THE PHYTOPATHOGEN

Evolution and Adaptation

THE PHYTOPATHOGEN

Evolution and Adaptation

Edited by

Abhijeet Ghatak, PhD
Mohammad Ansar, PhD

Apple Academic Press Inc.	Apple Academic Press Inc.
3333 Mistwell Crescent	9 Spinnaker Way
Oakville, ON L6L 0A2 Canada	Waretown, NJ 08758 USA

© 2017 by Apple Academic Press, Inc.

First issued in paperback 2021

No claim to original U.S. Government works

ISBN-13: 978-1-77463-702-9 (pbk)
ISBN-13: 978-1-77188-406-8 (hbk)

Library and Archives Canada Cataloguing in Publication

The phytopathogen : evolution and adaptation / edited by Abhijeet Ghatak, PhD, Mohammad Ansar, PhD.

Includes bibliographical references and index.
Issued in print and electronic formats.
ISBN 978-1-77188-406-8 (hardcover).--ISBN 978-1-77188-407-5 (pdf)
1. Plant diseases. 2. Plant viruses. I. Ghatak, Abhijeet, author, editor II. Ansar, Mohammad, author, editor

SB731.P59 2016	632'.3	C2016-905356-3	C2016-905357-1

Library of Congress Cataloging-in-Publication Data

Library of Congress Cataloging-in-Publication Data
Names: Ghatak, Abhijeet, editor. | Ansar, Mohammad, editor.
Title: The phytopathogen : evolution and adaptation / editors: Abhijeet Ghatak, PhD, Mohammad Ansar, PhD.
Description: Waretown, NJ : Apple Academic Press, 2017. | Includes bibliographical references and index.
Identifiers: LCCN 2016047918 (print) | LCCN 2016054347 (ebook) | ISBN 9781771884068 (hardcover : alk. paper) | ISBN 9781315366135 (ebook)
Subjects: LCSH: Plant diseases. | Plant viruses.
Classification: LCC SB731 .P54 2017 (print) | LCC SB731 (ebook) | DDC 632/.3--dc23
LC record available at https://lccn.loc.gov/2016047918

Apple Academic Press also publishes its books in a variety of electronic formats. Some content that appears in print may not be available in electronic format. For information about Apple Academic Press products, visit our website at **www.appleacademicpress.com** and the CRC Press website at **www.crcpress.com**

ABOUT THE EDITORS

Abhijeet Ghatak, PhD

Abhijeet Ghatak, PhD, is an Assistant Professor of Plant Pathology at Bihar Agricultural University (BAU), India, where his research currently covers fungal phyto-pathosystems and implications of nanotechnology as an option for plant disease management. His expertise lies in epidemiology and in helping to set strategy for plant disease management. He is a member of the core research group and research monitoring committee at BAU, helping to provide appropriate direction and to prioritize research. BAU has recognized Dr. Ghatak as the coordinator of host parasite inter-action research. His interests focus on host/organ-specificity, pathogen population variability and disease dynamics, and plant health manage-ment. He currently has a number of projects running under his supervision. Dr. Ghatak is also engaged in undergraduate and postgraduate teaching, covering courses related to principles of plant pathology, field and medicinal crop diseases, epidemiology, and forecasting. He received his postgraduate degree in Mycology and Plant Pathology from Banaras Hindu University, Varanasi, India, and acquired his doctorate in Plant Pathology from G. B. Pant University of Agriculture and Technology, Pantnagar, India. Dr. Ghatak has earned experience at the International Rice Research Institute, Philip-pines, while working on his PhD thesis research and has addressed several important issues related to neck blast epidemics in rice.

Mohammad Ansar, PhD

Mohammad Ansar, PhD, is an Assistant Professor at the Department of Plant Pathology at Bihar Agricultural University (BAU), India. At BAU, he acts as a coordinator of virology research and is engaged in teaching courses related to plant virology, seed health technology, and vege-table diseases. Apart from this, Dr. Ansar is employed as vegetable pathologist at the university. His major research activities focus on vegetable viruses with special interest on characteriza-tion, diagnosis, and epidemiology of plant viruses. He is currently a principal investigator of a state government funded project on whitefly- and thrips-transmitted viral diseases in tomato and chili. He earned his postgraduate certificate in Plant Pathology from C. S. Azad University of Agriculture and Technology, Kanpur, India, and his PhD degree in Plant Pathology from G. B. Pant University of Agriculture and Technology, Pantnagar, India. He is an expert on plant diagnosis and phytovirology.

DEDICATION

To

[signature: Charles Darwin]

– Charles Robert Darwin –

CONTENTS

LIST OF CONTRIBUTORS

Shamshad Alam
Biotechnology Laboratory, Central Rainfed Upland Rice Research Station (CRURRS), ICAR, Hazaribag 825301, India

Mohammad Ansar
Department of Plant Pathology, Bihar Agricultural University, Sabour 813210, Bihar, India

Md. Arshad Anwer
Department of Plant Pathology, Bihar Agricultural University, Sabour 813210, Bhagalpur, Bihar, India

Rekha Balodi
Plant Pathology, COA, GB Pant University of Agriculture and Technology, Pantnagar 263145, Uttarakhand, India

Arun P. Bhagat
Department of Plant Pathology, Bihar Agricultural University, Sabour 813210, Bhagalpur, Bihar, India

Subhash C. Bhardwaj
ICAR-Indian Institute of Wheat and Barley Research, Regional Station, Flowerdale, Shimla 171002, Himachal Pradesh, India

Sunaina Bisht
Plant Pathology, COA, GB Pant University of Agriculture and Technology, Pantnagar 263145, Uttarakhand, India

Anirudha Chattopadhyay
Department of Plant Pathology, CP College of Agriculture, Sardarkrushinagar Dantiwada Agricultural University, SK Nagar, Gujarat 385506, India

D. Durgadevi
Department of Plant Pathology, Centre for Plant Protection Studies, Tamil Nadu Agricultural University, Coimbatore 641003, India

Erayya
Department of Plant Pathology, Bihar Agricultural University, Sabour 813210, Bihar, India

Om P. Gangwar
ICAR-Indian Institute of Wheat and Barley Research, Regional Station, Flowerdale, Shimla 171002, Himachal Pradesh, India

Abhijeet Ghatak
Department of Plant Pathology, Bihar Agricultural University, Sabour 813210, Bihar, India

Lajja Vati Ghatak
Plant Breeding and Genetics, Bihar Agricultural University, Sabour 813210, Bihar, India

Shankar L. Godara
Department of Plant Pathology, College of Agriculture, SK Rajasthan Agricultural University, Bikaner, India

S. Harish
Department of Plant Pathology, Centre for Plant Protection Studies, Tamil Nadu Agricultural University, Coimbatore 641003, India

Somnath K. Holkar
Division of Crop Protection, Indian Institute of Sugarcane Research, Rae Bareli Road, P.O. Dilkusha, Lucknow 226002, India

Jahangir Imam
Biotechnology Laboratory, Central Rainfed Upland Rice Research Station (CRURRS), ICAR, Hazaribag 825301, India

Pratibha Kaushal
Division of Crop Improvement, Indian Institute of Sugarcane Research, Rae Bareli Road, P.O. Dilkusha, Lucknow 226002, India

Hanif Khan
ICAR-Indian Institute of Wheat and Barley Research, Regional Station, Flowerdale, Shimla 171002, Himachal Pradesh, India

Bhupendra S. Kharayat
Division of Plant Pathology, ICAR-Indian Agricultural Research Institute, Pusa, New Delhi 110012, India

Amarendra Kumar
Department of Plant Pathology, Bihar Agricultural University, Sabour 813210, Bihar, India

Gagan Kumar
Department of Mycology and Plant Pathology, Institute of Agricultural Sciences, Banaras Hindu University, Varanasi 221005, India

Mahesh Kumar
Department of Molecular Biology and Genetic Engineering, Bihar Agricultural University, Sabour 813210, Bihar, India

Ravi R. Kumar
Department of Molecular Biology and Genetic Engineering, Bihar Agricultural University, Sabour 813210, Bihar, India

Sanjeev Kumar
Division of Crop Improvement, Indian Institute of Sugarcane Research, Rae Bareli Road, P.O. Dilkusha, Lucknow 226002, India

Santosh Kumar
Department of Plant Pathology, Bihar Agricultural University, Sabour 813210, Bihar, India

Subodh Kumar
ICAR-Indian Institute of Wheat and Barley Research, Regional Station, Flowerdale, Shimla 171002, Himachal Pradesh, India

Vinod Kumar
Department of Molecular Biology and Genetic Engineering, Bihar Agricultural University, Sabour 813210, Bihar, India

Chanda Kushwaha
Department of Plant Pathology, Bihar Agricultural University, Sabour 813210, Bhagalpur, Bihar, India

R. Manikandan
Department of Plant Pathology, Centre for Plant Protection Studies, Tamil Nadu Agricultural University, Coimbatore 641003, India

Anand K. Meena
Department of Plant Pathology, College of Agriculture, SK Rajasthan Agricultural University, Bikaner, India

Ashok K. Meena
Department of Plant Pathology, College of Agriculture, SK Rajasthan Agricultural University, Bikaner, India

Jai S. Patel
Department of Botany, Banaras Hindu University, Varanasi 221005, India

Pramod Prasad
ICAR-Indian Institute of Wheat and Barley Research, Regional Station, Flowerdale, Shimla 171002, Himachal Pradesh, India

Sumit K. Pundey
Department of Mycology and Plant Pathology, Institute of Agricultural Sciences, Banaras Hindu University, Varanasi 221005, India

T. Raguchander
Department of Plant Pathology, Centre for Plant Protection Studies, Tamil Nadu Agricultural University, Coimbatore 641003, India

Neha Rani
Department of Plant Pathology, Bihar Agricultural University, Sabour 813210, Bhagalpur, Bihar, India

Sanjoy Guha Roy
Department of Botany, West Bengal State University, Barasat, Kolkata 700126, India

Ankita Sarkar
Department of Mycology and Plant Pathology, Institute of Agricultural Sciences, Banaras Hindu University, Varanasi 221005, India

Abhishek Sharma
Department of Vegetable Science, Punjab Agricultural University, Ludhiana, Punjab, India

Birinchi K. Sarma
Department of Mycology and Plant Pathology, Institute of Agricultural Sciences, Banaras Hindu University, Varanasi 221005, India

Md. Shamim
Department of Molecular Biology and Genetic Engineering, Bihar Agricultural University, Sabour 813210, Bihar, India

Nandani Shukla
Plant Pathology, COA, GB Pant University of Agriculture and Technology, Pantnagar 263145, Uttarakhand, India

Harikesh B. Singh
Department of Mycology and Plant Pathology, Institute of Agricultural Sciences, Banaras Hindu University, Varanasi 221005, India

Kundan Singh
Department of Plant Pathology, Bihar Agricultural University, Sabour 813210, Bhagalpur, Bihar, India

Mahendra Singh
Department of Soil Science & Agricultural Chemistry, Bihar Agricultural University, Sabour 813210, Bihar, India

Rakesh K. Singh
Department of Mycology and Plant Pathology, Institute of Agricultural Sciences, Banaras Hindu University, Varanasi 221005, India

Yogendra Singh
Department of Plant Pathology, Collage of Agriculture, G.B. Pant University of Agriculture & Technology, Pantnagar, Udham Singh Nagar, Uttarakhand, India

Nimmy M. Subramanian
National Research Centre on Biotechnology, Pusa campus, New Delhi 110012, India

Ram S. Upadhyay
Department of Mycology and Plant Pathology, Institute of Agricultural Sciences, Banaras Hindu University, Varanasi 221005, India

Mukund Variar
Biotechnology Laboratory, Central Rainfed Upland Rice Research Station (CRURRS), ICAR, Hazaribag 825301, India

LIST OF ABBREVIATIONS

%	percentage
ABA	abscisic acid
ACC	1-aminocyclopropane-1-carboxylate
AFLP	amplified fragment length polymorphism
AICRPS	All India Coordinated Research Project on Sugarcane
AMF	arbuscular mycorrhizal fungi
AMV	alfalfa mosaic virus
ARDRA	amplified ribosomal DNA restriction analysis
Avr	avirulence
BBTV	*Banana bunchy top virus*
BCA	biological control agents
BIC	biotrophic interfacial complex
B-lectin	binding lectin
BuYV	*Burdock yellows virus*
BVY	*Blackberry virus* Y
BYDV	*Barley yellow dwarf virus*
BYV	*Beet yellows virus*
CaCV	*Capsicum chlorosis virus*
CAD	cinnamyl alcohol dehydrogenase
CaMV	*Cauliflower mosaic virus*
CBEL	cellulose binding elicitor lectin
CBSV	*Cassava brown streak virus*
CFA	coronafacic acid
CFU	colony forming units
CLPP	community level physiological profiling
CMV	*Cucumber mosaic virus*
COMT	caffeic acid O-methyltransferase
COR	coronatine
CP	coat protein
CRNs	crinklers
CRURRS	Central Rainfed Upland Rice Research Station
CT	cycle threshold
CTV	*Citrus tristeza virus*
CuR	copper resistance

CWDE	cell wall degrading enzyme
DGGE	denaturing gradient gel electrophoresis
dHG	desoxyhemigossypol
DMFE	distribution of mutational fitness effects
EIHM	extra invasive hyphal membrane
ENC	effective number of codons
EPA	environmental protection agency
EPG	electrical penetration graph
ETI	effector triggered immunity
FAME	fatty acid methyl ester
FAO	Food and Agriculture Organization
Fe^{+3}	ferric iron
FHV	*Flock house virus*
G	gossypol
G×E	genotype-by-environment
GA	gibberellic acid
GBNV	*Groundnut bud necrosis virus*
GIP	glucanase inhibitor protein
GLRaV	*Grapevine leafroll associated virus*
GYSV	*Groundnut yellow spot virus*
H_2O_2	hydrogen peroxide
HCSV	*Hibiscus chlorotic spot virus*
HG	hemigossypol
HGT	horizontal gene transfer
HPLC	high performance liquid chromatography
HR	hypersensitivity response
hsv	host-specific virulence
IAA	indole acetic acid
ICEs	integrative conjugative elements
ICMV	*Indian cassava mosaic virus*
IGS	intergenic spacer
IH	invasive hyphae
IRRI	International Rice Research Institute
ISSR	inter simple sequence repeat
ITS	internal transcribed spacer
IYSV	*Iris yellow spot virus*
JIRCAS	Japan International Rice Research Center for Agricultural Sciences
LAV	*Leafhopper A virus*
LC	liquid chromatography

LGT	lateral gene transfer
LOX	lipoxygenase
LTH	Lijiangxintuanheigu
LTR	long terminal repeat
lx	probability of survival to age *x*
M. grisea	*Magnaporthe grisea*
M. oryzae	*Magnaporthe oryzae*
MAPK	mitogen-activated protein kinase
MS	mass spectrometry
MSV	*Maize streak virus*
mx	number of progeny produced by an individual of age *x*
N	population density
NBS-LRR	nucleotide binding site-leucine rich repeat
nec1	necrogenic toxin protein
NETU	nematode transmitted tubular viruses
NGS	next generation sequencing
NIPs	necrosis-inducing proteins
NOS	nitric oxide signaling
NRPS	non-ribosomal peptide synthase
OBAC	oat, bran, agar, cellobiose
OBB	oat bran broth
ORF	open reading frame
PAIs	pathogenicity associated islands
PAL	phenylalanine ammonia-lyase
PAMPs	pathogen associated molecular patterns
PBGC	pyochelin biosynthetic gene cluster
PCR	polymerase chain reaction
pd	potential drops
PDA1	pisatin demethylase gene
PGPR	plant growth-promoting rhizobacteria
PLFA	phospholipid fatty acid analysis
PLRV	*Potato leaf roll virus*
POX	peroxidases
PPO	polyphenol oxidase
PPV	*Plum pox virus*
PR	pathogenesis-related
PRRs	pattern recognition receptors
PSV	*Peanut stunt virus*
PTI	PAMP-trigerred immunity
PVY	*Potato virus Y*

Q-PCR	quantitative polymerase chain reaction
QTL	quantitative trait loci
R	resistance
R_0	basic reproductive number
RAPD	random amplified polymorphic DNA
RCR	rolling circle replication
RDV	*Rice dwarf virus*
RFLP	restriction fragment length polymorphism
RISA	ribosomal intergenic spacer analysis
RNAi	RNA interference
ROS	reactive oxygen species
RSGP	reverse sample genome probing
RT	reverse transcriptase
SA	salicylic acid
SAR	systemic acquired resistance
SCMV	*Sugarcane mosaic virus*
SCSU	sole-carbon-source utilization
SCYLV	*Sugarcane yellow leaf virus*
SLCMV	*Sri Lankan cassava mosaic virus*
SNPs	single nucleotide polymorphisms
SOD	sudden oak death
sp.	specie
SPMMV	*Sweet potato mild mottle virus*
spp.	species
SqVYV	*Squash vein yellowing virus*
SSCP	single strand conformation polymorphism
ssDNA	single-stranded DNA
TAV	*Tomato aspermy virus*
TE	transposable element
TEV	*Tobacco etch potyvirus*
TGGE	temperature gradient gel electrophoresis
ToLCD	tomato leaf curl disease
T-RFLP	terminal restriction fragment length polymorphism
TSWV	Tomato spotted wilt virus
TTSS	type three secretion system
txt A,B, C	Thaxtomin A,B,C
UBN	uniform blast nursery
UPP	undecaprenyl pyrophosphate phosphatase
UV	ultra violet
vir	virulence

WBNV	*Watermelon bud necrosis virus*
WYLV	*Watercress yellow spot virus*
YLD	yellow leaf disease
β	average number of offspring produced
γ	time

PREFACE

Plant pathology is an exciting area of plant science, creating unprecedented opportunities. Several plant pathogens have emerged in the changed climatic scenario and are causing severe damage when compared with previous records. The emergence of new pathogens in a niche needs evolutionary adaptation for their action. A bulk of study is on quantitative aspects of host–phytopathogen linkage resulting in emergence of phytopathogens aiding evolved aggressiveness, an important concern of the present era. Consequently, evolution and adaptation and recently emerged members of phytopathogens from several cropping systems are examined in this edition and focused on as the research priority for the book.

The book consists of various chapters that cover major thematic areas, such as growth and reproduction of phytopathogens, diversity and population dynamics, modification in host and pathogenesis, host govern specificity in pathosystem, and development.

One of the important behaviors in pathogens is reproduction; how phytopathogens evolve their approaches and strengthen their fecundity to survive in adverse situations is elaborated in the book. Moreover, variability in reproductive fitness and diversification in slow and rapid multiplying pathogens are also covered. Microbial communities and their population dynamics constitute important parameters in understanding soil-borne or root-infecting diseases. The emergence of exotic or newly introduced strains may occur with a crop having a specific resistance against the disease; sudden or a slow breaking down is a major challenge in the production of economically important crops. Therefore, the resistance-breaking issue is a central point, and thus, many concerns have been raised. *Phytophthora* spp. poses a great threat to different plant species in the tropics, sub-tropics, and temperate regions of the world. The chapter on pathogenesis within *Phytophthora* spp. of various host origins has been explained in great length in order to convey the infection biology.

The book covers a multitude of issues on diversity in major insect-transmitted virus infecting various crops in detail. Investigations of the conception of plant viruses may not be adequate to explain evolution in nature. Anthropogenic actions such as crop domestication, long-distance movement, and exploitation of natural habitats appear to influence plant virus spread and evolution in cultivating areas. The formation of new vector–virus relationships, the mechanisms involved, and reproductive strategies that play significant roles are presented in the book.

Courses on plant pathology are offered at various levels of undergraduate and postgraduate studies. The content of this book provides a good understanding of phytopathogens in context to evolution. The information presented on every aspects of evolutionary adaptation can be incorporated as a specific study in postgraduate programs. The book includes labeled examples and images that will be beneficial to generate new idea among the students and researchers.

The intention of this introduction on the evolution and adaptation in phytopathogens was to focus attention on discussions in several aspects of fascinating field and targeting related scientific community. The information provided in this book directly or indirectly helps the epidemiologist in formulating prediction models by using population dynamics and the structure of pathogens. Moreover, brief information on virological aspects provides an overview of the diversity and versatility of these very small but greatly vector-transmitted phytopathogens. The importance of diseases and strategies of insects for transmission of viruses are outlined; however, that will be incorporated as a developing epidemiological model in insect-transmitted virus.

We would highly appreciate receiving your comments, suggestions, and research contributions that relate to different themes of the book. It will be helpful to us in the preparation of a revised edition and future volumes.

— **Abhijeet Ghatak**
Mohammad Ansar

ACKNOWLEDGMENT

The editors are grateful to Dr. Mohammed Wasim Siddiqui, Founding Editor of the *Journal of Post Harvest Technology*, Sabouor, for helping in this project. Support from Bihar Agricultural University, Sabour, is thankfully acknowledged.

PART I
Evolutionary Process in Phytopathogens

CHAPTER 1

EVOLUTION AND ADAPTATION IN PHYTOPATHOSYSTEMS

ABHIJEET GHATAK[*]

Department of Plant Pathology, Bihar Agricultural University, Sabour 813210, Bihar, India

[*]*Corresponding author. E-mail: ghatak11@gmail.com*

CONTENTS

ABSTRACT

Modification in capability of a phytopathogen to sustain in a new area is necessarily essential. This is determined by the adaptation of phytopathogen to the abiotic condition at first, and then to the biotic adjustments leading evolution. Indeed, the process of evolution is started after genetical adaptation with its host. Therefore, this chapter is designed to extrapolate the role of variables influencing virulence, impact of population size on evolution, and host evolution participating in adaptation and evolution of phytopathogen. The virulence evolution in a phytopathogen is greatly dealt; whereby pathogen fitness, host existence, and survival capability are covered. It also touches the evolution and adaptation parts to mistletoe and phytoplasma, as phytopathogens that are not greatly studied. This chapter also highlights the difficulties during evolution. In short, every change is liable to diminutive units of adaptation, which ultimately leads to evolution of a pathosystem.

1.1   INTRODUCTION

Novel phytopathosystems are developed when introduction of plant species or phytopathogen is commenced in a newer area. In this connection, determination of phytopathosystem depends on the genetic compatibility of plant and pathogen. This is further decided by the pre-adaptation of the pathogen and succeeding evolutionary changes over a period. During an evolutionary process, which is an ever-changing process, a phytopathosystem may undergo in epidemics for many occasions that is contributed by variable dispersal, changed life-history, and evolutionary adjustment. Such modification may lead to alteration in aggressiveness/virulence of the pathogen and resistance of the host plant and to an environment which is conduce (Lajja Vati & Ghatak, 2015). The ecological outcome of this interaction is governed by the ecological responses and evolutionary influences (Parker & Gilbert, 2004).

Indeed, the introduced phytopathogen interacts with an exclusive atmosphere and a novel host. At first, the phytopathogen must have adaptive capacity for a new area against abiotic conditions that include extreme range of temperature, humidity, and most importantly the ultraviolet (UV) irradiation (Gilbert, 2002). Upon surviving these challenges, a phytopathogen must surpass the impediments of a biotic interaction with the host plant, starting from the infection process in the host, reproduction (production of spores or dispersal units), and finally dissemination to keep themselves "alive" from generation to generation (Zadoks & Schein, 1979). For each step, a phytopathogen needs to represent a strong selection influence, which may further direct the rapid adaptation to newer abiotic conditions and plant, and evolution with the host (Heitman et al., 2013). The adaptability of a phyto-pathogen toward its host is chiefly governed by the genetic compatibility between them. Further, the bio-chemical adjustments are taken place. After all these amendments the evolutionary process is generally begin.

The co-evolution of plants with microorganisms could be exemplified with symbiotic fungal association (Barrow et al., 2008; Biere & Tack, 2013) and intracellular symbiosis with bacteria (Markmann et al., 2008) with the plant roots found in the establishment process of early land plants on the earth. Therefore, such interaction is a result of a complex set of evolutionary influences (Alexander et al., 1996; Simms, 1996). A typical epiphytotic is generated due to a successful interaction by the host and phytopathogen, when the latter poses dominancy over the host. The degree of the epiphy-totic depends on the resistance of the host for a virulent phytopathogen (McDonald, 2004).

Generally two fundamental processes are involved in the gene-for-gene principle governed plant disease resistance: (a) anticipation of infection and then (b) plant responses are triggered to restrict the disease. The anticipation mechanism is led by the receptors, which are highly sensitive to react against the pathogen strains (Ellis et al., 2000). These strains can be encoded by the genes operated for disease resistance. Polymorphism of resistance (R) genes is mostly linked with loci available as tandem arrays of numerous replicas forming a cluster. The paralogs present in these clusters often display intri-cate evolutionary associations—as a result of loose genetic exchanges (Young, 2000). Upon infection or receiving signal from the phytopathogen, a plant is underway for physiological and morphological changes and forms secondary metabolites, which inhibit the progress of infection (Ansar et al., 2016). Moreover, the infection leads to the synthesis of pathogenesis-related (PR) proteins or defense secondary metabolites like phytoalexin.

In a phytopathogen, genetic modification usually happens during evolu-tion. In several biotrophs like rusts and powdery mildews, evolution to biot-rophy is a common feature. This is also an ordinary incidence in lower fungi, causing white rust (blister) and downy mildews. A report says wheat powdery mildew fungus *Blumeria graminis* shows a close connection between gene losses and the biotrophic mode of life cycle during primary and secondary metabolisms (Spanu et al., 2010). A similar interpretation on an obligate pathosystem (white rust of *Arabidopsis thaliana*) was made by Kemen et al. (2011). They conclude that the variation in gene is a common phenomenon of obligate pathosystem during evolution. A few biosynthetic pathways may not be operated or their chances of selection might be reduced during evolution of a closely linked host and its (obligate) phytopathogen. The flow in genetic variation imparts to outline the Fig Mosaic Virus (FMV) populations (Walia et al., 2014).

1.2   INFECTION CAPABILITY OF PHYTOPATHOGEN

A few factors are associated with plant and its pathogen's life-history and population structure that definitely influence the genetic structure of the populace and also the population size. It is therefore usually envisaged that a new epiphytotic appears due to the maladapted phytopathogens. The virulence of these phytopathogens should decrease over time. In support, a classic work on animal pathogens by Anderson and May (1982) suggests the evolution of increased virulence would be limited. This is due to the chances for transmission to a new vulnerable host. However, Kover and Clay (1998) admitted for decreased virulence under a condition when chances of vertical transmission are high. Indeed, the evolutionary mechanisms, under natural selection, lead to increased or decreased virulence of a pathogen. This is determined by the complex tradeoffs between reproduction rates and other life-history traits along with the mode of transmission (Bull, 1994; Ewald, 1993; Lenski & May, 1994; Williams & Hesse, 1991). The effects of host and phytopathogen generated by Vanderplank (1984), who determined using the analysis of variance that host resistance could be organized into vertical (race specific, governed by single gene) or horizontal (race nonspecific, governed by many genes), and respective pathogenic ability could be recognized by virulence (qualitative character) and aggressiveness (quantitative character). In the stated work, vertical resistance and virulence were resulted from the interaction effect among several lines of host and pathogen races; however, horizontal resistance and aggressiveness were defined as the main effects. The inheritance pattern (easily inheritable) and high potential protection ability up to the limit when a phytopathogen are not able to break the barrier of resistance even after evolving themselves to a certain extent; vertical resistance is regularly incorporated in the new varieties developed through breeding manipulation (Hovmøller, 2007; Wellings, 2007). Horizontal resistance has also been used for several crops because of its increased durability compared with vertical resistance (Andrivon et al., 2003; Chen & Line, 1995; McDonald & Linde, 2002; Niks & Rubiales, 2002; Roumen, 1994; Singh et al., 2005; Talukder et al., 2004; Vaz Patto et al., 2006; Zenbayashi et al., 2002). In general, little infection may take place in varieties operating horizontal resistance, but the varieties operating vertical resistance are free from infection. Therefore, if a race of phytopathogen is emerged (through evolution) and adapted to the system, it can result in a "boom-and-burst phenomenon" (Eversmeyer & Kramer, 2000; Keller et al., 2000; Mundt, 2014; Papaix et al., 2014).

Both host and environment influence aggressiveness of an isolate (Andrivon, 1993). Greater aggressiveness under elevated temperature was observed in the isolates of wheat stripe rust pathogen, *Puccinia striiformis* f. sp. *tritici*, suggesting the pathogen is well adapted to the "new" regime of temperature and can cause severe damage at warmer part of the globe, which was previously non-conducive to this fungus (Milus et al., 2009). Vander-plank (1968) very rightly marked that aggressiveness is positively related with fitness to survive, meaning the increase in aggressiveness increases of chances to survive, which is confirmed by the universal law of Darwin—survival of the fittest (von Sydow, 2014). Reproductive fitness inside the host is known as aggressiveness. In the case of the above-mentioned phyto-pathogen, the famous Vanderplank observation is likely to be factual. In the USA (Markell & Milus, 2008) and in Australia (Wellings, 2007), the newly emerged isolates adapted to these locations replacing the old isolates and repeatedly appeared resulting in heavy yield damage. This is an instance of population shift rather than new combinations of virulence: replacement was attributed by the superior (increased aggressiveness) isolates or races (Luig, 1985). There are several phytopathosystems supporting the theory that increased aggressiveness is liable for long-distance dispersal of the fungal spores. As for the wheat and rust fungus interaction, the races of stripe rust fungus (*P. striiformis* f. sp. *tritici*) and leaf rust fungus (*P. triticina*, formerly *P. recondita* f. sp. *tritici*), being superiorly aggressive, became widespread across America (Chen, 2005) and Australia (Park et al., 1995), respectively.

Tissue-dependent infection pattern determines aggressiveness in several phytopathosystems (Dufresne & Osbourn, 2001; Tooley & Kyde, 2007). *Phytophthora cactorum* infecting European beech (*Fagus sylvatica*) trees on stem and root tissues did not differ for aggressiveness, which was varied from leaf tissues (Weiland et al., 2010). The tissue-adapted aggressiveness of *Phytophthora* spp. is also verified in other trees (Hansen et al., 2005; Tooley & Kyde, 2007). Variability in the level of infection was measured with different monocyclic parameters of aggressiveness on leaf and neck tissues of rice plant using cross inoculation test, whereby the isolates obtained from leaves and necks were inoculated on these organs (Ghatak et al., 2013). The study was hypothesized on adaptation of *Magnaporthe oryzae* isolates on the type of organ they infect. This implicates the isolates originated from leaves can infect necks. Isolates originating from necks were more aggressive on both leaves and necks. Infection on necks by the leaf-originating isolates was probably due to the most aggressive leaf-isolate-infected necks. A hint on variation in the resistance expressed on leaf and the resistance expressed on neck could be attained by a work of Bonman et al. (1989), who detected the difference in resistance for the two organs on several rice genotypes in lowland environment. Tissue-adapted invasion strategies in rice blast fungus were also discussed by Marcel et al. (2010), whereby the workers have proposed for systemic mode of infection from seed to seedling. They concluded with fungal proliferation from the roots to vascular tissue progresses in a biotrophic approach.

1.3   VARIABLES INFLUENCING EVOLUTION OF VIRULENCE

Apart from the factors described in the earlier discussion, there are a few more critical variables that influence the evolution of virulence in the phytopathogen. They are as follows:

1.3.1   FITNESS OF THE PATHOGEN

For successful infection, a phytopathogen must be well qualified for all physiological parameters. Hence, fitness of phytopathogen, particularly the fungi, determines the adaptability to their host and environment. Because factors like host and environment determine the reproduction ability considering genetic recombination and dissemination, the evolution of a phytopathogen is a dependent phenomenon. Several attempts have been taken for resistance-associated studies, but fitness-associated investigation still needs special attention and still becomes a subject of understanding and exploration (Burns et al., 2012). Vanderplank (2012) very correctly envisaged that the basis of natural selection and phenotypic evolution of any organism is "fitness."

If the transmission ability of the phytopathogen in a new host is reduced, the virulence evolution will be limited (Laine & Barres, 2013). The reduced virulency may be fetched by the reduction in the fecundity or limited access of correct host(s). Therefore, lower transmission ability leads to poor virulence evolution capacity in a phytopathogen. It means a "healthy" race will have more chances and ability for evolution than an ordinary race. Alongside, the ability to explore its host and potentiality to adapt in such a situation, for example, adapting a new host, will definitely encourage the evolution.

1.3.2 EXISTING OF HOSTS

The existence of an alternate host for a phytopathogen allows the off-season survivability. This helps to regulate the biological processes and may become highly aggressive on the main host availability. Such groups of phytopathogens, having an alternate host, pose high threat to global produce and can cause epidemic under a favorable condition, for example, wheat rust. Under the conditions of the USA, the pathogen (*Puccinia graminis* f. sp. *tritici*) responsible for stem rust of wheat exists round the year on barberry, that is the alternate host of this pathogen (Agrios, 2005). In the wheat-growing season, the pathogen infects wheat crop and regularly makes problems to wheat cultivation. In the Indian scenario, importance of an alternative host comes into play as described in the Indian stem rust rules (Nagarajan & Singh, 1975). Barberry (*Berberis vulgaris*) bushes are not commonly found in India. The stem rust pathogen of wheat survives in self-sown wheat generally grown in the nearby fields of the path. They usually develop from the fallen grains of harvested ears, during the transportation from fields to the storage and milling centers. Apart from this, another major source of inoculum is the wheat cultivated in the Nilgiri hill and Palani hill regions of Tamil Nadu (a south Indian state), mostly in the period when the north- and central-Indian environment is unsuitable for wheat cultivation. The importance of long-distance dispersal of wheat stem rust has been elaborated (Nagarajan & Singh, 1990). The ability of wheat stem rust pathogen to survive in an alternate host or alternative host provides ample chances for evolution. A race of *P. graminis* f. sp. *tritici* from Uganda (Ug99) has been an emerging problem to the cultivated wheat (Nagarajan, 2012; Singh et al., 2015; Yu et al., 2014). In short, the presence of a host ensures round-the-year availability of active inoculum, leading to consistent infection potentiality, and thus, assisting in evolution.

A phytopathogen having low virulence level can reproduce on an alternative host. Likewise, such a phytopathogen attains high level of virulence on a few hosts if the former maintains its population size on the sympatric host. For example, the sudden oak epidemic pathogen *Phytophthora ramorum* can reproduce abundantly on tolerant *Arbutus menziesii* but is highly damaging on *Lithocarpus* and *Quercus*, where the fungus reproduce at a very low rate (Rizzo & Garbelotto 2003).

1.3.3   SAPROPHYTIC SURVIVAL

Such kind of survivability is associated with the facultative parasites: The parasite which is saprophyte by nature but turns into parasitic mode upon host availability, for example, *Pythium*, *Sclerotium*, etc. (Agrios, 2005). The phytopathogen pertaining to saprophytic survival pre-colonizes the substrate before it is colonized by the other microorganisms. In other words, it has the ability to live without its original or susceptible host. The organism survives well under saprophytic condition and must have strong saprophytic ability, that is production of antibiotics, increased aggressiveness, etc. This means such organisms may be high in virulence, and strong adaptability for these organisms generates the options for evolution. Variation of response for saprophytic survival among the strains of wheat take-all fungus *Gaeumanno-myces graminis* var. *tritici* has been detected under changing environmental pH (Daval et al., 2013). As discussed, the parasitic mode can be shifted into saprophytic mode. Therefore, the saprophytic survival allows evolution of phytopathogens with increased level of virulence without fitness of the pathogen. In the absence of cultivated host, the fungal phytopathogens can survive saprophytically by three means:

1.3.3.1   SOIL INHABITANTS

The organisms survive in the soil as saprophytes for a very long time (indefinite period), while the main host plant is absent, for example, *Pythium*, *Rhizoctonia*, and *Sclerotium*. These organisms can survive in soil for a relatively longer time.

1.3.3.2   ROOT INHABITANTS

Some more specialized parasites present in the soil adopt a close association with the host roots, for example, vascular wilt causing fungi: *Fusarium*, *Verticillium*, and cotton root rot fungus *Phymatotrichum omnivorum*.

1.3.3.3   RHIZOSPHERE COLONIZER

Being relatively higher tolerant to soil antagonism, these organisms can survive on the degraded roots (already dead) and live for a long period, for example, *Cladosporium fulvum* causing leaf mold in tomato.

1.3.4 SURVIVAL STRUCTURES

To escape the disease, a plant is dependent on various factors that affect the survival, degree of infection, reproduction, and spread (Agrios, 2005). One of the most important causes is deficit growth of the pathogen at susceptible phase of plant for poor infection or non-significant loss (disease). Many phytopathogenic fungi can overwinter or over summer producing survival structures (Baker & Snyder, 1965). A survival structure may be a fruiting body or a spore (resting spore). In general, an ordinary fungal spore germinates immediately after attaining maturation; however, a resting spore requires a phase of dormancy to germinate. Several resting spores are produced in phytopathogenic fungi; for instance Arthrospore, chlamydospore, and sclerotia prominently in asexual fungi, zygospore in advanced group of fungus (Zygomycota), and oospore in lower fungi having cellulosic wall. Zygospore and oospore are developed as a result of sexual reproduction, and remaining spores are produced through asexual reproduction. The soil inhabitants, for example, *Fusarium* (chlamydospore), *Rhizoctonia* (sclerotium), *Pythium* (oospore), etc., are able to survive for a very long period in the absence of their host. These members are unspecialized parasites having a large number of hosts to be parasitized, and therefore, they are more adaptive to their environment, thus, having enormous possibility for evolution. Mayer et al. (2001) elaborated several survival mechanisms of phytopathogens. In general, understanding the phytopathogens may survive prominently within the (infected) host and outside the host. Numerous kinds of hosts like cultivated or main host, alternate host, collateral host, and alternative host are examples for within-host survival. Outside-the-host survival is considered for the saprophytic mode of survival for various phytopathogens. A list of fungal phytopathogens surviving in different plant organs and soil is given in Table 1.1.

Bacterial phytopathogens do not form resting spores or similar structures and select shelters to sustain in nature (Agrios, 2005; Singh, 2009). Many bacterial phytopathogens survive in the seeds for a long period (over 10 years, *Xanthomonas axonopodis* pv. *phaseoli* and *Pseudomonas syringae* pv. *tomato*) and some others for a short period (up to 3 years, *X. oryzae* pv. *oryzae* and *X. axonopodis* pv. *malvacearum*). A few bacterial phytopathogens can survive in the crop residues, for example, *Clavibacter michiganensis* subsp. *insidiosum* in dry alfalfa stem residues. Likewise, *C. michiganensis* subsp. *michiganensis* and *X. axonopodis* pv. *malvacearum* can survive in the arid soil but may not survive under wet condition. The dreaded citrus canker pathogen *X. axonopodis* pv. *citri* shelters in the cankers of tree bark. In a similar way, the fire blight pathogen of pear and apple, *Erwinia amylovora*, passes the dormant phase in the dead twigs and remains as gemmiplane (survive in buds) during the dormancy of the tree. Some bacterial phyto-pathogens can survive in the rhizoplane (colonization in roots), for example, *E. carotovora* subsp. *atroseptica* (potato black leg), *P. syringae* pv. *lachry-mans* (angular spot of cucurbits), etc. (Table 1.2). Apart from these plant shelters, a few bacterial phytopathogens hide in the insect system and pass a phase of life. Most importantly, the pathogen of Stewart's wilt of corn survives in the corn flea beetle (*Chaetocnema pulicaria*). Other bacterial diseases that transmit through insect vector are given in Table 1.3.

TABLE 1.1   The Fungal Phytopathogen and Their Shelter.

Disease	Fungal phytopathogen	Shelter
Brown rot of pome and stone fruit	*Monilinia* spp.	Cankers and dead twigs
Powdery mildew of apple	*Podoshaera leucotricha*	Dormant buds, under scale
Powdery mildew of grapevines	*Uncinula necator*	Dormant buds, under scale
Downy mildew of grapevines	*Plasmopara viticola*	Dormant buds, under scale
Red rot of sugarcane	*Colletotrichum falcatum*	Cane cuttings
Leaf spot of rice	*Dreschlera oryzae*	Soil, seed
Rice blast	*Pyricularia oryzae*	Soil, seed
Powdery mildew of cucurbits	*Erysiphe cichoracearum*	Shaded or cooler place
	Sphaerotheca fuliginea	Shaded or cooler place
Powdery mildew of pea	*Erysiphe polygoni*	Shaded or cooler place
Viral diseases of cucurbits	*Virus*	Shaded or cooler place

TABLE 1.2   The Bacterial Phytopathogen and Their Shelter.

Disease	Bacterial phytopathogen	Shelter
Citrus canker	*Xanthomonas axonopodis* pv. *citri*	Cankers
Bacterial canker of stone fruits	*Pseudomonas syringae* pv. *syringae*	Cankers, buds, and systemic infection with no symptom
Fire blight of pear and apple	*Erwinia amylovora*	Dead twigs, and as gemmiplane during the dormancy of the tree
Fuscus blight of bean	*X. axonopodis* pv. *phaseoli* var. *fuscans*	Phylloplane of primary leaves
Bacterial blight of cotton	*X. axonopodis* pv. *malvacearum*	Seed cotyledon, rhizoplane
Pepper blight	*X. vesicatoria*	Rhizoplane
Angular spot of cucurbits	*P. syringae* pv. *lachrymans*	Rhizoplane
Potato black leg	*E. carotovora* subsp. *atroseptica*	Rhizoplane

TABLE 1.3 Insect Vectors of Bacterial Phytopathogens.

Disease	Bacterial phytopathogen	Insect vector
Potato black leg	*E. carotovora* subsp. *atroseptica*	Seed corn maggot (*Hylemya platuria*)
Soft rot of cucurbits	*Pectobacterium carotovora*	Cabbage maggot (*H. brassicae*)
Stewart wilt of corn	*Pantoea stewartii* subsp. *stewartii*	Corn flea beetle (*Chaetocnema pulicaria*)
Vascular wilt of cucurbit	*E. tracheiphila*	Cucumber beetle (*Acalymna vittatum* and *Diabrotica undecimpunctata*)
Fruit diseases	*Erwinia* spp.	Fruit flies (*Drosophila melanogaster*)

1.4 POPULATION SIZE AND EVOLUTION

Population size of a phytopathogen is a major factor that determines the degree of evolution (McDonald & Linde, 2002). Being cosmopolitan, the phytopathogens have a wide array of distribution. Some factors are responsible for the increase in effective population size, while some other factors are responsible for the reduced size of population (Fig. 1.1). The phytopathogens specific to their host are unable to increase their population. Moreover, the phytopathogens produced sexually have ample opportunity to increase their population size and for evolution. Spores dispersed distally are effective to increase their population size by introducing themselves in a new area.

1.5 RESISTANCE EVOLUTION IN THE HOST

Like the pathogens' evolution and population size through its biology (Fig. 1.1), the host evolution is also a (relatively slow) continuous process (Lion & Gandon, 2015). The co-evolutionary dynamics of hosts with their pathogen influence the novel host–parasite interaction through the shifting of the host from a place to another. Both the host and virulence of the pathogen are part of such co-evolutionary dynamics. Nevertheless, for host and pathogen, the rate of evolution of an attribute is comparative to (a) the strength of selection and (b) extent of genetic variation of the attribute. The disease-resistance genes in plants are highly polymorphic (Pryor & Ellis, 1993). Upon conducting many investigations, it is considered that manipulation in

DNA may lead to evolution of the genes and, thereby, in a species. This condition allows the plant for generating a novel resistance gene that can combat with the newly developed pathogen, that is with improved virulence. Richter et al. (1995) observed recombination at the disease-resistance gene, which was associated with construction of new resistance phenotypes—this work supports the hypothesis that altered DNA develops resistance in plants. Even, over 20 years of the work done by Richter et al. (1995), the molecular confirmation regarding the evolution of disease-resistance genes is not greatly attained.

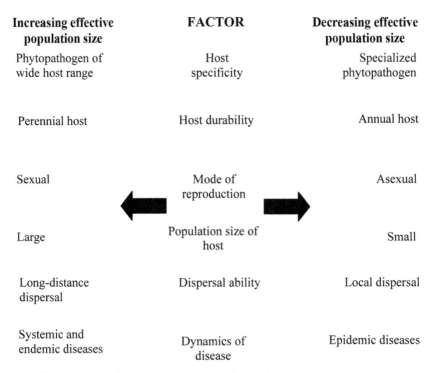

Increasing effective population size	FACTOR	Decreasing effective population size
Phytopathogen of wide host range	Host specificity	Specialized phytopathogen
Perennial host	Host durability	Annual host
Sexual	Mode of reproduction	Asexual
Large	Population size of host	Small
Long-distance dispersal	Dispersal ability	Local dispersal
Systemic and endemic diseases	Dynamics of disease	Epidemic diseases

FIGURE 1.1   Factors determining population size of phytopathogen.

1.6   EVOLUTION OF PLANT IMMUNITY

A highly developed plant (through evolution) has a great defense network against the pathogen infection. Primarily the immune response is subjected to pathogen assisted molecular pattern (PAMP)-triggered immunity (PTI) that has wide adaptability to recognize common identity of a phytopathogen. Under the co-evolution of plant along with its pathogen, effector proteins are delivered into the plant cell to suppress PTI that allows infection, colonization of the pathogen, and as a result emergence of disease (Ansar et al., 2016). To reply the effector proteins (from pathogen), the plants have R-proteins to observe the pathogen's effector proteins. Thus, the evolution of plant immunity is driven through pathogen's evolution in virulence.

1.7 EVOLUTION AND ADAPTATION IN RELATIVELY LESS-STUDIED ORGANISMS

1.7.1 MISTLETOE

The above-ground parasitic adaptation in mistletoe has occurred and evolved independently for four or five times (Nickrent, 2001). However, an excellent instance of convergent evolution is morphological similarities between *Viscaceae* and *Loranthaceae* (Barlow, 1983). A large number of investigations on mistletoe dispersal have dominated over highly specialized birds, associating physiological and behavioral trends with obligate frugivory and evaluation of co-evolutionary potential of these interactions (Reid, 1987, 1991; Snow, 1971).

1.7.2 PHYTOPLASMA

The sequence of genome in phytoplasma revealed the survival approaches and parasitism adaptation. Genome of this bacterium encodes a few metabolic functions that support intracellular parasitism, indicating the result of reductive evolution (Oshima et al., 2013). The genome analysis of phytoplasma exhibited that the reductive evolution as well as the virulence factors are unique to phytoplasma diseases. Additionally, the phytoplasma genome encodes a gene synthesis folate, which allows this bacterium to adapt in diverse and adverse environmental conditions, and also in various insects (Oshima et al., 2004). There are several evidences on putative motive units (PMUs) imparting a major role in genome evolution of phytoplasma (Hogenhout et al., 2008).

1.8 DIFFICULTIES DURING EVOLUTION

During a typical plant–pathogen interaction, deviation in local adaptation may affect virulence of the phytopathogen or resistance of the host, which ultimately leads to a complicated interaction between R genes and recessive alleles at Avr loci. Evolution of phytopathogen may be complex and unsuccessful under local adaptations for those organisms that can disseminate for miles. A fused gene flow may occur for local adaptation, particularly for those phytopathogens liable to long-distance dispersal through wind (Aylor, 2003; Brown & Hovmøller, 2002). This hypothesis is supported by the work of Parker (1991), who observed that the most virulent phytopathogen was found on their original hosts. However, other workers have advocated for absence of "home-host advantage" (Davelos et al., 1996; Zhan et al., 2002). Therefore, the inconsistent pattern of local adaptation may mislead the validation of virulence evolution (Zhan et al., 2002).

1.9 CONCLUSION

The evolution is a continuous process and thus the interacting organisms are always in active phase. The phytopathogens are shifting at the virulence level, and many other non-dispersive phases of their life cycle are directly or indirectly involved in the evolution and adaptation of a phytopathosystem. Alongside, it is always kept in the focal point that their hosts (plants) are also at the evolutionary phase, but possibly the (evolutionary) process is operating at a lower pace. Every change is linked to evolution and finally adaptation!

KEYWORDS

- adaptation
- evolution
- phytopathogen,
- phytopathosystem
- survival
- virulence

REFERENCES

Andrivon, D. Nomenclature for Pathogenicity and Virulence: The Need for Precision. *Phytopathology*. **1993**, *83,* 889–890.

Andrivon, D.; Lucas, J. M.; Ellisseche, D. Development of Natural Late Blight Epidemics in Pure and Mixed Plots of Potato Cultivars with Different Levels of Partial Resistance. *Plant Pathol.* **2003**, *52,* 586–594.

Ansar, M.; Ghatak, A.; Ghatak, L. V.; Srinivasaraghavan, A.; Balodi, R.; Raj, C. Secondary Metabolites in Pathogen-Induced Plant Defense. In *Plant Secondary Metabolites: Their Roles in Stress Ecophysiology;* Siddiqui, M. W., Bansal, V., Eds.; CRC Press: Florida, 2016, in press; Vol. 3.

Aylor, D. E. Spread of Plant Disease on a Continental Scale: Role of Aerial Dispersal of Pathogens. *Ecology.* **2003**, *84,* 1989–1997.

Baker, K. F.; Snyder, W. C. *Ecology of Soil-Borne Plant Pathogens: Prelude to Biological Control.* University of California Press: California, 1965.

Barlow, B. A. Biogeography of Loranthaceae and Viscaceae. In *The Biology of Mistletoes*; Calder, M., Bernhardt, P., Eds.; Academic Press Australia: Samford Valley, 1983; pp 19–46.

Barrow, J. R.; Lucero, M. E.; Reyes-Vera, I.; Havstad, K. M. Do Symbiotic Microbes Have a Role in Plant Evolution, Performance and Response to Stress? *Commun. Integr. Biol.* **2008**, *1*(1), 69–73.

Biere, A.; Tack, A. J. M. Evolutionary Adaptation in Three-Way Interactions between Plants, Microbes and Arthropods. *Funct. Ecol.* **2013**, *27,* 646–660.

Bonman, J. M.; Estrada, B. A.; Bandong, J. M. Leaf and Neck Blast Resistance in Tropical Lowland Rice Cultivars. *Plant Dis.* **1989**, *73,* 388–390.

Boonstra, H.; Duran, V.; Gamble, V. N.; Blumenthal, P.; Dominguez, L.; Pies, C. The "Boom and Bust Phenomenon": The Hopes, Dreams, and Broken Promises of the Contraceptive Revolution. *Contraception.* **2000**, *61*(1), 9–25.

Brown, J. K. M.; Hovmøller, M. S. Aerial Dispersal of Pathogens on the Global and Continental Scales and Its Impact on Plant Disease. *Science.* **2002**, *297,* 537–541.

Chen, X. M. Epidemiology and Control of Stripe Rust [*Puccinia striiformis* f.sp. *tritici*] on Wheat. *Can. J. Plant Pathol.* **2005**, *27,* 314–337.

Chen, X. M.; Line, R. F. Gene Action in Wheat Cultivars for Durable, High-Temperature, Adult-Plant Resistance and Interactions With Race-Specific, Seedling Resistance to Stripe Rust Caused by *Puccinia striiformis. Phytopathology.* **1995**, *85,* 567–572.

Daval, S.; Lebreton, L.; Gracianne, C.; Guillerm-Erckelboudt, A.Y.; Boutin, M.; Marchi, M.; Gazengel, K.; Sarniguet, A. Strain-Specific Variation in A Soil-borne Phytopathogenic Fungus for The Expression of Genes Involved in pH Signal Transduction Pathway, Pathogenesis and Saprophytic Survival in Response to Environmental pH Changes. *Fungal Genet. Biol.* **2013**, *61,* 80–89.

Davelos, A. L.; Alexander, H. M.; Slade, N. A. Ecological Genetic Interactions Between a Clonal Host Plant (*Spartina pectinata*) and Associated Rust Fungi (*Puccinia seymouriana* and *Puccinia sparganioides*). *Oecologia.* **1996**, *105,* 205–213.

Dufresne, M.; Osbourn, A. E. Definition of Tissue-Specific and General Requirements for Plant Infection in a Phytopathogenic Fungus. *Mol. Plant Microbe Interact.* **2001**, *14,* 300–307.

Ellis, J.; Dodds, P.; Pryor, T. Structure, Function and Evolution of Plant Disease Resistance Genes. *Curr. Opin. Plant Biol.* **2000**, *3,* 278–284.

Eversmeyer, M. G.; Kramer, C. L. Epidemiology of Wheat Leaf and Stem Rust in the Central Great Plains of the USA. *Annu. Rev. Phytopathol.* **2000,** *38,* 491–513.

Ghatak, A.; Willocquet, L.; Savary, S.; Kumar, J. Variability in Aggressiveness of Rice Blast (*Magnaporthe oryzae*) Isolates Originating from Rice Leaves and Necks: A Case of Pathogen Specialization? *PLoS ONE.* **2013,** *8*(6), e66180.

Hansen, E. M.; Parke, J. L.; Sutton, W. Susceptibility of Oregon Forest Trees and Shrubs to *Phytophthora ramorum*: A Comparison of Artificial Inoculation and Natural Infection. *Plant Dis.* **2005,** *89,* 63–70.

Heitman, J.; Sun, S.; James, T. J. Evolution of Fungal Sexual Reproduction. *Mycologia.* **2013,** *105*(1), 1–27.

Hogenhout, S. A.; Oshima, K.; Ammar, E.; Kakizawa, S.; Kingdom, H. N.; Namba, S. Phytoplasmas: Bacteria That Manipulate Plants and Insects. *Mol. Plant Pathol.* **2008,** *9*(4), 1–21.

Hovmøller, M. S. Sources of Seedling and Adult Plant Resistance to *Puccinia striiformis* f.sp. *tritici* in European Wheats. *Plant Breed.* **2007,** *126,* 225–233.

Keller, B.; Feuillet, C.; Messmer, M. Genetics of Disease Resistance: Basic Concepts and Application in Resistance Breeding. In *Mechanisms of Resistance to Plant Diseases;* Slusarenko, A. J., Fraser, R. S., van Loon, L.C., Eds.; Kluwer: The Netherlands, 2000; pp 101–160.

Laine, A. L.; Barres, B. Epidemiological and Evolutionary Consequences of Life-History Trade-Offs in Pathogens. *Plant Pathol.* **2013,** *62*(1), 96–105.

Lajja Vati; Ghatak, A. Phytopathosystem Modification in Response to Climate Change. In *Climate Dynamics in Horticultural Science: Impact, Adaptation, and Mitigation*; Choudhary, M. L., Patel, V. B., Siddiqui, M. W., Verma. R. B., Eds.; CRC Press: Florida, 2015; pp 161–178.

Lion, S.; Gandon, S. Evolution of Spatially Structured Host–Parasite Interactions. *J. Evol. Biol.* **2015,** *28,* 10–28.

Luig, N. H. Epidemiology in Australia and New Zealand. In *The Cereal Rusts*; Roelfs, A. P., Bushnell, W. R., Eds.; Academic Press: New York, 1985;Vol. 2, pp 301–328.

Marcel, S.; Sawers, R.; Oakeley, E.; Angliker, H.; Paszkowski, U. Tissue-Adapted Invasion Strategies of the Rice Blast Fungus *Magnaporthe oryzae*. *Plant Cell.* **2010,** *22,* 3177–3187.

Markell, S. G.; Milus, E. A. Emergence of A Novel Population of *Puccinia striiformis* f.sp. *tritici* in Eastern United States. *Phytopathology.* **2008,** *98,* 632–639.

Markmann, K.; Giczey, G.; Parniske, M. Functional Adaptation of a Plant Receptor-Kinase Paved the Way for the Evolution of Intracellular Root Symbioses with Bacteria. *PLoS Biol.* **2008,** *6*(3), e68.

Mayer, A. M.; Staples, R. C.; Gil-ad, N. L. Mechanisms of Survival of Necrotrophic Fungal Plant Pathogens in Hosts Expressing the Hypersensitive Response. *Phytochemistry.* **2001,** *58*(1), 33–41.

McDonald, B. A.; Linde, C. Pathogen Population Genetics, Evolutionary Potential, and Durable Resistance. *Annu. Rev. Phytopathol.* **2002,** *40,* 349–379.

Mundt, C. C. Durable Resistance: A Key to Sustainable Management of Pathogens and Pests. *Infect. Genet. Evol.* **2014,** *27,* 446–455.

Nagarajan, S. Is *Puccinia graminis* f.sp. *tritici* Virulence Ug99 a Threat to Wheat Production in North Western Plain Zone of India? *Indian Phytopathol.* **2012,** *65,* 219–226.

Nagarajan, S.; Singh, D. V. Long-Distance Dispersion of Rust Pathogens. *Annu. Rev. Phytopathol.* **1990,** *28,* 139–153.

Nagarajan, S.; Singh, H. Indian Stem Rust Rules an Epidemiological Concept on the Spread of Wheat Stem Rust. *Plant Dis. Rep.* **1975,** *59,* 133–136.

Nickrent, D. L. Mistletoe Phylogenetics: Current Relationships Gained from Analysis of DNA Sequences. In *Proceeding of West. Int. For. Dis. Work Conference*. Geils, B., Mathiasen, R., Eds.; USDA For. Serv.: Kona, HI, 2001.

Niks, R. E.; Rubiales, D. Detection of Potentially Durable Resistance Mechanisms in Plants to Specialized Fungal Pathogens. *Euphytica*. **2002**, *124*, 201–216.

Oshima, K.; Kakizawa, S.; Nishigawa, H.; Jung, H. Y.; Wei, W.; Suzuki, S.; Arashida, R.; Nakata, D.; Miyata, S.; Ugaki, M.; Namba, S. Reductive Evolution Suggested from the Complete Genome Sequence of a Plant-Pathogenic Phytoplasma. *Nat. Genet.* **2004**, *36*, 27–29.

Oshima, K.; Maejima, K.; Namba, S. Genomic and Evolutionary Aspects of Phytoplasmas. *Front. Microbiol.* **2013**, *4*, 1–8.

Papaix, J.; Burdon, J. J.; Lannou, C.; Thrall, P. H. Evolution of Pathogen Specialisation in a Host Metapopulation: Joint Effects of Host and Pathogen Dispersal. *PLoS Comput. Biol.* **2014**, *10*(5), e1003633.

Park, R. F.; Burdon, J. J.; McIntosh, R. A. Studies on the Origin, Spread, and Evolution of and Important Group of *Puccinia recondita* f.sp. *tritici* Pathotypes in Australasia. *Eur. J. Plant Pathol.* **1995**, *101*, 613–622.

Pryor, T.; Ellis, J. The Genetic Complexity of Fungal Resistance Genes in Plants. *Adv. Plant Pathol.* **1993**, *10*, 281–305.

Reid, N. The Mistletoe Bird and Australian Mistletoes: Co-evolution or Coincidence? *Emu*, **1987**, *87*, 130–131.

Reid, N. Coevolution of Mistletoes and Frugivorous Birds? *Aust. J. Ecol.* **1991**, *16*, 457–469.

Richter, T. E.; Pryor, T. J.; Bennetzen, J. L.; Hulbert, S. H. New Rust Resistance Specificities Associated with Recombination in the *Rpl* Complex in Maize. *Genetics*. **1995**, *141*, 373–381.

Rizzo, D. M.; Garbelotto, M. Sudden Oak Death: Endangering California and Oregon Forest Ecosystems. *Front. Ecol. Environ.* **2003**, *1*, 197–204.

Roumen, E. C. A Strategy for Accumulating Genes for Partial Resistance to Blast Disease in Rice within a Conventional Breeding Program. In *Rice Blast Disease*; Zeigler, R. S., Leong, S. A., Teng. P. S., Eds.; CAB International: Wallingford, UK, 1994; pp 245–266.

Singh, R. P.; Hodson, D. P.; Jin, Y.; Lagudah, E. S.; Ayliffe, M. A.; Bhavani, S.; Rouse, M. N.; Pretorius, Z. A.; Szabo, L. J.; Huerta-Espino, J.; Basnet, B. R.; Lan, C.; Hovmøller, M. S. Emergence and Spread of New Races of Wheat Stem Rust Fungus: Continued Threat to Food Security and Prospects of Genetic Control. *Phytopathology*. **2015**, *105*(7), 872–884.

Singh, R. P.; Huerta-Espino, J.; William, H. M. Genetics and Breeding for Durable Resistance to Leaf and Stripe Rusts of Wheat. *Turk. J. Agric. For.* **2005**, *29*, 121–127.

Singh, R. S. *Introduction to Principles of Plant Pathology*, 4th Edn; Oxford and IBH Publishing Company: New Delhi, 2009.

Snow, D. W. Evolutionary Aspects of Fruit Eating by Birds. *Ibis*. **1971**, *113*, 194–202.

Spanu, P. D.; Abbott, J. C.; Amselem, J.; Burgis, T. A.; Soanes, D. M., et al. Genome Expansion and Gene Loss in Powdery Mildew Fungi Reveal Tradeoffs in Extreme Parasitism. *Science*. **2010**, *330*, 1543–1546.

Talukder, Z. I.; Tharreau, D.; Price, A. H. Quantitative Trait Loci Analysis Suggests that Partial Resistance to Rice Blast is Mostly Determined by Race-Specific Interactions. *New Phytol.* **2004**, *162*, 197–209.

Tooley, P. W.; Kyde, K. L. Susceptibility of Some Eastern Forest Species to *Phytophthora ramorum*. *Plant Dis.* **2007**, *91*, 435–438.

Vanderplank, J. E. *Disease Resistance in Plants;* Academic Press: New York, 1968.

Vanderplank, J. E. *Disease Resistance in Plants;* Academic Press: New York, 1984.

Vanderplank, J. E. *Disease Resistance in Plants;* 2nd Edn; Elsevier: New York, 2012.

Vaz Patto, M. C.; Ferna´ndez-Aparicio, M.; Moral, A.; Rubiales, D. Characterization of Resistance to Powdery Mildew (*Erysiphe pisi*) in a Germplasm Collection of *Lathyrus sativus*. *Plant Breed.* **2006**, *125*(3), 308–310.

Von Sydow, M. 'Survival of the Fittest' in Darwinian Metaphysics: Tautology or Testable Theory? In *Reflecting on Darwin;* Voigts, E., Schaff,B., Pietrzak-Franger, M., Eds.; Taylor Francis Ltd.: London, U.K., 2014; pp 199–222.

Walia, J. J.; Willemsen, A.; Elci, E.; Caglayan, K.; Falk, B. W.; Rubio, L. Genetic Variation and Possible Mechanisms Driving the Evolution of Worldwide Fig Mosaic Virus Isolates. *Phytopathology.* **2014**, *104*, 108–114.

Weiland, J. E.; Nelson, A. H.; Hudler, G. W. Aggressiveness of *Phytophthora cactorum*, *P. citricola* I, and *P. plurivora* from European Beech. *Plant Dis.* **2010**, *94*, 1009–1014.

Wellings, C. R. *Puccinia striiformis* in Australia: A Review of the Incursion, Evolution and Adaptation of Stripe Rust in the Period 1979–2006. *Aust. J. Agric. Res.* **2007**, *58*, 567–575.

Young, N. D. The Genetic Architecture of Resistance. *Curr. Opin. Plant Biol.* **2000**, *3*, 285–290.

Yu, L. X.; Barbier, H.; Rouse, M. N.; Singh, S.; Singh, R. P.; Bhavani, S.; Huerta-Espino, J.; Sorrells, M. E. A Consensus Map for Ug99 Stem Rust Resistance Loci in Wheat. *Theor. Appl. Genet.* **2014**, *127* (7), 1561–1581.

Zadoks, J. C.; Schein, R. D. *Epidemiology and Plant Disease Management;* Oxford University Press: New York, 1979.

Zenbayashi, K.; Ashizawa, T.; Tani, T.; Koizumi, S. Mapping of the QTL (Quantitative Trait Locus) Conferring Partial Resistance to Leaf Blast in Rice Cultivar Chubu 32. *Theor. Appl. Genet.* **2002**, *104*, 547–552.

Zhan, J.; Mundt, C. C.; Hoffer, M. E.; McDonald, B. A. Local Adaptation and Effect of Host Genotype on the Rate of Pathogen Evolution: An Experimental Test in a Plant Pathosystem. *J. Evol. Biol.* **2002**, *15*, 634–647.

CHAPTER 2

THE PROCESS OF COEVOLUTION OF PLANTS AND THEIR PATHOGENS

BHUPENDRA S. KHARAYAT[1] and YOGENDRA SINGH[2]

[1]Division of Plant Pathology, ICAR-Indian Agricultural Research Institute, Pusa, New Delhi 110012, India

[2]Department of Plant Pathology, Collage of Agriculture, G.B. Pant University of Agriculture & Technology, Pantnagar 263145, Udham Singh Nagar, Uttarakhand, India

CONTENTS

ABSTRACT

Coevolution is an evolutionary process that prompts the genetic adaptation of a species in response to the natural selection imposed by another interacting species and the effects might be reciprocal. Host-parasite coevolution is defined as the reciprocal genetic change of two species by selective reciprocal influences. The process of coevolution between plants and the surrounding biota including viruses, fungi, bacteria, nematodes, insects, and mammals is considered by many biologists to have generated much of the earth's biological diversity. In natural ecosystem, variation in the genetic structure of pathogen population and the respective host is determined by a specific gene-for-gene coevolution. It is a form of reciprocal genetic change occurring in the two ecologically interacting species: the pathogen and its host. Mutation, sexual recombination and selection are important evolutionary forces acting at various stages of the interaction. Invasion and attack by a pathogen results in reduced plant fitness. Diffuse coevolution is more common, denoting interactions in which related pest species attack a range of plants that share similar chemical defenses. The distinction between pairwise and diffuse coevolution becomes more complex when genes involved in 'gene-for-gene' interactions are considered. Geographic variation in the structure of coevolutionary selection, e.g., ranging from mutualism to parasitism (selection mosaics, or genotype by genotype by environment, or G×G×E interactions); geographic variation in the intensity of coevolutionary selection, resulting in coevolutionary hotpots and coldspots; and trait remixing resulting from metapopulation dynamics, random genetic drift, gene flow. The concept of gene-for-gene coevolution is a major model for research on disease resistance in crop plants. Under the 'trench warfare' model, the same alleles are maintained over long time scales while in the 'arm race' model novel alleles are recurrently driven to fixation. Long-term evolutionary dynamics occurring between enzymes and inhibitors in plant–pest interactions have also been examined. In plant class I chitinase, nonsynonymous substitution rates exceed synonymous rates in Arabis and other dicots. Understanding of plants and their pathogen's coevolution in natural systems continues to develop as new theories at the population and species level are increasingly informed by studies unraveling the molecular basis of interactions between individual plants and their pathogens.

2.1   INTRODUCTION

Coevolution is defined, in evolutionary biology, as reciprocally induced evolutionary change between two or more species or populations. It is the mutual influence exerted by species on one another in the evolutionary process. Coevolutionary interactions, such as those between host and parasite, predator and prey, or plant and pollinator, evolve subject to the genes of both interactors. For example, that the evolution of pollination strategies can only be understood with knowledge of both the pollinator and the pollinated (Little et al., 2010). Coevolution can be broadly divided into three types: predator–prey coevolution, mutualism, and host–parasite coevolution (Simms, 1996). Coevolution is an evolutionary process that prompts the genetic adaptation of a species in response to the natural selection imposed by another interacting species and the effects might be reciprocal (Woolhouse et al., 2002). Coevolution can occur between any interacting populations: prey and predator, plant and herbivore, competitors, or mutualists, but it is expected to be particularly important in host–pathogen systems because of the intimate nature of the association and the strong selective pressures that each can exert on the other (Woolhouse et al., 2002). Host–parasite coevolution is defined as the reciprocal genetic change of two species by selective reciprocal influences. The process of coevolution between plants and the surrounding biota including viruses, fungi, bacteria, nematodes, insects, and mammals is considered by many biologists to have generated much of the earth's biological diversity (Occhipinti, 2013).

Hypotheses for the course of coevolution include: (a) escalating arm races in which plants relentlessly add to their chemical arsenals, whereas pathogens follow suit with new mechanisms for overriding those plant defenses, (b) cyclical selection in which highly defended plants are favored in times when virulent pathogens exert a severe pressure, but which gradually decline in prevalence because of costs associated with resistance traits when the plants are not under attack, and (c) a stasis that entails little evolutionary change in either plants or their pathogens because a paucity of genetic variation or the presence of specific constraints limit the opportunities for evolution (Kareiva, 1999). A process analogous to coevolution occurs in agricultural systems, in which natural enemies adapt to crop resistance introduced by breeding or genetic engineering. Because of this similarity, the investigation of resistance mechanisms in crops is helping to elucidate the workings of coevolution in nature, while evolutionary principles, including those derived from investigation of coevolution in nature, are being applied in the management of resistance in genetically engineered crops (Rausher, 2001).

2.2   NATURE OF COEVOLUTION

The coevolution can be defined as the process of reciprocal, adaptive genetic change in two or more species (Fig. 2.1). This simply means that changes in gene frequencies as a result of selection acting on one population create selection for changes in gene frequencies in the other population(s), although the kinds of population genetic processes that result can be different. Signatures of coevolution are pre-eminent in host–pathogen systems because of the strong selective pressures that the pathogen and host exert on each other. This results in the evolution of new defenses mechanism by the host. This ever-rising strategy of continued, fresh attacks and counter attack by parasites and formation of new defensive strategy by host evolve in the process (Knisken & Rausher, 2001). This "biological arm race" (Woolhouse et al., 2002) of fight of dominance by pathogens and plants continues in natural ecosystem.

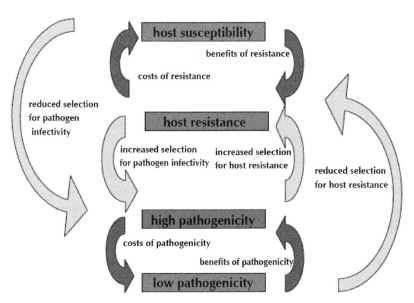

FIGURE 2.1   Coevolution of host–pathogen: Focusing reciprocity in those changes in allele frequencies due to strong selection in one species imposes selection resulting in marked changes in allelic frequencies in the other species. Green arrows indicate intra-specific selection; yellow arrows indicate inter-specific selection (Reprinted by permission from Nature Publishing Group: Nature Genetics, Woolhouse, M. E. J.; Webster, J. P.; Domingo, E.; Charlesworth, B.; Levin, B. R. Biological and Biomedical Implications of the Coevolution of Pathogens and Their Hosts. **2002**, *32*, 569–577. © 2002.).

2.2.1   EVOLUTIONARY FORCES CONTRIBUTING IN PLANT–PATHOGEN COEVOLUTION

In natural ecosystem, variation in the genetic structure of pathogen population and the respective host is determined by a specific gene-for-gene coevolution. It is a form of reciprocal genetic change occurring in the two ecologically interacting species: the pathogen and its host. Mutation, sexual recombination, and selection are important evolutionary forces acting at various stages of the interaction. Invasion and attack by a pathogen result in reduced plant fitness. Within the diseased plant population, mutation or recombination produces new receptor allele that can recognize the pathogen elicitor. In the presence of constant severe disease pressure, plants with the new resistance allele will be favorably selected (to produce more offspring) and increase in frequency. The increase in the resistance allele in plant population will in turn reduce pathogen's fitness and consequently, the overall level of disease in the plant population also decreases (McDonald & Linde, 2002). In natural populations, changes in the frequency of *avirulence* (*avr*) genes, driven by host population resistance structure, are complemented by genetic drift, which can generate marked differences between small neighboring demes as a consequence of chance survival or extinction of individual pathogen strains when population sizes crash (Anderson et al., 2010).

Variation by mutation or sexual recombination in the pathogen causing the loss or alteration of the *avr* allele will help the pathogen evade or suppress the plant resistance system. Since this increases the fitness advantage for the new variants or matching type, selection will favor and increase their frequency in the existing pathogen population. Population size, directional selection, mating system, and gene flow are well documented evolutionary forces responsible for the emergence and spread of new genetic variants or loss of defeated alleles (i.e., mutations) (McDonald & Linde, 2002). On the other hand, gene flow or migration from other pathogen populations may lead to the founding of new pathogen populations or the introduction of novel virulence combinations into existing populations (Anderson et al., 2010).

2.3   COEVOLUTION IN NATURAL ECOSYSTEM

Hosts and their pathogens have an intimate and antagonistic relationship; therefore, they are expected to influence each other's evolution. When a plant evolves a defense against a pathogen, then the pathogen may evolve counter-adaptations to its host's defense (Penn, 2001). Plant host–pathogen systems involve coevolutionary conflict where hosts are selected to evolve new resistance and pathogen to evolve new pathogenicity. The "theories of coevolution" of plant host and pathogen had given rise to two different schools of thought among the scientific community. One school emphasized the "Trench Warfare Model" while the other school believed that both the plant and pathogen undergo antagonistic coevolution supporting the "Arm Race Model" (Stahl et al., 1999). The followers of "trench warfare" believed that the advances and retreats of resistance (R) gene allele frequency maintained variation for resistance as an age-old dynamic polymorphic process (Jayakar, 1940). On the other hand, the proponents of "Arm Race Model" predicted the transient variation in disease resistance in plants to be the outcome of short-term ecological dynamics. This school also supported the R loci in host population to be monomorphic in nature (Stahl et al., 1999). Nevertheless, both the followers coalesced in accepting the "Geographical Mosaic of Coevolution Theory" which explicitly brought into account the inherent complexity of local and distant interacting communities, temporal and spatial species variation and the heterogeneity in magnitude, and direction of reciprocal selection across localities (Thompson, 2005; Tellier & Brown, 2007). Many studies have been devoted to the hypothesis of an endless "arm race" between antagonistic species, in which each species develops escalating attack and defense traits to counteract the adaptive responses of the other species. In reference to Lewis Carroll's book "Through the Looking Glass," Van Valen (1973) named this model of coevolution "the Red Queen Hypothesis" (RQH) because, even in a constant physical environment, interacting species must evolve continuously to maintain their position. The RQH was initially developed in the context of multiple species interactions, to account for the constant probability of species extinction. However, as modeling the coevolution of many species involves a number of difficulties, later studies based on the RQH have mostly been limited to interactions between pairs of species. From these restricted situations, two main predictions can be made: (a) a geographic view of the coevolutionary process suggests a dynamic mosaic structure, with local and temporary "hot spots" of antagonistic species coevolution and (b) arm races induce an advantage of rare genotypes of which resistance or virulence is more efficient and thus, frequency-dependent fluctuations of resistance and virulence may be predicted (Thompson, 1994). There is considerable evidence of genetic variation for resistance, but parasite-driven genetic change in resistance has never been observed directly (Little et al., 2010). Defense is thought to carry costs when it is not needed, that is, the fitness of resistant individuals in the absence of enemies is reduced compared to susceptible individuals.

Following this hypothesis, the arm race is constrained by the resistance cost and thus resistance and virulence of antagonistic species reach fixed levels that can be considered optimal. This hypothesis may result in a low level of genetic polymorphism of resistance and virulence in spatially restricted areas (Little et al., 2010). In contrast, resistance and virulence are expected to be polymorphic on a larger spatial scale if, because of ecological factors, the densities of antagonistic species are spatially heterogeneous. However, although resistance costs have often been demonstrated, expected geographical patterns of resistance and virulence are not observed: Antagonistic species densities are not clearly spatially associated with resistance and virulence variability levels (Little et al., 2010).

2.4   CLASSICAL MODELS OF COEVOLUTION

2.4.1   DIFFUSE COEVOLUTION

The term "diffuse coevolution" was introduced by Janzen (1980) in his note "When is it coevolution?" to describe the idea that selection on traits often reflects the actions of many community members, as opposed to pairwise interactions between species (Strauss et al., 2005). Diffuse coevolution is more common, denoting interactions in which related pest species attack a range of plants that share similar chemical defenses. The distinction between pairwise and diffuse coevolution becomes more complex when genes involved in "gene-for-gene" interactions are considered (Penn, 2001). Fox (1988) suggested that plants might exhibit generalized adaptations to cope with a suite of attacking herbivores, rather than having traits that were result of a one-on-one coevolutionary arm race. Gould (1988) further clarified a variety of ecological and genetic mechanisms that might lead to diffuse coevolution in response to selection from multiple species. Diffuse or multi-species coevolution in which many species, such as a suite of parasites and a suite of hosts, respond to each other's evolution (Thompson, 1989). This is probably the most realistic type of host–parasite coevolution since evolutionary adaptations to one parasite are likely to affect resistance to other parasites. However, to be considered coevolution, it is not enough for evolutionary changes in one lineage to cause changes in another lineage without the reverse. Karasov et al. (2014) identified a naturally interacting R gene and effectors pair in *Arabidopsis thaliana* and its facultative plant bacterium,

Pseudomonas syringae. The protein encoded by the *R* gene RPS5 recognizes an avrPphB homolog (avrPphB2) effectors and exhibits a balanced poly-morphism that has been maintained for over 2 million years. The presence of an ancient balanced polymorphism, the *R* gene confers a benefit when plants are infected with bacterium *P. syringae* carrying avrPphB2 but also incurs a large cost in the absence of pathogen infection. RPS5 alleles are maintained at intermediate frequencies in *Arabidopsis* populations globally, suggesting ubiquitous selection for resistance. These results and an associated model suggest that the *R* gene polymorphism in *A. thaliana* may not be main-tained through a tightly coupled interaction involving a single coevolved *R* gene and effectors pair. More likely, the stable polymorphism is maintained through complex and diffuse community-wide interactions.

2.4.2 GEOGRAPHIC MOSAIC THEORY

Coevolution among species and the impacts of geography on evolution have always been major research areas within evolutionary biology (Gomulkie-wicz et al., 2007). The geographic mosaic theory of coevolution hypothesizes that three processes are the primary drivers of coevolutionary dynamics: (a) geographic variation in the structure of coevolutionary selection, for example, ranging from mutualism to parasitism (selection mosaics, or geno-type by genotype by environment, or G×G×E interactions); (b) geographic variation in the intensity of coevolutionary selection, resulting in coevolu-tionary hotpots and coldspots; and (c) trait remixing resulting from meta-population dynamics, random genetic drift, gene flow (Thompson, 1994, 2005). The term "selection mosaic" refers to variability in the functions that describe reciprocal fitness interactions across space. The mosaics are likely to increase in complexity as pairwise interactions diversify into multi-specific interactions. It is not enough that the strength of coevolutionary selection varies between populations, it is also necessary that the direction of that selection varies, so that the outcomes of coevolution are different in different populations, depending on their environment. In other words, the costs and benefits to both partners of any particular adaptation are dependent not only on the adaptations of their partner, but also on the environment in which the interaction occurs (Nash, 2008).

The strength of coevolution varies between populations of interacting organisms. For a pair of interacting species, a "hot spot" is a location where the fitness of both species is affected by the distribution of traits in the other species or coevolutionary selection is intense. In contrast, in a "cold spot" the fitness of at least one of the species is unaffected by the other or the interacting species evolve independently of each other (Gomulkiewicz et al., 2000). Finally, in order for the coevolutionary process to work, there must be a mechanism that allows traits that have evolved in one population to be transferred to and mixed with traits that have evolved in other areas for example trait mixing (Gomulkiewicz et al., 2000). "Trait remixing" includes gene flow across landscapes, random genetic drift within populations, extinction and recolonization of local populations, and mutation (Gomulkiewicz et al., 2007).

The geographic mosaic theory predicts that these three processes lead to three observable patterns: spatial variation in the traits mediating an interspecific interaction, trait mismatching among interacting species (local maladaptation), and a few species-level coevolved traits (Thompson, 2005). These patterns are often used as evidence for the presence of a geographic mosaic of coevolution, but they can also result from other non-coevolutionary processes (Gomulkiewicz et al., 2000).

2.5 RECENT MODELS OF COEVOLUTION IN PLANT–PATHOGEN INTERACTION SYSTEM

2.5.1 GENE-FOR-GENE SYSTEM

The concept of gene-for-gene coevolution is a major model for research on disease resistance in crop plants (Thrall & Burdon, 2002). The gene-for-gene model refers to a specific genetic interaction between a host and its pathogen. It states that a resistance gene (*R* gene) in the host and an *avr* gene in the pathogen must be present for the host to be resistant. These models assume that single plant resistance genes interact specifically with single pathogen *avr* genes (Best et al., 2008). It also assumes multiple alleles at *R* gene and *avr* gene loci are maintained over long stretches of evolutionary time, which results in the maintenance of high genetic diversity linked to these loci (Wichmann et al., 2005). It is now clear that a strict gene-for-gene model of coevolutionary interactions between one *R* gene and one *avr* gene is overly simplistic; a single *avr*-locus may recognize multiple *R* genes, single *R* genes can control resistance to several pest species, and tight genomic clustering of *R* genes might influence the evolutionary dynamics of resistance (Michelmore & Meyers, 1998).

In this model, the outcome of infection is based on the interaction of dominant *R* genes in the host and dominant *avr* genes in the pathogen.

This genetic interaction is termed as effectors triggered immunity (Jones & Dangl, 2006), where R proteins constitute the recognition component of the plant immune system and detect the presence of specific pathogen effector (Avr) proteins and trigger a defense response that prevents infection. Simple gene-for-gene models predict stable polymorphisms through frequency-dependent selection, provided there are costs associated with resistance and virulence (Leonard, 1997).

2.5.2 ARM RACE

The ongoing coevolutionary struggle between hosts and pathogens, with hosts evolving to escape from pathogen infection and pathogens evolving to escape from host defenses, can generate an "arm race," that is, the occurrence of recurrent selective sweeps that each favors a novel resistance or virulence allele that goes to fixation (Aguileta et al., 2009). The coevolutionary arm race between pathogen and plant host has prompted diverse pathogen virulence strategies and modified plant surveillance and defense systems. The plant host must defend itself against pathogenic virulence in order to survive, while the pathogen has to evade or suppress host immune response in order to proliferate (Voigt, 2014). An "arm race" model of coevolution predicts that selective sweeps of adaptation between host R gene and pathogen (avr gene) should reduce both the number of alleles and the genetic diversity at these loci (Bergelson et al., 2001). The evolution of R and avr genes is the result of an arm race going on between host and pathogen, where the host tries to prevent the infection, while the pathogen tries to bypass the host defenses to cause disease (Maor & Shirasu, 2005; Van der dose & Rep, 2007). In the arm race model, a pathogen evolves an effector allele that enhances its fitness through the manipulation of host physiology and/or the suppression of host defense. The frequency of such an allele rapidly increases in the pathogen population, eventually replacing the older allele. In turn, a new allele of a gene for a host effectors target or a counter-defense protein that helps the host to evade the effect of the pathogen effector evolves, and its allele frequency rapidly increases to finally fix in the population. This cycle is repeated indefinitely.

The coevolutionary arm race model for plant–pathogen interactions implies that R genes are relatively young and monomorphic. However, recent reports show R gene longevity and co-existence of multiple R genes in natural populations. This indicates that R genes (Fig. 2.2) are maintained by balancing selection, which occurs when loss of the matching avr gene in the pathogen is associated with reduced virulence (Terauchi & Yoshida, 2010). The characteristics of the arm race selection are: genes encoding for the proteins involved exhibit rapid evolution resulting in a larger amount of amino acid replacements between the species, and a selective sweep reduces the genetic variation (intra-species polymorphism) in the regions tightly linked to the selected sites and renders extensive linkage disequilibrium (LD) around the sites (Van der Hoorn et al., 2002). Within bacteria, selection for a single gene will result in selection for the entire bacterial chromosome harboring that gene (a phenomenon known as genetic hitchhiking), due to the fact that the bacterial chromosome comprises a single linkage group. Hence, strong selection for a single bacterial gene (a selective sweep) may cause an entire chromosome to become predominant in the population, leading to a population wide reduction in genetic variation (Smith & Haigh, 1974).

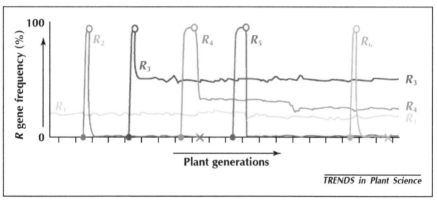

FIGURE 2.2 Model for selection of R genes in natural plant populations showing the frequency of various R genes (R1–R6) in the plant population over a long period of time. X-axis representing several plant generations. Circles marks indicating the point at which new R genes are generated (filled circles) or when pathogens circumvent recognition by a mutation in the matching *avr* gene (open circles). Strong selection pressure after these events results in the rise and fall, respectively, of the R gene frequency. Crosses (x) indicating incidental extinction of non-functional R genes. Recognition of the *avr* gene product by R1, R3, and R4 gene products cannot be circumvented without a reduction in virulence of the pathogen, resulting in maintenance of these genes by balancing selection (horizontal lines). By contrast, recognition of *avr* factors that match R2, R5, and R6 can be circumvented by mutations in the *avr* gene without causing reduction in virulence of the pathogens. Consequently, in the pathogen population, the matching R genes are likely to become rare (R5) or extinct (R2 and R6) because the modified *avr* genes will replace the original *avr* genes present in existing pathogen population (Reprinted from *Trends Plant Sci.* **2002**, *7*(2), Van der Hoorn, R. A.; Pierre, J. G. M, De, Wit; Jooste, M. H. A. J., Balancing Selection Favors Guarding Resistance Proteins, pp. 67–71. © 2002, with permission from Elsevier.).

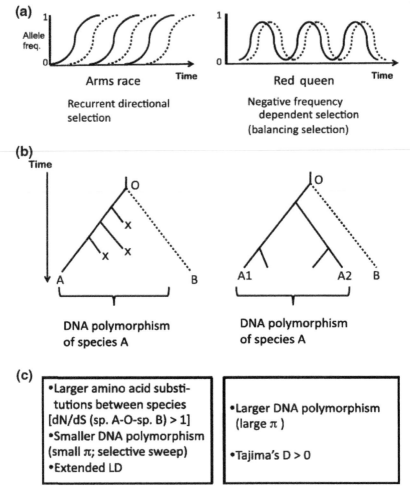

FIGURE 2.3 Effects of arm race and trench warfare or Red Queen on DNA polymorphisms and divergence under evolutionary scenarios of host–pathogen interactions. (a) Models of allele frequency changes in arm race (left) and Red Queen (right) host–pathogen interactions. The frequencies of the pathogen effector allele are shown by solid line and the frequencies of the host defense allele are shown by dotted line. (b) Gene genealogy of a locus involved in arm race or Red Queen interactions between the host and pathogen. Solid lines are indicating the genealogy within a species, whereas the dotted lines are for a closely related species. In the arm race scenario (left), a single allele (indicated as A) is selected when it confers a higher fitness to the carrier organism. Other alleles are selected out from the population as indicated by "x." In the Red Queen scenario (right), two or more alleles (indicated as A1 and A2) are maintained in the species for a long time. (c) Typical DNA sequence signatures resulting from arm race (left) and Red Queen (right) scenarios (Source: Ryohei Terauchi and Kentaro Yoshida, "Towards population genomics of effector-effector target interactions," in New Phytologist, 2010 Sep;187(4):929-39. With permission from John Wiley.)

2.5.3 TRENCH WARFARE OR RED QUEEN

Hosts and pathogens are engaged in a never-ending struggle, hosts evolving to escape pathogen infection and pathogens evolving to escape host defenses, as illustrated by the Red Queen tale (Aguileta et al., 2009). In the Red Queen model, which is also known as the "Trench-warfare" model (Stahl et al., 1999), two or more alleles of effector genes exist in the pathogen population, and the allele with the largest contribution to pathogen fitness increases in frequency. This model represents negative frequency-dependent selection, and the allele frequencies of the matching genes in the pathogen and host oscillate over time. Red Queen selection may maintain the same set of alleles in the population for a longer time. The two types of coevolutionary processes may leave contrasting patterns of DNA polymorphisms and divergence in the genomes of the organisms involved and these can be detected using statistical tests for DNA variation developed for molecular population genetics studies (Fig. 2.3 b, c). The coevolutionary "arm race" is a widely accepted model for the evolution of host–pathogen interactions. This model predicts that variation for disease resistance will be transient, and that host populations generally will be monomorphic at disease-resistance (R gene) loci. However, plant populations show considerable polymorphism at R gene loci involved in pathogen recognition. Under the "trench warfare" model, the same alleles are maintained over long time scales while in the "arm race" model novel alleles are recurrently driven to fixation. A high diversity of phenotypes has in fact long been observed regarding host resistance or pathogen infection ability depending on host/pathogen genotypes (Salvaudon et al., 2008). Stahl et al. (1999) tested the arm race model in *Arabidopsis thaliana* by analyzing sequences flanking *Rpm1*, a gene conferring the ability to recognize *Pseudomonas* pathogens carrying *AvrRpm1* or *AvrB*. Stahl et al. (1999) rejected the arm race hypothesis: Resistance and susceptibility alleles as this locus have coexisted for millions of years. These analyses support a "trench warfare" hypothesis, in which advances and retreats of resistance-allele frequency maintain variation for disease resistance as a dynamic polymorphism.

2.6 EVIDENCES FOR PLANT–PATHOGEN COEVOLUTION

Proof of occurring coevolution requires evidence of genetic changes in the field for both hosts and pathogens. Nature offers a number of examples in which organisms evolve defenses to attacks of parasites. These parasites in turn, find ways to circumvent these defenses and attack the organisms with greater vigor (Occhipinti, 2013).

2.6.1   PATHOGEN INFECTION AND HOST DEFENSES RECIPROCALLY AFFECT THE FITNESS OF HOST AND PATHOGEN

During the course of coevolution, plants have developed defense mechanisms against pathogenic intruders such as fungi, bacteria, viruses, and nematodes. Plants and pathogens are involved in an intimate detrimental physiological and ecological interaction. The strength of the selective pressure depends on the virulence of pathogen, the driving force in host–parasite coevolution. Evolutionary biologists define the virulence of a parasite as the reduction in host fitness caused by infection (Little et al., 2010). Virulence is a specific product of plant–pathogen interaction; virulence does not depend on parasite or plant alone (Ebert & Hamilton, 1996). The successful pathogen infection is determined by the combination of host and pathogen genotypes. Pathogen infection reduces their fitness while hosts have developed different defense strategies to avoid or limit infection and to compensate for its costs (Ebert & Hamilton, 1996). However, virulence can also include reductions in host fecundity (Best et al., 2008), which is relevant for some parasites such as castrating parasites and obligate killers (Ebert et al., 2004). In plants, the two major defense mechanisms are resistance (defined as the ability of the host to limit parasite multiplication) and tolerance (defined as the ability of the host to reduce the damage caused by parasite infection) (Clarke, 1986). Host defenses may have a negative effect on parasite fitness (Fig. 2.4). Hence, hosts and parasites may modulate the dynamics and genetic structure of each other's populations (Fig. 2.4), and hosts and pathogens may coevolved, defining coevolution as the process of reciprocal, adaptive genetic change in two or more species (Knisken & Rausher, 2001). Within the tolerance framework, however, there is room for more comprehensive measures of host fitness traits, and for full consideration of the consequences of coevolution. For example, the evolution of tolerance can also result in changed selection on parasite populations, which should provoke parasite evolution despite the fact that tolerance is not directly antagonistic to parasite fitness (Little et al., 2010). Clay and Kover (1996) suggested that the evolution of tolerance dampens coevolution by not directly reducing parasite numbers, but we posit that any host evolution that knocks parasites away from their evolutionary optimum (or creates a new one) will be countered by parasite evolution.

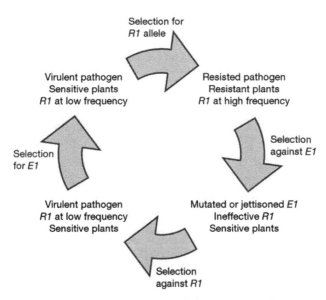

FIGURE 2.4 Coevolution of host *R* genes and the pathogen effectors complement. A pathogen carries an effector gene (*E1*) that is recognized by a rare *R1* allele (top). This results in selection for an elevated frequency of *R1* in the population. Pathogens in which the effector is mutated are then selected, because they can grow on *R1* containing plants (right). *R1* effectiveness erodes, and, because at least some *R* genes have associated fitness costs, plants carrying *R1* can have reduced fitness (bottom), resulting in reduced *R1* frequencies. The pathogen population will still contain individuals with *E1*. In the absence of *R1*, *E1* will confer increased fitness, and its frequency in the population will increase (left). This will lead to resumption of selection for *R1* (top). In populations of plants and pathogens, this cycle is continuously turning, with scores of effectors and many alleles at various *R* loci in play (Reprinted with permission from Nature Publishing Group: Nature. **2006,** 444, pp323–329. Jones, J. D. G.; Dangl, J. L. The Plant Immune System. © 2006.)

Woolhouse et al. (2002) point out three conditions that are required for host–pathogen coevolution: (a) reciprocal effects of the relevant traits of the interaction (e.g., defense and pathogenicity) on the fitness of the two species (pathogens and hosts), (b) dependence of the outcome of the host–pathogen interaction on the combinations of host and pathogen genotypes involved, and (c) genetic variation in the relevant host and pathogen traits. If aforesaid three conditions are met, changes in genotypic frequencies in both the host and the pathogen populations in the field demonstrating coevolution.

During antagonistic coevolution between viruses and their hosts, viruses have a major advantage by evolving more rapidly. To counter viruses, plants have developed defense systems such as gene-for-gene resistance and RNA silencing. Viruses need to evade recognition by resistance genes and encode suppressors of RNA silencing for successful infection. This suggests that and their hosts coexist and have coevolved, although the processes remain largely unknown (Ishibashi et al., 2012).

Although it is currently assumed that plants coevolved with their pathogens, this evidence is available for a few plant–pathogen systems and derives from analysis of the interaction of plants with bacteria, fungi, and oomycetes in their natural habitat (Salvaudon et al., 2008; Burdon & Thrall, 2009). Also taking account into the resistance cost hypothesis is a useful addition to Red Queen or arm race model, this theoretical view of coevolutionary process does not provide predictions that fit well with most situations observed in the field conditions (Little, 2002; Woolhouse & Webster, 2000).

2.6.2 PATHOGEN ATTACK AND HOST DEFENSE

The pathogen attack–plant defense, counter attack, counter-defense, and counter-counter-defense are continuous processes under host–pathogen coevolution. During plant–pathogen coevolution, pathogens have evolved various strategies to attack the plants, which include toxins, cell wall degrading enzymes, *avr* genes to overcome host resistance. Similarly, plants have evolved a range of defense mechanisms (a complex array of chemicals and enzymatic defenses, both constitutive and inducible, *R* genes etc.) to prevent ingress and colonization by potential pathogens, which derive from diverse phyla and include fungi, oomycetes, animals, bacteria, and viruses (Voigt, 2014). These wide chemical arrays involve pathogenesis related (PR)-proteins, phytoalexins, phytoanticipins etc. Plants produce a variety of defense enzymes that attack the polysaccharides or peptidoglycans in pathogen cell walls. These glucanhydrolases are often part of the PR response, and include chitinases and β-1,3 endoglucanases. Apart from direct protection by inhibiting pathogen growth, glucanhydrolases produce oligosaccharides that are elicitors of plant PR response and systemic acquired resistance (Occhipinti, 2013). Pathogens can counteract these enzymes with various inhibitors. Long-term evolutionary dynamics occurring between enzymes and inhibitors in plant–pest interactions have also been examined. In plant class I chitinase, nonsynonymous substitution rates exceed synonymous rates in *Arabis* and other dicots (Bishop et al., 2000). Plant chitinases are believed to defend against fungal infection by attacking chitin, a principal component of fungal cell walls. Class I chitinases in *Arabis* species often exhibit higher rates of nonsynonymous than synonymous substitutions, a hallmark of adaptive evolution (Rausher, 2001). Moreover, amino acid substitutions are concentrated in the molecule's active site, a pattern not usually seen in enzyme evolution. Because the structure of chitin does not evolve, the most reasonable interpretation of this pattern is that plant chitinases are coevolving with pathogen chitinase inhibitors, small carbohydrate, or protein molecules that competitively inhibit the breakdown of chitin by chitinase (Rausher, 2001).

2.7 CONCLUSION

Plants are engaged in a continuous coevolutionary struggle for dominance with their pathogens. The outcomes of these interactions are of particular importance to human activities, as they can have dramatic effects on agricultural systems. In natural ecosystem, variation in the genetic structure of pathogen population and the respective host is determined by a specific gene-for-gene coevolution. It is a form of reciprocal genetic change occurring in the two ecologically interacting species: the pathogen and its host. Hosts and their pathogens have an intimate and antagonistic relationship; therefore, they are expected to influence each other's evolution. Diffuse coevolution is more common, denoting interactions in which related pest species attack a range of plants that share similar chemical defenses. Diffuse or multispecies coevolution in which many species, such as a suite of parasites and a suite of hosts, respond to each other's evolution. This is probably the most realistic type of host–parasite coevolution since evolutionary adaptations to one parasite are likely to affect resistance to other parasites. The geographic mosaic theory predicts that these three processes lead to three observable patterns: spatial variation in the traits mediating an interspecific interaction, trait mismatching among interacting species (local maladaptation), and a few species-level coevolved traits.

The gene-for-gene model refers to a specific genetic interaction between a host and its pathogen. It states that a *R* gene in the host and an *avr* gene in the pathogen must be present for the host to be resistant. The coevolutionary "arm race" is a widely accepted model for the evolution of host–pathogen interactions. This model predicts that variation for disease resistance will be transient, and that host populations generally will be monomorphic at disease-resistance (*R* gene) loci. However, plant populations show considerable polymorphism at *R* gene loci involved in pathogen recognition. Under the "trench warfare" model, the same alleles are maintained over long time scales while in the "arm race" model novel alleles are recurrently driven to fixation.

Understanding of plants and their pathogen's coevolution in natural systems continues to develop as new theories at the population and species-level are increasingly informed by studies unraveling the molecular basis of interactions between individual plants and their pathogens.

KEYWORDS

- **avirulence**
- **coevolution**
- **gene-for-gene model**
- **geographic mosaic theory**
- **trench-warfare model**

REFERENCES

Aguileta, G.; Refregier, G.; Yockteng, R.; Fournier, E.; Giraud, T. Rapidly Evolving Genes in Pathogens: Methods for Detecting Positive Selection and Examples among Fungi, Bacteria, Viruses and Protists. *Infect. Genet. Evol.* **2009,** *9,* 656–670. doi:10.1016/j.meegid.2009.03.010.

Anderson, J. P.; Cynthia, A.; Gleason, R. C.; Foley, A.; Thrall, P. H.; Burdon, J.; Singh, K. Plants Versus Pathogens: An Evolutionary Arms Race. *Funct. Plant. Biol.* **2010,** *37*(6), 499–512. doi:10.1071/FP09304.

Bergelson, J.; Kreitman, M.; Stahl, E. A.; Tian, D. C. Evolutionary Dynamics of Plant R-Genes. *Science.* **2001,** *292,* 2281–2285.

Best, A.; White, A.; Boots, M. Maintenance of Host Variation in Tolerance to Pathogens and Parasites. *Proc. Natl. Acad. Sci. U. S. A.* **2008,** *105,* 20786–20791.

Bishop, J. G.; Dean, A. M.; Mitchell-Olds, T. Rapid Evolution in Plant Chitinases: Molecular Targets of Selection in Plant–Pathogen Coevolution. *Proc. Natl. Acad. Sci. U. S. A.* **2000,** *97,* 5322–5327.

Burdon, J. J.; Thrall, P. H. Coevolution of Plants and Their Pathogens in Natural Habitats. *Science.* **2009,** *324*(5928), 755–756. doi: 10.1126/science.1171663.

Clarke, D. D. Tolerance of Parasites and Disease in Plants and Its Significance in Host–Parasite Interactions. *Adv. Plant Pathol.* **1986,** *5,* 161–198.

Clay, K.; Kover, P. X. The Red Queen Hypothesis and Plant/Pathogen Interactions. *Annu. Rev. Phytopathol.* **1996,** *34,* 29–50.

Ebert, D.; Carius, H. J.; Little, T. J.; Decaestecker, E. The Evolution of Virulence When Parasites Cause Host Castration and Gigantism. *Am. Nat.* **2004,** *164,* 19–32.

Ebert, D.; Hamilton, W. D. Sex against Virulence: The Coevolution of Parasitic Diseases. *Trends Ecol. Evol.* **1996,** *11,* 79–82.

Fox, L. R. Diffuse Coevolution within Complex Communities. *Ecology.* **1988,** *69,* 906–907.

Gomulkiewicz, R.; Thompson, J. N.; Holt, R. D.; Nuismer, S. L.; Hochberg, M. E. Hot Spots, Cold Spots, and the Geographic Mosaic Theory of Coevolution. *Am. Nat.* **2000,** *156,* 156–174.

Gomulkiewicz, R.; Drown, D. M.; Dybdahl, M. F.; Godsoe, W.; Nuismer, S. L.; Pepin, K. M.; Ridenhour, B. J.; Smith, C. I.; Yoder, J. B. Dos and Don'ts of Testing the Geographic Mosaic Theory of Coevolution. *Heredity.* **2007,** *98,* 249–258.

Gould, F. Genetics of Pairwise and Multispecies Plant–Herbivore Coevolution. In *Chemical Mediation of Coevolution;* Spencer, K. C, Ed.; Academic Press: San Diego, CA, 1988; pp 13–55.

Ishibashi, K.; Mawatari, N.; Miyashita, S.; Kishino H.; Meshi T.; Ishikawa M. Coevolution and Hierarchical Interactions of Tomato Mosaic Virus and the Resistance Gene Tm-1. *PLoS Pathog.* **2012,** *8*(10), e1002975. doi: 10.1371/journal.ppat.1002975.

Janzen, D. H. When is it Coevolution? *Evolution.* **1980,** *34*(3), 611–612.

Jayakar S. D. A Mathematical Model for Interaction of Gene Frequencies in a Parasite and Its Host. *Theor. Popul. Biol.* **1940,** *1,* 140–164.

Jones, J. D. G.; Dangl, J. L. The Plant Immune System. *Nature.* **2006,** *444,* 323–329.

Karasov, T. L.; Kniskern, J. M.; Gao, L.; Brody, J.; De Y.; Ding, J.; Ullrich, D.; Lastra, R. O.; Nallu, S.; Roux, F.; Innes, R. W.; Barrett, L. G.; Hudson, R. R.; Bergelson, J. The Long-Term Maintenance of a Resistance Polymorphism through Diffuse Interactions. *Nature.* **2014,** *512,* 436–440.

Kareiva, P. Coevolutionary Arms Races: Is Victory Possible? *Proc. Natl. Acad. Sci. U. S. A.* **1999,** *96,* 8–10.

Knisken, J.; Rausher, M. D. Two Modes of Host–Enemy Coevolution. *Popul. Ecol.* **2001,** *43,* 3–14.

Leonard, K. Modelling Gene Frequency Dynamics. In *The Gene-for-gene Relationship in Plant–Parasite Interactions;* Crute, I. R., Holub, E. B., Burdon, J. J., Eds.; CAB International: Wallingford, 1997; pp 211–230.

Little, T. J. The Evolutionary Significance of Parasitism: Do Parasite-Driven Genetic Dynamics Occur Ex Silico? *J. Evol. Biol.* **2002,** *15,* 1–9.

Little, T. J.; Shuker, D. M.; Colegrave, N.; Day, T.; Graham, A. L. The Coevolution of Virulence: Tolerance in Perspective. *PLoS Pathog.* **2010,** *6* (9): e1001006. doi: 10.1371/journal. ppat.1001006.

Maor, R.; Shirasu, K. The Arms Race Continues: Battle Strategies between Plants and Fungal Pathogens. *Curr. Opin. Microbiol.* **2005,** *8*(4), 399–404.

McDonald, B. A.; Linde, C. Pathogen Population Genetics, Evolutionary Potential, and Durable Resistance. *Annu. Rev. Phytopathol.* **2002,** *40,* 349–379.

Michelmore, R. W.; Meyers, B. C. Clusters of Resistance Genes in Plants Evolve by Divergent Selection and a Birth-and-Death Process. *Genome Res.* **1998,** *8,* 1113–1130.

Nash, D. R. Process Rather than Pattern: Finding Pine Needles in the Coevolutionary Haystack. *J Biol.* **2008,** *7,* 14. doi: 10.1186/jbiol75.

Occhipinti, A. Plant Coevolution: Evidences and New Challenges. *J. Plant. Interact.* **2013,** *8*(3), 188–196. doi: 10.1080/17429145.2013.816881.

Penn, D. J. Host–Parasite Coevolution. In *Handbook of Plant Sciences;* Keith, R., Ed.; John Wiley & Sons, Ltd.: England, 2007; Vol. 2, pp 808–815.

Rausher, M. D. Co-evolution and Plant Resistance to Natural Enemies. *Nature.* **2001,** *411,* 857–864.

Salvaudon, L.; Giraud, T.; Shykoff, J. A. Genetic Diversity in Natural Populations: A Fundamental Component of Plant–Microbe Interactions. *Curr. Opin. Plant Biol.* **2008,** *11,* 135–143.

Simms, E. L. The Evolutionary Genetics of Plant–Pathogen Systems. *Bioscience.* **1996,** *46,* 136–143.

Smith, J. M.; Haigh, J. Hitch–Hiking Effect of a Favorable Gene. *Genet. Res.* **1974,** *23,* 23–35.

Stahl, E. A.; Dwyer, G.; Mauricio, R.; Kreitman, M.; Bergelson, J. Dynamics of Disease Resistance Polymorphism at the *Rpm1* Locus of *Arabidopsis. Nature.* **1999,** *400,* 667–671.

Strauss, S. Y.; Sahli, H.; Conner, J. K. Toward a More Trait-Centered Approach to Diffuse (Co) Evolution. *New Phytol.* **2005,** *165,* 81–90.

Tellier, A.; Brown, J. K. M. Stability of Genetic Polymorphism in Host–Parasite Interactions. *Proc. R. Soc. Lond. B. Biol. Sci.* **2007,** *274,* 809–817.

Terauchi, R.; Yoshida, K. Towards Population Genomics of Effector-Effector Target Interactions. *New Phytol.* **2010,** *87*(4), 929–39.

Thompson J. N. *The Geographic Mosaic of Coevolution;* University of Chicago Press: Chicago, 2005; pp 443.

Thompson, J. N. Concepts of Coevolution. *Trends Ecol. Evol.* **1989,** *4,* 179–183.

Thompson, J. N. *The Coevolutionary Process;* The University of Chicago Press: Chicago, 1994; pp 345.

Thrall, P. H.; Burdon, J. J. The Evolution of Gene-for-Gene Interactions: Consequences of Variation in Host and Pathogen Dispersal. *Plant Pathol.* **2002,** *51,* 169–184.

Van der dose, H. C.; Rep, H. Virulence Gene and the Evolution of Host Specificity in Plant Pathogenic Fungi. *Mol. Plant Microbe Interact.* **2007,** *20,* 1175–1182.

Van der Hoorn, R. A.; Pierre, J. G. M, De, Wit; Jooste, M. H. A. J. Balancing Selection Favors Guarding Resistance Proteins. *Trends Plant Sci.* **2002,** *7*(2), 67–71.

Van Valen L. A New Evolutionary Law. *Evol. Theor.* **1973,** *1,* 1–30.

Voigt, C. A. Callose-Mediated Resistance to Pathogenic Intruders in Plant Defense-Related Papillae. *Front. Plant Sci.* **2014,** *5*(168), 1–6. doi: 10.3389/fpls.2014.00168

Wichmann, G.; Ritchie, D.; Kousik, C. S.; Bergelson, J. Reduced Genetic Variation Occurs among Genes of the Highly Clonal Plant Pathogen *Xanthomonas axonopodis* pv. Vesicatoria, Including the Effector Gene *avrBs2. Appl. Environ. Microbiol.* **2005,** *71*(5), 2418–2432.

Woolhouse, M. E. J.; Webster, J. P. In Search of the Red Queen. *Parasitol. Today.* **2000,** *16,* 506–508.

Woolhouse, M. E. J.; Webster, J. P.; Domingo, E.; Charlesworth, B.; Levin, B. R. Biological and Biomedical Implications of the Coevolution of Pathogens and their Hosts. *Nat. Genet.* **2002,** *32,* 569–577.

CHAPTER 3

REPRODUCTIVE FITNESS OF FUNGAL PHYTOPATHOGENS: DERIVING CO-EVOLUTION OF HOST– PATHOGEN SYSTEMS

REKHA BALODI[1*], LAJJA VATI GHATAK[2], SUNAINA BISHT[1], and NANDANI SHUKLA[1]

[1]Plant Pathology, GB Pant University of Agriculture and Technology, Pantnagar 263 145, Uttarakhand, India

[2]Plant Breeding and Genetics, Bihar Agricultural University, Sabour 813 210, Bihar, India

*Corresponding author. E-mail: balodirekha30@gmail.com

CONTENTS

ABSTRACT

Plant-pathosystems are complex systems composed of host, pathogen, environment, and human beings. Plants (hosts) are affected are many types of pathogens belonging to prokaryotes, eukaryotes, or viruses, both host and pathogen remain in continuous arm race for their survival. Pathogens belonging to kingdom fungi are notorious, destructive, and are known for epidemics and starvation they have caused globally. With the changing management practices and climate change, fungal pathogens pose a great threat to the food security. It is important to understand the mechanism of evolution and adaptation in these pathogens to understand their increased fitness. For both evolution and adaptation, reproduction plays foundation role, thus understanding of reproductive fitness will help in better understanding of epidemiology and spread of these pathogens. This chapter is an attempt to explain reproductive fitness and various methods for estimation of reproductive fitness of fungal phytopathogens. Further, mechanisms of generation of variation and adaptation are elaborated by *Magnaporthe* pathosystem, which is the most problematic pathogen of important cereal hosts.

3.1   INTRODUCTION

Host–pathogen systems are composed of dynamic and ever-evolving components, which interact with each other and during this interaction, they evolve mechanisms to survive and thrive well in presence of their counterparts. Successful infection and colonization of host by the pathogen leads to yield losses and economic damage. The yield losses caused by the phytopathogens are of economic concern and become the cause of starvation in developing countries (Strange & Scott, 2005; Mikaberidze et al., 2014). Agro-ecosystems are under continuous intervention by humans, these interventions may facilitate emergence of new pathogens or speed up the process of evolution. Therefore, several approaches involving re-introduction of combination of environmental, species, and genetic heterogeneity are adopted to slow down these processes and prevent emergence of new pathogens (Davies & Davies, 2010).

Among all phytopathogens, fungi are the most destructive and cause huge losses in terms of productivity and environmental deterioration (Daverdin et al., 2012). Fungi have been the causal agents of many widespread epiphytotics, which have changed the outlook and course of human civilization toward the science of plant pathology (Fisher et al., 2012). Fungi are reported to be the most important cause of loss of biodiversity and have emerged as devastating pathogens in various taxa (Fisher et al., 2012). Some of the emerging diseases caused by fungi in different taxonomic groups are enlisted in Table 3.1. Thus, the importance of fungal diseases is not limited to field of plant pathology but has a wider scope in current scenario. Also, humans have played a vital role in dissemination of fungal pathogens, and thereby, affected the habitat flexibility and preferred mode of perpetuation. Many fungi have multiple ways of genetic recombination, hybridization, or horizontal gene transfer, thus, with the human intervention, their evolution is accelerated and new lineages are formed; which result in novel pathogens (Farrer et al., 2011; Fisher et al., 2012).

TABLE 3.1 Emerging Fungal Diseases of Importance Reported from Plant/Animal system (Source: Fisher et al., 2012).

S. N.	Crop/host	Disease	Pathogen	Reference
1.	Rice, wheat, barley, and other members of Poaceae	Blast or grey spot	*Magnaporthe grisea, M. oryzae*	Fisher et al. (2012)
2.	Wheat	Rust	*Puccinia graminis*	Fisher et al. (2012)
3.	Sea corals	Aspergillosis	*Aspergillus sydowii*	Kim and Harvell (2004)
4.	Bees	Colony collapse disorder	*Nosema* sp.	Cameron et al. (2011)
5.	Humans	Blood stream infection	*Candida* sp.	Pfaller and Diekema (2004)

Further, among the various measures adopted to control fungal pathogens use of chemicals (fungicides) (Waard et al., 1993) and genetic resistance (Hammond-Kosack & Kanyuka, 2007) is considered as most suitable method. Development of resistance among fungal pathogens toward the fungicides used and development of new virulent races against the major genes incorporated in the crop plants are major threat to the strategies adopted for the management of plant diseases. Both of these phenomena are regulated by the genetic composition of the fungal pathogens and evolution of the new genes in the genomes of these organisms by various mechanisms (Stukenbrock & McDonald, 2008; McDonald & Linde, 2002). Development of new strains/pathotypes of pathogens has led to emergence of novel diseases and diseases of minor importance have become destructive and difficult to manage (Oliver & Solomon, 2008). In addition, host specialization in biotrophic fungi coupled with high virulence and broad host range is a deleterious combination. For example, the oomycetous pathogen *Phytophthora ramorum* (causing sudden Oak death and *Ramorum* blight) is known to infect 109 host species (Grunwald et al., 2008). Blast/grey spot caused by *Magnaporthe* spp. is another example of devastating disease where host specialization coupled with high virulence and genetic variability of the pathogen make it very destructive to most members of family Poaceae (Couch et al., 2005).

The fitness of these fungi plays a crucial role in making them successful pathogens and adapting their hosts. Fitness of the pathogens is defined variously by different authors (Antonovics & Alexander, 1989; Pringle & Taylor, 2002), and has remained the central theme of understanding the pathogen population composition with respect to host resistance and variable response from the host to the pathogen. An essential adaptation of the pathogen is successful reproduction, which is necessary for the survival and spread of the pathogen. Various hosts and environmental factors determine the extent of reproduction of the pathogen and hence play an important role in the evolution of the pathogen populations. Among all the factors affecting reproductive fitness, host population affecting the rate and line of evolution of the pathogen population is the most significant and determines the path of evolution of the host–pathogen system as a whole. The studies on the development of resistance/resistance mechanisms or tolerance in host populations in response to the ever-evolving pathogens are many, but the effect of the host on the reproductive fitness of the pathogens is still a matter of understanding and exploration (Bruns et al., 2012). In this chapter, we have tried to elaborate the concept of reproductive fitness with respect to changing host population and environmental factors and its implications in the field of plant pathology.

3.2   CONCEPT OF FITNESS

Fitness of any organism forms the basis of natural selection and results in phenotypic evolution. In plant-pathosystems, the term "fitness" is used in a variety of ways, and is often related to the effect of host resistance on pathogen population (Leonard, 1977; Parlevliet, 1981) and evolution of pathogen population with respect to the host. The evolution of pathogen population occur by acquisition of new virulence genes and potential loss of previously present virulence genes in absence of resistant varieties (Jones & Dangl, 2006; Nelson, 1979) resulting in defeat of bred resistance in host population (Vertifolia effect, Vander Plank, 2012). Growth and reproduction within the host are critical factors which govern the overall fitness of the pathogen (Antonovics & Alexander, 1989). Measurement of the fitness in plant-pathosystems is difficult because of involvement of more than one component and one component cannot be excluded while measuring fitness of other. There are various methods of assessing fitness of a pathogen ranging from determination of changes in adaptive markers to change in virulence and aggressiveness of the pathogen and ascertaining general vigor and spore production by the mycelium (Braiser, 1999).

3.2.1 REPRODUCTIVE FITNESS OF THE FUNGI

Fungi are ubiquitous organisms with complex life cycles, their ecology influence their evolution and have an impact on the evolution of related organisms. Diversity and complexity in the life cycle of fungal kingdom have affected the development of formal theory of evolution and adaptation for these organisms (Braiser 1999; Pringle & Taylor, 2002). Gilchrist et al. (2006) proposed patch array model for life-history strategies of asexually reproducing saprobic fungi (Fig. 3.1), but this model can be adapted to the commensal/parasitic or mutualistic fungi. In this model, a fungal spore settles from the spore pool to the appropriate resource patch, then germinates and reproduces asexually within it. The spore will produce numerous new spores creating a spore pool and this spore pool will continuously colonize the un-colonized resource patch. Both colonized and un-colonized resource patches are removed from the system at a constant rate by natural phenomenon and there is a constant rate of spore decay in the system. Within this model, there is a critical density of un-colonized patches for the spread/colonization of any new strain of the pathogen. A new fungus can replace the already "existent species" of the pathogen if it has more capacity to colonize the un-colonized substrate and process will result in competitive exclusion. Another model postulated by the same author is within patch model, which elaborate the life-history strategy of an asexual form in saprobic and filamentous fungus within the defined patch of the resources.

Adaptation to biotic or abiotic stress in the environment may play a key role in increasing the fitness of pathogens. For their survival, they need to adapt by increasing virulence and pathogenicity toward their hosts and being resistant/tolerant to the management strategies employed. In fungal pathogens, the transport systems may have diverse functions like efflux of fungicides/biocides out of the system for increasing virulence. For example, in case of *Magnaporthe oryzae* (*M. oryzae*), ABC 1 transporter protects the fungus against azole group of fungicides and rice phytoalexin sakuranetin

(Morris et al., 2009). Further, the life-history strategies adopted by the fungal pathogens increases their adaptation on respective hosts for example, *M. oryzae* follows both biotrophic and necrotrophic phases in the lifecycle. Initial infection occurs as biotroph but later on in the infection cycle, living cells are killed and the pathogen sporulates. Further, it was observed by Marcel et al. (2010) that the pathogen follows completely biotrophic mode of infection as root-infecting pathogen, suggesting diverse life-history strategies adopted by the pathogens, and their adaptation to the environment make them destructive pathogens.

Patch Array Model

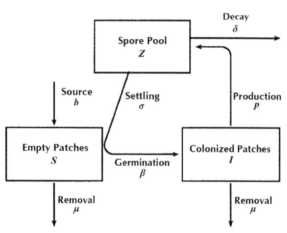

FIGURE 3.1 Patch array model for fungus and array processes. Where S is number of un-colonized patches at a time t, I is density of colonized patches at time t with fungus age a, Z is number of spores in the spore pool at time t, p is rate of spore production of age a, b is un-colonized patch production rate, σ spore settling rate from the spore pool, β is probability of successful spore settlement and germination, μ is patch removal rate, δ rate of spore decay within spore pool (Source: Michael A. Gilchrist, Deborah L. Suisky, and Anne Pringle, "Identifying Fitness and Optimal Life-History Strategies for an Asexual Filamentous Fungus," in Evolution. 2006 May; 60(5):970-9. With permission from John Wiley.)

Reproduction is the innate property of any living organism, which allows production of similar/identical individuals as the parents. Sexual mode of reproduction is wide spread among eukaryotes with only a small proportion reproducing by asexual means (Otto & Lenormand, 2002), though many species reproduce by both means. Reproduction helps in perpetuation of species by replacing dying/old individuals by more fit and vigorous individuals. Further, in case of pathogens, rate and success of reproduction along with the type of reproduction (i.e., either sexual or asexual) determines the fitness of pathogen (Fig. 3.2). In case of fungi also, sexual reproduction is the chief method of producing heterogeneity in the population, though for increasing the size of population, asexual mode of reproduction is preferred which is more common (Levin & Bergstrom, 2000; Saleh et al., 2012; Xu, 2004). Sexual reproduction in fungi is based on the recognition of opposite mating types through a pheromone/receptor system and triggering cell–cell fusion followed by meiosis. The opposite mating types may be present on different individuals (i.e., heterothallic) or on same individual (i.e., homothallic). However, in case of *Candida albicans* self-mating is reported despite it carrying only one mating type. Another key component of sexual reproduction in fungi is "female fertility," as in female organs only meiosis occurs; in the absence of female fertile strains, sexual reproduction in fungi becomes difficult. Mutations responsible for female-sterile phenotypes have already been identified in such loci in fungi: *Agaricus bisporus*, *Podospora* sp., *Fusarium* sp., *Neurospora crassa*, and *Podospora anserina*. Epigenetic effects may also be involved in sterile phenotypes. Since sexual reproduction in fungi is controlled by multiple factors, mutations or changes in the expression pattern of the genes involved may cause rapid loss of sexual reproduction (Saleh et al., 2012).

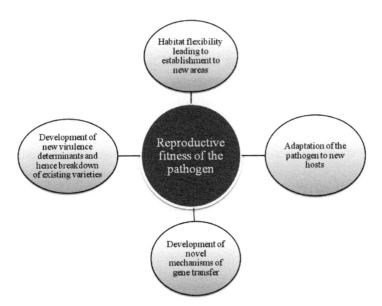

FIGURE 3.2 Reproductive fitness affects multiple variables of pathogen fitness and thus, plays a very important role in evolution and adaptation of pathogens.

Understanding of the mechanisms, which govern the reproductive fitness, will help in answering many unanswered questions; like mechanism of natural selection in plant pathogens, reasons of prevalence of asexual reproduction in many plant pathogens, development of virulence in natural populations (Pringle & Taylor, 2002), and will help in development of strategies for effective plant disease management.

3.2.2 MEASUREMENT OF REPRODUCTIVE FITNESS

Measurement of reproductive fitness is difficult because of the involvement of the host factors. The measurement becomes relative due to qualitative or quantitative resistance offered by the host and tolerance of the host toward the pathogen. Pathogen fitness lies in the causation of the disease, which depends on the host genotype, population structure, and phenology of the host along with the environmental factors affecting both host and pathogen. Thus, it is a complex phenomenon, which includes both qualitative and quantitative characters. For measurement of the fitness of the pathogen on the host, it is appropriate to use basic reproductive number and reproductive rate, as both indicate proliferation of the pathogen on host.

3.2.2.1 BASIC REPRODUCTIVE NUMBER

A pathogen's reproductive fitness can be quantified by the basic reproductive number "R_0." The R_0 is defined as "the average number of secondary cases of infection generated by one primary case in a susceptible host population (Anderson & May, 1986)." In other words, R_0 can be measured as a function of disease gradient from initial source of inoculum and is strongly influenced by spatial distribution of host as well as pathogen population. Mathematically, $R_0 = \beta N/\gamma$, where βN is average number of offsprings produced by a population with population density N and γ is the time units (Heesterbeek, 2002). If $R_0 > 1$ the disease will spread in the population, and, at $R_0 < 1$ the pathogen will eventually die. Hence, the basic reproductive number is a quantity with a clear biological meaning that characterizes reproductive fitness of the pathogen. Therefore, R_0 can be used to estimate the critical proportion of the host population that needs to be immunized (i.e., vaccinated) in order to eradicate the disease (Anderson & May, 1991).

It is reported that R_0 is affected by the field size and shape; it increases in small fields as well as in square fields, whereas in large fields it saturates to a constant value and decreases in elongated fields (Mikaberidze et al., 2014). The merit of this method lies in establishing relation between R_0, field size, and field dimensions and understanding maximum extent of dispersal of the disease. For example, similar approach used for wheat stripe rust caused by *Puccinia striiformis* is an important example where disease gradients were thoroughly measured over large distances (Sackett & Mundt, 2005, 2009; Cowger et al., 2005; Wellings, 2011) and thus estimation was done for R_0 as a function of the field size and shape.

3.2.2.2 REPRODUCTIVE RATE

The reproductive rate of a pathogen is the sum of both transmission between hosts and growth, and reproduction within host (May & Anderson, 1983). Within host, processes directly affect the pathogen virulence and aggressiveness, and hence, affect the path of pathogen evolution. Three pathogen life-history stages are essential to within host reproduction; (a) infection, (b) the latent period of growth within the host, and finally, (c) propagule production (Pariaud et al., 2009). Together, these life-history stages determine a pathogen's overall reproductive success within the host; however, the contribution of individual life-history stages to overall reproduction can vary among pathogen genotypes (Pariaud et al., 2009a) and across ecological contexts (Woodhams et al., 2008). The phenomenon is important because competition is thought to select for greater pathogen aggressiveness (May & Nowak, 1995). If the intensity of pathogen competition varies across a host population, a diverse host population will modulate and slow the evolution of increased pathogen aggressiveness.

3.3 APPROACHES FOR MEASURING FITNESS OF PATHOGENS

Plant pathology deals with populations rather than individual, but evolutionary processes resulting in the adaptation may occur at individual or group level. For example, in case of biotrophic pathogens, the increase in virulence should be to the level that they do not eradicate the population of their hosts. This may be achieved by selection, operating at individual level, where initially individuals with high virulence are favored but as the host population decreases, the individuals with less virulence become prominent. In contrast, when selection operates at group level, populations with intermediate level of virulence (generated due to variation in individuals) are favored (Antonovics

& Alexander, 1989). Whether at individual or group level, adaptations result in increased fitness of the pathogens, but it is easier to estimate the fitness at the individual level rather than at group level, as individuals are clearly defined units compared to populations or group of individuals, though it is true that natural selection operates at group level as well.

There are two methods used for the measurement of the fitness; (a) based on the contribution of an individual in a generation, known as "Predicted fitness method" and (b) based on the multi generation changes in phenotype frequency and/or gene frequency where these observed changes are translated to single generation fitness which is known as "Realized fitness method." Both approaches are elaborated in the following sections.

3.3.1 PREDICTED FITNESS METHODOLOGY

The contribution of an individual, genotype, or phenotype is measured by assessing fitness components that is, the factors which contribute to the next generation of the organisms and their fitness for example, survival, mating, genetic transmission etc. Therefore, one factor of the life cycle, which is directly related to the fitness (referred as "fitness component") is used for assessment (Antonovics & Alexander, 1989). Estimation of the fitness by using one fitness component reduces overall complexity of the procedure and is simpler. For example, while estimating resistance toward a pathogen, estimation of pollen tube growth rate may not be a fitness component, but yield of the plant may serve as a fitness component. Thus, the fitness components are determined by the trait under consideration and purpose of the estimation. In the simplest measurement of the fitness for the phytopathogens, measurement of reproductive rate (Antonovics & Alexander, 1989; Pringle & Taylor, 2002) can be used as fitness component. Reproductive rate is given as follows:

$$R = \sum l_x m_x$$

where l_x is the probability of survival to age x and m_x is the number of progeny produced by an individual of age x. For this measurement, two assumptions are made. One, both male and female contribute equally to the zygote formation and the other that the generations do not overlap with the phenotypes differing in their age-specific survival and reproduction.

Though these components were designed for higher plants and animals, they hold true for the fungal pathogens as well. For a fungal pathogen, probability of inoculum reaching the host plant will serve as the initial fitness component followed by infection efficiency of the pathogen on host (Antonovics & Alexander, 1989; Nelson 1979), which will be determined by the reproductive fitness of the pathogen in host. The analysis of fitness components is an intermediate step in determining overall fitness of the pathogen, but is required for evaluation of contribution of each component toward total fitness of the pathogen (Gilchrist et al., 2006). Further, from the disease management point of view, it is important to understand the effect of plant characteristics on the overall fitness of fungal pathogens (Antonovics & Alexander, 1989; Parlevliet, 1981).

3.3.2 REALIZED FITNESS METHODOLOGY

The methodology based on the single fitness measures is criticized as it is possible that some important components of the fitness may have been left. For rectifying this error, an alternative approach may be taken, in which actual observed changes in the genotype or phenotype are measured in one or more generations. These estimates of fitness render realized fitness. Further, these estimates are affected by environmental conditions and type of inheritance governing the trait. Both of them will have determinative effect on the trait in question. Thus, the drawback of this method is, if assumptions made for estimating the fitness of the microbe (e.g., constant effect of environment) are not considered; situation becomes complicated. In such cases, pathogen fitness become different from biological fitness and estimation of growth rates of the pathogen from disease progress curve may be an alternative. In realized fitness methodology, choice of model is critical which depend on both biological knowledge and model structure. For example, in situation where disease estimates are made, logistic model is more appropriate whereas in situations where absolute numbers are measured at the time of initiation of epidemic, exponential models are more appropriate (Antonovics & Alexander, 1989).

3.4 ENVIRONMENTAL EFFECTS ON FITNESS

Fitness of any phenotype depends upon the environmental factors surrounding the individual, along with genetic attributes. The understanding of environmental influences on the trait under consideration will help in making predications. For example, if the same phenotype has different fitness in different environments than with the understanding of mode of inheritance, evolutionary predictions can be made. On the other hand, if a phenotype has same fitness in different environmental conditions, then it will help in ecological prediction of the character under consideration. Under field conditions, with high reproductive rates and evolutionary pressure for infecting the host, the pathogen population may evolve to produce individuals with high pathogenicity toward host within one or two generations. But, this do not happen because of the physiological and environmental constraints limiting the life cycle traits and trade-offs for high pathogenicity may compromise other fitness attributes of the pathogen and hence, delayed response is resulted. For example, in case of *Mycosphaerella graminicola*, life cycle consist of biotrophic and necrotrpohic phases. The biotrophic phase hyphal proliferation occurs extracellularly and pathogen derives their nutrition from apoplastic cells whereas necrotrophic phase is associated with the appearance of lesions and production of spores on the host surface. The trigger mechanism from biotrophic to nectrotrophic phase is not known, but existence of two different types of life styles in the life cycles creates a check on the rate of evolution of the pathogen. Biotrophy promotes the virulence of the pathogen but it cannot proliferate in this phase. Thus, by separating the sporulation from the virulence of the pathogen, a trade-off is created which control the virulence, rate of reproduction and production of genetic variation of the pathogen, and keep both under the limits of evolution (Ponomarenko et al., 2011; Heraudet et al., 2008).

3.5 *MAGNAPORTHE* PATHOSYSTEM: UNDERSTANDING THE INFLUENCE OF REPRODUCTIVE FITNESS IN PATHOGEN EVOLUTION

Magnaporthe spp. is the causal agent of grey spot or blast disease in many members of family Poaceae. Blast disease in rice and other economically important crops is caused by *M. oryzae*, recently separated from *Magnaporthe grisea* (*M. grisea*) on the basis on molecular evidences (Couch & Kohn, 2002). The blast fungus has emerged as a model system of studying fungal biology due to the economic importance of the crops affected by the pathogen. Along with the economic importance, advances and development of new techniques for study the pathogen have facilitated the investigations on *Magnaporthe* as model phytopathogenic fungus. Blast pathogen displays remarkable morphogenetic and biochemical specialization to its pathogenic lifestyle and is an efficient and devastating agent of disease (Talbot, 2003,

Skamnioti & Gurr, 2008, Scheuermann et al., 2012). In addition to rice species, *Magnaporthe* infect other hosts as well causing huge losses to the members of Poaceae. Finger millet (*Eleusine coracana*), which is an important crop in India and South Asia, is also affected by this pathogen which causes huge losses (Ekwamu, 1991, Kuyek, 2000). This fungus on wheat causing blast is an emerging problem in many Latin American countries like Brazil (Duveiller et al., 2007; Igarashi et al., 1986).

Magnaporthe is highly variable pathogen and development of resistant cultivars is considered important for management of the pathogen (Jeon et al., 2013). But, new fungal strains develop in nature to overcome the resistance of the cultivars (Jeon et al., 2013; Kang & Lee, 2000) (Fig. 3.3). The mechanisms involved in the breakdown of resistance may be insertion of the transposons in the avirulence (Avr) gene (Kang et al., 2001), mutation or selective advantage to a subpopulation in a diverse population of the pathogen against a particular resistance gene (Jeon et al., 2013). Explanation of rapid breakdown of the resistance in field was attributed to hyper-variability of the pathogen (Ou, 1979), which led to the idea of development of partial resistance/durable resistance for the blast pathogen (Bonman et al., 1991).

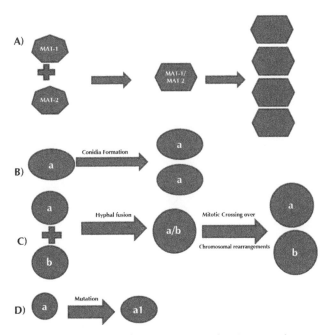

FIGURE 3.3 Mechanisms of genomic rearrangements in *Magnaporthe grisea*: (A) sexual reproduction facilitated by MAT1-2 mating types, (B) asexual reproduction, (C) parasexual recombination, and (D) mutation.

3.5.1   CLONAL PROPAGATION OF M. ORYZAE MAKE IT HIGHLY DESTRUCTIVE PATHOGEN

Asexual stage of rice-infecting *M. grisea* is described as *Pyricularia oryzae* (previously *P. grisea*). The conidia are produced on the conidiophores, are three-celled, and have a distinct stalk (Fig. 3.4). These spores behave as air-borne inoculum for primary infection by the fungus and cause secondary spread of the disease in the field through wind dispersal. These conidia are produced on all infected parts of the plant which gives lesions a dusty grey appearance. In the absence of sexual reproduction, conidia are the main source of spread and perpetuation of the pathogen in nature and this is the most frequently observed and widespread mode of reproduction for the pathogen.

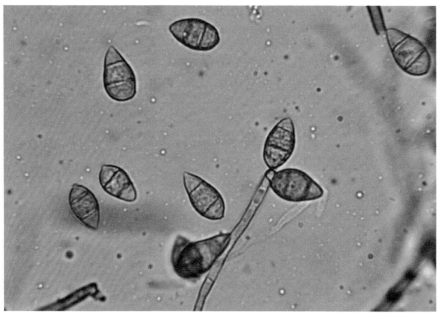

FIGURE 3.4   Conidia of *Magnaporthe grisea*. Source: forestry images (From Donald Groth, Louisiana State University, Ag Center, Bugwood.org. Used with permission.).

Under favorable environmental condition, population of the pathogen increase by asexual reproduction and spread to susceptible host, making it a highly destructive pathogen even in the absence of sexual reproduction. Further, as reported by Saleh et al. (2012), in clonal population ability of sexual reproduction may eventually be lost, as the pathogen does not require much variation and the asexually produced conidia can infect the diverse host population. Thus, asexual reproduction conserves the energy and time required for the sexual reproduction and makes this pathogen better adapted.

3.5.2 SEXUAL REPRODUCTION: CREATES VARIABILITY IN M. ORYZAE POPULATION

Magnaporthe is a filamentous, heterothallic, ascomycetous fungus, having two mating types MAT-1 and MAT-2 determined by two idiomorphs on chromosome 7 (Zhu et al., 1999). *M. oryzae* isolates can be classified into hermaphrodite, male fertile, or sterile, but crossing occurs only between fertile isolates from different mating types, and when at least one isolate is hermaphrodite (Zeigler, 2000). Under favorable conditions for sexual reproduction, perithecia are produced (Valent & Chumley, 1991). Under natural conditions, populations of *M. grisea* are host-restricted and show differences in the fertility (Couch et al., 2005). For example, in strains of *M. oryzae* infecting rice, fertile strains are rarely recovered from field (Saleh et al., 2012). Artificially fertile strains of *M. oryzae* are produced by successively backcrossing rice pathogenic isolates with isolates of weeping lovegrass (*Eragrostis curvularia*), or finger millet, which show greater fertility (Talbot, 2003; Valent et al., 1991).

As discussed previously, two mating types of *Magnaporthe* are found in nature but individual mating type appears to be restricted spatially. For example, in Argentina and Korea, isolates were of primarily MAT1-1 mating type (Consolo et al., 2005, Park et al., 2008) whereas in Africa, 29% of isolates were found to be MAT1-1 and 71% to MAT2-2, but none of the isolates was hermaphrodite, avoiding chances of crossing between them (Takan et al., 2011). Further, Kumar et al. (1999) found that in the Indian Himalayan region, 22% of the isolates belonged to MAT1-1 and 43% to MAT1-2, and male fertile and hermaphrodite isolates were also detected suggesting a possible occurrence of sexual recombination in this region. It is important because Himalayan regions are considered as one of the centers of origin for this pathogen (Saleh et al., 2014). Further, it has been established that this pathogen losses ability of sexual reproduction as it moves out of the center of origin.

The reason for loss of sexual reproduction may be founder effect that is, when rice cultivar was introduced to new areas, only one mating type established in the area (Zeigler, 2000). Further, it is possible that MAT locus could be linked to an Avr gene, with no compatibility with the rice cultivars grown in the area; this may represent a drawback in the fitness of the pathogen and is eliminated in time (Notteghem & Silue, 1992). Chromosomal rearrangements due to mobile genetic elements may be responsible for the erosion of sexual reproduction without causing fitness penalty (Chuma et al., 2011). One more reason could be that the fungus is able in the environment as a dormant mycelium and production of dormant sexual structures is not required for the survival of the pathogen during fellow season; thus, nature prefers the loss of sexual reproduction in the pathogen to reduce the cost of metabolism or simply weather conditions are not suitable for sexual reproduction of the fungus (Zeigler, 2000).

In a study conducted by Saleh et al. (2012), the fitness of female sex in *M. oryzae*, which is clonally propagated in the field by asexual reproduction, was evaluated. They studied whether the loss of sexual reproduction is controlled by genetic factors or determined epigenetically, and they monitored changes in female fertility in the absence of sexual reproduction in an *in vitro* experiment. They demonstrated by experimental evolution that female fertility is lost in *M. oryzae* population reproducing asexually, and loss of female fertility is controlled by genetic phenomenon rather than any epigenetic phenomenon. Genetic mechanism causing loss in female fertility may be mutation, recombination, and chromosomal rearrangements. Further, the experiment was done under experimental evolution condition and thus represents a close proximity to the natural conditions. Previously, Tharreau et al. (1997) found that spontaneous mutation in *M. grisea* isolates infecting finger millet result in loss in ability for perithecial development. These results suggest that more than one mechanism control the evolution of blast pathogen in nature.

3.5.3. ALTERNATIVE TO SEXUAL REPRODUCTION: PARASEXUAL REPRODUCTION

Sexual reproduction is important source for generation of variability in the pathogens and adaptation of the pathogens to their hosts, but in fungal kingdom, about 20% of the members reproduce by methods other than sexual reproduction (Seidl & Thomma, 2014).

As an alternative to the sexual cycle (described above), parasexual cycle may occur in *Magnaporthe*, this cycle involves mycelial anastomosis (haploid hyphal fusion), karyoga my results in the formation of heterozygous nature of the nuclei in conidia, which can be ascertained by estimating size of conidia, size of nucleus, or DNA content (Noguchi et al., 2006). Karyogamy is followed by mitotic recombination followed by *haploidization* due to aneuploidy that is, successive loss of a chromosome from each homologous pair (Crawford et al., 1986, Noguchi, 2011). Prerequisite for parasexual recombination is hyphal anastomosis, which occurs in vegetative compatible hyphae, followed by translocation of one or two nuclei in the fused cells and formation of compatible *heterokaryotic* state. Parasexual recombination generates genetic variability, but could also be a mechanism to restore the genome for preventing the mutation accumulation that endangers pathogen fitness, but it remains unproven till date. Parasexual recombination in *Magnaporthe* was first suggested by Yamasaki and Niizeki (1965) and observed by Zeigler et al. (1997); they observed growing of mycelium tufts between co-cultured isolates, followed by appearance of new haplotypes, indicating recombination between isolates. Since isolates of *Magnaporthe* from same or different population can easily recombine, it is suggested that vegetative compatibility is weakly controlled (Crawford et al., 1986). Noguchi et al. (2006) gave the first direct evidence that genetic exchange in *M. grisea* can be governed by parasexual recombination. In their study, they demonstrated that conidia generated through co-culturing of the parental isolates resulted in colonies with conidia having one nucleus per cell, with haploid genome content as determined by nuclear staining and flow cytometry, suggesting these were produced by the parasexual recombination and were stable. Further, these recombinants were carrying gene for antibiotic resistance though their parent strain carried only one gene for resistance (either Bialaphos resistant or Blasticidin S resistant). Virulence spectrum of these isolates suggests that they may become stably virulent on rice, though their parents were avirulent on rice (Noguchi et al., 2007).

3.5.4 MUTATION IN M. ORYZAE GENOME RESULT IN RAPID EVOLUTION OF AVR GENES

M. oryzae isolates are reported to have high genetic instability to the characters related to morphology, fertility, and pathogenicity when maintained in the laboratory conditions (Valent & Chumley, 1991). Jia et al. (2000) reported that interaction between rice and *M. oryzae* follows gene-for-gene system and that mutations in Avr genes can be decisive for interaction. Many important Avr genes are located in telomeric and sub-telomeric region in the genome of *M. oryzae*, for example, Avr-Pita (Orbach et al., 2000),

AvrMedNoï-1; AvrKu86-1 (Dioh et al., 2000), Avr-Piz (Luo et al., 2002), Avr-Pii (Yasuda et al., 2006), Avr-Pi15 (Ma et al., 2006) and Avr-Pit, and Avr-Pia (Chen et al., 2007). Deletions or alternations in these regions may lead to disruption of Avr gene function and hence gain in virulence. Further, it is reported 9.7% of the genome of *M. oryzae* is composed of transposons (Dean et al., 2005) and highest transposon concentration (24%) is observed in the telomeric region (Rehmeyer et al., 2006). Transposons are reported as hot spot for recombination and it is reported that maximum concentration of transposons in the telomeric region may lead to recombination in the Avr genes (Rehmeyer et al., 2006). In addition, flanking transposons are reported to be the cause of variation in Avr genes, for example, in the case of Avr Co39, RETRO5, and REP1, transposons flank the genes and cause variation (Farman et al., 2002). In Avr-Pita, Pot3 transposon inserts in the regulatory region of the gene and causes functional disruption (Kang et al., 2001). Transposons can move throughout the genome and can take the Avr gene to same chromosome, different chromosome, or on supernumerary chromosome suggesting they play important role in the evolution of Avr genes in *M. oryzae*.

It has been reported that gene for melanin biosynthesis has a mutation rate of 20% and this gene play essential part in penetration process, suggesting mutation may be an important mechanism for Avr gene manipulation in blast fungus (Chumley & Valent, 1990).

Spontaneous mutations or mutations mediated by transposons can help in the evolution of pathogen population as they produce variation to adapt quickly to the newly produced cultivar.

3.6 CONCLUSION

Reproductive fitness of the pathogen is a complex character and is dependent on many variables. Due to complexity of the nature of this character, it is difficult to interpret and measure. In case of *M. oryzae*, pathogen has evolved different life strategies to adapt to changing requirements of the host. On the one hand, clonal propagation (asexual reproduction of the pathogen) in favorable environmental conditions makes it a destructive pathogen. Genetic variability in the genome of the pathogen is created by mechanisms of evolution like sexual reproduction, parasexuality, and mutation.

KEYWORDS

- **fitness**
- **reproductive fitness**
- **evolution**
- **plant-pathosystem**
- **adaptation**
- *Magnaporthe* **spp.**

REFERENCES

Anderson, R. M.; May, R. M. The Invasion, Persistence and Spread of Infectious Diseases within Animal and Plant Communities. *Philos. Trans. R Soc. Lond. B Biol. Sci.* **1986,** *314,* 533–70.

Anderson, R. M.; May, R. M. *Infectious Diseases of Humans*; Oxford University Press: Oxford, 1991.

Antonovics, J.; Alexander, H. M. The Concept of Fitness in Plant Fungal Pathogen Systems. In *Disease Resistance and Host Fitness; McGraw-Hill:* New York, 1989.

Bonman, J. M.; Estrada, B. A.; Kim, C. K.; Ra, D. S.; Lee, E. J. Assessment of Blast Disease and Yield Loss in Susceptible and Partially Resistant Rice Cultivars in Two Irrigated Lowland Environments. *Plant Dis.* **1991,** *75,* 462–466.

Braiser, C. M. *Fitness, Continuous Variation and Selection in Fungal Populations: An Ecological Perspective*; Kluwer Academic Publishers: Dordrecht, The Netherlands, 1999; pp 307–340.

Bruns, E.; Carson, M.; May, G. Pathogen and Host Genotype Differently Affect Pathogen Fitness through their Effects on Different Life-History Stages. *BMC Evol. Biol.* **2012,** *12,* 135.

Cameron, S. A. Patterns of Widespread Decline in North American Bumble Bees. *Proc. Natl. Acad. Sci. U.S.A.* **2011,** *108,* 662–667.

Chen, Q. H.; Wang, Y. C.; Li, A. N.; Zhang, Z. G.; Zheng, X. B. Molecular Mapping of Two Cultivar-Specific Avirulence Genes in the Rice Blast Fungus *Magnaporthe grisea. Mol. Genet. Genom.* **2007,** *277*(2), 139–148.

Chuma, I.; Isobe, C.; Hotta, Y.; Ibaragi, K.; Futamata, N.; Kusaba, M.; Yoshida, K.; Terauchi, R.; Fujita, Y.; Nakayashiki, H.; Valent, B. Multiple Translocation of the AVR-Pita Effector Gene among Chromosomes of the Rice Blast Fungus *Magnaporthe oryzae* and Related Species. *PLoS Pathog.* **2011,** *7*(7), 1–20.

Chumley, F. G.; Valent, B. Genetic Analysis of Melanin-Deficient, Nonpathogenic Mutants of *Magnaporthe grisea. Mol. Plant Microbe Interact.* **1990,** *3*(3), 135–143.

Couch, B. C.; Fudal, I.; Lebrun, M.; Tharreau, D.; Valent, B.; Kim, P.; Notteghem, J.; Kohn, L. M. Origins of Host-Specific Populations of the Blast Pathogen *Magnaporthe oryzae* in

Crop Domestication with Subsequent Expansion of Pandemic Clones on Rice and Weeds of Rice. *Genetics.* **2005,** *170,* 613–630.

Couch, B. C.; Kohn, L. M. A Multilocus Gene Genealogy Concordant with Host Preference Indicates Segregation of a New Species, *Magnaporthe oryzae,* from *M. grisea. Mycologia.* **2002,** *94*(4), 683–693.

Cowger, C.; Wallace, L. D.; Mundt, C. C. Velocity of Spread of Wheat Stripe Rust Epidemics. *Phytopathology.* **2005,** *95,* 972–82.

Crawford, M. S.; Chunley, F. G.; Weaver, C. G. Characterization of the Heterokaryotic and Vegetative Diploid Phases of *Magnaporthe grisea. Genetics.* **1986,** *114*(4), 1111–1129.

Consolo, V. F.; Cordo, C. A.; Salerno, G. L. Mating-Type Distribution and Fertility Status in *Magnaporthe grisea* Populations from Argentina. *Mycopathologia.* **2005,** *160*(4), 285–290.

Davies, J.; Davies, D. Origins and Evolution of Antibiotic Resistance. *Microbiol. Mol. Biol. Rev.* **2010,** *74*(3), 417–433. http://Doi.Org/10.**1128**/MMBR.**00016**-10.

Daverdin, G.; Rouxel, T.; Gout, L.; Aubertot, J. N.; Fudal, I. Genome Structure and Reproductive Behaviour Influence the Evolutionary Potential of a Fungal Phytopathogen. *PLoS Pathog.* **2012,** *8*(11), E100*3*020. Doi:10.1371/Journal.Ppat.1003020.

Dean, R. A.; Talbot, N. J.; Ebbole, D. J.; Farman, M. L.; Mitchell, T. K.; Orbach, M. J.; Thon, M.; Kulkarni, R.; Xu, J-R.; Pan, H.; Read, N. D.; Lee, Y-H.; Carbone, I.; Brown, D.; Oh, Y. Y.; Donofrio, N.; Jeong, J. S.; Soanes, D. M.; Djonovic, S.; Kolomiets, E.; Rehmeyer, C.; Li, W.; Harding, M.; Kim, S.; Lebrun, M-H.; Bohnert, H.; Coughlan, S.; Butler, J.; Calvo, S.; Ma, L-J.; Nicol, R.; Purcell, S.; Nusbaum, C.; Galagan, J. E.; Birren, B. W. The Genome Sequence of the Rice Blast Fungus *Magnaporthe grisea. Nature.* **2005,** *434*(7036), 980–986.

Dioh, W.; Tharreau, D.; Notteghem, J. L.; Orbach, M. J.; Lebrun, M. H. Mapping of Avirulence Genes in the Rice Blast Fungus, *Magnaporthe grisea,* with RFLP and RAPD Markers. *Mol. Plant Microbe Interact.* **2000,** *13*(2), 217–227.

Duveiller, E.; Singh, R. P.; Nicol, J. M. The Challenges of Maintaining Wheat Productivity: Pests, Diseases, and Potential Epidemics. *Euphytica.* **2007,** *157,* 417–430.

Ekwamu, A. Influence of Head Blast Infection on Seed Germination and Yield Components of Finger Millet (*Eleusine coracana* L. Gaertn) Trop. *Pest Manag.* **1991,** *37,* 122–23.

Farman, M. L.; Eto, Y.; Nakao, T.; Tosa, Y.; Nakayashiki, H.; Mayama, S.; Leong, S. A. Analysis of the Structure of the AVR1-CO39 Avirulence Locus in Virulent Rice-Infecting Isolates of *Magnaporthe grisea. Mol. Plant Microbe Interact.* **2002,** *15*(1), 6–16.

Farrer, R. A.; Weinert, L. A.; Bielby, J.; Garner, T. W.; Balloux, F.; Clare, F.; Fisher, M. C. Multiple Emergences of Genetically Diverse Amphibian-Infecting Chytrids Include a Globalized Hypervirulent Recombinant Lineage. *Proc. Natl. Acad. Sci. U.S.A.* **2011,** *108*(46), 18732–18736.

Fisher, M. C.; Henk, D. A.; Briggs, C. J.; Brownstein, J. S.; Madoff, L. C.; McCraw, S. L.; Gurr, S. J. Emerging Fungal Threats to Animal, Plant and Ecosystem Health. *Nature.* **2012,** *484,* 186–194.

Gilchrist, M. A.; Sulsky, D. L.; Pringle, A. Identifying Fitness and Optimal Life-History Strategies for an Asexual Filamentous Fungus. *Evolution.* **2006,** *60*(5), 970–979.

Grunwald, N. J.; Goss, E. M.; Press, C. M. *Phytophthora ramorum*: A Pathogen with a Remarkably Wide Host Range Causing Sudden Oak Death on Oaks and Ramorum Blight on Woody Ornamentals. *Mol. Plant Pathol.* **2008,** *9,* 729–740.

Hammond-Kosack, K. E.; Kanyuka, K. Resistance Genes (R Genes) in Plants. In *Encyclopedia of Life Sciences (ELS);* John Wiley & Sons, Ltd.: Chichester, 2007. Doi: 10.1002/9780470015902.A0020119.

Heesterbeek, J. A Brief History of R_0 and a Recipe for its Calculation. *Acta Biotheor.* **2002,** *5,* 189–204.

Heraudet, V.; Salvaudon, L.; Shykoff, J. A. Trade-Off between Latent Period and Transmission Success of a Plant Pathogen Revealed by Phenotypic Correlations. *Evol. Ecol. Res.* **2008,** *10,* 913–924.

Igarashi, S.; Utiamada, C. M.; Igarashi, L. C.; Kazuma, A. H.; Lopes, R. S. Pyricularia in Wheat. 1. Occurrence of *Pyricularia* sp. in Parana State. *Fitopatol. Bras.* **1986,** *11,* 351–352.

Jeon, J.; Choi, J.; Lee, G. W.; Dean, R. A.; Lee, Y. H. Experimental Evolution Reveals Genome-Wide Spectrum and Dynamics of Mutations in the Rice Blast Fungus, *Magnaporthe oryzae. PLoS One.* **2013,** *8*(5), E65416. Doi:10.1371/Journal.Pone.0065416.

Jia, Y.; Mcadams, S. A.; Bryan, G. T.; Hershey, H. P.; Valent, B. Direct Interaction of Resistance Gene and Avirulence Gene Products Confers Rice Blast Resistance. *EMBO J.* **2000,** *19*(15), 4004–4014.

Jones, J. D. G.; Dangl, J. L. The Plant Immune System. *Nature.* **2006,** *444,* 323–329. Doi:10.1038/Nature05286.

Kang, S.; Lee, Y. H. Population Structure and Race Variation of the Rice Blast Fungus. *Plant Pathol. J.* **2000,** *16,* 1–8.

Kang, S.; Lebrun, M. H.; Farrall, L.; Valent, B. Gain of Virulence Caused by Insertion of a Pot3 Transposon in a *Magnaporthe grisea* Avirulence Gene. *Mol. Plant Microbe Interact.* **2001,** *14,* 671–674. Doi: 10.1094/Mpmi.2001.14.5.671.

Kim, K.; Harvell, C. D. The Rise and Fall of a Six-Year Coral-Fungal Epizootic. *Am. Nat.* **2004,** *164,* S52–S63.

Kuyek, D. Blast, Biotech and Big Business. Implications of Corporate Strategies on Rice Research in Asia. Retrieved from Http://Www.Grain.Org/Publications/Reports/Blast.Htm. 2000.

Kumar, J.; Nelson, R. J.; Zeigler, R. S. Population Structure and Dynamics of *Magnaporthe grisea* in the Indian Himalayas. *Genetics.* **1999,** *152*(3), 971–984.

Leonard, K. J. Selection Pressures and Plant Pathogens. *Ann. N. Y. Acad. Sci.* **1977,** *287,* 207–222.

Levin, B. R.; Bergstrom, C. T. Bacteria Are Different: Observations, Interpretations, Speculations, and Opinions about the Mechanisms of Adaptive Evolution in Prokaryotes. *Proc. Natl. Acad. Sci. U.S.A.* **2000,** *97,* 6981–6985.

Luo, C. X.; Hanamura, H.; Sezaki, H.; Kusaba, M.; Yaegashi, H. Relationship Between Avirulence Genes of the Same Family in Rice Blast Fungus *Magnaporthe grisea. J. Gen. Plant Pathol.* **2002,** *68*(4), 300–306.

Ma, J. H.; Wang, L.; Feng, S. J.; Lin, F.; Xiao, Y.; Pan, Q. H. Identification and Fine Mapping of Avrpi15, a Novel Avirulence Gene of *Magnaporthe grisea. Theor. Appl. Genet.* **2006,** *113*(5), 875–883.

Marcel, S.; Sawers, R.; Oakeley, E.; Angliker, H.; Paszkowski, U. Tissue-Adapted Invasion Strategies of the Rice Blast Fungus *Magnaporthe oryzae. The Plant Cell.* **2010,** *22*(9), 3177–3187. Doi: Http://Dx.Doi.Org/10.**1105**/Tpc.110.**078**048.

May, R.; Nowak, M. Coinfection and the Evolution of Parasite Virulence. *Proc. R. Soc. Lond. B Biol. Sci.* **1995,** *26,* 209–215.

May, R. M.; Anderson, R. M. Epidemiology and Genetics in the Coevolution of Parasites and Hosts. *Proc. R. Soc. Lond. B Biol. Sci.* **1983,** *219,* 281–313.

McDonald, B. A.; Linde, C. The Population Genetics of Plant Pathogens and Breeding for Durable Resistance. *Euphytica.* **2002,** *12,* 163–180.

Mikaberidze, A.; Mundt, C. C.; Bonhoeffer, S. The Effect of Spatial Scales on the Reproductive Fitness of Plant Pathogens. **2014**. http://arxiv.org/pdf/1410.0587.pdf.

Morris, C. E.; Bardin, M.; Kinkel, L. L.; Moury, B.; Nicot, P. C. Expanding the Paradigms of Plant Pathogen Life History and Evolution of Parasitic Fitness Beyond Agricultural Boundaries. *PLoS Pathog.* **2009,** *5*(12), E1000693. Doi: 10.1371/Journal.Ppat.1000693.

Nelson, R. R. The Evolution of Parasitic Fitness. In *Plant Disease: An Advanced Treatise*; Horsfall, J. G., Cowling, E. B., Eds.; Academic Press: London, UK, 1979; pp 23–46.

Noguchi, M. T. Parasexual Recombination in *Magnaporthe oryzae. Japan Agric. Res. Q.* **2011,** *45*(1), 39–45.

Noguchi, M. T.; Yasuda, N.; Fujita, Y. Evidence of Genetic Exchange by Parasexual Recombination and Genetic Analysis of Pathogenicity and Mating Type of Parasexual Recombinants in Rice Blast Fungus, *Magnaporthe oryzae. Phytopathology.* **2006,** *96*, 746–750.

Noguchi, M. T.; Yasuda, N.; Fujita, Y. Fitness Characters in Parasexual Recombinants of the Rice Blast Fungus, *Pyricularia oryzae. Jpn Agric. Res. Q.* **2007,** *41*(2), 123–131.

Notteghem, J. L.; Silue, D. Distribution of the Mating Type Alleles in *Magnaporthe grisea* Populations Pathogenic on Rice. *Phytopathology.* **1992,** *82*(4), 421–424, ISSN 0031–949X.

Oliver, R. P.; Solomon, P. S. Recent Fungal Diseases of Crop Plants: Is Lateral Gene Transfer a Common Theme? *Mol. Plant Microbe Interact.* **2008,** *21*(3), 287–293. Doi: 10.1094/MPMI –21–3–0287.

Orbach, M.; Farrall, L.; Sweigard, J. A.; Chumley, F. G.; Valent, B. A Telomeric Virulence Gene Determines Efficacy for the Rice Blast Resistance Gene Pi-Ta. *The Plant Cell.* **2000,** *12*(11), 2019–2032.

Otto, S. P.; Lenormand, T. Resolving the Paradox of Sex and Recombination. *Nat. Rev. Genet.* **2002,** *3*, 252–261.

Ou, S. H. *Breeding Rice for Resistance to Blast-A Critical Review*, Proceedings, Rice Blast Workshop, International Rice Research Institute, PO Box 933, Manila: Philippines, 1979; pp 81–137.

Pariaud, B.; Ravigne, V.; Halkett, F.; Goyeau, H.; Carlier, J.; Lannou, C. Aggressiveness and its Role in the Adaptation of Plant Pathogens. *Plant Pathol.* **2009a,** *58*, 409 –424.

Pariaud, B.; Robert, C.; Goyeau, H.; Lannou, C. Aggressiveness Components and Adaptation to a Host Cultivar in Wheat Leaf Rust. *Phytopathology.* **2009,** *99*, 869–878.

Parlevliet, J. E. Stabilizing Selection in Crop Pathosystems: An Empty Concept or a Reality? *Euphytica.* **1981,** *30*, 259–269.

Park, S. Y.; Milgroom, M. G.; Han, S. S.; Kang, S.; Lee, Y. H. Genetic Differentiation of *Magnaporthe oryzae* Populations from Scouting Plots and Commercial Rice Fields in Korea. *Phytopathology.* **2008,** *98*(4), 436–442.

Pfaller, M. A.; Diekema, D. J. Rare and Emerging Opportunistic Fungal Pathogens: Concern for Resistance Beyond *Candida albicans* and *Aspergillus fumigates. J. Clin. Microbiol.* **2004,** *42*(10), 4419–4431. DOI: 10.1128/JCM.42.10.4419–4431.2004.

Ponomarenko, A.; Goodwin, S. B.; Kema, G. H. J. Septoria Tritici Blotch (STB) of Wheat. *Plant Health Instr.* **2011**. Doi:10.1094/PHI-I-2011-0407-01.

Pringle, A.; Taylor, J. W. The Fitness of Filamentous Fungi. *Trends Microbiol.* **2002,** *10*(10), 474–481. Doi:10.**1016/S0966**-842X(02)**02447**-2.

Rehmeyer, C.; Li, W.; Kusaba, M.; Kim, Y. S.; Brown, D.; Staben, C.; Dean, R.; Farman, M. Organization of Chromosome Ends in the Rice Blast Fungus, *Magnaporthe oryzae. Nucleic Acids Res.* **2006,** *34*(17), 4685–4701.

Sackett, K. E.; Mundt, C. C. Primary Disease Gradients of Wheat Stripe Rust in Large Field Plots. *Phytopathology.* **2005,** *95*, 983–991.

Sackett, K. E.; Mundt, C. C. Effect of Plot Geometry on Epidemic Velocity of Wheat Yellow Rust. *Plant Pathol.* **2009**, *58*, 370–377.

Saleh, D.; Milazzo, J.; Adreit, H.; Fournier, E.; Tharreau, D. South-East Asia is the Center of Origin, Diversity and Dispersion of the Rice Blast Fungus, *Magnaporthe oryzae*. *New Phytol.* **2014**, *201*, 1440–1456. Doi: 10.1111/Nph.12627.

Saleh, D.; Milazzo, J.; Adreit, H.; Tharreau, D.; Fournier, E. Asexual Reproduction Induces a Rapid and Permanent Loss of Sexual Reproduction Capacity in the Rice Fungal Pathogen *Magnaporthe oryzae*: Results of *In Vitro* Experimental Evolution Assays. *BMC Evol. Biol.* **2012**, *12*, 42. Http://Www.Biomedcentral.Com/**1471-2148**/12/42

Scheuermann, K. K.; Raimondi, J. V.; Marschalek, R.; Andrade, A. D.; Wickert, E. *Magnaporthe oryzae* Genetic Diversity and its Outcomes on the Search for Durable Resistance, the Molecular Basis of Plant Genetic Diversity, Prof. Mahmut Caliskan (Ed.), ISBN: 978-953-51-0157-4, Intech, 2012. Available From: Http://Www.Intechopen.Com/Books/ The -Molecular-Basis-Of-Plant-Geneticdiversity/Magnaporthe-Oryzae-Genetic-Diversity-And-Its -Outcomes-On-The-Search-For-Durable-Resistance

Seidl, M. F.; Thomma, B. P. H. J. Sex or No Sex: Evolutionary Adaptation Occurs Regardless. *Bioessays.* **2014**, *36*, 335–345.

Skamnioti, P.; Gurr, S. J. Against the Grain: Safeguarding Rice from Rice Blast Disease. *Trends Biotechnol.* **2008**, *27*(3), 141–150.

Strange, R. N.; Scott, P. R. Plant Disease: A Threat to Global Food Security. *Annu. Rev. Phytopathol.* **2005**, *43*, 83–116.

Stukenbrock, E. H.; Mcdonald, B. A. The Origins of Plant Pathogens in Agro-Ecosystems. *Annu. Rev. Phytopathol.* **2008**, *46*, 75–100.

Takan, J. P.; Chipili, J.; Muthumeenakshi, S.; Talbot, N. J.; Manyasa, E. O.; Bandyopadhyay, R.; Sere, Y.; Nutsugah, S. K.; Talhinhas, P.; Hossain, M.; Brown, A. E.; Sreenivasaprasad, S. *Magnaporthe oryzae* Populations Adapted to Finger Millet and Rice Exhibit Distinctive Patterns of Genetic Diversity, Sexuality and Host Interaction. *Mol. Biotechnol.* **2011**. Doi: 10.1007/S*12*033-011-9429-Z.

Talbot, N. J. On the Trail of a Serial Killer: Exploring the Biology of *Magnaporthe grisea*. *Annu. Rev. Microbiol.* **2003**, *57*, 177–202. Doi: 10.1146/Annurev.Micro.57.030502.090957.

Tharreau, D.; Notteghem, J. L.; Lebrun, M. H. Mutations Affecting Perithecium Development and Sporulation in *Magnaporthe grisea*. *Fungal Genet. Biol.* **1997**, *21*, 206–213. Doi: 10.1006/Fgbi.1996.0951.

Valent, B.; Chumley, F. G. Molecular Genetic Analysis of the Rice Blast Fungus *Magnaporthe grisea*. *Annu. Rev. Phytopathol.* **1991**, *29*, 443–67.

Valent, B.; Farrall, L.; Chumley, F. G. *Magnaporthe grisea* Genes for Pathogenicity and Virulence Identified through a Series of Backcrosses. *Genetics.* **1991**, *127*, 87–101.

Vander Plank, J. E. *Disease Resistance in Plants*; 2nd ed.; Academic Press/Elsevier: New York, 2012; pp 83.

Waard, M. A.; Georgopoulos, S. G.; Hollomon, D. W.; Ishii, H.; Leroux, P.; Ragsdale, N. N.; Schwinn, F. J. Chemical Control of Plant Diseases: Problems and Prospects. *Annu. Rev. Phytopathol.* **1993**, *31*(1), 403–421.

Wellings, C. R. Global Status of Stripe Rust: A Review of Historical and Current Threats. *Euphytica.* **2011**, *179*, 129–141.

Woodhams, D. C.; Alford, R. A.; Briggs, C. J.; Johnson, M.; Louise, A.; Woodhams, C.; Alford, A.; Briggs, J. Life-History Trade-Offs Influence Disease in Changing Climates: Strategies of an Amphibian Pathogen. *Ecology.* **2008**, *89*, 1627–1639.

Xu, J. P. The Prevalence and Evolution of Sex in Microorganisms. *Genome.* **2004**, *47*, 775–780.

Yamasaki, Y.; Niizeki, H. Studies on Variation of the Rice Blast Fungus, *Pyricularia oryzae* Cav. I. Karyological and Genetic Studies on Variation. *Bull. Natl. Inst. Agric. Sci. (Jpn).* **1965**, *13*, 231–273.

Yasuda, N.; Tsujimoto-Noguchi, M.; Fujita, Y. Partial Mapping of Avirulence Genes AVR-Pii and AVR-Pia in the Rice Blast Fungus *Magnaporthe oryzae. Can. J. Plant Pathol.* **2006**, *28*(4), 494–498.

Zeigler, R. S.; Correa, F. J. Applying *Magnaporthe grisea* Population Analyses for Durable Rice Blast Resistance. APSnet Features. 2000. Online. DOI: 10.1094/Apsnetfeature-2000 –0700A.

Zeigler, R. S.; Scott, R. P.; Leung, H.; Bordeos, A. A.; Kumar, J.; Nelson, R. J. Evidence of Parasexual Exchange of DNA in the Rice Blast Fungus Challenges its Exclusive Clonality. *Phytopathology.* **1997**, *87*(3), 284–294.

Zhu, H.; Blackmon, B. P.; Sasinowski, M.; Dean, R. A. Physical Map and Organization of Chromosome 7 in the Rice Blast Fungus, *Magnaporthe grisea. Genome Res.* **1999**, *9*(8), 739–750.

CHAPTER 4

EVOLUTION IN BIO-CONTROL AGENTS AND ITS ADAPTATION TO SUPPRESS SOIL-BORNE PHYTOPATHOGEN

Md. ARSHAD ANWER and KUNDAN SINGH

Department of Plant Pathology, Bihar Agricultural University, Sabour 813 210, Bhagalpur, India

CONTENTS

ABSTRACT

In the evolutionary process, both bio-control agents and pathogen evolved with time but it is evident that biological control has primacy over phyto-pathogens. Bio-control agents have evolved some mechanism to suppress the pathogen such as competition for space and nutrients, production of antimicrobial substances or antibiosis, parasitism and predation, induction of defense response in plants and metabolism of germination stimulants. This chapter describes about the evolution of biological control agents with detailed mode of action and their mechanism against plant pathogens espe-cially soil borne.

4.1   INTRODUCTION

In early 1930s, potential of *Trichoderma* species as biocontrol agents against plant diseases was first recognized (Weindling, 1932), and in following years, many diseases has been controlled (Aluko & Hering, 1970; Bliss, 1951; Chet, 1987; Elad & Kapat, 1999; Harman, 2000; Howell, 1982; Lifshitz et al., 1986; Lumsden et al., 1992; Sharon et al., 2001; Wellset al., 1972; Yedidia et al., 1999; Zhang et al., 1996). This has resulted in the commercial production of several *Trichoderma* species for disease protection and growth enhancement of a number of crops in the US (McSpadden & Fravel, 2002), and in the production of *Trichoderma* species and combination of species in India, Israel, Sweden and New Zealand.

Biological control of plant diseases is recent in comparison to the biological control of insects. In the year 1979, the first bacterium, *Agro-bacterium radiobacter* K 84 was registered with the US environmental protection agency (EPA) for the control of crown gall. Ten years later, in 1989 the first fungus *Trichoderma harzianum* ATCC 20476 was registered with the same agency for the control of plant diseases. Up to year 2005, a total of 14 bacteria and 12 fungi have been registered with the EPA for the control of plant diseases (Fravel, 2005) and sold commercially as one or more products. Sixty five percent of the registered organisms have been registered in the EPA within the past 10 years while the remaining 36% registered over the past five years. Many problems related to technologies were overcome and shifts in thinking occurred for these products to reach the shelves.

At present about 700 different microbial products are available worldwide. In India, about 16 *Bacillus thuringiensis*-based commercial preparations, 38 fungal-based formulations on *Trichoderma*, *Beauveria*, and *Metarhizium*, about 45 baculovirus-based formulations of *Spodoptera* and *Helicoverpa* are available. Microbials are expected to replace at least 20% of the chemical pesticides. Biological control agents (BCA) are being supplied by about 128 units in the country (80 private companies), Central Integrated Pest Management centers (30), SAUs (10), ICAR institutes (8), and parasitoid producing laboratories (4) are also supplying natural enemies (Wahab, 2003, 2004, 2009).

William Roberts in 1874 introduced the term antagonism and demonstrated the antagonistic action of microorganisms in liquid culture between *Penicillium glaucum* and bacteria. The term biological control of plant disease management was invented for the first time by C. F. Von in 1914. Since then various bio-control products have been found to be very effective managing the plant disease. Sanford (1926) observed that the potato scab was suppressed by green manuring antagonistic activities. In Germany, two hundred years ago in Persoon (1794) proposed first time *Trichoderma* as a genus. He showed similar fungi (microscopically) described as appearing like mealy powder enclosed by a hairy covering. Persoon proposed four species that is *T. viride*, *T. nigroscens*, *T. aureum*, and *T. roseum*. However, the *Trichoderma* is considered as *Trichoderma* Pers. Ex. Fr. In India, Thakur and Norris isolated *Trichoderma* first time during the year 1928 from Madras (Chennai). Weindling in 1932 first time reported the potential value of the genus *Trichoderma* as bioagents. He reported *Trichoderma lignorum* against several plant pathogens. Wright (1952–1957), Grossbard (1948–1952) and others demonstrated that antibiotics were produced in soil by *Penicillium*, *Aspergillus*, *Trichoderma*, *streptomyces* ssp. Kloepper et al. (1980), demonstrated the importance of siderophores produced by *Erwinia carotovora*. Howell et al. (1993) reported P and Q strains of *Trichoderma* sp. A large number of biocontrol agents have been investigated to exploit their beneficial effects on crop productivity. BCA are primarily fungal and bacterial in origin. They basically work through parasitism, competition, antibiosis, and induced systemic resistance (ISR) against plant pathogens (Papavizas, 1985; Stirling, 1993; Khan et al., 2006, 2009, 2010, 2011; Khan & Anwer, 2007, 2008, 2012; Bora et al., 2010; Anwer, 2014). The other important genera of biocontrol fungi which have been tested against plant pathogenic fungi and nematodes include *Chaetomium*, *Neurospora*, *Fusarium* (saprophytic), *Rhizoctonia*, *Dactylella*, *Arthrobotrys*, *Catenaria*, *Paecilomyces*, *Pochonia*, and *Glomus*. Whereas, among bacteria, *Pseudomonas* spp. (Gram negative bacteria, Weller, 1988) and Gram-positive bacteria such as *Bacillus* spp. (Von Der Weid et al., 2005), *Paenibacillus* spp. (Da Mota et al., 2008), and

Actinomycetes (Hamby & Crawford, 2000; El-Tarabily & Sivasithamparam, 2006; de Vasconcellos & Cardoso, 2009) are dominant antagonists.

All the BCA coexist with pathogen along with other microorganisms in natural ecosystem and have been keeping pathogen in balance for a long time showing primacy. So many evidence for the evolution of pathogen have been proved by many scientists (Howell, 2003; Anwer & Khan, 2013; Bouizgarne, 2013; Anwer, 2014; Anwer and Bushra, 2015; Anwer, 2015), but despite of BCAs' promise, and evolutionary studies in the context of biological control can be difficult.

4.2 THE PRIMACY EVOLUTION OF BIOLOGICAL CONTROL AGENTS OVER PHYTOPATHOGENS

To understand the mechanisms involved in disease suppression in biological control study of different strains of fluorescent *Pseudomonads* contributed greatly. Many of these bacteria could prevent plant diseases by various mechanisms: antibiosis, competition, or parasitism. In the genus *Pseudomonas*, *Pseudomonas fluorescens* which are ubiquitous rhizosphere inhabitant bacteria are the most studied group (Weller, 1988). They were shown to have a higher density and activity in the rhizosphere than in bulk soil. When introduced on seed or planting material, they promote plant growth or control plant diseases by suppressing harmful rhizosphere microorganisms. They are able to compete aggressively for sites in the rhizosphere and prevent proliferation of plant pathogens by niche exclusion, production of antibiotics and siderophores, or inducing systemic resistance (De Weger et al., 1986; Krishnamurthy & Gnanamanickam, 1998; Haas & De´fago, 2005) by stimulating plant growth by helping either uptake of nutrients from soil (De Weger et al., 1986); or by producing certain plant growth promoting materials (Ryu et al., 2005; Spaepen et al., 2007). In many cases *Pseudomonas fluorescens* have been applied to suppress *Fusarium wilts* of various phytopathogens (Lemanceau & Alabouvette, 1993), and also against tomato bacterial canker, *Clavibacter michiganenis* subsp. *michiganensis* (Amkraz et al., 2010). Whereas, their presence is also correlated with disease suppression in some suppressive soils (Kloepper et al., 1980; Lemanceau et al., 2006). Examples of commercially available biocontrol products from *Pseudomonas* are Bio-save (*P. syringae*) and Spot-Less (*P. aureofaciens* Tx-1). Van Peer et al. (1991) reported protection of carnation from fusarosis due to phytoalexin accumulation upon treatment with *Pseudomonas* strain WCS417. Other works followed as well as the use of *P. fluorescens* as an inducing agent to prevent the spread of various phytopathogens (Maurhofer et al., 1994; Duijff et al., 1994; Leeman et al., 1995; Pieterse et al., 2000). *P. fluorescens* CHA0 showed ability to protect tobacco against the tobacco necrosis virus concomitant with a systemic accumulation of salicylic acid and associated with the induction of multiple acidic pathogenesis related proteins, including PR-1a, -1b, and -1c (Maurhofer et al., 1994). Inoculation of *Arabidopsis thaliana* by *P. fluorescens* WCS417r and of rice by WCS374r conducted to ISR respectively to *Pseudomonas syringae* pv. tomato (Pieterse et al., 2000) and to the leaf blast pathogen *Magnaporthe oryzae* (De Vleesschauwer et al., 2008).

Microbial antagonists of plant pathogens evolved to work in several ways, the most common mechanisms being parasitism and predation, competition for nutrients or space, production of antimicrobial substances and induced resistance. These mechanisms often involve the synergistic action of several mechanisms.

4.2.1 COMPETITION FOR SPACE AND NUTRIENTS

Competition occurs when two or more organisms require the same resource for growth and survival. The use of this resource by one organism reduces the amount available to the other. A soil-borne pathogen which infects only certain parts of the root may therefore be limited by competition for suitable colonisable sites or space on the root surface. The rhizosphere is a region of intense microbial activity where there may also be competition for oxygen. On the leaf surface, where nutrients are in short supply, competition for nutrients is thought to play a significant role in pathogen suppression. Competition for the same site is sometimes called 'site exclusion' and frequently takes place amongst organisms which are closely related taxonomically, for example, the fire blight bacterium *Erwinia amylovora* and the saprophytic species *E. herbicola*. In the case of competition for nutrients, the situation is similar. Antagonists with the same nutrient requirements as the pathogen are its most effective competitors.

Competition for the same carbon source between *Pythium ultimum*, a common cause of seedling damping-off and rhizosphere bacteria has resulted in effectively evolved to biological control of *P. ultimum* in several crops. Ethanol and acetaldehyde released from seeds of pea and soybean following imbibition stimulates hyphal growth from sporangia of *P. ultimum*. Treating seeds with a strain of *Pseudomonas putida*, which utilise ethanol as its sole carbon source in culture, reduce the concentration of volatiles released, lessens hyphal growth or sporangial germination of the pathogen and increases seedling emergence in *Pythium* infested soil. Similarly, ethanol-metabolizing strains of *Enterobacter cloacae* reduce the ability of cottonseed exudate and volatiles to stimulate sporangium germination of *P. ultimum*.

One of the best-documented examples of nutrient competition in BCA evolution involves competition for iron between *fluorescent pseudomonads* and soilborne fungal pathogens such as *Fusarium oxysporum*. Strains of bacteria including *Pseudomonas fluorescens* and *P. putida* produce sidero-phores, metabolic products of microorganisms that bind iron and facilitate its transport from the environment into the microbial cell. The siderophores *pyoverdine* and *pseudobactin* have a high affinity for the soluble ferric iron (Fe^{+3}) and inhibit the growth of pathogens by limiting the availability of iron. Evidence for involvement of siderophores in biological control has been obtained from experiments with well-characterized and genetically evolved bacterial strains. Pyoverdine production is correlated with biological control while pyoverdine deficient strains have reduced capabilities in biological control.

Evolution for the nutrient competition as a mechanism of BCA depends on the type of pathogen that is targeted. It is not evolved for suppressing biotrophs such as rusts and powdery mildews because they do not require exogenous nutrients to infect the host, but on the other hand, a necrotrophic pathogen such as *Botrytis cinerea* is directly affected. Such necrotrophic pathogens require some exogenous nutrients during a definite saprophytic phase prior to infecting the host and are therefore vulnerable to nutrient competition. There are many examples of bacteria and yeasts effectively evolved to reducing spore germination or germ tube growth in necrotrophic pathogens by competing for nutrients. For example, a *Pseudomonas* species inhibits the germination of conidia of *Botrytis cinerea* by competing for amino acids.

4.2.2 ANTIBIOSIS: PRODUCTION OF ANTIMICROBIAL SUBSTANCES

In due course of evolution, most BCA (microorganisms) have been producing secondary metabolites. These compounds are generally produced in a phase subsequent to growth and are not essential mediators to the central metabolism. They generally have unusual structures and are toxic to pathogens (e.g. antibiotics and mycotoxins). They may be volatile such as hydrogen cyanide and ammonia or non-volatile compounds viz., oxalic acid, viridin, gliovirin, harzonic acid, bacillomycin, phenazine etc. *Bacillus* spp. and *Pseudomonas* spp. produces anti fungal lipopeptides such as fengycin, surfactin, iturin, and bacillomycin against fungal pathogen and causes pores in to fungal pathogens in the natural environment.

4.2.3 PARASITISM AND PREDATION

A very good example of primary evolution of BCA over pathogen is parasitism of one fungus by another (hyperparasitism or mycoparasitism) is well documented and is manifested as morphological disturbance, direct penetration of hyphae and hyphal lysis. The degree of hyperparasitism is affected by environmental factors such as temperature, light and pH, but nutrient status, especially the C:N ratio, is also very important. There are some species available commercially and are used to control fungal pathogens in the soil and on aerial plant surfaces. This mycoparasite penetrates resting structures such as sclerotia or may parasitize growing hyphae by coiling round them. *Trichoderma* spp. degrades fungal cell walls by the lytic action of glucanases and chitinases, while other species also produce cellulase. In this regard, the genus shows highest evolution (Fig. 4.1). *Trichoderma* species evolved such a way that they become useful in integrated disease management programs with enhanced pesticide-tolerant, hyperparasitic, and lytic capabilities. Another successful mycoparasite, *Ampelomyces quisqualis*, can control powdery mildew, an obligate parasite in glasshouse-grown cucumber. Other mycoparasites include *Sporidesmium sclerotiorum* and *Coniothyrium minitans*, which are antagonists of sclerotial fungi, and *Gliocladium* spp., which parasitize a range of soil-borne pathogens. Mycoparasitic Pythium spp. appears to have some affinity for plant-parasitic members of the same genus. On the phylloplane, several fungi, including *Verticillium lecanii*, *Sphaerellopsis filum*, and *Cladosporium* spp., are known to attack rust fungi.

Some bacteria which occur on the phylloplane and in the rhizosphere are also evolved to parasitize plant pathogens. For instance, *Bdellovibrio bacteriovorus* is widespread in soil and can attack other bacteria, particularly Gram-negative species, by attaching to and penetrating the bacterial cell wall and lysing the cell. There are also many instances of bacteria causing lysis of fungal spores. Isolates of *Bacillus* obtained from the surface of cereal rust urediniospores can lyse germ tubes arising from urediniospores. When bacterial suspensions are sprayed onto cereal leaves, pustule development is reduced.

A discussion on biological control evolution of plant pathogens would not be complete without considering predation of pathogens by other organisms.

In soil, numerous invertebrates, including mites, springtails, protozoans, earthworms and free-living nematodes, by virtue of their varied feeding habits, may contribute to some form of general biological suppression of pathogens. Unfortunately, specific information on their relative importance is not enough. Some amoebae ingest yeasts and small spores or bore holes in fungal hyphae and then enter them to rapidly clean out the cytoplasm of the cell. Large numbers of bacteria in the soil and rhizosphere are consumed by bacterial feeding nematodes and several species of fungal feeding nematodes are common in soil. Many are omnivores and may not have a preference for particular plant pathogenic fungi or bacteria.

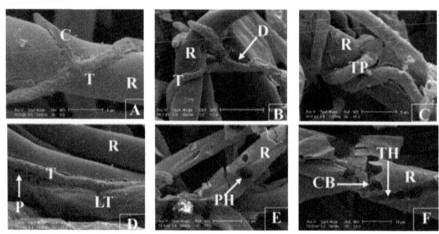

FIGURE 4.1   A. *Trichoderma* fungal strands coil (C) around the *Rhizoctonia*; B. Initial stages of degradation (D) as a result of *Trichoderma* generated enzymes; C. Penetration of the *Rhizoctonia commences*. D. Further enzyme activity causes cell degradation, Compare the shape and consistency of the upper *Rhizoctonia* to the lower area (LT); E. *Trichoderma* is pulled away to expose hole where penetration has taken place (PH) and *Trichoderma* has forced its way into the *Rhizoctonia*; F. Enzyme action has destroyed the *Rhizoctonia* pathogen to the extent of total breakdown.

T = *Trichoderma*, R = *Rhizoctonia*, TP = *Trichoderma* penetrating (Source: http://www.plant-health.co.za/eco-t_mode-of-action.html).

4.2.4 INDUCTION OF DEFENCE RESPONSE IN PLANTS

Another evolution has been noticed in the mechanism to manage pathogen by biological control agent especially *Trichoderma* species is that of induction of resistance in the host plant by treatment with the biological control agent. This concept is supported by the work of Yedidia et al. (1999), who demonstrated that inoculating roots of seven-day-old cucumber seedlings in an aseptic hydroponic system with *T. harzianum* (T-203) spores to a final concentration of 10^5/ml initiated plant defence responses in both the roots and leaves of treated plants. They also demonstrated that hyphae of the biological control fungus penetrated the epidermis and upper cortex of the cucumber root. The plant response was noticed by an increase in peroxidase activity (often associated with the production of fungitoxic compounds), an increase in chitinase activity, and the deposition of callose-enriched wall appositions on the inner surface of cell walls. Increased enzyme activities were observed in both roots and leaves. Interestingly, the plant defence became muted with time and began to resemble a symbiotic mycorrhizal association. Later, Yedidia et al. (2000) showed that inoculation of cucumber roots with *T. harzianum* (T-203) induced an array of pathogenesis-related proteins, including a number of hydrolytic enzymes. Plants treated with a chemical inducer (2,6-dichloroisonicotinic acid) of disease resistance displayed defence responses that were similar to those of plants inoculated with the biocontrol agent. In another study, Howell et al. (2002) demonstrated that seed treatment of cotton with biocontrol preparations of *T. virens* (G-6, G-11, G6-5) or application of *T. virens* culture filtrate to cotton seedling radicles induced synthesis of much higher concentrations of the terpenoids desoxyhemigossypol (dHG), hemigossypol (HG), and gossypol (G) in developing roots than those found in untreated controls. Gossypol was toxic only at high levels, but the pathway intermediates HG and dHG were strongly inhibitory to *R. Solani,* the cotton seedling pathogen, at much lower concentrations. *Trichoderma* species were found much more resistant to cotton terpenoids than were seedling disease pathogens. Biocontrol activity against *R. solani* was highly correlated with induction of terpenoid synthesis in cotton roots by *Trichoderma* species, even among strains of *T. virens* that were deficient for antibiotic production and mycoparasitism. In addition to terpenoid synthesis, treatment of cotton roots with *T. virens* also induced significantly higher levels of peroxidase activity than in control roots. Peroxidase activity and terpenoid levels in seedling hypocotyls were not significantly different from those found in the controls. In this case, plant defence responses appeared to be confined to the root system.

4.2.5   METABOLISM OF GERMINATION STIMULANTS

A recently discovered, evolution in mechanism employed by *Trichoderma* species to effect biological control of pre-emergence damping-off of cotton seedlings incited by *P. ultimum* and/or *Rhizopus oryzae*, Howell (2002) found that control by *T. virens* (G6-5, G6) or protoplast fusants of *T. virens/T. longibrachiatum* (Tvl-30, Tvl-35) was due to metabolism of germination stimulants released by the cotton seeds. These compounds usually induced pathogen propagules to germinate. Disease control could be affected by wild-type strains or by mutant strains that were efficient for mycoparasitism, induction of terpenoid synthesis and antibiotic production, in cotton roots. If, however, pathogen propagules were induced to germinate by artificial means, none of the above treatments gave effective control of the disease. The importance of metabolism of stimulatory compounds by the biocontrol agent is further supported by the fact that cotton cultivars that do not produce pathogen propagules stimulants during germination are virtually immune to the disease. Again, artificial induction of pathogen propagule germination caused to be these cotton cultivars susceptible to disease. Also, cultured of dead *T. virens* with seed exudates biocontrol preparation stimulated pathogen propagules to germinate, while exudates cultured with live biocontrol preparation did not (Fig. 4.2).

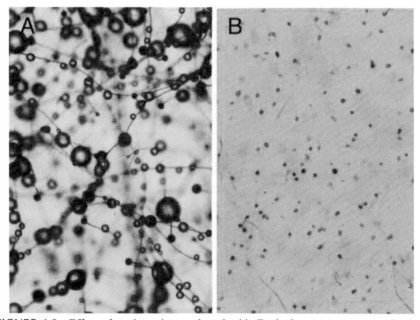

FIGURE 4.2   Effect of seed exudates cultured with *Trichoderma virens* preparations on germination of *Rhizopus oryzae* sporangiospores: (A) exudate cultured with dead *T. virens* preparation stimulates spore germination and mycelial growth of *R. oryzae*, while (B) exudate cultured with live *T. virens* preparation does not stimulate spore Germination (Source: Howell, 2003. Courtesy of The American Phytopathological Society, 2003.)

4.2.6 ADDITIONAL MECHANISMS

This is perhaps not of primary importance as mechanisms in biological control, most plant growth promoting BCA including *Trichoderma* species exhibit other characteristics during interactions with host plants that may contribute to tolerance or disease resistance. These characteristics visible themselves by increases in plant root and shoot growth, resistance to biotic and abiotic stresses, and changes in the nutritional status of the plant. These phenomena have been thoroughly reviewed by Harman (2000), who showed that seed treatment of corn planted in low nitrogen soil with *T. harzianum* T-22 resulted in plants that were larger and greener in the early part of the growing season. At maturity, the treated plants had larger stem diameters and increased yields of grain and fodder. In fields where nitrogen levels were sufficient, treatment with *T. harzianum* T-22 did not result in increased yields of grain or silage. However, the nitrogen required for optimum yields was lower for T-22 treated plants than for the untreated controls. Later Harman (2001) reported a strong interaction between T-22 and the nitrogen-fixing bacterium *Bradyrhizobium japonicum*. Treatment of soybeans with the combination stimulated root growth more effectively than did either alone. Theoretically, the combination of a nitrogen-fixing bacterium and a fungus that enables the plant to utilize nitrogen more efficiently should decrease even further the nitrogen fertilizer requirements of the crop. More recently, Yedidia et al. (2001) showed that treatment of cucumber plants in soil with *T. harzianum* (T-203) resulted in large increases in root area and cumulative root length, and significant increases in dry weight, shoot length, and leaf area over that of the untreated control. In hydroponic culture, T-203 inoculated cucumber roots contained significant increases in P, Cu, Fe, Mn, Zn, and Na. In the shoots of these plants, the concentrations of P, Zn, and Mn were increased. The authors considered that improvement of plant nutrition was directly related to the general beneficial growth effect on the root system of inoculation with *T. harzianum*. In another study, Anwer and Khan (2013) found that *Aspergillus niger* AnC2 enhances the tomato fruit quality and promotes the plant health.

4.2.7 TRICHODERMA EVOLUTION

In the evolutionary process, *Trichoderma* (Hypocreaceae, Hypocreales, and Ascomycota) is relatively young compared with its teleomorph genus *Hypocrea*. *Trichoderma*, commonly applied morphological identification of its species was, and still is, difficult because there are only a few relatively invariable morphological characteristics, leading to overlap among species (Samuels, 2006). Therefore, the not very correct species names to isolates were very common before DNA markers were developed. Scientists have provided the methodological framework for molecular identification of *Trichoderma* species by means of DNA barcoding and that resulted in the characterization of species using molecular data (Samuels, 2006; Druzhinina et al., 2006). Since then, diversity studies have become more meaningful and extensive sampling worldwide (Chaverri et al., 2003a; Hoyos-Carvajal et al., 2009; Migheli et al., 2009; Gal-Hamed et al., 2010; Jaklitsch, 2009, 2011) introduced a new period of *Trichoderma* taxonomy underlain by evolutionary concepts. Now a days, the availability of solid taxonomy within the genus and samples from various habitats and substrates from all over the world has covered the way to ecological genomics of *Trichoderma*, in which an understanding of several available genomes relies on the generalization of genus-wide traits and on the detection of unique features of selected species (Druzhinina & Kubicek, 2013).

The ultimate level of evolutionary resolution is offered by the analysis of whole genomes by Atanasova et al., 2013. In this, phylogenetic analysis of 100 orthologous protein sequences available from the three genomes sequenced of *T. reesei* (http:// genome.jgi-psf.org/Trire2/Trire2.home.html, Martinez et al., 2008), *T. virens* and *T. atroviride* (http://genome.jgi-psf. org/Trive1/Trive1.home.html and http://genome.jgi-psf.org/ Triat1/Triat1. home.html, respectively; Kubicek et al., 2011), representing the three well defined infraspecific groups of the genus (section *Longibrachiatum, Virens* clade and section *Trichoderma*, respectively), had revealed that the mycoparasitic species *T. atroviride* occupied an ancestral position relative to the mycoparasitic and phytostimulating species *T. virens,* whereas the moderate antagonist of other fungi and a superior producer of cellulases and moderate mycoparasite *T. reesei* holded the most derived position of the three (Fig. 4.3). This finding indicates that mycoparasitism is the natural property of the genus, which may be either powered by additional features (like in *T. virens*) or reduced (like in *T. reesei*) in a course of evolution and ecological specialization (Druzhinina et al., 2011). Kubicek et al. (2011) complemented their three-species phylogram with an *rpb2*-based tree for 100 species of the genus. The *rpb2* phylogram and the tree based on 100 orthologous protein sequences for genome sequence species were largely in agreement with the postulation that *T. atroviride* represents the oldest state. The same topology was also seen from the phylogram inferred (Atanasova et al., 2013). These analyses suggested that *T. virens* and *T. reesei* are evolutionarily more derived.

In another study, the genomes of four other *Trichoderma* species have been sequenced by DOE JGI and made publically available in 2012: *T. harzianum* sensu stricto CBS 226.95 (http:// genome.jgi-psf.org/Triha1/ Triha1.homehtml), *T. asperellum* CBS 433.97 (http://genome.jgi.doe.gov/ Trias1/Trias1.home.html), *T. longibrachiatum* ATCC 18648 (http://genome. jgi.doe.gov/Trilo1/Trilo1.home.html), and *T. citrinoviride* (http://genome. jgi.doe.gov/Trici1/Trici1.home.html). Sequencing of the *T. koningii* genome was reported by the Shanghai Institutes for Biological Sciences (China), although it is not accessible via the internet.

The inclusion of the above-mentioned species in a phylogram in Figure 4.3 certainly explained evolutionary relationships between and within respective clades. It should not, however, bring major changes to the genus tree because all novel genomes are closely related to the first three species for which genomes were compared by Kubicek et al. (2011).

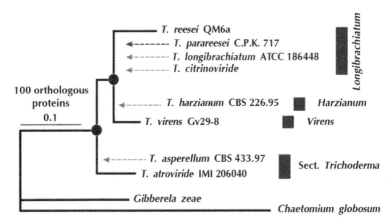

FIGURE 4.3 Bayesian phylogram based on the analysis of amino acid sequences of 100 orthologous syntenic proteins (MCMC, 1 million generations, 10,449 characters) in *T. reesei*, *T. virens*, *T. atroviride*, *Gibberella zeae*, and *Chaetomium globosum* (Kubicek et al., 2011). Circles above nodes indicate 100% posterior probabilities and significant bootstrap coefficients. Arrows indicate putative positions of the five other *Trichoderma* species for which complete genomes have been sequenced (From Atanasova, L; Druzhinina, I. S.; Jaklitsch, W. M. Two Hundred *Trichoderma* Species Recognized on the Basis of Molecular Phylogeny. In *Trichoderma Biology and Applications*; Mukherjee, P. K., Horwitz, B. A., Singh, U. S., Mukherjee, M., Schmoll, M., Eds.; CABI: USA, 2013; pp10–42. Used with permission from CAB International, Wallingford UK.)

4.3   BIOLOGICAL CONTROL OF SOIL–BORNE PHYTOPATHOGENS

Biocontrol involves harnessing disease-suppressive microorganisms to improve plant health. Disease suppression by biocontrol agents is the sustained demonstration of interactions among the plant, pathogen, biocontrol agent, microbial community on and around the plant, and the physical environment (Fig. 4.4).

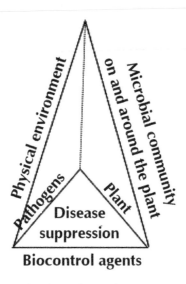

FIGURE 4.4   Disease suppression is the interactions among the plant, pathogen, BCA, microbial community on and around the plant, and the physical environment.

Even in model *in vitro* systems, the study of biocontrol involves interactions among a minimum of three organisms. Therefore, despite its potential in agricultural applications, biocontrol is one of the most poorly understood areas of plant and microbe interactions. The complexity of these systems has influenced the acceptance of biocontrol as a means of controlling plant diseases in two ways. First, practical results with biocontrol have been variable. Thus, despite some stunning successes with biocontrol agents in agriculture, there remains a general skepticism born of past failures (Weller, 1988; Cook & Baker, 1983). Second, progress in understanding an entire system has been slow. Recently, however, substantial progress has been made in a number of biocontrol systems through the application of genetic and mathematical approaches that contain the complexity. Biocontrol of soil-borne diseases is particularly complex because these diseases occur in the dynamic environment at the interface of root and soil, the rhizosphere, which is defined as the region surrounding a root that is affected by it. The rhizosphere is characterized by rapid change, intense microbial activity, and high populations of bacteria compared with non-rhizosphere soil. Plants release metabolically active cells from their roots and deposit as much as 20% of the carbon allocated to roots in the rhizosphere, suggesting a highly evolved relationship between the plant and rhizosphere microorganisms. The rhizosphere is subject to dramatic changes on a short temporal scale-rain events and daytime drought can result in fluctuations in pH, salt concentration, osmotic potential, water potential, and soil particle structure. Over longer temporal scales, the rhizosphere can change due to root growth, interactions with other soil biota, and weathering processes. It is the dynamic nature of the rhizosphere that makes it an interesting setting for the interactions that lead to disease and biocontrol of disease (Hawes, 1991; Waisel et al., 1991; Rovira, 1965, 1969, 1991). There are some well-studied examples of successful biocontrol agents.

4.3.1 FUNGAL DISEASES

Fungi are eukaryotes and constitute a group of plant pathogens that incite the most economically significant diseases of agricultural crops. They infect all types of crops including cereals, vegetables, legumes, and ornamentals. Important diseases caused by fungi are rusts (*Puccinia* spp., *Hemileia* spp.), smuts (*Ustilago* spp., *Tilletia* spp.), seed-rot (*Pythium* spp.), damping-off (*Pythium* spp.), root rot (*Rhizoctonia* spp.), wilt (*Fusarium* spp.), blight (*Phytophthora* spp.), powdery mildew (*Erysiphe* spp., *Sphaerotheca* spp.), and downy mildew (*Plasmopara* spp., *Peronospora* spp.), which attack crops under a varied range of agroclimatic conditions (Agrios, 2005). Among them some are soil borne viz. damping-off (*Pythium* spp.), root rot (*Rhizoctonia* spp.), and wilt (*Fusarium* spp.). Generally, moderately cooler climates with higher relative humidity are favourable for pathogenesis of fungi (Khan et al., 2010). Numerous studies have been conducted to test the effect of BCA, and on several occasions, their application has proved quite effective in controlling fungal-induced plant diseases (Khan & Anwer, 2011). The effects of the BCA have been tested on fungal pathogens in pot and field conditions (Table 4.1).

TABLE 4.1 Effect of Different Biological Control Agents On Soil Borne Fungal Pathogens Infecting Agricultural Crops.

Biocontrol agent	Pathogen/disease managed	Host plant	Reference(s)
Pot condition			
Trichodrma spp.	Seed rot fungi	Many	Elad and Chet (1987)
Penicillium oxalicum	*Pythium ultimum*	Seeds, seedling roots, corns, bulbs, and tubers	Elad and Chet (1987)
Aspergillus niger AN 27 (Kalisena SD)	*Fusarium oxysporum melonis*	Muskmelon seedlings	Radhakrishna and Sen (1986), Angappan et al. (1996), Kumar and Sen (1998)
Trichoderma spp.	*Fusarium* spp.	Chickpea and pigeon pea	Khan (2005)
A. niger AnC2	*Rhizoctonia solani*	Eggplant	Khan et al. (2006a)
Trichodrma spp.	*Rhizoctonia* spp	Many crops	Olson and Denson (2007)
Trichodrma spp.	*Macrophomina phaseolina*		Khan and Gupta (1998)
Trichodrma spp.	*Pythium* spp.		Pill et al. (2009)
T. asperellum T34	*Fusarium* spp.	Carnation	Sant et al. (2010)
T. harzianum + *T. asperellum*	*Fusarium solani*	Bean	Ibrahimov et al. (2009)
Trichoderma spp.	*Botrytis cinerea*	Chickpea	Khan et al. (2011)
Field condition			
T. harzianum	*Fusarium ciceri*	Chickpea	Khan et al. (2004), Singh et al. (2003), Prasad et al. (2002)
T. harzianum and *T. viride*	*Fusarium ciceri*	Chickpea	Gurha (2001)
T. harzianum	*Sclerotium* root rot	Sugar beet	Upadhyay and Mukhopadhyay (1986)
T. harzianum	*S. rolfsii* (collar rot)	Mint	Singh and Singh (2004)
T. virens	*F. oxysporum* f. sp. *lycopersici*	Tomato	Khan and Akram (2000)
T. koningi, T. hamatum, and *T. virens*	*F. oxysporum* f. sp. *lycopersici*	Tomato	Cipriano et al. (1989)
T. harzianum, T. viride, and *T. koningi*	*Macrophomina phaseolina* (root rot)	Egg plant	Khan and Gupta (1998)

TABLE 4.1 *(Continued)*

Biocontrol agent	Pathogen/disease managed	Host plant	Reference(s)
T. virens	Damping-off	Tomato	De and Mukhopadhyay (1994)
T. harzianum or *P. lilacinus*	Wilt	Tomato	Shahida and Gaffar (1991)
T. harzianum	*R. solani* (root rot)	Chickpea	Khan and Rehman (1997)
T. virens	*R. solani* (damping-off)	Many crops	Papavizas and Lewis (1989)
T. harzianum, T. viride, and *T. virens*	*F. oxysporum* f. sp. *ciceri*	Chickpea	Dhedhi et al. (1990)
T. harzianum	root rot (*F. solani, F. oxysporum,* and *M. phaseolin*)	Grape vines	Riad et al. (2010)
A. niger	*Fusarium* spp., *R. solani,* and *Pythium* spp.	Muskmelon	Chattopadhyay and Sen (1996)
A. niger	*Fusarium* spp.	Chickpea and tomato	Anwer (2014)
A. niger	Damping-off (*P. aphanidermatum* and *R. solani*)	Fruit and vegetable crops	Majumdar and Sen (1998)
A. niger	charcoal rot (*Macrophomina phaseolina*)	Potato	Mondal (1998)
A. niger	*Macrophomina phaseolina*	Sorghum	Das (1998)

4.3.2 BACTERIAL DISEASE

Management of plant pathogenic bacteria with the application of BCA is limited; however, a few studies conducted thus far have shown that bacterial diseases of plants can be successfully managed with bacterial BCA. For example, bacterial crown gall has been controlled by treating seeds or nursery stock with bacteriocin-producing strain *Agrobacterium radiobacter* K-1026 (Jindal, 1990). Some information is also available on management of bacterial plant pathogens with fungal BCA. Treatments of tubers and seeds with fungal antagonists has proved effective against plant pathogenic bacteria but was not effective under field conditions (Agrios, 2005). Kalita et al. (1996)

reported 47.5% reduction in citrus canker incidence (*Xanthomonas campestris* pv. *citri*) after application of a strain of *Aspergillus terreus*. Bacterial wilt of tomato (*Ralstonia solanacearum*) and soil populations of the pathogen were reduced by soil application of *Glomus mosseae* together with *P. fluorescens* (Kumar & Sood, 2002). Effects of bacterial biocontrol agents have been evaluated against different diseases (Table 4.2).

4.3.3 NEMATODE DISEASES

Bio-control of nematodes may be achieved with two kinds of microorganisms, that is, classical parasites or predators, and plant growth-promoting microorganisms (PGPM; Khan et al., 2009). Classical parasites or predators such as *Paecilomyces lilacinus, Dactylaria candida,* and *Pasteuria penetrans* have been used in nematode control during last few decades and to reduce nematode population by direct action (De Bach, 1964). PGPM may suppress rhizoshpheric nematode populations by promoting host growth, inducing systemic resistance and/or producing nematoxic metabolites like bulbiformin (Brannen, 1995), phenazine (Toohey, 1965), and pyoluteorin (Howell & Stepinovic, 1980). In recent years, considerable research has been carried out on use of BCA to control nematode populations in soil. Effects of microorganisms have been evaluated against different nematodes under pot and field conditions. Some important BCA managing nematode diseases are listed in Table 4.3.

4.4   CONCLUSION

During the process of evolution, BCA have supremacy over pathogens such an extent that in any undisturbed ecology there is less diseases. The mechanisms employed by biocontrol agents to effect biological control of plant diseases are many and complex, and their use varies with the kind of biocontrol agent, pathogen, and host plant involved in the interaction. Mechanisms are also influenced by the soil type, by the temperature, pH, and moisture of the plant and soil environment, and by other members of the microflora. What we observe and define as biocontrol may be the result of a number of different mechanisms working synergistically to achieve disease control. Our knowledge of the complexity of these systems is currently limited by our ability to perceive them, and a great deal of research will have to be undertaken in order to understand exactly what is taking place during the biocontrol process.

TABLE 4.2 Effect of Different Bacterial Biological Control Agents on Soil-Borne Pathogen Infecting Agricultural Crops.

Biocontrol agent	Produced lytic enzyme	Pathogen managed	Host plant	Reference(s)
Biocontrol bacteria producing lytic enzyme				
Aeromonas caviae	Chitinases	*Rhizoctonia solani* and *Fusarium oxysporum f. sp. vasinfectum*	Cotton	Inbar and Chet (1991)
		Sclerotium rolfsii	Bean	Inbar and Chet (1991)
Arthrobacter sp.		*Fusarium*	Carnation	Koths and Gunner (1967), Sneh (1981)
Arthrobacter		*Fusarium moniliforme var subglutinans*	Southern pines	Barrows-Broaddus and Kerr (1981)
Enterobacter agglomerans and Bacillus cereus		*Rhizoctonia solani*	Cotton	Chernin et al. (1995, 1997), Pleban et al. (1997)
Bacillus circulans and Serratia marcescens		*Phaeoisariopsis personata*	Peanut	Kishore et al. (2005)
Enterobacter agglomerans and Bacillus cereus		*Rhizoctonia solani*	Cotton	Chernin et al. (1995, 1997), Pleban et al. (1997)
Paenibacillus illinoisensis		*Rhizoctonia solani*	Cucumber	Jung et al. (1990)
Pseudomonas		*Fusarium oxysporum f. sp. cucumerinum*	Cucumber	Sneh et al. (1984)
Serratia plymuthica		*Sclerotinia sclerotiorum*	Cucumber	Kamensky et al. (2003)
Serratia marcescens		*Sclerotium rolfsii*	Beans	Chet et al. (1990)
		R. solani	Cotton	Chet et al. (1990)
Streptomyces lydicus		*Pythium and Aphanomyces*		Mahadevan and Crawford (1997)
Streptomyces sp.	Glucanases	*Phytophthora fragariae*	Raspberry	Valois et al. (1996)

TABLE 4.2 (Continued)

Biocontrol agent	Produced lytic enzyme	Pathogen managed	Host plant	Reference(s)
Pseudomonas cepacia		*R. solani, Sclerotium rolfsii,* and *Pythium ultimum*		Fridlender et al. (1993)
Actinoplanes philippinensis and *Micromonospora chalcea*		*Pythium aphanidermatum*	Cucumber	El-Tarabily (2006)
Lysobacter enzymogenes		*Pythium*	Sugar beet	Palumbo et al. (2005)
Serratia marcescens, Streptomyces viridodiasticus, and *Micromonospora carbonacea*	Chitinases and glucanases	*Sclerotinia minor*	Lettuce	El-Tarabily et al. (2000)
Streptomyces sp. and *Paenibacillus sp.*		*F. oxysporum* f. sp. *cucumerinum*	Cucumber	Singh et al. (1999)
Bacillus subtilis, Erwinia herbicola, Serratia plymuthica, and *Actinomycete*	Chitinases, proteases, and cellulases	*Eutypa lata*	Grapevine	Schmidt et al. (2001)
Stenotrophomonas maltophilia	Proteases	*Pythium ultimum*	Sugar beet	Dunne et al. (1997, 1998)
Biocontrol agent	**Produced antibiotics**	**Pathogen managed**		**Reference(s)**
Biocontrol bacteria producing antibiotics				
Bacillus	Bacillomycin	*Aspergillus flavus*		Moyne et al. (2001)
Bacillus cereus	Kanosamine	*Phytophthora medicaginis*		Milner et al. (1996)
B. cereus and *B. thuringiensis*	Zwittermicin A	*Phytophthora*		Silo-Suh et al. (1998)
Bacillus spp.	Zwittermicin A	*Sclerotinia sclerotiorum*		Zhang and Fernando (2004)
B. cereus	Zwittermicin A	*Phytophthora parasitica* var. *nicotianae*		He et al. (1994)

TABLE 4.2 *(Continued)*

Biocontrol agent	Produced lytic enzyme	Pathogen managed	Host plant	Reference(s)
B. subtilis	Iturin	*P. ultimum, R. solani, F. oxysporum, S. sclerotiorum,* and *M. phaseolina*		Constantinescu (2001)
B. subtilis	Iturin A and Surfactin	*R. solani*		Asaka and Shoda (1996)
Burkholderia cepacia	Pyrrolnitrin	*R. solani*		El-Banna and Winkelmann (1988)
Pseudomonas fluorescens	Pyrrolnitrin	*Gaeumannomyces graminis* var. *tritici*		Tazawa et al. (2000)
	Pyrrolnitrin	*R. solani*		Howell and Stipanovic (1979)
Enterobacter agglomerans	Pyrrolnitrin	*Agrobacterium tumefaciens, Clavibacterium michiganese, Xanthomonas campestris,* and *Pseudomonas syringae* pv. *Syringae*		Chernin et al. (1996)
Pseudomonas	2,4-DAPG	*Xanthomonas oryzae* pv. *Oryzae*		Velusamy et al. (2006)
P. fluorescens	Pyoluteorin	*P.ultimum*		Howell and Stipanovic (1980)
P. fluorescens	Phenazines	*Gaeumannomyces graminis* var. *tritici, and Botrytis cinerea*		Weller and Cook (1983), Brisbane and Rovira (1988), Khan and Anwer (2011)
P. fluorescens	Phenazines-1-carboxylate	*Gaeumannomyces graminis* var. *tritici*		Thomashow et al. (1990)
P. aureofaciens	Phenazines-1-carboxylate	*Sclerotinia homeocarpa*		Powell et al. (2000)

TABLE 4.2 *(Continued)*

Biocontrol agent	Produced lytic enzyme	Pathogen managed	Host plant	Reference(s)
P. chlororaphis	Phenazines-1-car-boxylate	*Fusarium oxysporum* f. sp. *radicis-lycopersici*		Chin-A-Woeng et al. (1998)
P. fluorescens	Viscosinamide	*R. solani*		Nielsen et al. (2002)
P. fluorescens	Viscosinamide	*P. ultimum*		Thrane et al. (2000)
P. fluorescens	Amphisin	*P. ultimum and R. solani*		Andersen et al. (2003)
Streptomyces hygroscopicus var. *geldonus*	Geldanamycin	*R. solani*		Rothrock and Gottlieb (1984)
Streptomyces libani	Oligomycin A	*Botrytis cinerea*		Kim et al. (1999)
Streptomyces cacaoi	Polyoxin D	*R. solani*		Isono et al. (1965)
Streptomyces kasugaensis	Kasugamycin	*P. oryzae*		Umezawa et al. (1965)

TABLE 4.3 Effect of Biological Control Agents on Different Nematodes Infecting Agricultural Crops.

Biocontrol agent	Pathogen /disease managed	Host plant	Reference(s)
Pot condition			
Paecilomyces lilacinus	*Rotylenchulus reniformis*	Chickpea	Ashraf and Khan (2008)
P. chlamydosporia	*H. avenae*	Wheat	Bhardwaj and Trivedi (1996)
T. harzianum, T. viride, T. koningii, T. reesei, and *T. hamatum*	*M. javanica*	Eggplant and cv black beauty	Bokhari (2009)
P. chlamydosporia	*M. hapla*	Tomato	De Leij et al. (1993)
A. niger	*M. incognita*	Eggplant	Khan et al. (2006b), Khan and Anwer (2007, 2008)
T. harzianum (T014)	*M. incognita*	Gladiolus	Khan and Mustafa (2005)
T. virens and *Burkholderia cepacia*	Root knot nematodes	Bell pepper	Meyer et al. (2000)
T. harzianum, P. lilacinum, and *A. niger*	*M. incognita*	Chickpea	Pant and Pandey (2001)
A. niger	*M. javanica*	Chilli	Shah et al. (1994)
T. harzianum	*M. arenaria*	Corn	Windham et al. (1989)
Field condition			
P. chlamydosporia	*H. avenae*	Cereal	Kerry et al. (1982a, 1982b)
P. chlamydosporia +P. lilacinus	*M. incognita*	Brinjal	Cannayane and Rajendran (2001)
A. niger	*Rotylenchulus reniformis*	Caster bean	Das (1998)
P. Chlamydosporia + P. fluorescens	*M. incognita*	Brinjal	Dhawan et al. (2008)
Pochonia chlamydosporia var. *catenulate*	Different plant parasitic nematodes	Vegetable crops	Garcia et al. (2004)
P. lilacinus	*M. javanica*	Tobacco	Hewlett et al. (1988)
T. harzianum	Root know nematode	Cardamom	Anonymous (1995)
P. lilacinus	*Meloidogyne* spp.	Various	Jatala (1986)
Penicillium notatum	*Globodera* spp.	Potato	Jatala (1986)
P. lilacinus and *T. virens*	*M. incognita*	Tomato	Khan and Akram (2000)

TABLE 4.3   *(Continued)*

Biocontrol agent	Pathogen /disease managed	Host plant	Reference(s)
P. fluorescens, T. virens, or *P. lilacinus*	Root knot nematodes	Tomato	Akram and Khan (2006)
Paecilomyces lilacinus	*Meloidogyne* spp.	Okra	Khan and Ejaz (1997)
P. lilacinus and *P. chlamydosporia*	*M. incognita*	Mungbean	Khan and Kaunsar (2000)
Paecilomyces lilacinus	*Meloidogyne* spp.	Tomato	Khan and Tarannum (1999), Pal and Gardener (2006), Schenek (2004)
Bacillus subtilis or *Pseudomonas stutzeri*	Root knot nematodes	Tomato	Khan and Tarannum (1999)
T. harzianum and *P. chlamydosporia*	*M. incognita*	Chickpea	Khan et al. (2005a)
A. niger	*M. incognita*	Tomato	Khan et al. (2007)
P. chlamydosporia, P. fluorescens, and *B. subtilis*	*M. incognita*	Hollyhock, petunia, and poppy	Khan et al. (2005b)
P. fluorescens or *B. subtilis*	Root knot nematodes	Green gram	Khan et al. (2007)
P. fluorescens	*M. incognita*	Chickpea	Khan et al. (2012)
P. chlamydosporia	Root knot nematodes	Papaya	Kumar (2009)
P. lilacinus	*R. reniformis*	Tomato	Lysek (1966)
T. pseudokoningii, T. viride, P. lilacinus, A. niger, and *G. mosseae*	*M. incognita*	Soybean	Oyekanmi et al. (2008)
T. harzianum and *P. lilacinus*	*M. incognita*	Chickpea	Pant and Pandey (2002)
P. chlamydosporia + *P. lilacinus*	*M. javanica*	Acid lime	Rao (2005)
T. harzianum and *P. chlamydosporia*	*Globodera rostochiensis* and *G. pallid*	Potato	Saifullah (1996a, 1996b)
A. niger	*M. incognita*	Okra	Sharma et al. (2005)
Trichoderma asperellum-203 and *Trichoderma atroviride*	*M. javanica*	*In vitro*	Sharon et al. (2009)
T. harzianum and *P. chlamydosporia*	*H. cajani*	Pigeonpea	Siddiqui and Mahmood (1996)

TABLE 4.3 *(Continued)*

Biocontrol agent	Pathogen /disease managed	Host plant	Reference(s)
T. harzianum with *P. fluorescens*	*M. javanica*	Tomato	Siddiqui and Shaukat (2004)
P. chlamydosporia	*M. hapla*	Tomato	Siddiqui et al. (2009)
P. chlamydosporia	*G. pallida*	Potato	Siddiqui et al. (2009)
Aspergillus niger	*Meloidogyne* spp.	Tomato	Singh et al. (1991)
T. harzianum	*M. arenaria*	Corn	Windham et al. (1989)
T. harzianum and *P. lilacinus*	*Meloidogyne javanica*	Okra	Zareen et al. (2001)
T. atroviride	*R. similis*	Banana	Zum Felde et al. (2006), Pocasangre Enamorado et al. (2007)

The results of many years of research with *Trichoderma* species as biocontrol agents has shown that not all the mechanisms and characteristics deemed necessary for optimum biocontrol are found in the same organism. Very often those strains that have the capacity to produce antibiotics and enzymes that are associated with biocontrol are not the ones that have good storage qualities or function well at temperature and moisture levels where pathogens flourish. Therefore, hybridization of different strains or species will be required in order to combine these beneficial characteristics. The production of hybrids from *Trichoderma* species, many of which have no known sexual stage, will entail the use of transformation or protoplast fusion in order to obtain strains with optimal sets of characteristics. Once produced and screened, hybrids may yield strains with expanded host, temperature, and moisture parameters, and they may yield strains with better storage qualities than those exhibited by the parents.

Many BCA control efficiently soil born pathogens in pot as well as natural field conditions. Time has come to exploit the potential biocontrol agents specifically against each and every disease to manage them with the philosophy that nature is having balancing power against each menace.

KEYWORDS

- *Aspergillus*
- growth-promoting microorganisms
- mycoparasitic
- phytopathogen
- *Trichoderma*

REFERENCES

Agrios, G. N. *Plant Pathology*. Elsevier Academic Press: San Diego, CA, 2005.

Akram, M.; Khan, M. R. Interaction of *Meloidogyne incognita* and *Fusarium oxysporum* f. sp. *lycopersici* on Tomato. *Ann. Plant Prot. Sci.* **2006,** *14*(2), 448–451.

Aluko, M. O.; Hering, T. F. The Mechanism Associated with the Antagonistic Relationship between *Corticium solani* and *Gliocladium virens*. *Trans. Br. Mycol. Soc.* **1970,** *55*, 173–179.

Amkraz, N.; Boudyach, E. H.; Boubaker, H.; Bouizgarne, B.; Ait Ben Aoumar, A Screening for Fluorescent Pseudomonades, Isolated from the Rhizosphere of Tomato, for Antagonistic Activity toward *Clavibacter michiganensis* subsp. *michiganensis*. *World J. Microbiol. Biotechnol.* **2010,** *26*, 1059–1065.

Andersen, J. B.; Koch, B.; Nielsen, T. H.; Sørensen, D.; Hansen, M.; Nybroe, O.; Christophersen, C.; Sorensen, J.; Molin, S.; Givskov, M. Surface Motility in *Pseudomonas* sp. DSS73 is Required for Efficient Biological Containment of the Root-Pathogenic Microfungi *Rhizoctonia solani* and *Pythium ultimum*. *Microbiology*. **2003,** *149*, 1147–1156.

Angappan, K.; Dureja, P.; Sen, B. Multiprong Actions of Biocontrol Agent, *Aspergillus niger* AN27. In *Second International Crop Science Congress on Crop Productivity and Sustainability – Shaping the Future*, Abstract of Poster Session, Nov. 17–24; New Delhi: India, 1996; pp 301.

Anonymous. Indian Institute of Spices Research (IISR), *Annual Report,* 1994–95; India: Calicut, 1995; pp 89.

Anwer, M. A.; Bushra, A. *Aspergillus niger*-A Novel Heavy Metal Bio-Absorbent and Pesticide Tolerant Fungus. *Res. J. Chem. Environ.* **2015,** *19*(2), 57–66.

Anwer, M. A.; Khan, M. R. *Aspergillus niger* As Tomato Fruit (*Lycopersicum esculentum* Mill.) Quality Enhancer and Plant Health Promoter. *J. Post-Harvest Technol.* **2013,** *01*(01), 36–51.

Anwer, M. A. *Aspergillus niger as a Noble Biological Control Agent*. Scholar's Press: Germany, 2014; pp 367.

Anwer, M. A. *Aspergillus niger as a Noble Biological Control Agent*. Scholar's Press: Germany, 2014; pp 367. (ISBN: 978–3–639–70222–4).

Anwer, M. A. Advances in Post Harvest Disease Control in Vegetables. In *Postharvest Biology and Technology of Horticultural Crops: Principles and Practices for Quality Maintenance;* Siddiqui, M. W., Ed.; Apple Academic Press Inc.: New Jersey, USA, 2015; pp 505–541.

Asaka, O.; Shoda, M. Biocontrol of *Rhizoctonia solani* Damping-Off of Tomato with *Bacillus subtillis* RB14. *Appl. Environ. Microbiol.* **1996,** *62*, 4081–4085.

Ashraf, M. S.; Khan, T. A. Biomanagement of Reniform Nematode, *Rotylenchulus reniformis* by Fruit Wastes and *Paecilomyces lilacinus* on Chickpea. *World J. Agric. Sci.* **2008,** *4*, 492–494.

Atanasova, L; Druzhinina, I. S.; Jaklitsch, W. M. Two Hundred *Trichoderma* Species Recognized on the Basis of Molecular Phylogeny. In *Trichoderma Biology and Applications*; Mukherjee, P. K., Horwitz, B. A., Singh, U. S., Mukherjee, M., Schmoll, M., Eds.; CABI: USA, 2013; pp10–42.

Barrows-Broaddus, J.; Kerr, T. K. Inhibition of *Fusarium moniliforme* var. *subglutinans*, The Casual Agent of Pitch Canker, By the Soil Bacterium *Arthrobacter* sp. *Can. J. Microbiol.* **1981,** *27*, 20–27.

Bhardwaj, P.; Trivedi, P. C. Biological Control of *Heterodera avenae* on Wheat Using Different Inoculum Levels of *Verticillium chlamydosporium*. *Ann. Plant Prot. Sci.* **1996,** *4*, 111–114.

Bliss, D. E. The Destruction of *Armillaria mellea* in Citrus Soils. *Phytopathology*. **1951,** *41*, 665–683.

Bokhari, F. M. Efficacy of Some *Trichoderma* Species in the Control of *Rotylenchulus reniformis* and *Meloidogyne javanica*. *Arch. Phytopathol. Plant Prot.* **2009,** *42*(4), 361–369.

Bora, B. C.; Gogoi, B. B.; Khan, M. R.; Khan, U.; Khan, M. M.; Anwer, M. A. Nematode Infestation in Jute and other Baste Fiber Crops. In *Nematode Infections Part II: Industrial Crops*; Khan, M. R., Jairajpuri, M. S., Eds.; The National Academy Of Sciences: Allahabad, India, 2010; pp 289–304.

Bouizgarne, B. Bacteria for Plant Growth Promotion and Disease Management. In *Bacteria In Agrobiology: Disease Management*; Maheshwari, D. K., Ed.; Springer-Verlag: Berlin, Heidelberg. 2013; pp 15–47.

Brisbane, P. G.; Rovira, A. D. Mechanisms of Inhibition of *Gaeumannomyces graminis* var. *tritici* by Fluorescent Pseudomonads. *Plant Pathol.* **1988,** *37*, 104–111.

Cannayane, I.; Rajendran, G. Application of Biocontrol Agents and Oil Cakes for the Management of *Meloidogyne incognita* in Brinjal (*Solanum melongena* L.). *Curr. Nematol.* **2001,** *12*(2), 51–55.

Chattopadhyay, C.; Sen, B. Integrated Management of *Fusarium* Wilt of Muskmelon Caused by *Fusarium oxysporum*. *Indian J. Mycol. Plant Pathol.* **1996,** *26*, 162–170.

Chaverri, P.; Castlebury, L. A.; Overton, B. E.; Samuels, G. J. *Hypocrea/Trichoderma*, Species with Conidiophore Elongations and Green Conidia. *Mycologia.* **2003,** *95*, 1100–1140.

Chernin, L.; Brandis, A.; Ismailov, Z.; Chet, I. Pyrrolnitrin Production by an *Enterobacter agglomerans* Strain with a Broad Spectrum of Antagonistic Activity towards Fungal and Bacterial Phytopathogens. *Curr. Microbiol.* **1996,** *32*, 208–212.

Chernin, L.; Ismailov, Z.; Haran, S.; Chet, I. Chitinolytic *Enterobacter agglomerans* Antagonistic to Fungal Plant Pathogens. *Appl. Environ. Microbiol.* **1995,** *61*, 1720–1726.

Chernin, L. S.; Fuente, L. D. L.; Sobolov, V.; Haran, S.; Vorgias, C. E.; Oppenheim, A. B.; Chet, I. Molecular Cloning, Structural Analysis, and Expression in *Escherichia coli* of a Chitinase Gene from *Enterobacter agglomerans*. *Appl. Environ. Microbiol.* **1997,** *63*, 834–839.

Chet, I.; Ordentlich, A.; Shapira, R.; Oppenheim, A. Mechanisms of Biocontrol of Soil-borne Plant Pathogens by Rhizobacteria. *Plant Soil*. **1990,** *129*, 85–92.

Chet, I. *Trichoderma*-Application, Mode of Action, and Potential as a Biocontrol Agent of Soil-borne Pathogenic Fungi. In *Innovative Approaches to Plant Disease Control*. Chet, I., Ed.; John Wiley & Sons: New York, USA, 1987; pp137–160.

Chin-A-Woeng, T. F. C.; Bloemberg, G. V.; Van Der Bij, A. J.; Van Der Drift, K. M. G. M.; Schripsema, J.; Kroon, B.; Scheffer, R. J.; Keel, C. ; Bakker, P. A. H. M.; Tichy, H. V.; De Bruijn, F. J.; Thomas-Oates, J. E.; Lugtenberg, B. J. J. Biocontrol by Phenazine-1-Carbox-amide-Producing *Pseudomonas chlororaphis* PCL1391 of Tomato Root Rot Caused By *Fusarium oxysporum* f. sp. *radicis-lycopersici*. *Mol. Plant Microbe. Interact.* **1998**, *11*, 1069–1077.

Cipriano, T.; Cirvilleri, G.; Cartia, G. *In-Vitro* Activity of Antagonistic Microorganisms against *Fusarium oxysporum* f. sp. *lycopersici*, the Causal Agent of Tomato Crown Root-Rot. *Inf. Fitopatol.* **1989**, *39*(5), 46–48.

Constantinescu, F. Extraction and Identification of Antifungal Metabolites Produced by Some *B. subtilis* Strains. Analele Institutului De Cercetari Pentru Cereale Protectia Plantelor 2001, *31*, 17–23.

Cook, R. J.; Baker, K. F. The Nature and Practice of Biological Control of Plant Pathogens. APS Press: St. Paul, MN, USA, 1983.

Da Mota, F. F.; Gomes, E. A.; Seldin, L. Auxin Production and Detection of the Gene Coding for the Auxin Efflux Carrier (AEC) Protein In *Paenibacillus polymyxa*. *J. Microbiol.* **2008**, *56*, 275–264.

Das, I. K.; Kalisena, A. Novel Biopesticide for Disease Free Crop and Better Yield. Proceedings of the National Symposium on Development of Microbial Pesticides and Insect Pest Management, November 12th and 13th, 1998; Pune: BARC Mumbai and Hindustan Antibiotics Ltd., 1998.

De Leij, F. A. A. M.; Kerry, B. R.; Denneby, J. A. *Verticillium chlamydosporium* as a Biological Control Agent for *Meloidogyne incognita* and *M. hapla* in Plant and Microplot Tests. *Nematologica.* **1993**, *39*, 115–126.

De Vasconcellos, R. L. F.; Cardoso, E. J. B. N. Rhizospheric *Streptomycetes* as Potential Biocontrol Agents of *Fusarium* and *Armillaria* Pine Rot and as PGPR For *Pinus taeda*. *Biocontrol.* **2009**, *54*, 807–816.

De Weger, L. A.; Van Boxtel, R.; Van Der Burg, B.; Gruters, R. A.; Geels, F. P.; Schippers, B.; Lugtenberg, B. Siderophores and Outer Membrane Proteins of Antagonistic, Plant-Growth-Stimulating, Root colonizing *Pseudomonas* sp. *J. Bacteriol.* **1986**, *165*, 585–594.

De, R. K.; Mukhopadhyay, A. N. Biological Control of Tomato Damping off by *Gliocladium virens*. *J. Biol. Control.* **1994**, *8*(1), 34–40.

Dhawan, S. C.; Singh, S.; Kamra, A. Bio-Management of Root-Knot Nematode, *Meloidogyne incognita* by *Pochonia chlamydosporia* and *Pseudomonas fluorescens* on Brinjal In Farmer's Field. *Indian J. Nematol.* **2008**, *38*(1), 110–111.

Dhedhi, B. M.; Gupta, O.; Patel, V. A. Influence of Metabolites of Microorganisms on the Growth of *Fusarium oxysporm* f. sp. *ciceri*. *Indian J. Mycol. Plant Pathol.* **1990**, *20*(1), 66–69.

Druzhinina, I. S.; Kubicek, C. P. Ecological Genomics of *Trichoderma*. In *Ecological Genomics of Fungi*; Martinn, F., Ed.; Wiley-Blackwell: Oxford, UK, 2013, in press.

Druzhinina, I. S.; Seidl-Seiboth, V.; Herrera-Estrella, A.; Horwitz, B. A.; Kenerley, C. M.; Monte, E.; Mukherjee, P. K.; Zeilinger, S.; Grigoriev, I. V.; Kubicek, C. P. *Trichoderma*: The Genomics of Opportunistic Success. *Nat. Rev. Microbiol.* **2011**, *9*, 749–759.

Duijff, B. J.; Bakker, P. A. H. M.; Schippers, B. Suppression of *Fusarium* Wilt of Carnation by *Pseudomonas putida* WCS358 at Different Levels of Disease Incidence and Iron Availability. *Biocontrol Sci. Technol.* **1994**, *4*, 279–288.

Dunne, C.; Crowley, J. J.; Moe¨Nne-Loccoz, Y.; Dowling, D. N.; De Bruijn, F. J.; O'Gara, F. Biological Control of *Pythium ultimum* by *Stenotrophomonas maltophilia* W81 is Mediated by an Extracellular Proteolytic Activity. *Microbiology.* **1997**, *143*, 3921–3391.

Dunne, C.; Moe¨Nne-Loccoz, Y.; Mccarthy, J.; Higgins, P.; Powell, J.; Dowling, D. N.; O'Gara, F. Combining Proteolytic and Phloroglucinol-Producing Bacteria for Improved Biocontrol of Pythium-Mediated Damping-Off of Sugar Beet. *Pathology.* **1998**, *47*, 299–307.

Elad, Y.; Chet, I. Possible Role of Competition for Nutrient in Biocontrol of *Pythium* Damping-Off By Bacteria. *Phytopathology.* **1987**, *77*, 190–195.

Elad, Y.; Kapat, A. The Role of *Trichoderma harzianum* Protease in the Biocontrol of *Botrytis cinerea. Eur. J. Plant Pathol.* **1999**, *105*, 177–189.

El-Banna, N.; Winkelmann, G. Pyrrolnitrin from *Burkholderia cepacia*: Antibiotic Activity against Fungi and Novel Activities against *Streptomycetes. J. Appl. Microbiol.* **1988**, *85*, 69–78.

El-Tarabily, K. A. Rhizosphere-Competent Isolates of *Streptomycete* and Non-Streptomycete *Actinomycetes* Capable of Producing Cell-Wall Degrading Enzymes to Control *Pythium aphanidermatum* Damping-Off Disease of Cucumber. *Can. J. Bot.* **2006**, *84*, 211–222.

El-Tarabily, K. A.; Sivasithamparam, K. Non-Streptomycete *Actinomycetes* as Biocontrol Agents of Soil-borne Fungal Plant Pathogens and as Plant Growth Promoters. *Soil Biol. Biochem.* **2006**, *38*, 1505–1520.

El-Tarabily, K. A.; Soliman, M. H.; Nassar, A. H.; Al-Hassani, H. A.; Sivasithamparam, K.; Mckenna, F.; Hardy, G. E. S. J. Biological Control of *Sclerotinia minor* Using a Chitinolytic Bacterium and *Actinomycetes. Plant Pathol.* **2000**, *49*, 573–583.

Fravel, D R. Commercialization and Implementation of Bio Control. *Annu. Rev. Phytopathol.* **2005**, *43*, 337–359.

Fridlender, M.; Inbar, J.; Chet, I. Biological Control of Soil-borne Plant Pathogens by a ß-1, 3 Glucanase Producing *Pseudomonas cepacia. Soil Biol. Biochem.* **1993**, *25*, 1211–1221.

Gal-Hamed, I.; Atanasova, L.; Komon´-Zelazowska, M.; Druzhinina, I. S.; Viterbo, A.; Yarden, O. Marine Isolates of *Trichoderma* sp. As Potential Halotolerant Agents of Biological Control for Arid-Zone Agriculture. *Appl. Environ. Microbiol.* **2011**, *77*, 5100–5109.

Garcia, L.; Bulnes, C.; Melchor, G.; Vega, E.; Ileana, M.; De Oca, N. M.; Hidalgo, L.; Marrero, E. Safety of *Pochonia chlamydosporia* var. *catenulata* in Acute Oral and Dermal Toxicity/Pathogenicity Evaluations in Rats and Rabbits. Vet. *Hum. Toxicol.* **2004**, *46*(5), 248–250.

Gurha, S. N. Effect of Some *Trichoderma* sp. On *Fusarium oxysporum* sp. *ciceri* in vitro. *Ann. Plant Prot. Sci.* **2001**, *9*, 332–334.

Haas, D.; De´Fago, G. Biological Control of Soil-borne Pathogens by Fluorescent Pseudomonades. *Nat. Rev. Microbiol.* **2005**, *3*, 307–319.

Hamby, M. K.; Crawford, D. L. The Enhancement of Plant Growth by Selected *Streptomyces* Species. In *100th General Meeting Of American Society For Microbiology,* Los Angeles, CA. Abstract 567, 2000.

Harman, G. E. Myths and Dogmas of Biocontrol: Changes in Perceptions Derived from Research on *Trichoderma harzianum* T- 22. *Plant Dis.* **2000**, *84*, 377–393.

Harman, G. E. Microbial Tools to Improve Crop Performance and Profitability and to Control Plant Diseases. In *Int. Sympos. Biol. Control Plant Dis. New Century-Mode Action Application Technol.* 2001, pp 4-1–4-14.

Harman, G. E. Myths and Dogmas of Biocontrol. Changes in Perceptions Derived from Research on *Trichoderma harzianum* T-22. *Plant Dis.* **2000,** *84,* 377–393.

Hawes, M. C. Living Plant Cells Released from the Root Cap: A regulator of Microbial Populations in the Rhizosphere? In *The Rhizosphere and Plant Growth;* Keister, D. L., Cregan, P. B., Boston, M. A., Eds.; Kluwer Academic Publishers: The Netherlands, 1991; pp 51–59.

He, H.; Silo-Suh, L. A.; Handelsman, J.; Clardy, J.; Zwittermicin, A. An Antifungal and Plant Protection Agent from *Bacillus cereus. Tetrahedron lett.* **1994,** *35,* 2499–2502.

Hewlett, T. E.; Dickson, D. W.; Mitchell, D. J.; Kannwischer-Mitchell, M. E. Evaluation of *Paecilomyces lilacinus* as a Biocontrol Agent of *Meloidogyne javanica* on Tobacco. *J. Nematol.* **1988,** *20,* 578–584.

Howell, C. R., Stipanovic, R. D. Control of *Rhizoctonia solani* on Cotton Seedlings with *Pseudomonas fluorescens* and with an Antibiotic Produced by the Bacterium. *Phytopathology.* **1979,** *69,* 480–482.

Howell, C. R.; Stipanovic, R. D. Suppression of *Pythium ultimum*-Induced Damping Off of Cotton Seedlings by *Pseudomonas fluorescens* and its Antibiotic, Pyoluteorin. *Phytopathology.* **1980,** *70,* 712–715.

Howell, C. R. Cotton Seedling Preemergence Damping-Off Incited by *Rhizopus oryzae* and *Pythium* sp. and it's Biological Control with *Trichoderma* sp. *Phytopathology.* **2002,** *92,* 177–180.

Howell, C. R. Effect of *Gliocladium virens* on *Pythium ultimum, Rhizoctonia solani,* and Damping-Off of Cotton Seedlings. *Phytopathology.* **1982,** *72,* 496–498.

Howell, C. R.; Stipanovic, R. D.; Lumsden, R. D. Antibiotic Production by Strains of *Gliocladium virens* and Its Relation to the Biocontrol of Cotton Seedling Diseases. *Biocontrol Sci. Technol.* **1993,** *3,* 435–441.

Howell, C. R. Mechanisms Employed by *Trichoderma* Species in the Biological Control of Plant Diseases: The History and Evolution of Current Concepts, *Plant Dis.* **2003,** *87*(1), 4–10.

Hoyos-Carvajal, L.; Orduz, S.; Bissett, J. Genetic and Metabolic Biodiversity of *Trichoderma* from Colombia and Adjacent Neotropic Regions. *Fungal Genet. Biol.* **2009,** *46,* 615–631.

Ibrahimov, A. S.; Zafari, D. M.; Valizadeh, E.; Akrami, M. Control of Fusarium Rot of Bean by Combination of by *Trichoderma harzianum* and *Trichoderma asperellum* in Greenhouse Condition. *Agric. J.* **2009,** *4,* 121–123.

Inbar, J.; Chet, I. Evidence that Chitinase Produced by *Aeromonas caviae* is Involved in the Biological Control of Soil-borne Plant Pathogens by this Bacterium. *Soil Biol. Biochem.* **1991,** *23,* 973–978.

Isono, K.; Nagatsu, J.; Kawashima, Y.; Suzuki, S. Studies on Polyoxins, Antifungal Antibiotics. Part I. Isolation and Characterization of Polyoxins A and B. *Agric. Biol. Chem.* **1965,** *29,* 848–854.

Jaklitsch, W. M. European Species of *Hypocrea* Part I. The Green-Spored Species. *Stud Mycol.* **2009,** *63,* 1–91.

Jaklitsch, W. M. European Species of *Hypocrea* Part II: Species with Hyaline Ascospores. *Fungal Divers.* **2011,** *48,* 1–250.

Jatala, P. Biological Control of Plant Parasitic Nematodes. *Annu. Rev. Phytopathol.* **1986,** *24,* 453–491.

Jindal, K. K. Emerging Bacterial Disease of Temperate Fruits in Himachal Pradesh. *Plant Dis. Res. (Special).* **1990,** *5,* 151–155.

Jung, W. J.; An, K. N.; Jin, Y. L.; Park, R. D.; Lim, K. T.; Kim, K. Y.; Kim, T. H. Biological Control of Damping Off Caused by *Rhizoctonia solani* Using Chitinase Producing *Paenibacillus illinoisensis* KJA-424. *Soil Biol. Biochem.* **2003,** *35,* 1261–1264.

Kalita, P.; Bora, L. C.; Bhagabati, K. N. Phylloplane Microflora of Citrus and their Role in Management of Citrus Canker. *Indian Phytopathol.* **1996,** *49,* 234–237.

Kamensky, M.; Ovadis, M.; Chet, I.; Chernin, L. Soil-Borne Strain IC14 of *Serratia plymuthica* with Multiple Mechanisms of Antifungal Activity Provides Biocontrol of *Botrytis cinerea* and *Sclerotinia sclerotiorum* Diseases. *Soil Biol. Biochem.* **2003,** *35,* 323–331.

Kerry, B. R.; Crump, D. H.; Mullen, L. A. Studies of the Cereal Cyst Nematode *Heterodera avenae* under Continuous Cereals. 1974–1978. I. Plant Growth and Nematode Multiplication. *Ann. Appl. Biol.* **1982a,** *100,* 477–487.

Kerry, B. R.; Crump, D. H.; Mullen, L. A. Studies of the Cereal Cyst Nematode *Heterodera avenae* under Continuous Cereals. 1975–1978. II. Fungal Parasitism of Nematode Females and Eggs. *Ann. Appl. Biol.* **1982b,** *100,* 489–499.

Khan, M. R. ; Anwer, M. A. Fungal Based Bioinoculants for Plant Disease Management. In *Microbes and Microbial Technology: Agricultural and Environmental Applications*; Pichtel, J. (USA), Ahmad, I. (India), Eds.; Springer: USA, 2011; pp 447–489.

Khan, M. R.; Anwer, M. A. Molecular and Biochemical Characterization of Soil Isolates of *Aspergillus niger* Aggregate and an Assessment of their Antagonism against *Rhizoctonia solani*. *Phytopathologia mediterranea,* **2007,** *46,* 304–315.

Khan, M. R.; Anwer, M. A. DNA and Some Laboratory Tests of Nematode Suppressing Efficient Soil Isolates of *Aspergillus niger*. *Indian Phytopathol.***2008,** *61*(2), 212–225.

Khan, M. R.; Altaf, S.; Mohidin, F. A.; Khan, U.; Anwer, M. A. Biological Control of Plant Nematodes with Phosphate Solubilizing Microorganisms. In *Phosphate Solubilizing Microbes for Crop Improvement;* Khan, M. S., Zaidi, A., Eds.; Nova Science Publisher, Inc.: New York, USA, 2009; pp 395–426.

Khan, M. R.; Anwer, M. A.; Shahid, S. Management of Grey Mould of Chickpea, *Botrytis cinerea* with Bacterial and Fungal Biopesticides Using Different Modes of Inoculation and Application. *Biol. Control.***2011,** *57*(1), 13–23. (NAAS Score: 7.92).

Khan, M. R.; Ashraf, S.; Shahid, S.; Anwer, M. A. Response of Chickpea Cultivars to Foliar, Seed and Soil Inoculations with *Botrytis cinerea*. *Phytopathol. Mediterr.* **2010,** *49* (3), 275–286.

Khan, M. R.; Khan, M. M.; Anwer, M. A.; Haque, Z. Laboratory and Field Performance of Some Soil Bacteria Used as Seed Treatments on *Meloidogyne incognita* in Chickpea. *Nematol. Mediterr.* **2012,** *40,* 143–151.

Khan, M. R.; Anwer, M. A.; Khan, S. M.; Mohiddin, F. A. A Characterization of Some Isolates of *Aspergillus niger* and Evaluation of their Effects against *Meloidogyne incognita*. *Tests of Agrochemicals and Cultivars,* **2006,** *27,* 33–34. (NAASScore: 8.15).

Khan, M. R. *Biological Control of Fusarial Wilt an Root-Knot of Legumes*. Department of Biotechnology, Ministry of Science and Technology: New Delhi, 2005.

Khan, M. R.; Akram, M. Effect of Certain Antagonistic Fungi and Rhizobacteria on Wilt Disease Complex Caused by *Meloidogyne incognita* and *Fusarium oxysporium* f.sp. *lycopersici* on Tomato. *Nematol. Mediterr.* **2000,** *28,* 139–144.

Khan, M. R.; Anwer, M. A. DNA and Some Laboratory Tests of Nematode Suppressing Efficient Soil Isolates of *Aspergillus niger*. *Indian Phytopathol.* **2008,** *61*(2), 212–225.

Khan, M. R., Ejaz, M. N. Effect of Neem Leaves and *Paecilomyces lilacinus* on Root-Knot Nematode Disease of Okra. *Vasundhara.* **1997,** *2,* 1–5.

Khan, M. R.; Gupta, J. Antagonistic Effects of *Trichoderma* Species against *Macrophomina phaseolina* on Eggplant. *J. Plant Dis. Prot.* **1998,** *105*, 387–393.

Khan, M. R.; Kounsar, K. Effects of Certain Bacteria and Fungi on the Growth of Mungbean and Reproduction of *Meloidogyne incognita. Nematol. Mediterr.* **2000,** *28*, 221–226.

Khan, M. R.; Mustafa, U. Root-Knot Nematode Problem in Gladiolus Cultivars and their Management. *Int. J. Nematol.* **2005,** *15*, 59–63.

Khan, M. R.; Rehman, Z. Bio-management of Root Rot of Chickpea Caused by *Rhizoctonia. Vasundhara.* **1997,** *3*, 22–26.

Khan, M. R.; Tarannum, Z. Effects of Field Application of Various Microorganisms on *Meloidogyne incognita* on Tomato. *Nematol. Mediterr.* **1999,** *27*, 233–238.

Khan, M. R.; Khan, S. M.; Mohiddin, F. A. Biological Control of Fusarial Wilts of Chickpea through Seed Treatment with the Commercial Formulations of *Trichoderma harzianum* and/ or *Pseudomonas fluorescens. Phytopathol. Mediterr.* **2004,** *43*, 20–25.

Khan, M. R.; Khan, S. M.; Mohiddin, F. A. Root-Knot Problem of Some Winter Ornamental Plants and Its Bio-management. *J. Nematol.* **2005b,** *37*, 198–206.

Khan, M. R.; Mohiddin, F. A.; Khan, S. M., and Khan, B. Effect of Seed Treatment with Certain Biopesticides on Root-Knot of Chickpea. *Nematol. Mediterr.* **2005a,** *2*, 107–112.

Kim, B. S.; Moon, S. S.; Hwang, B. K. Isolation, Identification and Antifungal Activity of a Macrolide Antibiotic, Oligomycin A, Produced by *Streptomyces libani. Can. J. Bot.* **1999,** *77*, 850–858.

Kishore, G. K.; Pande, S.; Podile, A. R. Biological Control of Late Leaf Spot of Peanut (*Arachis hypogaea*) with Chitinolytic Bacteria. *Phytopathology.* **2005,** *95*, 1157–1165.

Kloepper, J. W.; Leong, J.; Teintze, M.; Schroth, M. N. *Pseudomonas* Siderophores: A Mechanism Explaining Disease Suppression in Soils. *Curr. Microbiol.* **1980,** *4*, 317–320.

Koths, J. S.; Gunner, H. R. Establishment of a Rhizosphere Microflora on Carnation as a Means of Plant Protection in Steamed Greenhouse Soils. *Am. Soc. Hortic. Sci.* **1967,** *91*, 617–626.

Krishnamurthy, K.; Gnanamanickam, S. S. Induction of Systemic Resistance and Salicylic Acid Accumulation in *Oryza sativa* L. In the Biological Suppression of Rice Blast Cause by Treatments with *Pseudomonas* sp. *World J. Microbiol. Biotechnol.* **1998,** *14*, 935–937.

Kubicek, C. P., et al. Comparative Genome Sequence Analysis Underscores Mycoparasitism as the Ancestral Life Style of *Trichoderma. Genome Biol.* **2011,** *12,* R40.

Kumar, S. Management of Root-Knot Nematode in Papaya by *Pochonia chlamydosporia. Indian J. Nematol.* **2009,** *39*(1), 48–56.

Kumar, S.; Sen, B. Kalisena, A. Novel Biopesticide for Disease Free Crop and Better Yield. Proceedings of the National Symposium on Development of Microbial Pesticides and Insect Pest Management, November 12th and 13th, 1998; Pune: BARC Mumbai and Hindustan Antibiotics Ltd., 1998.

Kumar, S.; Sood, A. K. Management of Bacterial Wilt of Tomato with VAM and Bacterial Antagonists. *Indian Phytopathol.* **2002,** *55*(4), 513.

Leeman, M.; Van Pelt, J. A.; Den Ouden, F. M.; Heinsbroek, M., Bakker, P. A. H. M.; Schippers, B. Induction of Systemic Resistance by *Pseudomonas fluorescens* in Radish Cultivars Differing in Susceptibility to *Fusarium* Wilt, Using a Novel Bioassay. *Eur. J. Plant Pathol.* **1995,** *101*, 655–664.

Lemanceau, P.; Alabouvette, C. Suppression of *Fusarium* Wilts by Fluorescent Pseudomonas: Mechanisms and Applications. *Biocontrol. Sci. Technol.* **1993,** *3*, 219–234.

Lemanceau, P.; Maurhofer, M.; De´Fago, G. Contribution of Studies on Suppressive Soils to the Identification of Bacterial Biocontrol Agents and to the Knowledge of their Modes of Action. In Plant-Associated Bacteria, Gnanamanickam, S. S.; Springer, New York, 2006; pp 231–267.

Lifshitz, R.; Windham, M. T.; Baker, R. Mechanism of Biological Control of Preemergence Damping-Off of Pea by Seed Treatment with *Trichoderma* sp. *Phytopathology.* **1986,** *76,* 720–725.

Lumsden, R. D.; Locke, J. C.; Adkins, S. T.; Walter, J. F.; Ridout, C. J. Isolation and Localization of the Antibiotic Gliotoxin Produced by *Gliocladium virens* from Alginate Prill in Soil and Soilless Media. *Phytopathology.* **1992,** *82,* 230–235.

Lysek, H. Study of Biology of Geohelminths. II. The Importance of Some Soil Microorganisms for the Viability of Geohelminth Eggs in the Soil. *Acta Univ. Palacki. Olomuc.* **1966,** *40,* 83–90.

Mahadevan, B.; Crawford, D. L. Properties of the Chitinase of the Antifungal Biocontrol Agent *Streptomyces lydicus* WYEC108. *Enzyme Microb. Technol.* **1997,** *20,* 489–493.

Majumdar, S.; Sen, B. Kalisena, A. Novel Biopesticide for Disease Free Crop and Better Yield. Proceedings of the National Symposium on Development of Microbial Pesticides and Insect Pest Management, November 12th And 13th, 1998; Pune: BARC Mumbai and Hindustan Antibiotics Ltd., 1998.

Martinez, D., et al. Genome Sequencing and Analysis of the Biomass-Degrading Fungus *Trichoderma reesei* (Syn. *Hypocrea jecorina*). *Nat. Biotechnol.* **2008,** *26,* 553–560.

Maurhofer, M.; Hase, C.; Meuwly, P.; Metraux, J. P.; Defago, G. Induction of Systemic Resistance of Tobacco to Tobacco Necrosis Virus by the Root-Colonizing *Pseudomonas fluorescens* Strain CHA0: Influence of the Gaca Gene and of Pyoverdine Production. *Phytopathology.* **1994,** *84,* 139–146.

Mcspadden, G. B. B.; Fravel, D. R. Biological Control of Plant Pathogens: Research, Commercialization, and Application in The USA. *Plant Health Prog.* Online, Publication Doi: 10.1094/PHP-2002- 0510–01-RV.

Meyer, S. L. F.; Massoud, S. I.; Chitwood, D. J.; Roberts, D. P. Evaluation of *Trichoderma virens* and *Burkholderia cepacia* for Antagonistic Activity against Root-Knot Nematode *Meloidogyne incognita*. *Nematology.* **2000,** *2,* 871–879.

Migheli, Q.; Balmas, V.; Komon´-Zelazowska, M.; Scherm, B.; Fiori, S., Kopchinskiy, A. G.; Kubicek, C. P. Druzhinina, I. S. Soils of a Mediterranean Hot Spot of Biodiversity and Endemism (Sardinia, Tyrrhenian Islands) are Inhabited by Pan-European, Invasive Species of *Hypocrea/Trichoderma*. *Environ. Microbiol.* **2009,** *11,* 35–46.

Milner, J. L.; Silo-Suh, L.; Lee, J. C.; He, H.; Clardy, J.; Handelsman, J. Production of Kanosamine by *Bacillus cereus* UW85. *Appl. Enviorn. Microbiol.* **1996,** *62,* 3061–3065.

Mondal, G. *In Vitro* Evaluation of *Aspergillus niger* AN 27 against Soil borne Fungal Plant Pathogens and Field Testing against *Macrophomina phaseolina* on Potato. Ph.D. Thesis, New Delhi: I.A.R.I, 1998; pp 117–XLVIII.

Moyne, A. L.; Shalby, R.; Cleveland, T. E.; Tuzun, S. Bacillomycin, D, An Iturin with Antifungal Activity against *Aspergillus flavus*. *J. Appl. Microbiol.* **2001,** *90,* 622–629.

Nielsen, T. H.; Sorensen, D.; Tobiasen, C.; Andersen, J. B.; Christophersen, C.; Givskov, M.; Sorensen, J. Antibiotic and Biosurfactant Properties of Cyclic Lipopeptides Produced by Fluorescent *Pseudomonas* sp. From the Sugar Beet Rhizosphere. *Appl. Environ. Microbiol.* **2002,** *68,* 3416–3423.

Olson, H. A.; Benson, D. M. Induced Systemic Resistance and the Role of Binucleate *Rhizoctonia* and *Trichoderma hamatum* 382 in Biocontrol of *Botrytis* Blight in Geranium. *Biol. Control.* **2007,** *42*(2), 233–241.

Oyekanmi, E. O.; Coyne, D. L.; Fawole, B. Utilization of the Potentials of Selected Microorganisms as Biocontrol and Biofertilizer for Enhanced Crop Improvement. *J. Biol. Sci.* **2008,** *8,* 746–752.

Pal, K. K.; Gardener, B. M. Biological Control of Plant Pathogens. *Plant Health Instructor.* Doi: 10.1094/PHI–A–2006–1117–02.

Palumbo, J. D.; Yuen, G. Y.; Jochum, C. C.; Tatum, K.; Kobayashi, D. Y. Mutagenesis of Beta-1,3-Glucanase Genes in *Lysobacter enzymogenes* Strain C3 Results in Reduced Biological Control Activity toward *Bipolaris* Leaf Spot of Tall Fescue and *Pythium* Damping-Off of Sugar Beet. *Phytopathology.* **2005,** *95,* 701–707.

Pant, H.; Pandey, G. Efficacy of Bio-Control Agents for the Management of Root-Knot Nematode on Chickpea. *Ann. Plant Prot. Sci.* **2001,** *9*(1), 117–170.

Pant, H.; Pandey, G. Use of *Trichoderma harzianum* and Neem Cake Alone and in Combination on *Meloidogyne incognita* Galls in Chickpea. *Ann. Plant Prot. Sci.* **2002,** *10,* 134–178.

Papavizas, G. C. Biological Control of Soil Borne Diseases. *Summa. Phytopathol.* **1985,** 11, 173–179.

Papavizas, G. C.; Lewis, J. A. Effect of *Gliocladium* and *Trichoderma* on Damping Off of Snapbean Caused by *Sclerotium rolfsii* in the Greenhouse. *Plant Pathol.* **1989,** *38,* 277–286.

Persoon, C. H. Neuer Veersuch Einer Systematischen Eintheilung Der Schwamme (Dispositio Methodica Fungorum). *Romer's News Mag. Bot.* **1794,** *1,* 63–128.

Pieterse, C. M. J.; Van Pelt, J. A.; Ton, J.; Parchmann, S.; Mueller, M. J.; Buchala, A. J.; Me´Traux, J-P.;Van Loon, L. C. Rhizobacteria-Mediated Induced Systemic Resistance (ISR) in *Arabidopsis* Requires Sensitivity to Jasmonate and Ethylene But is Not Accompanied by an Increase in Their Production. *Physiol. Mol. Plant Pathol.* **2000,** *57,* 123–134.

Pill, W. G.; Collins, C. M.; Goldberger, B.; Gregory, N. Responses of Non-Primed or Primed Seeds of 'Marketmore 76' Cucumber (*Cucumis sativus* L.) Slurry Coated with *Trichoderma* Species to Planting in Growth Media Infested with *Pythium aphanidermatum. Sci. Hortic.* **2009,** *121*(1), 54–62.

Pleban, S.; Chernin, L.; Chet, I. Chitinolytic Activity of an Endophytic Strain of *Bacillus cereus. Lett. Appl. Microbiol.* **1997,** *25,* 284–288.

Pocasangre Enamorado, L. E.; Zum Felde, A.; Cañizares, C.; Muñoz, J.; Suarez, P., Jimenez, C., Riveros, A. S., Rosales, F. E., And Sikora, R. A. Field Evaluation of the Antagonistic Activity of Endophytic Fungi towards the Burrowing Nematode, *Radopholus similis*, in Plantain. In *Latin America Pro Musa Symposium*; White River (ZAF): Greenway Woods Resort, 2007; pp 10–14.

Powell, J. F., Vargas, J. M., Nair, M. G., Detweiler, A. R.; Chandra, A. Management of Dollar Spot on Creeping Bentgrass with Metabolites of *Pseudomonas aureofaciens* (TX-1). *Plant Dis.* **2000,** *84,* 19–24.

Prasad, R. D.; Rangeshwaran, R.; Anuroop, C. P.; Rashmi, H. J. Biological Control of Wilt and Root-Rot of Chickpea under Field Condition. *Ann. Rev. Plant Prot. Sci.* **2002,** *10*(1), 72–75.

Radhakrishna, P.; Sen, B. Efficacy of Different Methods of Inoculation of *Fusarium oxysporum* and *F. solani* for Inducing Wilt in Muskmelon. *Indian Phytopathol.* **1986,** *38,* 70–73.

Rao, M. S. Management of *Meloidogyne javanica* on Acid Lime Nursery Seedlings by Using Formulations of *Pochonia chlamydosporia* and *Paecilomyces lilacinus* [*Citrus aurantifolia* (Christm. Et Panz.) Swingle; India]. *Nematol. Mediterr.* **2005,** *33*(2), 145–148.

Riad, S. R. E.; Ziedan, E. H.; Abdalla, A. M. Biological Soil Treatment with *Trichoderma harzianum* to Control Root Rot Disease of Grapevine (*Vitis vinifera* L.) In Newly Reclaimed Lands in Nobaria Province. *Arch. Phytopathol. Plant Prot.*2010, 43(1), 73–87.

Rothrock, C. S.; Gottlieb, D. Role of Antibiosis in Antagonism of *Streptomyces hygroscopicus* var. *geldanus* to *Rhizoctonia solani* in Soil. *Can. J. Microbiol.* **1984,** *30,* 1440–1447.

Rovira, A. D. Rhizosphere Research - 85 Years of Progress and Frustration. In *The Rhizosphere and Plant Growth*; Keister, D. L., Cregan, P. B., Eds.; Kluwer Academic Publishers: Boston, MA, 1991; pp 3–13.

Rovira, A. D. Interactions between Plant Roots and Soil Microorganisms. *Annu. Rev. Microbiol.* **1965,** *19,* 241–266.

Rovira, A. D. Plant Root Exudates. *Bot. Rev.* **1969,** *35,* 35–57.

Ryu, C-M.; Hu, C. H.; Locy, R. D.; Kloepper, J. W. Study of Mechanisms for Plant Growth Promotion Elicited by Rhizobacteria in *Arabidopsis thaliana. Plant Soil.* **2005,** *268,* 285–292

Saifullah. Fungal Parasitism of Young Females of *Globodera rostochiensis* and *G. pallida. Afro. Asian J. Nematol.* **1996a,** *6,* 17–22.

Saifullah. Killing Potato Cyst Nematode Males, A Possible Control Strategy. *Afro Asian J. Nematol.* **1996b,** *6,* 23–28.

Samuels, G. J. *Trichoderma*: Systematics, the Sexual State, and Ecology. *Phytopathology.* **2006,** *96,* 195–206.

Sanford, G. B., Some Factors Affecting the Pathogenicity of *Actinomycetes Scabies. Phytopathology.* **1926,** *16,* 525e547.

Sant, D.; Casanova, E.; Segarra, G.; Aviles, M.; Reis, M.; Trillas, M. I. Effect of *Trichoderma asperellum* Strain T34 on *Fusarium* Wilt and Water Usage in Carnation Grown on Compost-Based Growth Medium. *Biol. Control.* In Press, Available On Line, January 2010, Doi:10.1016/J.Biocontrol.2010.01.012.

Schenek, C. Control of Nematodes in Tomato with *Paecilomyces lilacinus.* Hawaii Agriculture Research Center Vegetable Report 5, 2004.

Schmidt, C. S.; Lorenz, D.; Wolf, G. A. Biological Control of the Grapevine Dieback Fungus *Eutypa Lata* I: Screening of Bacterial Antagonists. *J. Phytopathol.* **2001,** *149,* 427–435.

Shah, N. H.; Khan, M. I.; Azam, M. F. Studies on the Individual and Concomitant Effect of *Aspergillus niger, Rhizoctonia solonia*, and *M. javanica* on Plant Growth and Nematode Reproduction on Chilli (*Capsicum annuum* L). *Ann. Plant Prot. Sci.* **1994,** *1*(2), 75–78.

Shahida, P.; Gaffar, A. Effect of Microbial Antagonists in the Control of Root-Rot of Tomato. *Pak. J. Bot.* **1991,** *23*(2), 179–182.

Sharma, H. K.; Prasad, D.; Sharma, P. Compatibility of Fungal Bio-agents as Seed Dressers with Carbofuran in Okra against *Meloidogyne incognita*. National Symposium on Recent Advances and Research Priorities in Indian Nematology, 2005; pp 9–11.

Sharon, E.; Bar-Eyal, M.; Chet, I.; Herra-Estrella, A.; Kleifeld, O.; Spiegel, Y. Biological Control of the Root-Knot Nematode *Meloidogyne javanica* by *Trichoderma harzianum. Phytopathology.* **2001,** *91,* 687–693.

Sharon, E.; Chet, I.; Spiegel, Y. Improved Attachment and Parasitism of *Trichoderma* on *Meloidogyne javanica in vitro. Eur. J. Plant Pathol.* **2009,** 123, 291–299.

Siddiqui, I. A.; Shaukat, S. S. *Trichoderma harzianum* Enhances the Production of Biocontrol of *Meloidogyne javanica* by *Pseudomonas fluorescens* in Tomato. *Lett. Appl. Microbiol.* **2004,** *38*(2), 169–175.

Siddiqui, I. A.; Atkins, S. D.; Kerry, B. R. Relationship between Saprotrophic Growth in Soil of Different Biotypes of *Pochonia chlamydosporia* and the Infection of Nematode Eggs. *Ann. Appl. Biol.* **2009**, *155*(1), 131–141.

Siddiqui, Z. A.; Mahmood, I. Biological Control of *Heterodera cajani* and *Fusarium udum* on Pigeonpea by *Glomus mosseae, Trichoderma harzianum* and *Verticillium chlamydosporium. Isr. J. Plant Sci.* **1996**, *44*, 49–56.

Silo-Suh, L. A.; Stab, V. E.; Raffel, S. R.; Handelsman, J. Target Range of Zwittermicin A, An Aminopolyol Antibiotic From *Bacillus cereus. Curr. Microbiol.* **1998**, *37*, 6–11.

Singh, P. P.; Shin, Y. C.; Park, C. S.; Chung, Y. R. Biological Control of *Fusarium* Wilt of Cucumber by Chitinolytic Bacteria. *Phytopathology.* **1999**, *89*, 92–99.

Singh, A.; Singh, H. B. Control of Collar Rot in Mint (*Mentha* sp.) Caused by *Sclerotium rolfsii* Using Biological Means. *Curr. Sci.* **2004**, 87(3), 362–366.

Singh, H. B.; Singh, A.; Nautiyal, C. S. Commercializing of Biocontrol Agents: Problems and Prospects. In *Frontiers of Fungal Biodiversity in India*; Rao, G. P., Manoharachaty, C., Bhat, D. J., Rajak, R. C., Lakhanpal, T. N., Eds.; International Book Distributing Co.: Lucknow, 2003; pp 847–861.

Singh, S. M.; Azam, M. F.; Khan, A. M.; Saxena, S. K. Effect of *Aspergillus niger* and *Rhizoctonia solani* on Development of *Meloidogyne incognita* on Tomato. *Curr. Nematol.* **1991**, *2*, 163–166.

Sneh, B. Use of Rhizosphere Chitinolytic Bacteria for Biological Control of *Fusarium oxysporum* f. sp. *dianthi* in Carnation. *Phytopathol. Z.* **1981**, *100*, 251–256.

Sneh, B.; Dupler, M.; Elad, Y.; Baker, R. Chlamydospore Germination of *Fusarium oxysporum* f. sp. *cucumerinum* as Affected by Fluorescent and Lytic Bacteria From *Fusarium*-Suppressive Soil. *Phytopathology.* **1984**, *74*, 1115–1124.

Spaepen, S.; Vanderleyden, J.; Remans, R. Indole-3-Acetic Acid in Microbial and Microorganism-Plant Signaling. *FEMS Microbiol. Rev.* **2007**, *31*, 425–448.

Stirling, G. R. Biocontrol of Plantpathogenic Nematode And Fungus. *Phytopathology.* **1993**, *83*, 1525–1532.

Tazawa, J.; Watanabe, K.; Yoshida, H.; Sato, M. ; Homma, Y. Simple Method of Detection of the Strains of Fluorescent *Pseudomonas* sp. Producing Antibiotics, Pyrrolnitrin and Phloroglucinol. *Soil Microorg.* **2000**, *54*, 61–67.

Thomashow, L. S.; Weller, D. M.; Bonsall, R. F.; Pierson, L. S. Production of the Antibiotic Phenazine- 1-Carboxylic Acid by Fluorescent *Pseudomonas* Species in the Rhizosphere of Wheat. *Appl. Enviorn. Microbiol.* **1990**, *56*, 908–912.

Thrane, C.; Nielsen, T. H.; Nielsen, M. N.; Olsson, S.; Sorensen, J. Viscosinamide Producing *Pseudomonas fluorescens* DR54 Exerts Biocontrol Effect on *Pythium ultimum* In Sugar Beet Rhizosphere. *FEMS Microbiol. Ecol.* **2000**, *33*, 139–146.

Umezawa, H.; Okami, T.; Hashimoto, T.; Suhara, Y.; Hamada, M.; Takeuchi, T. A New Antibiotic, Kasugamycin. *J. Antibiot. Ser. A.* **1965**, 18, 101–103.

Upadhyay, J. P. Mukhopadhyay, A. N. Biological Control of *Sclerotium rolfsii* by *Trichoderma harzianum* in Sugar Beet. Trop. *Pest Manage.* **1986**, *32*, 215–220.

Valois, D.; Fayad, K.; Barasubiye, T.; Garon, T.; Dery, C.; Brzezinski, R.; Beaulieu, C. Glucanolytic *Actinomycetes* Antagonistic to *Phytophthora fragariae* var. *rubi,* The Causal Agent of Raspberry Root Rot. *Appl. Environ. Microbiol.* **1996**, *62*, 1630–1635.

Van Peer, R.; Niemann, G. J.; Schippers, B. Induced Resistance and Phytoalexin Accumulation in Biological Control of *Fusarium* Wilt of Carnation by *Pseudomonas* sp. Strain WCS417r. *Phytopathology.* **1991**, *81*, 728–734.

Velusamy, P.; Immanuel, J. E.; Gnanamanickam, S. S., Thomashow, L. Biological Control of Rice Bacterial Blight by Plant-Associated Bacteria Producing 2, 4-Diacetylphloroglucinol. *Can. J. Microbiol.* **2006**, *52*, 56–65.

Von DerWeid, I.; Artursson, V.; Seldin, L.; Jansson, J. K. Antifungal and Root Surface Colonization Properties of GFP Tagged *Paenibacillus brasilensis* PB177. *World J. Microbiol. Biotechnol.* **2005**, *2*(1), 1591–1597.

Wahab, S. Biotechnological Approaches in Plant Protection. In *Biopesticides and Pest Management;* Koul, O., Dhaliwal, G. S., Marwaha, S. S., Arora, J. K., Eds.; Campus Books International: New Delhi, 2003; Vol.1, pp 113–127.

Wahab, S. Biotechnological Approaches in the Management of Plant Pests, Diseases and Weeds for Sustainable Agriculture. In *Deep Roots, Open Skies: New Biology In India;* Basu, S. K., Batra, J. K., Salunke, D. M., Eds.; Narosa Publishing House: New Delhi, India, 2004; 200, pp 113–129.

Wahab, S. Biotechnological Approaches in the Management of Plant Pests, Diseases and Weeds for Sustainable Agriculture. *J. Biopesticides.* **2009**, 2(2), 115–134.

Waisel, Y.; Eshel, A.; Katkafl, U. *Plant Roots: The Hidden Half.* Yoav Wiesel, Ed.; Marcel Dekker: New York, 1991.

Weindling, R. *Trichoderma lignorum* as a Parasite of Other Soil Fungi. *Phytopathology.* **1932**, *22*, 837–845.

Weller, D. M. Biological Control of Soil-borne Plant Pathogens in the Rhizosphere with Bacteria. *Annu. Rev. Phytopathol.* **1988**, *26*, 379–407.

Weller, D. M.; Cook, R. J. Suppression of Take-All of Wheat by Seed Treatments with Fluorescent Pseudomonads. *Phytopathology.* **1983**, *73*, 463–469.

Wells, H. D., Bell, D. K., Jaworski, C. A. Efficacy of *Trichoderma harzianum* as a Biocontrol for *Sclerotium rolfsii. Phytopathology.* **1972**, *62*, 442–447.

Windham, G. I.; Windham, M. T.; Williams, W. P. Effects of *Trichoderma* sp. On Maize Growth and *Meloidogyne arenaria* Reproduction. *Plant Dis.* **1989**, *73*, 493–496.

Yedidia, I.; Benhamou, N.; Chet, I. Induction of Defense Responses in Cucumber Plants (*Cucumis Sativus* L.) by the Biocontrol Agent *Trichoderma harzianum. Appl. Environ. Microbiol.* **1999**, *65*, 1061–1070.

Yedidia, I.; Benhamou, N.; Kapulnik, Y.; Chet, I. Induction and Accumulation of PR Proteins Activity during Early Stages of Root Colonization by the Mycoparasite *Trichoderma harzianum* Strain T-203. *Plant Physiol. Biochem.* **2000**, *38,* 863–873.

Yedidia, I.; Srivastva, A. K.; Kapulnik, Y.; Chet, I. Effect of *Trichoderma harzianum* on Microelement Concentrations and Increased Growth of Cucumber Plants. *Plant Soil.* **2001**, *235*, 235–242.

Zareen, A.; Khan, N. J., Zaki, M. J. Biological Control of *Meloidogyne javanica* (Treub) Chitwood, Root Knot Nematodes of Okra [*Abelmoschus esculentus* (L) Moench]. *Pak. J. Biol. Sci.* **2001**, *4*, 990–994.

Zhang, Y.; Fernando, W. G. D. Zwittermicin A Detection in *Bacillus* sp. Controlling *Sclerotinia sclerotiorum* on Canola. *Phytopathology.* **2004**, *94*, S116.

Zhang, J.; Howell, C. R.; Starr, J. L. Suppression of *Fusarium* Colonization of Cotton Roots and *Fusarium* Wilt by Seed Treatments with *Gliocladium virens* and *Bacillus subtilis. Biocontrol Sci. Technol.* **1996**, *6*, 175–187.

Zum Felde, A.; Pocasangre, L. E.; Carnizares Monteros, C. A.; Sikora, R. A.; Rosales, F. E.; Riveros, A. S. Effect of Combined Inoculations of Endophytic Fungi on the Biocontrol of *Radopholus similis. Info. Musa.* **2006**, *15*(1–2), 12–17.

CHAPTER 5

EVOLUTION AND ADAPTATION OF PHYTOPATHOGENS IN PERSPECTIVE OF INTENSIFIED AGROECOSYSTEM

RAKESH K. SINGH[1*], ANIRUDHA CHATTOPADHYAY[2], and SUMIT K. PANDEY[1]

[1]*Department of Mycology and Plant Pathology, Institute of Agricultural Sciences, Banaras Hindu University, Varanasi 221005, India*

[2]*Department of Plant Pathology, CP College of Agriculture, Sardarkrushinagar Dantiwada Agricultural University, SK Nagar 385506, Gujarat, India*

Corresponding author. E-mail: rakesh_bhuin@outlook.com

CONTENTS

ABSTRACT

In recent times, increasing problem of evolution and adaptation of plant pathogens in an intensified agroecosystem have become evident from growing incidence of plant disease epidemics with associated yield loss occurring either in large scale or in small pockets of various parts of the world. The dynamics of evolution of phytopathogens is very complex and is influenced by various deriving forces like genetic homogeneity, change in agroecological factors which exert directional selection pressure in the favor of evolution, and emergence of new virulent races/ pathotypes. These new pathogenic variants are highly competent on invading new ecological niches, due to their tremendous genomic plasticity and higher adaptability to changing environmental conditions. Individual groups of pathogens take different paths to evolve and they also adopt different mechanisms in different time and space. In this content, we try to present the various evolutionary mechanisms adopted by different groups of plant pathogens in different agroecosystems and we also try to point out possible driving factors promoting the evolution and adaptation of plant pathogens in an intensified agroecosystem. Remodulation of existing agroecosystem by creating genetic heterogeneity and agroecological diversity would be the suitable approach to slow down the process of evolution. Introduction of improved monitoring system with advanced diagnostic facilities for the detection and spread of new variants and strengthening plant quarantine activity will also be helpful to prevent the entry and establishment of new pathogens or pathogenic variants in an uninvaded area. Therefore, appropriate integration of all these multidisciplinary approaches in groups are emphasized and justified to combat this situation.

5.1   INTRODUCTION

Evolution is the improvement of inherited characteristics in a population of an organism due to changes in genetic composition occurred through successive generations and leads to the change in the frequencies of alleles in the population. This evolutionary process gives rise to diversity in the biological population of individual organisms including plants as well as their pathogens. It is a co-evolutionary process in which host-plants and their pathogens are engaged in continuous battles in a plant–pathogen system with an outcome of improved plants showing enhanced defense to pathogen attack as well as more pathogens evolving to circumvent these host-plant defense. Although, in this co-evolutionary system, generally pathogens have an added advantage relative to their host due to their shorter generation time and larger population size, this response in pathogen population is totally dependent on host genotype (Zhan et al., 2014). This allows pathogens to generate more variants in the population within a fixed period of time (Thomas et al., 2010). Hence, there will be evolution in their pathogenic ability for responding to changes in host defense systems. This will ultimately exert a positive selection pressure in the favor of new variants of pathogen to maintain adaptation on its natural host. This followed by a tremendous increase in its population size is consistent with the increase in host population size. Therefore, there is a continuous increasing issue of plant pathogens causing problems for agricultural crops with the emergence of new pathogens and the continuous evolution of known pathogens. This increase in newly emerging infectious diseases has been partly linked to agricultural practices, climate change (which affects the survival and distribution of pathogens and vectors such as insect pests and nematodes), increased population densities enabling increased disease incidences, and subsequent crop losses.

5.2   EVOLUTIONARY CONSEQUENCES OF PHYTOPATHOGENS

Evolution of phytopathogen is a host-governed phenomenon. In a natural ecosystem, when a host plant is going to be attacked by a pathogen, the plant will activate some of the defense genes encoding non-specific receptor molecules called pattern recognition receptors (PRRs). PRRs will perceive some conserved signaling molecules of pathogen origin called pathogen-associated molecular patterns (PAMPs) that will ultimately activate the innate immunity response in plant and exert broad-spectrum activity against all genetic variants (strains/isolates/races) of a pathogen species. In this case, the plant can be called as non-host and pathogens are non-pathogens. Any microbial species can be recognized as a pathogen only when it can break down the basal defense system of the host plant. Evolution of microbial agents at this level can lead to an increase in the host range and this can be better recognized as host shift or host jump. In a natural population, host plants and their pathogens are involved in a continuous co-evolutionary race and in this co-evolutionary race, a trait of one species/organism evolves in response to a trait of another species/organism, which in turn evolves in response to the trait in the first species (Janzen, 1980). This a kind of reciprocal interaction which is also evident in a host–pathogen system, in which pathogen evolves in response to its host defense to circumvent defense mechanisms and host evolves in response to pathogen attack through enhanced protection. This co-evolution process is thought to occur in natural ecosystems where plant and pathogen exhibit a gene-for-gene relationship (McDonald, 2004).

According to our present understanding of gene-for-gene interactions, during the evolutionary race for survival of a host plant from pathogen attack, a new resistance gene (R1) is evolved by genetic modification of one or more resistance genes of the host plants. This resistant gene(s) encodes some specific receptor molecules to recognize elicitors or effector molecules encoded by avirulence (*avr1*) gene of the pathogen, which in turn initiates a cascade of signal-transduction pathway that ultimately results in a hypersensitive response to resist the infection of existing races of the pathogen. This will exert a tremendous selection pressure on existing pathogen population, and pathogens carrying this *avr1* gene cannot survive on such plant population carrying R1 genes. For survivability, pathogens will come out with transformed new genes governing its virulence to host. In the long run, the positive selection pressure will favor the multiplication of new matching virulent races of the pathogen, resulting in the breakdown of host resistance. This was previously explained by "arms-race" concept in which effectiveness of R gene is lost by pathogen *avr* gene mutation followed by the evolution of new resistance gene specificities. In this way, a dynamic co-evolutionary process involving both plant and pathogen is continuing in a meta-population. In a natural ecosystem, a balancing force is always working. Generally, a host meta-population comprises many micro populations. Each micro population is susceptible to different pathotypes of the pathogens to different degree. Any micro population that is under severe disease pressure will subsequently multiply at less number with fewer numbers of progeny in the following season resulting in selection pressure against the respective pathotypes with predominance over others. In this way, over the time, there are fluctuations in the frequencies of different genotypes in host and pathogen populations, but overall the level of disease remains constant. But, in "artificial" agricultural ecosystem, the intensification in crops and cropping pattern leads to the genetic uniformity in the host population which results in the continuous selection for new pathogens unless they are re-engineered to make them less conducive to pathogen emergence (Stukenbrock & McDonald, 2008).

5.3 MECHANISMS OF EVOLUTION AND EMERGENCE OF PHYTOPATHOGENS

There are different mechanisms adopted by different plant pathogens to evolve in natural as well as agroecosystem in different time scale. Comparative genomics, population genetics, evolutionary biology, and bioinformatics analyses may provide some light on the evolution, speciation, and adaptation of plant pathogen to address the origin of plant pathogens, that is, when and from where they came and how they have emerged and adapted on domesticated crops (Stukenbrock & McDonald, 2008).

5.3.1 MECHANISM OF EVOLUTION AND EMERGENCE OF FUNGAL PLANT PATHOGENS

The evolution and emergence of fungal plant pathogens is a common phenomenon in an intensified agroecosystem (Table 5.1). Although the rate of evolution in fungi is a slow process, their greater genomic diversity with pathogenicity to a large number of host and higher parasitic fitness created by a number of mechanisms helps them to evolve in a rapid pace in shorter time scale. Some of the mechanisms are discussed in the following section.

TABLE 5.1 Mechanism of Host-Governed Evolution and Adaptation of Fungal Plant Pathogens.

Evolutionary mechanism	Events	Plant–pathogen system	Reference
Host tracking/host domestication	Co-evolution of a pathogen with its host	*Mycosphaerella graminicola* (*Zymoseptoria tritici*) on wheat	Stukenbrock et al. (2007)
		Magnaporthe oryzae on rice	Couch et al. (2005)
		Phytophthora infestans on potato	Fry and Goodwin (1997)
		Ustilago maydis on maize	Munkacsi et al. (2007)

TABLE 5.1   *(Continued)*

Evolutionary mechanism	Events	Plant–pathogen system	Reference
Host shift and host jumping	Process by which a pathogen infects a new previously unaffected host species that is either a close relative of the former host (in host shift) or taxonomically very distant from the former host (in host jump)	*Phytophthora infestans* from wild *Solanum* species to potato	Grunwald and Flier (2005); Gomez-Alpizar et al. (2007)
		Magnaporthe oryzae from Setaria millet to rice	Couch et al. (2005)
		Rhynchosporium secalis from wild grasses to barley and rye	Zaffarano et al. (2006)
Horizontal gene transfer (transposable element, mitochondrial DNA; mycoviruses)	A process by which a genomic region is transferred to another organism, for example, by a transposon, plasmid, mycoviruses, etc.	ToxA from *Phaeosphaeria nodorum* into *Pyrenophora tritici-repentis*	Friesen et al. (2006)
		Host specific toxins in *Alternaria alternate*	Thomma (2003)
		PEP cluster in *Nectria haematococca*	Han et al. (2001)
Hybridization	Methods of combining two genetically divergent individuals from the same species (intraspecific hybridization) or from different species (interspecific hybridization) to create a new variant or hybrid	Hybrids of *Ophiostoma ulmi* and *O. novo-ulmi* on elm trees	Brasier (2001)
		Hybrid of *Phytophthora cambivora* and relative of *P. fragariae* on alder	Ioos et al. (2006)
		Phytophthora ramorum is a hybrid between two species of *Phytophthora*	Grunwald et al. (2012)

5.3.1.1   HOST DOMESTICATION OR HOST TRACKING

Host tracking can be defined as the co-evolution of a pathogen with its host during the process of host domestication. In this co-evolutionary process, domestication of a crop species associated with striking changes in both morphology, physiology, and genomic level of the cultivated species at the cost of loss of natural defense against pathogens in comparison to their wild relatives is occurring over longer time frames in natural ecosystems (Roy, 2001; Stukenbrock & Bataillon, 2012). This leads to the development of specific crop-based agroecosystem which favor the emergence and specialization of new pathogen species. In this case, usually the centre of origin of the pathogen coincides with the centre of origin of the host and during domestication, the selection of desirable host genotypes simultaneously select for pathogen genotypes that are adapted to the selected crop hosts in the specialized ecological niche (Stukenbrock & McDonald, 2008). The new agroecosystem is highly conducive for the propagation of selected pathogens with high fitness and imposes strong directional selection pressure on the pathogen population that helps to expand with the new agricultural host (Stukenbrock & Bataillon, 2012). In host tracking, the pathogen is likely to be younger than its host, in contrast to co-speciation where host and its pathogen have diverged simultaneously. This is evident from the co-evolution of blast pathogen *Magnaporthe oryzae* along with the domestication of rice in Asian centre of origin (Couch et al., 2005). Later, with the intensification and increasing acreage of rice cultivation, the propagation and global dispersal of clonal lineages of the rice-infecting blast pathogen is intensified. Similarly, the emergence of the pathogen *Mycosphaerella graminicola* causing Septoria leaf blotch of wheat coincides with the domestication of wheat and indicates the co-speciation of host and pathogen in the Fertile Crescent (Stukenbrock & McDonald, 2008) from where it later migrated to new areas in consistent with the intensive cultivation of wheat at global level (Banke & McDonald, 2005; Stukenbrock & McDonald, 2008). In case of potato late blight pathogen, the co-domestication of *Phytophthora infestans* with its host potatoes in the Andes valley of South America is inferred by Gomez-Alpizar et al. (2007) followed by an "out-of South America" migration. But in most of the cases, it has been found that evolution of pathogens with its host domestication is associated with the host specialization in pathogen population, but trade-off in the pathogenicity of pathogen from other host due to loss of the essential pathogenicity and avirulence genes (Stukenbrock & McDonald, 2008). The genetic isolation and divergence of pathogen population from other lineages may be due to rapid clonal propagation and strong selection mediated by the intensification of newly domesticated host.

5.3.1.2   HOST SHIFT AND HOST JUMPING

It can be defined as the process in which a pathogen infects a new, previously unaffected, host species that is either a genetically close relative of the former host (in host shift) or taxonomically very distant from the original host (in host jump). Usually after emergence, a new pathogen may adopt to a new host following a host shift or host jump. In this case, the geographical origin of the host needs not be the same as the geographical origin of the pathogen (Stukenbrock & McDonald, 2008). In an agroecosystem, a new pathogen may shift or jump into a new host from either a wild host population or from another crop species through a biological step-by-step process. At first, the new host species has to be exposed to the pathogen population followed by establishment of a compatible interaction between host and pathogen and finally, successful infection and dissemination of pathogen within the new host population (Woolhouse et al., 2005). The evidence of host shift and host jump was observed in *Rhynchosporium secalis* which originally emerged on rye and subsequently expanded host range to *Hordeum* spp. and *Agropyron* spp. (Zaffarano et al., 2006), *P. infestans* which shifted from wild *Solanum* species to cultivated potato (*Solanum tuberosum*) (Grunwald & Flier, 2005; Gomez-Alpizar et al., 2007), and rice-infecting lineages of *M. oryzae* which emerged and specialized on rice host following a host shift from Setaria millet and subsequent host jump to common rice weeds like cutgrass (*Leersia hexandra*) and torpedo grass (*Panicum repens*) (Couch et al., 2005). The close proximity between host plants facilitates either host shift or host jump to a new plant species. Hence, intensive agroecosystem and change in climatic condition played a major role in the emergence of new pathogen, their adaptation to new hosts, and their worldwide dissemination (Stukenbrock & McDonald, 2008). This process is influenced by various factors like ecology and behavior of the host species, their biology, mode of infection, survivability, and dissemination of the pathogen as well as the ecological factors. Generally, wild plant species like weeds growing within or adjacent to agricultural fields are likely to be the important source for infection and survivability of new pathogens. The emergence and adaptation of a new pathogen may be due to (a) changes in the host ecology and environment occurring with the introduction of new crop species into agroecosystem, like introduction of soybeans into the Amazon rainforest basin or wheat into the former subtropical forest of southern Brazil, (b) changes in the host behavior, and (c) movement of host species, that is, the global transportation of infected plant material also increases the chance of host shifts/

host jump of pathogens by introduction of co-evolved pathogens into native populations of related hosts. This became evident when infected Asian chestnuts were brought to North America leading to the chestnut blight epidemic (Anagnostakis, 1987).

5.3.1.3 HYBRIDIZATION

Hybridization or genetic recombination can be defined as the method of combining two genetically divergent individuals from the same species (intraspecific hybridization) or from different species (interspecific hybridization) to create a new variant or hybrid in nature. In the population of a fungal pathogen, genetic recombination is possible in two ways that is, sexual recombination and somatic recombination that can lead to changes in chromosome number, ploidy levels, or extensive genomic rearrangement (Schardl & Craven, 2003). Thus, hybridization is becoming the natural process for the evolution of new taxa and it helps in the emergence of evolution of plant pathogens (Olson & Stenlid, 2002). There are various reports on the increasing trends of hybridization between plant-pathogenic fungi and emergence of devastating new diseases on both cultivated and wild plants caused by their new hybrids. For example, interspecific hybridization between species of the Dutch elm disease pathogens *Ophiostoma ulmi* and *O. novo-ulmi* although results in the development of transient hybrids forming "genetic bridges" (Brasier et al., 1998; Brasier, 2001), allows unilateral gene flow from one species (*O. ulmi*) to the other (*O. novo-ulmi*) including pathogenicity gene (*Pat* 1) *via* gene introgression (Abdelali et al., 1999). Recently, some other new evidence of hybridization in pathogenic population has come to light. In the Netherlands, a new *Phytophthora* pathogen on *Primula* and *Spathiphyllum* has been reported that is a hybrid between *P. cactorum*, which is probably an endemic resident of the Netherlands, and *P. nicotianae*, an introduced species (Manin't Veldt et al., 1998). In addition, in Europe, a new, aggressive *Phytophthora* pathogen of alder was identified by Brasier et al. (1999) as a allopolyploid interspecific hybrid between the introduced *P. cambivora* (pathogen of hardwood trees but not a pathogen of Alnus) and *Phytophthora* species close to *P. fragariae* (pathogen of raspberry and strawberry). There is another documentation of the emergence of a new plant pathogen, *P. andina*, a common pathogen of Andean crops *S. betaceum*, *S. muricatum*, *S. quitoense*, and several wild *Solanum* spp. originated by hybridization between the potato late blight pathogen *P. infestans* and another *Phytophthora* species, belonging to *P.* clade 1c (Goss et al., 2011). In New Zealand and North America, a range of newly evolved inter-specific hybrids have been identified between different introduced *Melampsora* species on poplar trees (Spiers & Hopcroft, 1994; Frey et al., 1999; Newcombe et al., 2000) whereas, in the forests of north-eastern California, hybrids developed between the host-specialized "S" and "P" taxa of the conifer root pathogen *Heterobasidion* spp. (Garbelotto et al., 1996). In this way, the hybrids can act as a genetic bridge either between two species or sub-species, transferring pathogenicity traits from one species/ sub-species to the other via hybridization which is the potential tools for evolution of new "superpathogen" with new host ranges (Brasier, 1995).

5.3.1.4   HORIZONTAL GENE TRANSFER

It is a process by which the exchange of specific genes or genomic regions between species occurs without any sexual interaction. Horizontal (or lateral) gene transfer (HGT) is the non-sexual movement of genetic information between two organisms in the form of specific genes or genomic regions, transposable elements, plasmids, mycoviruses, etc. HGT differs from vertical gene transfer, which is the normal transmission process of genetic material from parent to offspring. The horizontally transferred genes or gene sequences can be detected based on atypical codon usage pattern and variation in guanine-cytosine (GC) composition along a chromosome indicates recent introduction of alien genetic material (Lawrence & Ochman, 1998). As all HGT detection methods have some limitations, Fitzpatrick (2009, 2011) emphasized to follow a total evidence approach where several independent methods are to be used and cross corroborated before inferring that a HGT event has occurred. This HGT can be made possible by following various mechanisms *viz.*, conjugation and transformation observed in the case of bacteria and *Saccharomyces cerevisiae* under specific artificial laboratory condition (Nevoigt et al., 2000), close ecological proximity between two organisms like a yeast wine strain (*S. cerevisiae* EC118) and *Zygosaccharomyces bailii*, a major contaminant of wine fermentations (Novo et al., 2009), sexual or somatic fusion with ancestral species as in case of *M. graminicola* acquiring eight novel dispensable chromosomes by this mechanism followed by degeneration and extensive recombination with core chromosome (Goodwin et al., 2011), formation of conidial anastomosis between the vegetatively incompatible strains of *C. lindemuthianum* and *C. gossypii* (Roca et al., 2004).

The HGT plays an important role in the evolution of a non-pathogen or a weak pathogen into a virulent one after acquiring genes needed for pathogenicity and virulence from a different species. This can be exemplified from interspecific transfer of virulence gene *ToxA*, encoding host-specific toxin from *Stagonospora nodorum* to *Pyrenophora tritici-repentis* resulting in serious emergence of tan spot disease on wheat (Friesen et al., 2006). Similarly acquisition of host-specific toxin encoding genes by various *Alternaria alternata* pathotypes through HGT results in the emergence of new pathotypes infecting previously uninfected host (Hatta et al., 2006). Comparative genomic study reveals that a cluster of genes including three virulence genes (*PEP1*, *PEP2*, and *PEP5*) and the pisatin demethylase gene (*PDA1*) on a conditionally dispensable chromosome of *Nectria haematococca* (anamorph *Fusarium solani*) determining pathogenicity to Pea could have been acquired by HGT (Han et al., 2001). A similar mobile chromosome that is, chromosome number 14 containing *Six1* (*Avr3*), *Six3* (*Avr2*) and oxidoreductase (*ORX1*) genes in *F. oxysporum* f. sp. *lycopersici* essential for pathogenicity on tomato plants was reported by Ma et al. (2010). There is also an evidence of cross-kingdom HGT between plant and fungi, fungi and bacteria, fungi and viruses, etc. occurring in evolutionary pathway. For example, the gene cluster involved in gibberellin biosynthesis pathway of *Gibberella fujikuroi*, a rice pathogen, was postulated to be acquired through HGT from plant or unknown organism (Niehaus et al., 2014).

5.3.2   MECHANISM OF EVOLUTION AND EMERGENCE OF BACTERIAL PLANT PATHOGENS

Bacterial plant pathogens continue to cause serious problems to major crops worldwide with the emergence of new pathogens and evolution of known pathogens. There are certain driving factors influencing their evolution and emergence in a modern agroecosystem (Table 5.2). Therefore, it is important to identify and to understand the mechanisms underlying genetic diversity and their functionality to regulate evolution and adaptation of bacterial plant pathogens.

TABLE 5.2 Mechanism of Evolution and Emergence of Bacterial Plant Pathogens.

Evolutionary mechanism	Events	Plant–pathogen system	Reference
Recombination	Exchange of genetic material between two individuals to form a new combination of genes on chromosome. In bacteria, the recombination takes place by (a) transformation, (b) transduction, and (c) conjugation	Genetic recombination between environmental strains and pathogenic strains of *Pseudomonas syringae* to introduce pathogenic traits and to maintain genetic variation	Monteil et al. (2013)
		Genetic recombination between *Pseudomonas viridiflava* and diverse, environmental population in weed communities	Goss et al. (2005)
		Inter-subspecific recombination between *Xylella fastidiosa* subsp. *fastidiosa* and *X. fastidiosa* subsp. multiplex helps in host shifting to mulberry	Nunney et al. (2014)
Mutation/deletion/ genome reduction	Sudden heritable changes in the nucleotide level results in the imperfect copying of genomic material from parent to offspring of plant viruses	Deletion or pseudogenization of regulation of pathogenicity factors (*rpf*) genes in specialized pathogen *Xanthomonas albilineans* (causing leaf scald disease in sugar cane) and *Xylella fastidiosa* in comparison to *X. axonopodis* pv. *vesicatoria*, a free-living pathogen	Jackson et al. (2011)
		Mutation in the regulatory gene *efpR* in three populations of *R. solanacearum* helps in the evolution and pathogenic fitness on beans, a distant host	Guidot et al. (2014)
Horizontal gene transfer (plasmids; conjugation; transposon)	Process by which a part of genomic region is transferred to another organism belonging to a different species or between different vegetatively incompatible lines	Horizontal gene transfer between *Pseudomonas syringae* pv. *tomato* and *Xanthomonas gardneri*, two tomato pathogens	Potnis et al. (2011)
		A large, mobile pathogenicity island containing thaxtomin biosynthetic pathway, *nec1*, a putative tomatinase gene, and many mobile genetic elements confers plant pathogenicity on *Streptomyces* species	Kers et al. (2005)
		Acquisition of pathogenicity and virulence genes by *Xanthomonas vesicatoria*, through horizontal gene transfer from multiple origins	Potnis et al. (2011)
		Horizontal transfer of copper resistance (Cu^R) operon on a large conjugative plasmid among the pathogenic population of *Xanthomonas axonopodis citrumelonis* and *X. vesicatoria*, and the epiphyte *Stenotrophomonas maltophilia* for adaptation and pathogenic fitness	Goss et al. (2013)

5.3.2.1   GENETIC RECOMBINATION

Genetic recombination can be defined as the exchange of genetic material between two individuals to form new combination of genes on chromosome. In bacteria, the recombination takes place by transferring genetic material (DNA) from one bacterial cell (donor cell) to the other (recipient cell) *via* (a) transformation, (b) transduction, and (c) conjugation, forming partial normal diploids, also called heterozygotes or merozygotes. Thus, the recipient cells are converted to heterozygotes or merozygotes containing either a fragment of the donor DNA (exogenote) or a complete recipient DNA (endogenote). The genetic recombination has been shown to increase the rate of pathogen adaptation in its host (Baltrus et al., 2008) and helps in the emergence of new plant pathogens through transfer of virulence factors (Friesen et al., 2006). This can be seen in case of *Xylella* spp. Inter-subspecific homologous recombination between two of the subspecies of *Xylella fastidiosa*, that is, *X. fastidiosa* subsp. *fastidiosa* and *X. fastidiosa* subsp. *multiplex* results in the emergence of a new subspecies, *X. fastidiosa subsp. morus* infecting mulberry (Nunney et al., 2014). Genetic recombination among agricultural pathogens occasionally produces new strains with increased fitness as evident in the population of *Pseudomonas syringae* (Goss et al., 2013). The environmental strain of *P. syringae* acts as the reservoirs of pathogenic traits and genetic variation for pathogenic strains introduced via genetic recombination.

5.3.2.2   MUTATION/ DELETION/ GENOME REDUCTION

Mutation, deletion, or reduction in genome size of plant pathogen sometime helps in the evolution of plant pathogens and their adaptation to new environment. There is some evidence of reduction in genome size associated with evolution of plant pathogens linked to restriction in host range and adaption to specified ecological niche. For example, deletion or pseudogenization of regulation of pathogenicity factors (*rpf*) genes and lack of type-III secretion system in the xylem-limited pathogen *Xanthomonas albilineans* (causing leaf scald disease in sugar cane) and *X. fastidiosa* in comparison to *X. axonopodis pv. vesicatoria*, a free-living pathogen, suggests their evolution and adaptation in restricted host range with the changed life style, especially in xylem of the plant. Similarly, in a multi-host experimental assay, it has been found that populations of *R. solanacearum* evolved on a distant host showing increased pathogenic fitness on both original and distant hosts.

The magnitude of pathogenic fitness was greater on distant hosts due to the mutation in the regulatory gene *efpR* (Guidot et al., 2014). This supports host-mediated evolution and adaptation of plant pathogens *in vivo*.

5.3.2.3   HORIZONTAL GENE TRANSFER (HGT)

HGT is an important mechanism for the evolution of plant pathogenic bacteria. The exchange of mobile genetic elements, *viz.*, pathogenicity islands and plasmid carrying antibiotic resistance genes by HGT in many bacterial pathogens contribute to rapid changes in virulence potential, genome evolution for pathogenic traits influencing aggressiveness, fitness, adaptation to new environment, and antibiotic resistance (Dobrindt et al., 2004). This is evident from the acquisition of effector protein encoding genes by *X. gardneri*, a tomato pathogen from *P. syringae pv. tomato via* HGT for increasing virulence (Potnis et al., 2011). The horizontal transfer of important pathogenicity genes like Type II and III protein secretion clusters and some core effectors including *avrBs2* from *X. gardneri* and *X. campestris pv. campestris*, and the core effector *xopL* from *X. citri K* into *X. vesicatoria* supports its multiple origins with rapid speciation and host range (Potnis et al., 2011). Conserved sequence identity and similarity in the organization of the copper resistance (Cu^{R-}) operon on a large conjugative plasmid in the pathogens *X. axonopodis citrumelonis* and *X. vesicatoria*, and the epiphyte *Stenotrophomonas maltophilia* supports the horizontal transfer of genes important for adaptation and pathogenic fitness of emerging strains in agricultural environment. Genetic mobility of a conserved pathogenicity island among various *Streptomyces* species supports the emergence of *S. acidiscabies* and *S. turgidiscabies* due to HGT from *S. scabies* to saprophytic species (Healy et al., 1999; Bukhalid et al., 2002; Kers et al., 2005).

5.3.3   MECHANISM OF EVOLUTION AND EMERGENCE OF PLANT VIRUSES AND VIRUS-LIKE AGENTS

With the recent intensification of agricultural practices and change in cropping ecology, the emergence of plant viruses become more evident due to tremendous changes in the genetic plasticity of viruses with increase in host range required for replication and adaptation (Table 5.3). This can be recognized from the expansion of ecological niche and geographical zone (Elena et al., 2011), accompanied with increased disease severity (Cleaveland et

TABLE 5.3 Mechanism of Host-Governed Evolution and Adaptation of Plant Virus.

Evolutionary mechanism	Events	Plant–pathogen system	Reference
Mutation	Sudden heritable changes in the nucleotide level results in the imperfect copying of genomic material from parent to offspring of plant viruses	Single mutational change in amino acid from Aspartic acid (Asp[27]) to Lysine (Lys[27]) of NIaPro determines the host-specificity of Papaya ringspot virus pathotypes P & host switching of cucurbit-infecting pathotype W to papaya	Chen et al. (2008)
Genetic recombination	Exchange of genetic material between two individuals to form new combination of genes on chromosome	Recombination has also been important in the evolution of the plant pararetroviruses like Caulimoviruses	Chenault and Melcher (1994)
		DNA recombination in the Geminiviruses	Padidam et al. (1999)
		Recombination between host and virus genome in Banana streak virus	Harper et al. (1999)
		A virulent recombinant was a hybrid of East African cassava mosaic virus and African cassava mosaic virus (ACMV) that overcame crop resistance to ACMV and decimated cassava production in Uganda in 1997 (113)	Pita et al. (2001)
Pseudo-recombination or reassortment	Method of reassortment or shuffling of genomic segments during encapsidation of genome segments of two strains of same viruses or different viruses	Pseudo-recombination increase the host range of MYMIV	Varma and Malathi (2003); Rouhibakhsh and Malathi (2005)
Host shift and host jump	Parasitic shift or jump to infect a new previously unaffected host species that is either a close relative of the former host (in host shift) or taxonomically very distant from the former host (in host jump)	Cotton leaf curl virus infecting wild cotton (*Gossypium barbadense*) in India	Varma and Malathi (2003)
		Mungbean yellow mosaic India virus (MYMIV) shifting to cowpeas	Varma and Malathi (2003)
		Maize rough dwarf virus infecting maize in Mediterranean region before introduction of American hybrid cotton	Thresh (2006)
Migration	Migration of viruses with vectors or through trans-boundary movement of infected plant material into a new geographical area	Emergence of Cacao swollen-shoot virus in West African countries due to introduction of exotic plants from South America	Otim-Nape et al. (2009)

al., 2007). Usually plant viruses follow various mechanisms and pathways to evolve or emerge which end up with their adaptation into its new host followed by their epidemiological spread in new host population (Elena et al., 2011). Some of these are discussed in the following sections.

5.3.3.1   MUTATION

Sudden heritable changes in the nucleotide level results in the imperfect copying of genomic material from parent to offspring of plant viruses known as mutation, which usually occurs during replication of nucleic acid. RNA viruses are notorious for having high mutation rates, due to replication with RNA-dependent polymerases lacking proof reading activity. The mutation of plant viruses are influenced by various external factors as well as host factors (Schneider & Roossinck, 2001). Hence, Roossinck (1997) hypothesized that the genetic variation in some plant RNA viruses is associated with their difference in host range. For example, cucumoviruses (CMV) have a very broad host range, both experimentally and naturally (Palukaitis et al., 1992), whereas the tobamoviruses have a much narrower natural host range, but can infect a large number of plants experimentally (Fraile et al., 1995). This suggests that the greater mutation or genetic variation in plant viruses has allowed them to infect and adapt to more plant host. This is evidenced by luteoviruses, like *Potato leaf roll virus* (PLRV) with little genetic variation having a very narrow host range, whereas *Beet western yellows virus* have much broader host range with much greater variation (DeMiranda et al., 1995). Thus, higher genetic diversity in plant virus has been recorded at their centers of origin, where the plant virus presumably initiated its adaptation to a new host (Fargette et al., 2004; Tomimura et al., 2004; Ohshima et al., 2010).

5.3.3.2   GENETIC RECOMBINATION

Recombination is the most common method of genetic variation in both DNA and RNA plant viruses. It can lead to changes *via* deletion, addition, or exchange of sequence between two genomes. It can be possible either by homologous recombination between two nearly identical RNAs, or non- homologous recombination between two RNAs that have a short anti-parallel stretch of complementarily (Simon & Bujarski, 1994). Recombination has played a very important role in the evolution of many plant viruses like luteoviruses (Gibbs, 1995), nepoviruses (Le Gall et al., 1995), bromoviruses (Allison et al., 1989), and CMV (Fraile et al., 1997), and also in plant pararetroviruses like Caulimoviruses (Chenault & Melcher, 1994). Thus, recombination may have led to the emergence of plant viruses and viroids with altered host range and virulence (Hammond et al., 1989; Padidam et al., 1999; Owens et al., 2000). Integration of virus genome with its host genome in *Banana streak virus* may have arisen by recombination during their evolution (Harper et al., 1999). This evidence suggests the significance of host factors in the evolution of plant viruses.

5.3.3.3   PSEUDO-RECOMBINATION

This is the method of reassortment or shuffling of genomic segments during encapsidation of genome segments of two strains of same viruses or different viruses. This has been found in the case of plant viruses having divided genome. This is the mechanism of introducing variation without any genetic changes, especially when mixed infections are most common among field isolates of plant viruses. But it has been demonstrated in only a few plant-virus systems like cucumovirus (White et al., 1995), tobravirus (Robinson et al., 1987), begomoviruses (Pita et al., 2001), etc. Although reassortment may not be a common event in nature, sometimes it has a dramatic impact on the evolution of new viral species, especially if the reassortant confers a selective advantage, such as an expanded host range as in case of CMV in Spain (Fraile et al., 1997) and Yellow mosaic viruses in legumes like especially mungbean, cowpea, etc. in India (Varma & Malathi, 2003; Rouhibakhsh & Malathi, 2005).

5.3.3.4   MIGRATION

Migration or movement of plant viruses occur with the help of migration of insect vectors or through trans-boundary movement of infected plant material into a new geographical area. Although, long-range movement of insect vectors carrying plant viruses is always helpful for emergence of plant viruses in a new area, intercontinental movement of some plant viruses either through true seeds or by infected vegetative propagules and planting materials of plants is the major concern for introduction of plant viruses and viroids.

5.3.4 MECHANISM OF EVOLUTION AND ADAPTATION OF PLANT MOLLICUTES

The comparative genomic study of genome sequences of phytoplasma and spiroplasma helps to decipher their rapid evolution and adaptation to a broad range of environments, from phloem of their host plant to gut lumen, hemolymph, saliva, and endocellular organelles of their insect vector (Hogenhout & Music, 2010). Most of the phytoplasma and spiroplasma continue their life cycle between plants and insects and require both of them for their survival and dispersal in nature. This "host switching" is an essential stage in the life cycle of pathogens. To adopt in this diverse bio-ecological niches, phytoplasma have evolved from its ancestral progenitor in various ways (Table 5.4). One of the hypothesized ways is phage-mediated gene exchange *via* infection of ancient phage that leads to the evolution of the phytoplasma and adaptation to trans-kingdom parasitic lifestyle (Wei et al., 2008). This is evident from the presence of sequence-variable mosaics (SVMs), that is, genes repetitively clustered in non-randomly distributed segments similar to the gene content and organization of prophage clusters, but absent in genomes of ancestral relatives including *Acholeplasma* spp.(Wei et al., 2008). Another possible way is the transfer of potential mobile units (PMUs) from one phytoplasma to another. PMU is a repetitive sequence containing genes for DNA replication and synthesis, recombination, a transcription factor, and other several genes encoding membrane-targeted and -secreted proteins for virulence as well as transposases which can easily exchange its position within genome, just like transposable element and helps in reorganization of genome through expansions and deletions of PMUs. This expansions or reductions of PMUs play a major role in the evolution and adaptation of phytoplasma to different environments (Bai et al., 2006). This reduction in genome size of phytoplasma and spiroplasma during their evolution from bacterial ancestors helps to keep minimal sets of genes required only for cellular life and parasitism in specified tissues of plants. Therefore, most of the phytoplasma are lacking genes for sugar metabolism as compared to their mycoplasma counterparts (Fraser et al., 1995; Bai et al., 2006). The loss of genes over time was accompanied with the gain of some new ability for evolution of parasitism and adaptation to nutrient-rich environments in host plant and insect vectors. The genetic recombination of plasmid with phytoplasma chromosome and intramolecular recombination among phytoplasma plasmids are also increasing biological diversities of phytoplasmas

(Nishigawa et al., 2002; Liefting et al., 2004; Bai et al., 2006). The extensive genetic diversity in the population of phytoplasma at strain level and their co-existence within the same host species in the same geo- and bio-ecological niches increases the opportunity of genetic recombination. Thus, it helps in rapid evolution and adaptation of phytoplasmas in various environments and the emergence of new phytoplasmal plant diseases (Zhao et al., 2010).

TABLE 5.4   Mechanism of Evolution and Adaptation of Plant Mollicutes.

Evolutionary mechanism	Events	Plant–pathogen system	Reference
Phage-mediated gene exchange	Leads the formation of sequence-variable mosaics (SVMs) containing phytoplasma strain-specific genes, generating genetic diversity	Difference in genome size architecture of "*Candidatus* Phytoplasma asteris"-related strains OY-M and AY-WB	Wei et al. (2008)
Horizontal transfer of potential mobile units(PMU)	Horizontal exchange of potential mobile units(PMU) among divergent phytoplasma lineages and its homologous recombination	Reductive evolution and reorganization of genome of "*Ca. Phytoplasma asteris*" OY-M and AY-WB	Bai et al. (2006)
Genetic recombination	Recombination of plasmid with phytoplasma chromosome and intramolecular recombination among phytoplasma plasmids	Intermolecular recombination between extra chromosomal DNAs in onion yellows (OY) phytoplasma wild (W) type and mild (M) strain	Nishigawa et al. (2002); Liefting et al. (2004); Bai et al. (2006) Oshima et al. (2004)

5.3.5   MECHANISMS OF EVOLUTION AND EMERGENCE OF PLANT PARASITIC NEMATODE

Similar to the plant-pathogenic fungi and bacteria, plant pathogenic nematodes also adopt some common path for its evolution of parasitism into host plant (Table 5.5). Host governs co-speciation *via* host jump host switching and also HGT, the basic mechanism of evaluation and emergence of plant parasitic nematodes.

TABLE 5.5   Mechanism of Evolution and Emergence of Plant Parasitic Nematode.

Evolutionary mechanism	Events	Plant–pathogen system	Reference
Horizontal gene transfer from bacteria and fungi	Non-sexual exchange of genetic materials between different species within kingdom or across the kingdom	Horizontal acquisition of endoglucanase genes by plant parasitic nematode of order Tylenchida for plant parasitism	Smant et al. (1998); Yan et al. (1998, 2001); Davis et al. (2000)
		Horizontal acquisition of genes by *Meloidogyne incognita* having homology to nod factor encoding genes of *Rhizobium leguminosarum* helps in ecological adaptation	Mccarter et al. (2003); Scholl et al. (2003).
Co-speciation	Obscured by host switching or jumping	Co-evolution of anguinid nematodes in accordance with their host plants of same or related systematic groups belonging to the Poaceae or Asteraceae	Subbotin et al. (2004)

5.3.5.1   CO-SPECIATION

Co-speciation is a common phenomenon within the population of nematode, which is often obscured by host switching or host jumping, and it is particularly common in some symbiotic or parasitic association between two organisms. A higher level of co-speciation in anguinid nematode with their grassy host species under Poaceae family is being studied by Subbotin et al. (2004) using comparative phylogenic information, where 60% chance of co-speciation was estimated. Co-speciation and adaption of different *Anguina* sp. was also evident from the evolutionary trends in gall formation in different host by anguinids.

5.3.5.2   HORIZONTAL GENE TRANSFER

HGT in place of It is a common path for evolution and adaptation of plant pathogenic fungi, bacteria, as well as plant parasitic nematodes. There are numerous evidences of HGT through which plant parasitic nematodes have acquired a large number of pathogenicity genes encoding cell wall degrading enzymes from diverse origin (Mitreva et al., 2009). The discovery of production and secretion of cellulase enzymes by plant parasitic nematode helps to hypothesize about the acquisition of pathogenicity genes by nematodes and subsequent evolution of parasitism on plant *via* HGT from soil bacteria to ancestral bacterivorous nematode (Keen & Roberts, 1998).

Later on, many more examples of acquisition of large number of pathogenicity genes by various plant parasitic nematodes in relation to their evolution and adaptation was evident from phylogenetic analysis of gene sequences from whole genome sequence or expressed sequence tag (EST) data base (Haegeman et al., 2011). Annotations of the whole genome sequence of few plant-parasitic nematodes have revealed the presence of a set of genes for plant parasitism which is the "signature" of evolution and adaptation of nematodes for plant parasitism. Some of these parasitism genes are acquired horizontally and have systematically been found in all plant-parasitic species. Whereas, some other genes encoding effector proteins and peptides mimicking plant hormones are important for plant-specific parasitism and have species-specific origin. Overall, plant parasitic nematodes showed convergent evolutionary pathway to adapt for plant parasitism (Bird et al., 2014).

5.4 ADAPTATION OF PLANT PATHOGENS TO NEW ENVIRONMENT

After evolution of any pathogen, it has to be acclimatized into a new environment. This adaptation is favored by positive directional selection in the favor of newly emerging pathogen population. The rate of adaptation sometimes may be very slow or fast. This adaptation potential can be expressed in terms of the basic reproduction number, R_0. If the R_0 is < 1, then a large proportion of infection will be acquired directly from the original source of host population and there is "an epidemic waiting to happen," whereas, if $R_0 > 1$, then most of the infection will be acquired and spread within the new host population and have a potentiality to cause major epidemic. The probability of successful adaptation depends on several factors, *viz.*, (a) the number of primary infections (I_0) (b) the initial R_0 of the infection in the new host population (c) the changes or progress in R_0 at each step, and (d) number of mutations and other genetic changes required (Woolhouse et al., 2005). The probability of adaptation is also influenced by availability of suitable host population and suitable environmental and ecological factors.

5.5   UNDERLYING FACTORS INFLUENCING THE EVOLUTION AND EMERGENCE OF PLANT PATHOGENS

In nature, the co-evolutionary dynamics of pathogen evolution is significantly influenced by some important factors *viz*., different host factors including the nature of the host plant (either annual or perennial), their distribution and introduction in new environment, their deployment in different cropping system, different factors related to plant pathogens including the nature of the pathogen (monocyclic, polycyclic, or polyetic), their survivability, dissemination, etc., and some ecological/environmental factors. In agroecosystem, the evolution and emergence of new pathogen is brought with the domestication of new plant species for cultivation providing a new ecological niche. It can also be favored by high densities of host individuals, genetic homogeneity within host populations, etc. The evolution of phenotypic traits of plant pathogens in response to host resistant genes and their deployment is also a common phenomenon (Lo Iacono et al., 2012). Some factors associated with plant pathogen also have significant contribution for their own evolution and adaptation. Some pathogenic traits like infectivity, virulence, and aggressiveness associated with some life-history traits like propagule production, survivability, dispersal and dissemination, etc. are influencing the emergence of new pathotypes and their adaptation in new environment.

Mode of reproduction, sexual or parasexual, in pathogen population also plays a very important role in the evolution and emergence of plant pathogenic variants by generation of new genetic variation through intragenic recombination (Ferreira et al., 2012), reshuffling of existing genes to create new gene combination, change in ploidy level, and elimination of deleterious mutations accumulated during asexual reproduction (Innocenti et al., 2011; Jaramillo et al., 2013). On the other hand, pathogen population in modern agroecosystems are regularly challenged by pesticides, resistance genes, crop rotation, and a variety of cultural practices aimed at reducing infections (Stukenbrock & McDonald, 2008). These intensified agricultural practices and new agricultural environment imposes strong directional selection pressure on the pathogen genome, notably on genes controlling the defeat of crop resistance genes and resistance toward pesticides (Stukenbrock & Bataillon, 2012). Speciation is another factor responsible for creating divergence and emergence of a new pathogen within population. It can occur either sympatrically or allopatrically (Stukenbrock & Bataillon, 2012). In sympatric speciation, a species evolves into a new species without a physical barrier and the genetic divergence generates within population inhibiting in the same geographical region. Thus, the ancestor species and the new species will live side by side during the speciation process, whereas the allopatric speciation occurs between populations that are geographically separated (Mayr, 1982). This is evident naturally in the case of potato late blight pathogen (*Phythophthora infestans*), rice blast pathogen (*M. oryzae*), and yellow rust pathogen (*Puccinia striformis*). The emergence of these three species was associated with a strong reduction in sexual recombination and the spread of only few specialized clonal lineages (Goodwin et al., 1994; Couch et al., 2005; Bahri et al., 2009). The global expansion of agricultural trade and transportation of germplasm coupled with inefficient quarantine and monitoring system also aggravate the spread and adaption of pathogen in previously uninvaded area. The movement of infected seed and planting materials like tubers, cuttings, etc. helps the entry and establishment of pathogens like panama wilt of banana, potato late blight, cassava mosaic virus, coconut cadang-cadang viroid, chestnut blight, oak wilt, etc. in a new ecological niche. This helps pathogen to adopt in diverse ecosystem and to become a cosmopolitan pathogen with large genetic diversity. As a result of increased genetic diversity, the pathogen population respond very rapidly to deployment of control measures like newly introduced resistance genes, new fungicides, and other disease management practices (Van Der Bosch & Gilligan, 2008; Stukenbrock & McDonald, 2008).

5.6   IMPROVED MANAGEMENT STRATEGIES TO COMBAT THIS PROBLEM IN AGROECOSYSTEM

Keeping concern on continuous emergence and adaptation of new plant pathogens in agroecosystem, there is an urgent need to take some measures to combat these emerging pathogens. These measures should be effective as well as sustainable in long run. Multidisciplinary collaboration is essential for the development of suitable management strategies. Stukenbrock and Bataillon (2012) emphasized on a close integration of evolutionary biology and genomic tools to predict the emergence, establishment, and adaptation of plant pathogens in agroecosystems that will be helpful for designing a novel and sustainable strategies. These strategies must be dynamic specially and temporally to slow down the emergence, adaptation, and spread of pathogens in various agroecosystem, keeping pace with constant change in the population dynamics of host and pathogen. Suitable integration of management strategies including generic approaches as well as specific approaches should be adopted based on crop and locality. No single static and specific approaches can be favored in long run and priority should be given to reduce the chances of development of new variants and to avoid their direct selection. Therefore, intelligent selection and suitable integration of various pest management practices will be effective on economical, ecological, and efficacious way. The common generic practices based on crop sanitation and cultural management is to be incorporated with specific strategies like use of fungicides, resistance cultivars, etc. The current approaches should emphasize on suitable utilization of existing strategies like deployment of resistance genes (gene pyramiding, multiline, varietal mixture, etc.) and agrochemicals (alternate use of pesticides, combination of pesticides etc.) that also largely limit the evolutionary consequences of plant pathogens in recent intensive agroecosystem. Although exploitation of host resistance is considered to be most effective and environmentally safest approach, durability in long run become risky. Therefore, focus is to be diverted toward utilization of polygenic resistance rather than monogenic one. The new breeding strategies should emphasize on the exploitation of wild genotypes and landraces for getting durable resistance in long run. On the other hand, modern cultivation practices like monoculture, crop homogeneity at field level, etc. enhances the genetic erosion. Therefore, genetic diversity within crop is narrowing day by day and we are losing our natural gene bank. Hence, some newer pathogens are spreading from associated weed host to main crop or other crop species in rapid speed *via* host shift or host jump. To overcome this undesirable host shifting or host jumping, agroecosystem will need to be re-engineered to prevent the continuous emergence of new pathogens (Stukenbrock & McDonald, 2008). Change in crop ecology and crop cultivation practices like creation of genetic heterogeneity within and between crops, cultivation of landraces, conservation of wild genotypes and changing in farming landscapes will alter the microclimate to make it less conducive to pathogen emergence. Adverse effects of new technologies like sprinkler irrigation, intensive cultivation in green houses, etc. have been examined in large scale. Improved monitoring system similar to the Global Rust Initiative should be developed to detect the emergence and movement of new pathogens or pathogenic variants (Stukenbrock & McDonald, 2008). Rapid detection and diagnosis of new virulent races/strains or "super races/strains" of the various deadly pathogens and their distribution pattern is also essential to develop suitable strategies in this respect. Advancement in molecular tools will also help in this case. Plant quarantine system in different countries should also be improvised with modern tools and techniques for detection of entry and establishment of new deadly pathogens in an uninvaded area and strict phytosanitary measures should also be regulated to prevent entry of new pest having quarantine importance.

5.7 CONCLUSION

Plant pathogens are continuously evolving in the present-day intensified agroecosystem. Significant increase in crop losses from various parts of the world and epidemic outbreak of minor pathogens as a major one are evidences of that. This evolutionary path is very complex as individual groups of pathogen, even within same group individuals are following different mechanisms. Identification of these mechanisms and their role in the evolution and adaptation can be elucidated by various experimental approaches. Advancement in evolutionary biology, ecology, and genomic tools with interdisciplinary approach including plant pathology provide some valuable information to understand the dynamics of evolution of plant pathogen in modified agroecosystem. Rapid development and easy accessibility of next generation sequencing technologies also give insight to understand evolutionary consequences of phytopathogens in different time scale. Although adequate knowledge on evolutionary principles is available to find out reason of disease outbreak, its application in designing a suitable and dynamic management system is still in its infancy stage.

The modern structure of agroecosystem is the main hotspot for emergence of plant pathogens as it exerts directional selection pressure in the favor of evolution and adaptation of phytopathogens due to various existing driving forces like introduction of new pathogens in new ecological niche, introduction of pathogenic variants, change in population structure of vectors, adoption of modern intensified farming techniques, change in climatic behavior, etc. In such situation, our target would be re-modulation of agroecosystem to slow down or to stop the emergence of plant pathogens by various approaches including creation of genetic heterogeneity within and between crops, modification of crop architecture and farming practices that favors less conducive cropping ecology for the emergence of phytopathogens. The genetic heterogeneity and ecological diversity can be created by growing multiline, using varietal mixture and varietal mosaics, and preferring species mixture or cultivating landraces with modern hybrids within sole cropping system and adopting mixed cropping system and combining agricultural crops with horticulture and forestry system.

Nowadays, farm mechanization in agriculture sector is progressing in rapid pace with adoption of new technologies like green house or polyhouse-based cultivation, and drip- or sprinkler-based irrigation system. Biotechnological approaches like recombinant DNA technology and tissue culture practices are also gaining momentum with the introduction and greater acceptance of transgenic crops raised for resistance to pathogens or other agronomic traits by tissue culture. But for gaining some short-term benefits, such new technologies have been exploited in large scale without any concern to its adverse effect and possible response of pathogenic population. Therefore, there is an urgency to work on this issue addressing the new evolutionary ideas coupled with recent advancement in technologies, for formulating the sustainable management strategies.

KEYWORDS

- agroecosystem
- ecology
- evolution
- plant pathogen

REFERENCES

Abdelali, E.; Brasier, C. M.; Bernier, L. Localization of a Pathogenicity Gene in *Ophiostoma ovo-ulmi* and Evidence that it may be Introgressed from *O. Ulmi. Molr. Plant Microbe Interac.* **1999,** *12,* 6–15.

Allison, R. F.; Janda, M.; Ahlquist, P. Sequence of Cowpea Chlorotic Mottle Virus Rnas 2 and 3 and Evidence of a Recombination Event during Bromovirus Evolution. *Virology.* **1989,** *172,* 321–330.

Anagnostakis,S. Chestnut Blight: The Classical Problem of an Introduced Pathogen. *Mycologia.* **1987,** *79,* 23–37.

Bahri, B.; Leconte, M.; Ouffroukh, A.; De, Vallavieille-Pope, C.; Enjalbert, J. Geographic Limits of a Clonal Population of Wheat Yellow Rust in the Mediterranean Region. *Mol. Ecol.* **2009,** *18,* 4165–4179.

Bai, X.; Zhang, J.; Ewing, A.; Miller, S. A.; Radek, A. J.; Shevchenko, D. V.; Tsukerman, K.; Walunas, T.; Lapidus, A.; Campbell, J. W.; Hogenhout, S. A. Living with Genome Instability: The Adaptation of Phytoplasmas to Diverse Environments of their Insect and Plant Hosts. *J. Bacteriol.* **2006,** *188,* 3682–3696.

Baltrus, D. A.; Guillemin, K.; Phillips, P. C. Natural Transformation Increases the Rate of Adaptation in the Human Pathogen *Helicobacter pylori. Evolution.* **2008,** *62,* 39–49.

Banke, S.; Mc Donald, B. A. Migration Patterns among Global Populations of then Pathogenic Fungus *Mycosphaerella graminicola. Mol. Ecol.* **2005**, *14,* 1881–1896.

Bird, D. M.; Jones, J. T.; Charles, H. O.; Kikuchi, T.; Danchin, E. G. J. Signatures of Adaptation o Plant Parasitism in Nematode Genomes. *Parasitol.* **2014**. Doi:10.1017/ S0031182013002163.2014.

Brasier, C. M. Episodic Selection as a Force in Fungal Microevolution with Special Reference to Clonal Speciation and Hybrid Introgression. *Can. J. Bot.* **1995**, *73,* 1213–1221.

Brasier, C. M. Rapid Evolution of Introduced Plant Pathogens via Inter Specific Hybridization. *Bio. Sci.* **2001**, *51*(2), 123–133.

Brasier, C. M.; Cooke, D.; Duncan, J. M. Origin of a New Phytophthora Pathogen through Interspecific Hybridization. *Proc. Natl. Acad. Sci.* **1999**, *96,* 5878–5883.

Brasier, C. M.; Kirk, S. A.; Pipe, N.; Buck, K. W. Rare Hybrids in Natural Populations of the Dutch Elm Disease Pathogens *Ophiostoma ulmi* and *O. Novo-ulmi. Mycol. Res.* **1998**, *102,* 45–57.

Bukhalid, R. A.; Takeuchi, T.; Labeda, D.; Loria, R. Horizontal Transfer of the Plant Virulence Gene, Nec1, and Flanking Sequences among Genetically Distinct *Streptomyces* Strains in the Diastatochromogenes Cluster. *Appl. Environ. Microbiol.* **2002**, *68,* 738–744.

Chen, K. C.; Chiang, C. H.; Raja, J. A.; Liu, F. L.; Tai, C. H.; Yeh, S. D. A Single Amino Acid of Nia Pro of Papaya Ringspot Virus Determines Host Specificity for Infection of Papaya. *Mol. Plant Microbe Interac.* **2008**, *21,* 1046–1057.

Chenault, K. D.; Melcher, U. Phylogenetic Relationships Reveal Recombination among Isolates of Cauliflower Mosaic Virus. *J. Mol. Evol.* **1994**, *39,* 496–505.

Cleaveland, S.; Haydon, D. T.; Taylor, L. Overviews of Pathogen Emergence: Which Pathogens Emerge, When and Why? *Curr. Top. Microbiol. Immunol.* **2007**, *315,* 85–111.

Couch, B. C.; Fudal, I.; Lebrun, M.-H.; Tharreau, D.; Valent, B. Origins of Host-Specific Populations of the Blast Pathogen *Magnaporthe oryzae* in Crop Domestication with Subsequent Expansion of Pandemic Clones on Rice and Weeds of Rice. *Genetics.* **2005**, *170,* 613–630.

Davis, E. L.; Hussey, R. S.; Baum, T. J.; Bakker, J.; Schots, A.; Rosso, M. N.; Abad, P. Nematode Parasitism Genes. *Annu. Rev. Phytopathol.* **2000**, *38,* 365–396.

De-Miranda, J. R.; Stevens, M.; De Bruyne, E.; Smith, H. G.; Bird, C.; Hull, R. Sequence Comparison and Classification of Beet Luteo Virus Isolates. *Arch. Virol.* **1995**, *140,* 2183–2200.

Dobrindt, U.; Hochhut, B.; Hentschel, U.; Hacker, J. Genomic Islands in Pathogenic and Environmental Microorganisms. *Nat. Rev. Microbiol.* **2004**, *2,* 414–424.

Elena, S. F.; Carrera, J.; Rodrigo, G. A. Systems Biology Approach to the Evolution of Plant–Virus Interactions. *Curr. Opin. Plant Biol.* **2011**, *14*(4), 372–377.

Fargette, D.; Pinel, A.; Abubakar, Z.; Traore, O.; Brugidou, C.; Fatogoma, S.; Hebrard, E.; Choisy, M.; Sere, Y.; Fauquet, C.; Konate, G. Inferring the Evolutionary History Of Rice Yellow Mottle Virus From Genomic, Phylogenetic, and Phylogeographic Studies. *J. Virol.* **2004**, *78,* 3252–3261.

Ferreira, R. C.; Briones, M. R. S. Phylogenetic Evidence Based on *Trypanosoma cruzi* Nuclear Gene Sequences and Information Entropy Suggest that Inter-Strain Intragenic Recombination is a Basic Mechanism Underlying the Allele Diversity of Hybrid Strains. *Infect. Genet. Evol.* **2012**, *12,* 1064–1071.

Fitzpatrick, D. A. Lines of Evidence for Horizontal Gene Transfer of a Phenazine Producing Operon into Multiple Bacterial Species. *J. Mol. Evol.* **2009**, *68,* 171–185.

Fitzpatrick, D. A. Horizontal Gene Transfer in Fungi. *FEMS Microbiol. Lett.* **2011**, *329*(1), 1–8.

Fraile, A.; Alonso-Prados, J. L.; Aranda, M. A.; Bernal, J. J.; Maplica, J. M.; Garcia-Arenal, F. Genetic Exchange by Recombination or Reassortment is Infrequent in Natural Populations of a Tripartite RNA Plant Virus. *J. Virol.* **1997**, *71*, 934–940.

Fraile, A.; Aranda, M. A.; Garcia-Arenal, F. Evolution of the Tobamo viruses. In *Molecular Basis of Virus Evolution;* Gibbs, A. J., Calisher, C. H., Garcia-Arenal, F., Eds.; Cambridge University Press: Cambridge, 1995; pp 338–350.

Fraser, C. M.; Gocayne, J. D.; White, O.; Adams, M. D.; Clayton, R. A.; Fleischmann, R. D.; Bult, C. J.; Kerlavage, A. R.; Sutton, G.; Kelley, J. M.; Fritchman, R. D.; Weidman, J. F.; Small, K. V.; Sandusky, M.; Fuhrmann, J.; Nguyen, D.; Utterback, T. R.; Saudek, D. M.; Phillips, C. A.; Merrick, J. M.; Tomb, J. F.; Dougherty, B. A.; Bott, K. F.; Hu, P. C.; Lucier, T. S.; Peterson, S. N.; Smith, H. O.; Hutchison, C. A.; Venter, J. C. The Minimal Gene Complement of *Mycoplasma genitalium. Science.* **1995**, *270*, 397–403.

Frey, P.; Gatineau, M.; Martin, F.; Pinon, J. *Molecular Studies of the Poplar Rust Melampsora medusae-populina, an Interspecific Hybrid between M. larici-populina and M. medusae*; 1999, September 13–17, Proceedings of the International Poplar Symposium II: Orleans, France pp 34.

Friesen, T. L.; Stukenbrock, E. H.; Liu, Z.; Meinhardt, S.; Ling, H. Emergence of a New Disease as a Result of Interspecific Virulence Gene Transfer. *Nat. Genet.* **2006**, *38*, 953–956.

Fry, W. E.; Goodwin, S. B. Resurgence of the Irish Potato Famine Fungus. *Bio Science.* **1997**, *47*, 363–371.

Garbelotto, M.; Ratcliff, A.; Bruns, T. D.; Cobb, F. W.; Otrosina, W. J. Use of Taxon Specific Competitive Priming PCR To Study Host Specificity, Hybridization and Inter-Group Gene Flow in Inter Sterility Groups of Hetero Basidionannosum. *Phytopathol.* **1996**, *86*, 543–551.

Gibbs, M. The Luteo Virus Super Group: Rampant Recombination and Persistent Partnerships. In *Molecular Basis of Virus Evolution*; Gibbs, A. J., Calisher, C. H., Garcia-Arenal, F., Eds.; Cambridge University Press: Cambridge, 1995; pp 351–368.

Gomez-Alpizar, L.; Carbone, I.; Ristaino, J. B. An Andean Origin of Phytophthora Infestans Inferred from Mitochondrial and Nuclear Gene Genealogies. *Proc. Nat. Acad. Sci.* U.S.A. **2007**, 104, 3306–3311.

Goodwin, S. B.; Ben, M.; Barek, S.; Dhillon, B.; Wittenberg, A. H. J.; Crane, C. F. Finished Genome of the Fungal Wheat Pathogen *Mycosphaerella graminicola* Reveals Dispensome Structure, Chromosome Plasticity, and Stealth Pathogenesis. *Plos Gene.* **2011**, *7*(6), E1002070. Doi:10.1371/Journal.Pgen.1002070.

Goodwin, S. B.; Cohen, B. A.; Fry, W. E. Panglobal Distribution of a Single Clonal Lineage of the Irish Potato Famine Fungus. *Proc. Nat. Acad. Sci. U.S.A.* **1994**, *91*, 11591–11595.

Goss, E. M.; Cardenas, M. E.; Myers, K.; Forbes, G. A.; Fry, W. E.; Restrepo, S.; Grunwald, N. J. The Plant Pathogen Phytophthora Andina Emerged Via Hybridization of an Unknown Phytophthora Species and the Irish Potato Famine Pathogen, P. Infestans. *PLoS ONE.* **2011**, *6*(9), E24543. Doi:10.1371/Journal.Pone.0024543.

Goss, E. M.; Kreitman, M.; Bergelson, J. Genetic Diversity, Recombination, and Cryptic Clades in *Pseudomonas viridiflava* Infecting Natural Populations of *Arabidopsis thaliana. Genetics.* **2005**, *169*, 21–35.

Goss, E. M.; Potnis, N.; Jones, J. B. Grudgingly Sharing their Secrets: New Insight into the Evolution of Plant Pathogenic Bacteria. *New Phytol.* **2013**, *199*, 630–632.

Grunwald, N. J.; Garbelotto, M.; Goss, E. M.; Heungens, K.; Prospero, S. Emergence of the Sudden Oak Death Pathogen *Phytophthora ramorum Trends Microbiol.* **2012,** *20(3),* *131–138.*

Grunwald, N. J.; Flier, W. G. The Biology of Phytophthora Infestans at its Centre of Origin. *Annu. Rev. Phytopathol.* **2005,** *43,* 171–190.

Guidot, A.; Jiang, W.; Ferdy, J. B.; Thebaud, C.; Barberis, P.; Gouzy, J.; Genin, S. Multi Host Experimental Evolution of the Pathogen *Ralstonia solanacearum* Unveils Genes Involved in Adaptation to Plants. *Mol. Biol. Evol.* **2014,** *31*(11), 2913–2928.

Haegeman, A.; Jones, J. T.; Danchin, E. G. J. Horizontal Gene Transfer in Nematodes: A Catalyst for Plant Parasitism. *Mol. Plant Microbe Interac.* **2011,** *24*(8), 879–887.

Hammond, R.; Smith, D. R.; Diener, T. O. Nucleotide Sequence and Proposed Secondary Structure of Columnea Latent Viroid: A Natural Mosaic of Viroid Sequences. *Nucleic Acids Res.* **1989,** *17*, 10083–10094.

Han, Y.; Liu, X.; Benny, U.; Kistler, H. C.; Van, Etten, H. D. Genes Determining Pathogenicity to Pea are Clustered on a Supernumerary Chromosome in the Fungal Plant Pathogen *Nectria haematococca. Plant J.* **2001,** *25,* 305–314.

Harper, G.; Julian, O. O.; Harrison, J. S. H.; Hull, R. Integration of Banana Streak Badana Virus into the Musa Genome: Molecular and Cytogenetic Evidence. *Virol.* **1999,** *255*, 207–213.

Hatta, R.; Shinjo, A.; Ruswandi, S.; Kitani, K.; Yamamoto, M. DNA Transposon Fossils Present on the Conditionally Dispensable Chromosome Controlling AF-Toxin Biosynthesis and Pathogenicity of *Alternaria alternata. J. Gen. Plant Pathol.* **2006,** *72,* 210–219.

Healy, F. G.; Bukhalid, R. A.; Loria, R. Characterization of an Insertion Sequence Element Associated with Genetically Diverse Plant Pathogenic *Streptomyces* sp. *J. Bacteriol.* **1999,** *181*, 1562–1568.

Hogenhout, S.; Music, M. S. Phytoplasma Genomics, from Sequencing to Comparative and Functional Genomics – What Have We Learnt? In *Phytoplasmas: Genomes, Plant Hosts and Vectors;* P. G. Weintraub, P. Jones, Eds.; CAB International: U.K., 2010; pp 19 –36.

Innocenti, P.; Morrow, E. H.; Dowling, D. K Experimental Evidence Supports a Sex-Specific Selective Sieve in Mitochondrial Genome Evolution. *Science.* **2011,** *332,* 845–848.

Ioos, R.; Andrieux, A.; Marcais, B.; Frey, P. Genetic Characterization of the Natural Hybrid Species *Phytophthora alni* Inferred from Nuclear and Mitochondrial DNA Analyses. *Fungal Genet. Biol.* **2006,** *43*, 511–529.

Jackson, R. W.; Johnson, L. J.; Clarke, S. R.; Arnold, D. L. Bacterial Pathogen Evolution: Breaking News. *Trends Genetics.* **2011,** *27*(1), 32–40.

Janzen, D. H. When is it Coevolution? *Evolution.* **1980,** *34*, 611–612.

Jaramillo, N.; Domingo, E.; Munoz-Egea, M. C.; Tabares, E.; Gadea, I. Evidence of Muller's Ratchet in Herpes Simplex Virus Type 1. *J. Gen. Virol.* **2013,** *94*, 366–375.

Keen, N. T.; Roberts, P. A. Plant Parasitic Nematodes: Digesting a Page from the Microbe Book. *Proc. Nat. Acad. Sci. U.S.A.* **1998,** *95*, 4789–4790.

Kers, O. A.; Cameron, K. D.; Joshi, M. V.; Bukhalid, R. A.; Morello, J. E.; Wach, M. J.; Gibson, D. M.; Loria, R. A Large, Mobile Pathogenicity Island Confers Plant Pathogenicity on *Streptomyces* Species. *Mol. Microbiol.* **2005,** *55*(4), 1025–1033.

Lawrence, J. G.; Ochman, H. Molecular Archaeology of the *Escherichia coli* Genome. *Proc. Nat. Acad. Sci. U.S.A.* **1998,** *95,* 9413–9417.

Le Gall, O.; Lanneau, M.; Candresse, T.; Dunez, J. The Nucleotide Sequence of the RNA-2 of an Isolate of the English Serotype of Tomato Black Ring Virus: RNA Recombination in the History of Nepoviruses. *J. Gen. Virol.* **1995,** *76,* 1279–1283.

Liefting, L. W.; Shaw, M. E.; Kirkpatrick, B. C. Sequence Analysis of Two Plasmids from the Phytoplasma Beet Leaf Hopper-Transmitted Virescence Agent. *Microbiol.* **2004,** *150,* 1809–1817.

Lo Iacono, G.; Van Den Bosch, F.; Paveley, N. The Evolution of Plant Pathogens in Response to Host Resistance: Factors Affecting the Gain from Deployment of Qualitative and Quantitative Resistance. *J. Theor. Biol.* **2012,** *304,* 152–163.

Ma, L. J. et al. Comparative Genomics Reveals Mobile Pathogenicity Chromosomes in *Fusarium. Nature.* **2010,** *464,* 367–373.

Manin't Veldt, W. A.; Veenbaas Rijks, W. J.; Ilieva, E.; De Cock, A. W. A. M.; Bonants, P. J. M.; Pieters, R. Natural Hybrids of *Phytophthora nicotianae* and *P. cactorum* Demonstrated by Isozyme Analysis and Random Amplified Polymorphic DNA. *Phytopathology.* **1998,** *88,* 922–929.

Mayr, E. *The Growth of Biological Thought*; Belknap: Cambridge (MA), 1982.

Mccarter, J. P.; Mitreva, M. D.; Martin, J.; Dante, M.; Wylie, T.; Rao, U.; Pape, D.; Bowers, Y.; Theising, B.; Murphy, C. V.; Kloek, A. P.; Chiapelli, B. J.; Clifton, S. W.; Bird, D. Mc, K.; Waterston, R. H., Analysis and Functional Classification of Transcripts from the Nematode Meloidogyne Incognita. *Genome Biol.* **2003,** *4,* R26.

Mcdonald, B. A. Population Genetics of Plant Pathogens. The Plant Health Instructor. **2004.** Doi: 10.1094/PHI-A-2004-0524-01.

Mitreva, M.; Smant, G.; Helder, J. Role of Horizontal Gene Transfer in the Evolution of Plant Parasitism among Nematodes. In *Horizontal Gene Transfer Genomes in Flux;* Gogarten, M. B., Gogarten, J. P., Olendzenski, L. C., Eds.; Humana Press: New York (NY) 2009; Vol. 532, pp 517–535.

Monteil, C. L.; Cai, R.; Liu, H.; Mechan, L.; Lontop, M. E.; Leman, S.; Studholme, D. J.; Morris, C. E.; Vinatzer, B. A. Non-agricultural Reservoirs Contribute to Emergence and Evolution of *Pseudomonas syringae* Crop Pathogens. *New Phytol.* **2013,** *199,* 800–811.

Munkacsi, A. B.; Georgiana May, S. S. *Ustilago maydis* Populations Tracked Maize through Domestication and Cultivation in the Americas. *Proc. R. Soc.* **2008,** *275,* 1037–1046.

Nevoigt, E.; Fassbender, A.; Stahl, U. *Cells of the Yeast Saccharomyces cerevisiae* are Transformable by DNA under Non-Artificial Conditions. *Yeast.* **2000,** *16,* 1107–1110.

Newcombe, G.; Stirling, B.; Mcdonald, S.; Bradshaw, J. R. Melampsora X Columbiana, a Natural Hybrid of *M. medusa M. occidentalis. Mycol. Res.* **2000,** *104,* 261–274.

Niehaus, E. M.; Jan Evska, S.; Von Bargen, K. W.; Sieber, C. M. K.; Harrer, H. Apicidin F: Characterization and Genetic Manipulation of a New Secondary Metabolite Gene Cluster in the Rice Pathogen Fusarium Fujikuroi. *PLoS ONE.* **2014,** 9(7), E103336. Doi:10.1371/Journal.Pone. 0103336.

Nishigawa, H.; Oshima, K.; Kakizawa, S.; Jung,H-Y.l Kuboyama, T.; Miyata, S.; Ugaki, M.; Namba, S. Evidence of Intermolecular Recombination between Extrachromosomal Dnas in Phytoplasma: A Trigger for the Biological Diversity of Phytoplasma. *Microbiology.* **2002,** *148,* 1389–1396.

Novo, M.; Bigey, F.; Beyne, E. Eukaryote-To Eukaryote Gene Transfer Events Revealed by the Genome Sequence of the Wine Yeast *Saccharomyces cerevisiae* EC1118. P. *Natl Acad. Sci. U.S.A.* **2009,** *106,* 16333–16338.

Nunney, L.; Schuenzel, E. L.; Scally, M.; Bromley, R. E.; Stouthamerc, R. Large-Scale Inter-Sub-Specific Recombination in the Plant-Pathogenic Bacterium *X. fastidiosa* Is Associated with the Host Shift to Mulberry. *Appl. Environ. Microbiol.* **2014,** *80*(10), 3025–3033.

Ohshima, K.; Akaishi, S.; Kajiyama, H.; Koga, R.; Gibbs, A. J. Evolutionary Trajectory of Turnip Mosaic Virus Populations Adapting to a New Host. *J. Gen. Virol.* **2010,** *91,* 788–801.

Olson, A.; Stenlid, J. Pathogenic Fungal Species Hybrids Infecting Plants. *Microbes Infect.* **2002,** *4*(13), 1353–1359.

Oshima, K.; Kakizawa, S.; Nishigawa, H.; Jung, H. Y.; Wei, W.; Suzuki, S.; Arashida, R.; Nakata, D.; Miyata, S.; Ugaki, M.; Namba, S. Reductive Evolution Suggested from the Complete Genome Sequence of a Plant–Pathogenic Phytoplasma. *Nat. Genet.* **2004,** *36,* 27–29.

Otim-Nape, G. W.; Serubombwe, W. S.; Alicai, T.; Thresh, J. M. Plant Virus Diseases in Sub-Saharan Africa: Impact, Challenges, and the Need For Global Action. In *Plant Virology in Sub-Saharan Africa*; International Institute of Tropical Agriculture (IITA), Nigeria, 2009; pp 299–311.

Owens, R. A.; Yang, G.; Gundersen-Rindal, D.; Hammond, R. W.; Candresse, T.; Bar-Joseph, M. Both Point Mutation and Recombination Contribute to the Sequence Diversity of Citrus Viroid III. *Virus Genes.* **2000,** *20*(3), 243–252.

Padidam, M.; Sawyer, S.; Fauquet, C. M. Possible Emergence of New Gemini Viruses by Frequent Recombination. *Virology.* **1999,** 265, 218–225.

Palukaitis, P.; Roossinck, M. J.; Dietzgen, R. G.; Francki, R. I. B. Cucumber Mosaic Virus. In *Advances in Virus Research*; Maramorosch, K., Murphy, F. A., Shatkin, A. J., Eds.; Academic Press: San Diego, 1992; Vol. 23, pp 281–348.

Pita, J. S.; Fondong, V. N.; Sangare, A.; Otim-Nape, G. W.; Ogwal, S.; Fauquet, C. M. Recombination, Pseudo Recombination and Synergism of Gemini Viruses are Determinant Keys to the Epidemic of Severe Cassava Mosaic Disease in Uganda. *J. Gen. Virol.* **2001,** *82,* 655–665.

Potnis, N.; Krasileva, K.; Chow, V.; Almeida, N. F.; Patil, P. B.; Ryan, R. P.; Sharlach, M.; Behlau, F.; Dow, J. M.; Momol, M. T.; et al. Comparative Genomics Reveals Diversity among *Xanthomonas* Infecting Tomato and Pepper. *BMC Genomics.* **2011,** *12,* 146.

Robinson, D. J.; Hamilton, W. D. O.; Harrison, B. D.; Baulcombe, D. C. Two Anomalous Tobra Virus Isolates: Evidence for RNA Recombination in Nature. *J. Gen. Virol.* **1987,** *68,* 2551–2561.

Roca, M. G.; Davide, L. C.; Davide, L. M. C.; Mendes-Costa, M. C.; Schwan, R. F.; Wheals, A. E. Conidial Anastomosis Fusion between Colletotrichum Species. *Mycol. Res.* **2004,** *108*(11), 1320–1326.

Roossinck, M. J. Mechanisms of Plant Virus Evolution. *Annu. Rev. Phytopathol.* **1997,** *35,* 191–209.

Rouhibakhsh, A.; *Malathi,* V. G. *Severe Leaf Crinkle Disease of Cowpea* – a New Disease of Cowpea in Northern India Caused by Mung Bean Yellow Mosaic India Virus and a Satellite DNA β. *New Dis. Rep.* **2005,** *10,* 19.

Roy, B. A. Patterns of Association between Crucifers and their Flower-Mimic Pathogens: Host Jumps are More Common than Coevolution or Co-speciation. *Evol.* **2001,** *55,* 41–53.

Schardl, C. L.; Craven, K. D. Interspecific Hybridization in Plant-Associated Fungi and Oomycetes: A Review. *Mol. Ecol.* **2003,** *12,* 2861–2873.

Schneider, W. L.; Roossinck, M. J. Genetic Diversity in RNA Virus Quasi-Species Is Controlled by Host–Virus Interactions. *J. Virol.* **2001,** *75,* 6566–6571.

Scholl, E. H.; Thorne, J. L.; Mccarter, J. P.; Bird, D. M. Horizontally Transferred Genes in Plant-Parasitic Nematodes: A High-Throughput Genomic Approach. *Genome Biol.* **2003,** 4, R39.

Simon, A. E.; Bujarski, J. J. RNA-RNA Recombination and Evolution in Virus-Infected Plants. *Annu. Rev. Phytopathol.* **1994,** *32,* 337–362.

Smant, G.; Stokkermans, J. P.; Yan, Y.; De Boer, J. M.; Baum, T. J.; Wang, X.; Hussey, R. S.; Gommers, F. J.; Henrissat, B.; Davis, E. L.; Helder, J.; Schots, A.; Bakker, J. Endogenous Cellulases in Animals: Isolation of Beta-1, 4-Endoglucanase Genes from Two Species of Plant–Parasitic Cyst Nematodes. *Proc. Natl. Acad. Sci. U .S. A.* **1998,** *95*(9), 4906–4911.

Spiers, A. G.; Hopcroft, D. H. Comparative Studies of Poplar Rusts *Melampsora medusae, M. larici-populina* and their Interspecific Hybrid *M. medusae-populina. Mycol. Res.* **1994,** *98,* 889–903.

Stukenbrock, E. H.; Banke, S.; Javan-Nikkhah, M.; Mcdonald, B. A. Origin and Domestication of the Fungal Wheat Pathogen *Mycosphaerella graminicolavia* Sympatric Speciation. *Mol. Biol. Evol.* **2007,** *24,* 398–411.

Stukenbrock, E. H. Bataillon T. A Population Genomics Perspective on the Emergence and Adaptation of New Plant Pathogens in Agro-ecosystems. *PLoS Pathog.* **2012,** *8*(9), E1002893. Doi:10.1371/Journal.ppat.1002893.

Stukenbrock, E. H.; Mcdonald, B. A. The Origins of Plant Pathogens in Agro-ecosystems. *Annu. Rev. Phytopathol.* **2008,** *46,* 75–100.

Subbotin, S. A.; Krall, E. L.; Riley, I. T.; Chizhov, V. N.; Staelens, A.; Loose, M. D.; Moens, M. Evolution of the Gall-Forming Plant Parasitic Nematodes (Tylenchida: Anguinidae) and their Relationships with Hosts as Inferred from Internal Transcribed Spacer Sequences of Nuclear Ribosomal DNA. *Mol. Phylogenet. Evol.* **2004,** *30,* 226–235.

Thomas, J. A.; Welch, J. J.; Lanfear, R.; Bromham, L. A Generation Time Effect on the Rate of Molecular Evolution in Invertebrates. *Mol. Biol. Evol.* **2010,** *27,* 1173–1180.

Thomma, B. P. H. J. *Alternaria* sp. from General Saprophyte to Specific Parasite. *Mol. Plant Pathol.* **2003,** *4*(4), 225–236.

Thresh, J. M. Plant Virus Epidemiology: The Concept of Host Genetic Vulnerability. *Adv. Virus Res.* **2006,** *67,* 89–125.

Tomimura, K.; Spak, J.; Katis, N.; Jenner, C. E.; Walsh, J. A.; Gibbs, A. J.; Ohshima, K. Comparisons of the Genetic Structure of Populations of Turnip Mosaic Virus in West and East Eurasia. *Virol.* **2004,** *330,* 408–423.

Van Der Bosch, F.; Gilligan, C. A. Models of Fungicide Resistance Dynamics. *Annu. Rev. Phytopathol.* **2008,** *46,* 123–147.

Varma, A.; Malathi, V. G. *Emerging Gemini Virus Problems*: A Serious Threat to Crop Production. *Ann. Appl. Biol.* **2003,** *142,* 145–164.

Wei, Davis, R. E.; Jomantiene, R.; Zhao, Y. Ancient, Recurrent Phage Attacks and Recombination Shaped Dynamic Sequence-Variable Mosaics at the Root of Phytoplasma Genome Evolution. *Proc. Natl. Acad. Sci. U.S.A.* **2008,** *105*(33), 11827–11832.

White, P. S.; Morales, F. J.; Roossinck, M. J. Interspecific Reassortment in the Evolution of a Cucumo Virus. *Virol.* **1995,** *207,* 334–337.

Woolhouse, M. E. J.; Haydon, D. T.; Antia, R. Emerging Pathogens: The Epidemiology and Evolution of Species Jumps. *Trends Ecol. Evol.* **2005,** *20*(5), 238–244.

Yan, Y.; Smant, G.; Davis, E. Functional Screening Yields a New Beta-1, 4-Endoglucanase Gene from *Heterodera glycines* that may be the Product of Recent Gene Duplication. *Mol. Plant Microbe Interact.* **2001,** *14,* 63–71.

Yan, Y.; Smant, G.; Stokkermans, J.; Qin, L.; Helder, J.; Baum, T.; Schots, A.; Davis, E. Genomic Organization of Four Beta-1, 4-Endoglucanase Genes Ii Plant- Parasitic Cyst Nematodes and its Evolutionary Implications. *Gene.* **1998,** *220*(1–2), 61–70.

Zaffarano, P. L.; Mcdonald, B. A.; Zala, M.; Linde, C. C. Global Hierarchical Gene Diversity Analysis Suggests the Fertile Crescent is not the Centre of Origin of the Barley Scald Pathogen *Rhynchosporium secalis. Phytopathology.* **2006,** *96*, 941–950.

Zhan, J.; Thrall, P. H.; Burdon, J. J. Achieving Sustainable Plant Disease Management through Evolutionary Principles. *Trends Plant Sci.* **2014,** *19*(9), 570–575.

Zhao, Y.; Wei, W.; Davis, R.; Lee, I. M. Recent Advances in 16S Rrna Gene-Based Phytoplasma Differentiation, Classification and Taxonomy. In *Phytoplasmas: Genomes, Plant Hosts and Vectors;* Weintraub, P. G., Jones, P., Eds.; CAB International: U.K., 2010; pp 64–92.

PART II
Host Selection by the Phytopathogen

CHAPTER 6

HOST-DERIVED SPECIFICITY IN *MAGNAPORTHE-*PHYTOPATHOSYSTEM

MUKUND VARIAR*, SHAMSHAD ALAM, and JAHANGIR IMAM

Biotechnology Laboratory, Central Rainfed Upland Rice Research Station (CRURRS), ICAR, Hazaribag 825301, India

Corresponding author. E-mail: mukund.variar@gmail.com

CONTENTS

ABSTRACT

Resistance genes are the most effective weapons against the onslaught of rice pathogens, especially, blast, caused by *Magnaporthe oryzae*. Resistance is induced by specific recognition of the pathogen secreted avirulence (*Avr*) gene products and activation of immune responses at the site of infection. Species and cultivar specificity is governed by the presence or absence of the *Avr* gene products called effectors or elicitors that accumulate in the biotrophic interfacial complex (BIC) and are delivered into plant cells (cytoplasmic effectors) or dispersed in the extracellular space between the fungal cell wall and the extra invasive hyphal membrane (EIHM) (apoplastic effectors). Pathogens secrete effectors to modulate host immunity and to facilitate parasitism but several among them are recognized by host resistance proteins eliciting host defense response. Molecular and genetic studies have shown that the ability to infect different rice cultivars or other host species may be conferred in many cases by single host/cultivar specificity genes, although the degree of pathogenecity, for example, number of lesions, lesion size, and spore yield may be modulated by multiple genes in the pathogen. More than 40 *Avr* genes of *M. oryzae* have been identified, 12 of them have been cloned, and characterized. Frequent presence/absence polymorphisms by deletions, nucleotide substitutions, and inactivation by transposable element (TE) insertions in the *Avr* genes, which is an adaptive advantage for the pathogen, renders the resistance genes ineffective, and allows the fungus to infect, and invade previously resistant genotypes. Virulence analyses with Eastern Indian isolates of *M. oryzae* on a set of monogenic differentials revealed that most of the 24 *R* genes except a few had a narrow resistance spectrum. The few genes (*Pi9*, *Piz-5* (*Pi2*), *Pita-2*, *Piz, and Pi1*) showing broad spectrum resistance to *M. oryzae* isolates had different specificities. A combination of *Pi9* and *Pita-2* excluded all the pathotypes collected from Eastern India and hence has potential for effective management of rice blast disease in this region.

6.1   INTRODUCTION

During the course of domestication, rice plant has been subjected to selection both by nature and man which led to reduction of diversity in the present rice species. Over the period, rice plant has encountered many biotic and abiotic stresses which might have tailored its physiological and molecular responses. Among biotic stress, rice blast caused by *Magnaporthe oryzae* is a serious threat to rice production causing yield losses to the tune of 157 million tons of rice per annum worldwide. The blast pathogen infect rice crop at every stages of its growth starting from seedling to grain filling stage. In recent years, rice blast epidemics have occurred in China, where 5.7 million hectares of rice were destroyed between 2001 and 2005, Korea, Japan, Vietnam, and the United States (Wilson &Talbot, 2009). This fungal disease is a major yield depressant in certain agro-ecological regions of India, especially the plateaus and hills of East, and North India during the South West (SW) monsoon season. According to one estimate, about 564,000 tons of rice is lost due to blast in Eastern India, of which nearly 50% (246,000 tons) is in the upland ecosystem (Widawsky et al., 1990). Rice is cultivated in these regions mostly under rainfed conditions, in undulating or terraced fields under different toposequences, and hydrologies. Rice cultivars grown and crop management practices followed, therefore, vary widely, necessitating incorporation of resistance genes into different cultivars locally preferred by farmers. Pathogen variability exacerbates disease complexity and poor inherent yields dissuade the farmers from investing in crop protection. N fertilizer application above the recommended doses (150 kg N/ha in parts of Andhra Pradesh), cooler climate in the table lands of Karnataka during the SW monsoon season and the second rice season in Tamil Nadu (North East monsoon) and preference for susceptible fine grained varieties pre-dispose the rice crop to blast in these states (Variar et al., 2009). Farmers cultivating quality rices in the North (Basmati types are highly susceptible) are constrained to limit N application to about 50 kg/ha to avoid severe losses from the disease.

Rice blast disease is caused by *M. oryzae Couch sp. Nov = Magnaporthe grisea sensu* Yaegashi and Udagawa, which was recently defined as a new species, separate from *M. grisea*, based on multilocus genealogy, and mating experiments (Couch & Kohn, 2002). Phylogenetic analysis divides *Magnaporthe* isolates, which are morphologically indistinguishable, into two distinct clades, one that is associated with *Digitaria* (crabgrass)-infecting isolates of the fungus named *M. grisea* and one that is associated with isolates capable of infecting rice, millets, and other grasses (species of *Oryza, Setaria, Lolium, Eragrostis,* and *Eleusine*), which was named *M. oryzae.* Although the collective host range of *Magnaporthe* spp. is extensive, individual isolates of the fungus are limited to infecting a small number of grass species.

The infection cycle of *M. oryzae* begins with the attachment of conidia to rice plants. Conidia elaborate spore tip mucilage for attachment to the host surface and germinate on hydrophobic host surface cues. The germ tubes produced from conidia form specialized cells called appressoria that generate pressure and physical force to rupture the cuticle and invade the different host tissues such as leaves, stems. Recent evidence suggests that appressorium development is controlled by cell cycle checkpoints and involves autophagy (Veneault-Fourrey et al., 2006). This culminates in programmed cell death of the fungal conidium and recycling of its contents to the appressorium. Mature appressorium generates turgor by accumulating high concentrations of compatible solutes which translates in to mechanical force by which the penetration peg forces its way through the leaf cuticle. The fungus then invades rice tissue using specialized filamentous invasive hyphae (IH), which successively occupy living rice cells and colonize tissue extensively. Ramification of the fungal hyphae through the plant tissue results in the disease lesions that are symptomatic of rice blast disease. The fungus sporulates profusely from disease lesions under conditions of high humidity, allowing the disease to spread rapidly to adjacent rice plants by wind, and dew drop splash. The process of conidia attachment, germ tube generation, appressoria formation, and penetration have been studied extensively in the rice blast pathosystem (Howard & Valent, 1996) but there is limited information on the molecular events that occur after penetration of the cell wall and the molecular mechanisms underlying the host–pathogen interactions.

6.2   EFFECTOR MEDIATED HOST SPECIFICITY

Plants have evolved sophisticated and multifaceted mechanisms to recognize and respond to infection to a number of pathogens in nature like non-host resistance, constitutive barriers, and race-specific resistance or host derived resistance (Liu et al., 2007). Generally, the non-host resistance, also called species-level resistance implies that any given pathogen can cause disease only on a limited range of plant species. The constitutive barrier includes cuticular wax layers, pre-formed antimicrobial enzymes, secondary metabolites, and toxic compounds which can block the attack of pathogens (Zeng et al., 2006). Race-specific resistance or host derived resistance is also known as gene for gene resistance in which host R genes can specifically recognize the cognate Avr genes from the pathogens (Flor, 1971). Such race-specific plant–pathogen interaction leads to resistance, while lack of functional products of either gene results in disease. Resistance (R) genes are the most effective weapons against pathogen invasion since they can specifically recognize the corresponding pathogen effectors to activate plant immune responses at the site of infection.

The rice blast fungus is considered a hemi-biotroph with an initial biotrophic invasion phase followed by a distinct phase of necrotrophic killing of host cells. The pathogen then secretes effector proteins into the cytoplasm of living host cells to block the host defenses and control host cellular processes needed for disease. Specificity of the blast fungus to certain graminaceous plants and cultivar specificity among rice isolates may result from the presence of these effectors previously described as avirulence/virulence factors, elicitors, and host/cultivar specific toxins. They serve to suppress immunity, modulate metabolism, and prevent recognition of the invading fungus. Although the functions of most of these effectors are still unknown, it is increasingly clear that fungal effectors are delivered both to the inside of host cells, as well as to the host–pathogen interface during infection. Molecular and genetic studies have shown that the ability to infect different rice cultivars or other host species may be conferred in many cases by single host/cultivar specificity genes, although the degree of pathogenecity, for example, number of lesions, lesion size, and spore yield may be modulated by multiple genes in the pathogen (Leong et al., 1994).

In the gene for gene model, also known as receptor-ligand model, the *R* protein is proposed to act as a receptor to recognize a corresponding pathogen *Avr* protein and form an *R-Avr* complex that leads to the activation of host defense response. This type of *R-Avr* interaction is well documented in rice-*M. oryzae*. The *Avr* genes encode effector molecules during normal growth and pathogenicity of the pathogen. Effectors facilitate disease development, but some (*Avr* effectors) also trigger the host's resistance gene-mediated hypersensitive response (HR) and block disease. This subset of pathogen effectors or *Avr* gene products is recognized by host resistance (*R*) gene products, resulting in a HR that blocks pathogen growth. The *Avr* genes determine the host range of pathogens, on the basis whether a pathogen carrying a set of *Avr* genes would be capable of producing disease on a particular host which contains complementary *R* genes or not. The natural population of *M. oryzae* has a common and interesting phenomenon of modification or shedding of *Avr* genes which give them the advantage of shifts in the host range (Sharma et al., 2012).

As with other biotrophic and hemibiotrophic pathosystems, rice blast effectors have been identified so far by their AVR activity. More than 40 *Avr* genes of *M. oryzae* have been identified, 12 of them have been cloned and characterized (*PWL1, PWL3, PWL4, PWL2, Avr1*-CO39, *Avr-Pita, ACE1, Avr-Piz-t, Avr-Pia, Avr-Pii, Avr-Pik/Km/Kp,* and *Avr-Pi54*), and two, *Avr-Pi15* and *Avr-Pi7* have been mapped (Kang et al., 1995; Sweigard et al., 1995; Farman & Leong, 1998; Orbach et al., 2000; Bohnert et al., 2004;

Li et al., 2009; Yoshida et al., 2009; Ray et al., 2011; Ma et al., 2006; Feng et al., 2007) (Table 6.1). Out of 12, nine have been cloned by map-based cloning approach and rest three by genome-wide association mapping. With the exception of Avirulence Conferring Enzyme 1 (*ACE1*) which encodes an appressorium specific polyketide synthase/non-ribosomal peptide synthetase (Bohnert et al., 2004), the cloned blast *Avr* genes *Avr-Pita1* (Orbach et al., 2000; Khang et al., 2008), *PWL1*, *PWL3*, *PWL4* (Kang et al., 1995), *PWL2* (Sweigard et al., 1995), and *Avr-CO39* (Peyyala & Farman, 2006) encode small, IH specific, secreted effector proteins. *ACE1* is a hybrid enzyme formed by a fusion between a polyketide synthase and a non-ribosomal peptide synthetase, enzymes that mediate the biosynthesis of secondary metabolites in fungi. Mutagenesis experiments indicated that *ACE1* biosynthetic activity is required for *Avr* suggesting that this enzyme probably mediates the production of an effector metabolite that triggers hypersensitivity after translocating to rice cells. The PWL effectors are small, glycine-rich proteins that are present in rice pathogens and function as *Avr* proteins in infection of weeping lovegrass and finger millet. Fungal strains expressing either gene are blocked from infecting weeping lovegrass. *Avr-Piz-t* functions to suppress pathogen-associated molecular pattern (PAMP)-triggered immunity by inhibiting the ubiquitin ligase activity of the rice RING E3 ubiquitin ligase APIP6 (Park et al., 2012). *Avr-Pita* encodes a putative neutral zinc metalloprotease (Jia et al., 2000) and it belongs to a gene family with at least two additional members (Khang et al., 2008).

TABLE 6.1 Characteristics of the Cloned Blast-Resistance Genes and Quantitative Trait Loci (QTL) in Rice.

R gene	Chromosome no.	Cognate Avr gene	Encoding protein	Expression	Cloning strategy	References
Pib	2	NA	NBS-LRR	Circadian, inducible by stress	Map based	Wang et al. (1999)
Pita	12	Yes	NBS-LRR	Constitutive	Map based	Bryan et al. (2000)
Pi9	6	NA	NBS-LRR	Constitutive	Map based	Qu et al. (2006)
Pi2	6	NA	NBS-LRR	Constitutive	Map based	Zhou et al. (2006)
Piz-t	6	Yes	NBS-LRR	Constitutive	Map based	Zhou et al. (2006)
Pid2	6	NA	Receptor kinase	Constitutive	Map based	Chen et al. (2006)

TABLE 6.1 *(Continued)*

R gene	Chromo-some no.	Cognate Avr gene	Encoding protein	Expression	Cloning strategy	References
Pi36	8	NA	CC-NBS-LRR	Constitutive	Map based	Liu et al. (2007)
Pi37	1	NA	NBS-LRR	Constitutive	Map based *in Silico*	Lin et al. (2007)
Pik-m	11	Yes	NBS-LRR	Constitutive	Map based	Ashikawa et al. (2008)
Pit	1	NA	CC-NBS-LRR	Transcriptionally inactive	Map based	Hayashi and Yoshida (2009)
Pi5	9	NA	CC-NBS-LRR	*Pi5-1* is pathogen dependent, *Pi5-2* is constitutive	Map based	Lee et al. (2009)
Pid3	6	NA	NBS-LRR	Constitutive	*In Silico* homology based	Shang et al. (2009)
Pi21*	4	NA	Proline-containing protein	Inducible by stress	Map based	Fukuoka et al. (2009)
Pbl	11	NA	CC-NBS-LRR	Transcriptionally inactive	Map based	Hayashi et al. (2010)
Pish	1		CC-NBS-LRR	Constitutive	Mutant Screening	Takahashi et al. (2010)
Pik	11	Yes	CC-NBS-LRR	Constitutive	Map based	Zhai et al. (2011)
Pik-p	11	Yes	CC-NBS-LRR	Constitutive	Map based *in Silico*	Yuan et al. (2011)
Pi54 (*Pik-h*)	11	NA	NBS-LRR	Constitutive	Map based	Sharma et al. (2005)
Pia	11	Yes	CC-NBS-LRR	Constitutive	Multifaceted genomics approach	Okuyama et al. (2011)
NLS1	11	NA	CC-NBS-LRR	Constitutive	Map based	Tang et al. (2011)
Pi25	12	NA	NBS-LRR	Constitutive	Map based	Chen et al. (2011)
Pi54rh	11	NA	CC-NBS-LRR	Constitutive	Allele mining approach	Das et al. (2012)

*Clone blast resistance QTLs.

M. oryzae effectors can be divided into cytoplasmic and apoplastic effectors based on their localization in plant cells (Zhang & Xu, 2014). Cytoplasmic effectors, including *Avr-Pita*, *PWL1*, *PWL2*, *Bas2*, and *Avr-Piz-t*, are preferentially accumulated in the biotrophic interfacial complex (BIC) before being delivered into plant cells. The BIC is a distinct plant-derived, membrane-rich structure developed at the tip of primary IH by *M. oryzae*. In each newly invaded rice cell, effectors are first secreted into BICs before delivery. The BICs are persistent and left behind when the primary IH differentiates into the secondary IH. In addition, the fungus continues to secrete effectors into BICs even after IH have grown extensively as pseudohyphae and invaded neighboring plant cells (Khang et al., 2010). On the other hand, apoplastic effectors such as *Bas4, Avr1-CO39,* and *Slp1* are not associated with the BIC. After secretion, they are dispersed in the extracellular space between the fungal cell wall and the extra invasive hyphal membrane (EIHM), surrounding the IH in rice cells. No specific protein motifs or sequences have been identified in the cytoplasmic or apoplastic effectors that are responsible for their localization in plant cells after being secreted. Therefore, it is impossible to predict whether an effector is apoplastic or cytoplasmic solely based on its amino acid sequence. Movement of cytoplasmic effectors from vein associated cells into neighboring cells was found to be rare and of low efficiency in comparison with effector movement among regular epidermal cells suggesting that effector trafficking depends on rice cell types. In addition, cell-to-cell effector translocation also depends on the protein size. These results indicate that Magnaporthe effectors may be moved symplastically through plasmodesmata (Khang et al., 2010).

6.3   INSTABILITY OF THE *AVR* GENES

Resistance to *M. oryzae* in rice follows a gene-for-gene specificity where major resistance *R* genes (*Pi* for *Pyricularia*) are effective in controlling infection by races of *M. oryzae* possessing corresponding *Avr* genes (Jia et al., 2000). The gene-for-gene hypothesis implies that a single plant *R* gene product recognizes a unique *Avr* protein. A strong and rapid immune response that follows this recognition which prevents further invasion. However, many of these resistance genes lost their effectiveness within a short period. Instability of resistance is ascribed to genetic changes that occur in the associated *Avr* gene. Previous studies have shown that *Avr* genes might be undergoing frequent mutational events, including spontaneous deletions, nucleotide substitutions, and inactivation by TE insertions, which can lead to loss of *Avr* (Valent & Khang, 2010). Huang et al. (2014), found frequent presence/absence polymorphisms, high levels of nucleotide variation, high non-synonymous to synonymous substitution ratios, and frequent shared non-synonymous substitutions in the *Avr* genes of 62 rice blast strains from different parts of China. When the genomic composition of the flanking sequences was surveyed in the area around *Avr* genes and non-*Avr* genes, a large number of repetitive sequences with several copies were found in the flanking sequences of three *Avr* genes while the flanking sequences surrounding the non-*Avr* genes were non-repetitive. Moreover, no haplotype diversity was detected in the non-*Avr* genes of the 62 blast strains but haplotype diversity ranged from 0.16 to 0.96 in *Avr* loci. These results partially explain why frequent presence/absence polymorphisms and translocations across the genome are detected in *Avr* genes. Instability of the *Avr* genes is thus an adaptive advantage for the pathogen that is known to reproduce asexually in nature.

Information on presence/absence of *Avr* genes in natural populations from diverse rice-growing regions and their interaction with *R* genes may be useful to breeders attempting to develop durably resistant varieties. A set of 63 blast pathogen isolates from Northeast and Eastern India were evaluated for the presence/absence of *Avr* genes and phenotyped for their virulence spectrum to know whether molecular detection of *Avr* genes would help to choose the resistance genes for deployment. Preponderance of many *Avr* genes in the population suggested that avirulent strains had greater fitness in the absence of corresponding resistance genes. However, virulence analyses revealed that many isolates possessing alleles of *Avr* genes were able to infect monogenic lines harboring cognate *R* genes suggesting that such isolates might possess alternate mechanisms to escape host surveillance (unpublished data).

6.4 POPULATION STRUCTURE

Strategies to incorporate the non-matching resistance genes against the existing pathogen population in the field require studies on virulence and molecular characterization. Characterization of pathogen populations of *M. oryzae* is done both by conventional pathotyping based on the reaction on a set of differentials and molecular tools such as DNA-fingerprinting. The diversity and structure of *M. oryzae* on rice were described in many countries using RFLP (Levy et al., 1991), rep-PCR markers (George et al., 1998), or RAPD (Ross et al., 1995; Chadha & Gopalakrishna, 2005). The

MGR586 element, a repeated and dispersed sequence related to DNA TEs, has been widely used as a probe for genetic analysis of the blast pathogen population. These studies showed that pathogen populations are composed of a limited number of groups/lineages of genetically similar individuals. Because of the difficulties involved in RFLP analysis of large samples of *M. oryzae*, George et al. (1998) developed a cost effective, simple, repetitive element-based polymerase chain reaction (rep-PCR) fingerprinting method, specific for monitoring *M. oryzae* populations. A close correspondence between the groupings of isolates based on *Pot2* rep-PCR and those obtained by MGR586 was demonstrated. These markers are useful for distinguishing genetic groups, or families, of the pathogen. However, the molecular markers used in population studies are not race-specific. Tharreau et al. (2009), geno-typed more than 1700 isolates of *M. Oryzae* from 40 countries with 13 SSR markers and found evidence of intercontinental introductions of the fungus in the North, South, and West African populations. They hypothesized that Himalayan foot hills are a centre of diversity of *M. oryzae* and that sexual reproduction was taking place in this area but the fungus lost the ability for sexual reproduction in other parts of the world due to migration and other founder effects. Recently, Saleh et al. (2014) provided new insights on the native areas, diversity reservoirs, and invasion routes of rice blast. The trans-portation of infected materials is majorly linked to the several independent events of intercontinental migrations. Plant domestication has also a role in shaping the population structure of plant pathogen. Saleh et al. (2014) found that genetic clustering of *M. oryzae* isolates was significantly associ-ated with the prevalence of varieties of *indica* or *japonica* type in the area sampled. Following rice domestication, *M. oryzae* possibly adapted inde-pendently to these two subspecies leading to differentiation in two clusters. Such a structure, probably accompanied by a specialization of the pathogen on the different rice subspecies, could be exploited to develop new strate-gies of deployment of resistance genes. We investigated the mating type distribution and fingerprint analysis in 63 *M. oryzae* isolates collected from Northeast and Eastern India to assess the sexual recombination and genomic diversity (Imam et al., 2014). The result on mating type distribution and lineage diversity indicated that sexual recombination might be one of the reasons for lineage diversity in *M. oryzae* isolates.

Molecular diversity was detected using *Pot2*-TIR and MGR586-TIR rep-PCR of a total of 63 single spore isolates of rice blast fungus *M. oryzae* collected during 2010–2013 from different cultivars in different rice growing regions of Northeast and Eastern India. DNA fingerprint data for both the primers yielded high lineage diversity. Cluster analysis of 63 isolates for both the primers showed high haplotypic diversity at all sites. Seven lineages were detected by *Pot2*-TIR and 11 by MGR586-TIR fingerprints. Among the seven lineages detected by *Pot2*-TIR, lineage C and G represented the isolates from Jharkhand and Assam, respectively; whereas lineage B was from Jharkhand and Odisha. The remaining four lineages contained isolates of mixed geographical origin. Isolates from Jharkhand were distributed in all the seven lineages. The MGR586-TIR DNA fingerprinting detected 11 lineages, out of which six (lineages F, G, H, I, J, K) were site specific but were represented only by single isolate. Lineages B and C contained isolates of Jharkhand only. The remaining two lineages contained isolates of mixed geographical origin. The lineage A was the largest representing 47 isolates from all the states. The DNA fingerprinting analysis of 63 isolates from Northeast and Eastern India clearly indicated that in general, the *M. oryzae* population cannot be delineated into region-specific groups.

6.5 VIRULENCE STRUCTURE OF *M. ORYZAE* ISOLATES IN EASTERN INDIA

Pathotyping or virulence analysis with a genetically well-defined set provides high degree of resolution for describing the virulence structure of a population. A new set of 26 differential varieties targeting 24 resistance genes in the genetic background of LTH developed by the collaboration of International Rice Research Institute (IRRI) and the Japan International Rice Research Center for Agricultural Sciences (JIRCAS) (Tsunematsu et al., 2000; Kobayashi et al., 2007), is a basic tool to understand pathogenic variability in *M. oryzae* and to classify pathogen population into races (Fukuta et al., 2010). Virulence analysis of 72 *M. oryzae* isolates from Eastern India at Hazaribag (Alam et al., 2013) against the set of monogenic differential varieties revealed matching virulence to all monogenic differentials carrying different resistant genes. The pathogen population comprised isolates capable of infecting minimum six to maximum 21 resistant genes out of 26 *R* genes evaluated. All isolates exhibited compatibility to LTH but none were virulent on Tetep, which was used as a resistant check. Virulence frequency of different *M. oryzae* isolates was found to range from 21.4 (Mo-ei-205) to 77.8% (Mo-ei-5 and Mo-ei-76). Isolates collected from Jharkhand were more virulent than the isolates from other regions. Out of 72 isolates phenotyped, the isolates originating from Jharkhand (Mo-ei-5, Mo-ei-43, and Mo-ei-103) had the highest virulence as they exhibited compatibility with 21 out of 26 resistance genes.

The frequency of virulence on 26 international differentials targeting 26 major genes viz., *Pi9*, *Piz5(Pi-2)*, *Pita-2*, *Pita-2*, *Piz*, *Pi1*, *Pi5*, *Pi7*, *Pii*, *Pi20(t)*, *Pi11(t)*, *Pik-h*, *Pik-m*, *Pik-s*, *Pi12(t)*, *Piz-t*, *Pish*, *Pik*, *Pib*, *Pi3*, *Pit*, *Pi19(t)*, *Pita*, *Pik-p*, *Pita(Pi-4)*, and *Pia* varied from 4.2 to 84.7%. *Pi9*, *Piz5(Pi2)*, *Pita-2*, *Pi1*, and *Piz* genes showed wide resistant spectrum as they exhibited compatibility to few isolates of Eastern India (Fig. 6.1). IRBL9w (*Pi9*) IRBLz5-CA *Piz5(Pi-2)*, IRBLta2-Pi (*Pita-2*), IRBLta2-Re (*Pita-2*), IRBLz-Fu (*Piz*) and IRBL-1CL (*Pi1*) were susceptible to 4.2%, 9.5%, 18.6%, 23%, 31.9% and 32.4% of isolates, respectively; which showed that *Pi9*, *Piz-5(Pi2)*, *Pita-2*, *Piz and Pi1* had broad spectrum of resistance to prevalent pathotypes and may be useful to prevent blast disease in Eastern India. Similar results were observed in Philippines where *Pi9*, *Piz5(Pi2)* and *Pita-2* were found highly effective (Kobayashi et al., 2007). Highest frequency of virulence was recorded on susceptible check LTH with all the isolates recording susceptible reaction. However, no isolate knocked down the resistance of Tetep. Tetep is known to harbor at least four major blast-resistance genes *Pi1*, *Pita*, *Pit* and *Pi54* (Inukai et al., 1995), which contribute to its broad spectrum resistance. Highest frequency of virulence after susceptible check was recorded on IRBLa-A (*Pia*) (84.72%). The virulence frequency on remaining monogenic differentials ranged from 43.7 to 83.3%. *Pita* gene, which was introgressed from two different donors in monogenic lines showed different frequency of virulence. Frequency of virulence on IRBLta-CT2 (*Pita*) was 83.3% and IRBLta-K1 (*Pita*) was 75%. The blast isolate of Eastern India shows higher virulence to *Pia, Pita, Pik-p, Pi19, Pi3, Pit, and Pib*. Compatibility of most of the isolates with one or the other major genes indicated that monogenic resistance may not provide durable resistance to prevalent pathotypes of Eastern India.

Knockdown of resistant gene in rice blast disease is quite common, especially when resistance is governed by a single gene (Hittalmani et al., 2000). Earlier studies indicated that rice varieties with durable resistance to rice blast usually have several major blast-resistance genes (Lei et al., 2013; Wang et al., 1994). IR64, a high yielding variety with durable resistance to blast, released in 1984 was identified to carry six specific blast resistant genes, five of which were identified earlier (Sallaud et al., 2003; Kobayashi et al., 2009; Sallaud et al., 2003). The other such examples are Moroberekan (Wang et al., 1994; Inukai et al., 1996; Naqvi et al., 1996; Chen et al., 1999), Suweon 365 (Ballini et al., 2008; Ahn et al., 2000), Teqing (Tabien et al., 2000), Sanhuangzhan 2 (Liu et al., 2004), Digu (Chen et al., 2006; Chen al et al., 2009; Chen et al., 2011), and Gumei 2 (Wu et al., 2005) each of which possess at least 3–5 blast *R* genes. Gene pyramiding of effective resistant genes or their combination is thus an important approach of developing broad spectrum durable resistant varieties (Ram et al., 2007; Jiang et al., 2012).

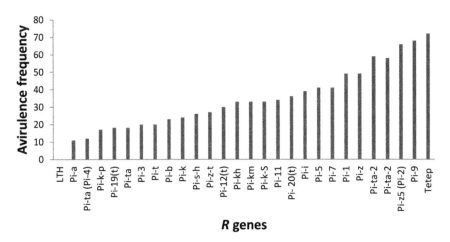

FIGURE 6.1   Virulence frequency of East Indian isolates on monogenic differentials.

Virulence analysis with Eastern Indian isolates revealed that most of the blast resistant genes except few (*Pi9, Piz5(Pi2), Pita-2, Piz, and Pi1*) expressed narrow resistant spectrum to the tested blast isolates. The effective *R* genes represent the potential source for blast-resistance breeding program for this region. Isolates virulent to almost all resistant genes were present in pathogen population, suggesting that most resistance genes were effective against a part of the pathogen population and it would be impossible to obtain desirable levels of resistance by the introgression of single resistant genes. Therefore, pyramiding of two or more resistant gene showing complementary resistant spectra will be needed to develop durable resistance. *Pi9* + *Pita-2* exhibited complementary resistance spectrum and can exclude all virulences of the pathogen population of Eastern India. Rathour et al. (2006) in Northwest Himalayan region of India also reported that combination of gene showing complementary resistance in single genetic background can provide durable resistance (Rathour et al., 2006). Gouda et al. (2012) incorporated *Pi1+Piz5* showing complementary resistance spectra into restorer line PRR78 which showed complete resistance against all the races. In other studies, genes having broad spectrum resistance and combining them with other major genes have been proposed as a means of achieving durable resistance against the pathogen population (Chen et al., 1996; Mekwatana-karn et al., 1999).

6.6   GENETICS OF RESISTANCE

Differences in virulence between isolates were first observed by Sasaki in 1922 and resistance genes of rice to blast disease were described soon after. So far 100 rice blast-resistance (R) genes have been identified (Sharma et al., 2012). Of the 100 blast-resistance genes identified, 45% are from *japonica* cultivars, and 51% from *indica* cultivars and the rest 4% from wild species of rice. These R genes are distributed throughout the 12 rice chromosomes except chromosome 3. Most of the blast-resistance genes and their analogues are located on chromosomes 4, 6, 11, and 12 but their efficiencies against different pathotypes are different, some exhibiting more broad spectrum resistance than others. Interestingly, most of them encode nucleotide binding site-leucine rich repeat (NBS-LRR) proteins except *Pid2*, which is reported to encode a receptor-like kinase protein with a predicted extracellular domain of a bulb-type mannose-specific binding lectin (B-lectin) and an intracellular serine–threonine kinase domain (Chen et al., 2006). The *Pi21* encodes a proline-rich protein that includes a putative heavy metal-binding domain and protein–protein interaction motifs (Fukuoka et al., 2009). Generally, most of these resistance genes are identified in landraces, cultivars, or wild rice collections by use of differential physiological races of *M. oryzae* (Tanksley et al., 1997). The traditional phenotypic screening method for blast resistance is time consuming, laborious, and requires specialized techniques. With the fine mapping and cloning of many blast-resistance genes, PCR-based markers have been developed to screen and identify R genes. DNA markers have a significant advantage of increasing the precision of identification and incorporation of blast-resistance genes (Wang et al., 2007). Among the 100 rice blast-resistance genes that have been identified, 22 have been cloned.

The cloning and phylogenetic analysis of NBS-LRR regions of R genes clearly revealed that *Pi2*, *Piz-t*, and *Pi9*, are present at the same locus, indeed allelic to each other and clustered in the same clade. This is obvious and expected as the three genes are tightly linked to each other (Zhou et al., 2006, 2007). The *Pi2/Pi9* locus, which is proximal to the centromere on chromosome 6, other than tightly linked *Piz-t*, also harbors *Piz*, *Pigm(t)*, *Pi26*, and *Pi40* genes (Liu et al., 2002; Hayashi et al., 2004; Deng et al., 2006; Jeung et al., 2007). These genes have broad spectrum resistance to *M. oryzae* isolates but with different specificities. The cloning and sequence analysis of *Pi2/ Piz-t* revealed that both these genes are true alleles to each other with only eight amino acids that are different and showed similar expression pattern (Zhou et al., 2006). The expression analysis results of the three genes *Pi2, Piz-t*, and *Pi9* revealed that their expression is constitutive and is not induced by blast infection. One striking point is that *Pi2, Piz-t*, and *Pi9* genes were identified from different varieties/species. *Pi2* was introduced from *indica* cultivar 5173, *Piz-t* from *indica* rice cultivar TKM1 and *Pi9* from *Oryza minuta*, a tetraploid wild species of *Oryza* genus (Nagai et al., 1970; Mackill & Bonman, 1992; Amante-Bordeos et al., 1992). The *Pi2/Pi9* locus is more dynamic within and among different wild rice species than the one within the cultivated rice haplotypes (Zhou & Wang, 2009). The clusters of these related NBS-LRR genes at a single locus can result in the generation of orthologues and paralogues with new *R* alleles of more resistance/specificities which may be able to combat the blast disease.

The *Pib* gene is a member of a small gene family, cloned by map-based cloning strategy (Wang et al., 1999). It confers resistance to most of the Japanese blast races and has been mapped to the distal end of the long arm of chromosome 2. The *Pib* gene encodes a polypeptide of 1251 amino acids; it contains an NBS region and C-terminal LRRs, but no distinct transmembrane domain. An NBS domain is a signaling motif shared by plant *R* gene products. Northern blot analysis of the *Pib* gene family members (*Pib, PibH8, HPibH8-1*, and *HPibH8-2*) revealed that their expression was regulated by altered environmental signals such as temperature, light, darkness, water, and chemical treatments, including jasmonic acid, salicylic acid (SA), ethylene, and probenazole.

The *Pik* cluster of *Pik, Pik-p, Pik-h, Pik-s*, and *Pik-m* alleles are tightly linked blast-resistance genes (Xu et al., 2008; Kiyosawa, 1972; McCouch et al., 1994), located in proximity to the telomeric region of the long arm of the chromosome 11. Phylogenetic analysis showed that *Pik-h* gene is far from *Pik-p, Pik-s*, and *Pik-m* alleles, which are clustered in two clades near to each other. In case of *Pik, Pik-p*, and *Pik-m*, two adjacent NBS-LRR class genes are essential for complete resistance to blast. The *Pik-m* locus has two NBS-LRR genes *Pik-m1-TS* and *Pik-m2-TS*. Complementarity tests showed that transgenic plant carrying either of the NBS-LRR genes did not confer *Pik-m* resistance specificity (Constanzo & Yulin, 2010). The expression analysis results revealed that both the genes express constitutively and are only marginally induced by blast infection. Since *Pik-p* is the allele that was found in wild relatives of rice, it was suggested that it is the ancestral allele and the other *Pik* alleles evolved after rice domestication (RoyChowdhury et al., 2012).

Pi5 gene, cloned by map-based cloning strategy, confers resistance to many *M. oryzae* isolates collected from Korea and the Philippines (Wang et al., 1994; Chen et al., 2000). This gene, similar to *Pik-m*, requires two candidate NBS-LRR gene *Pi5-1* and *Pi5-2* for conferring resistance (Lee et al., 2009). The *Pi5-1* and *Pi5-2* proteins contain a unique C terminus that is distinct from those of other R proteins and also these two have more introns (four and five, respectively) than other *R* genes (Chen et al., 2006). These distinct properties of *Pi5-1* and *Pi5-2* genes put them in a single clade different from other NBS-LRR genes. The molecular basis of these two R proteins is still unclear and the expression analysis work only indicates that the *Pi5-1* transcript level is high after pathogen challenge, whereas the *Pi5-2* gene is constitutively expressed (Jeon et al., 2003). Genetic transformation experiments of a susceptible rice cultivar revealed that neither the *Pi5-1* nor the *Pi5-2* gene was found to confer resistance to *M. oryzae*. Both the CC-NBS-LRR genes, *Pi5-1* and *Pi5-2*, are required to confer *Pi5*-mediated resistance to *M. oryzae* (Lee et al., 2009).

The rice blast-resistance gene *Pit* was identified in the Indonesian rice variety Tjahaja, which confers race-specific resistance against the fungal pathogen, *Magnaporthre grisea* (Kiyosawa, 1972). Map-based cloning revealed that *Pit* belongs to the CC-NBS-LRR family of *R* genes. Long terminal repeat (LTR) retrotransposon contributed to the evolution of the rice blast-resistance gene *Pit*. A comparison of the sequences of *Pit* alleles between a resistant cultivar, K59, and a susceptible cultivar, Nipponbare, revealed that four amino acids substitution and the LTR retrotransposon in the promoter region are associated with the resistance phenotype of the *Pit* gene (Hayashi & Yoshida, 2009). It was also clearly indicated that the promoter activity conferred by the renovator retrotransposon fragment, rather than the amino acid substitutions, plays an important role in *Pit* resistance by enhancing its transcription.

The *Pi37* gene, present on chromosome 1 was identified in the rice cultivar St. No. 1. It confers partial and complete resistance to Japanese and Chinese *M. oryzae* isolates (Ezuka et al., 1969a, b; Chen et al., 2005). *Pi37* encodes a 1290 amino acid long NBS-LRR protein product, and the presence of substitution at two sites in the NBS region (V239A and I247M) is associated with the resistance phenotype. *Pi37* gene has four paralogues and is more closely related to the maize rp1 complex than any blast *R* gene. *Pi37* thus appears to be the first representative of a cereal NBS-LRR gene lacking an intron (Lin et al., 2007). Expression analysis showed that *Pi37* is expressed constitutively and only slightly induced by blast infection.

The *Pita* gene is located near the centromere of chromosome 12 and along with *Pita-2* gene at the same locus. Katy was the first US cultivar reported to contain the *Pita* gene that was derived from the landrace *indica* variety Tetep (Moldenhauer et al., 1990). All *Pita-2* containing cultivars also contains *Pita* gene and are mostly inseparable. The reason for this can be the recombination suppression observed near the centromere region (Bryan et al., 2000). *Pi-ta* encodes a predicted 928-amino acid cytoplasmic receptor with a centrally localized nucleotide binding site. *Pita*, *Pi36*, and *Pid2* are single copy *R* genes where resistance specificity is determined by single amino acid. Both *Pita* and *Pi36* are NBS-LRR encoding *R* genes but *Pid2* is the only non-NBS-LRR encoding *R* gene, which encodes a receptor like kinase. The *indica* rice variety Kasalath carries a gene called as *Pi36*, resistant to Chinese *M. oryzae* isolates and is located on chromosome 8. The *Pi36* gene encodes a 1056 amino acid NBS-LRR protein. A single amino acid substitution event (Asp to Ser) at position 590 of *Pi36* distinguishes resistance and susceptibility. *Pi36* is more closely related to the barley powdery mildew resistance genes *Mla1* and *Mla6* and is constitutively expressed as in other blast-resistance genes (Liu et al., 2007).

Pid2, a dominant resistance gene, cloned by a map-based cloning strategy from Chinese *Oryza sativa* subs. *indica* variety Digu, confers gene-for-gene resistance to the Chinese blast strain, ZB15. *Pid2* encodes a receptor-like kinase protein with a predicted extracellular domain of a bulb-type mannose specific binding lectin (B-lectin) and an intracellular serine–threonine kinase domain. *Pid2* encodes a predicted RLK protein of 825 amino acids, consisting of an N-terminal hydrophobic signal peptide, a B-lectin domain, a PAN domain, a TM domain, and a C-terminal intracellular protein kinase domain and is the only non-NBS-LRR encoding *R* gene (Chen et al., 2006). *Pid2* is a single-copy gene that is a constitutively expressed gene. A single amino acid difference at position 441 of *Pid2* distinguishes resistant and susceptible alleles of rice blast-resistance gene *Pid2*. Because of its novel extracellular domain, *Pid2* represents a new class of plant resistance genes (Chen et al., 2006).

Pid3 gene was also cloned from the rice "Digu" (*indica*) by performing a genome-wide comparison of the NBS-LRR gene family between two genome-sequenced varieties, "9311" (*indica*) and "Nipponbare" (*japonica*) (Shang et al., 2009). The rice blast-resistance gene *Pid3* encodes a NBS-LRR protein. The *Pid3* gene in *japonica* was identified as a pseudogene because of a nonsense mutation at the nucleotide position 2208 starting from the translation initiation site. But this mutation was not found in any *indica*, or African cultivated rice varieties, or AA genome containing wild rice species (Shang et al., 2009) indicating that the mutation occurred after the divergence of *indica* and *japonica* rice. Q et al. (2013) performed functional analysis of *Pid3-A4*, an ortholog of *Pid3* revealed by allele mining in the common wild rice A4 (*Oryza rufipogon*). The predicted protein encoded by *Pid3-A4* shares 99.03% sequence identity with *Pid3*, with only nine amino acid substitutions. In wild rice plants, *Pid3-A4* is constitutively expressed, and its expression is not induced by *M. oryzae* isolate Zhong-10-8-14 infection. Therefore, *Pid3-A4* should be quite useful for the breeding of rice blast resistance, especially in Southwestern China. The same variety also has the *Pid2* gene and the technique used for the identification of *Pid3* gene is less labor intensive and precise. The *Pi25* was mapped on chromosome 6 at the same locus of *Pi2/Pi9*, which is the locus of resistance genes. The *Pi25* confers neck blast resistance to Chinese isolate 92-183 (race ZC15). Since the locus has a cluster of genes, it may be possible that two or more genes help in conferring neck blast resistance (Jian et al., 2004). The phylogenetic analysis revealed the similarity between the *Pid3* and *Pi25* genes which occupied the same clade.

The *Pi54* gene is cloned from *indica* rice line, Tetep on chromosome 11. It is slightly distant from the *Pik* locus but its sequence similarity clearly reveals that it is very similar to *Pik-h* gene (Sharma et al., 2005; Sharma et al., 2010). The *Pi54* protein contains a NBS-LRR domain in addition to a small zinc finger domain (Sharma et al., 2005). Functional complementation test confirmed its ability for conferring a stable and high level of resistance against geographically diverse strains of *M. oryzae* from different parts of India (Rai et al., 2011). The *Pi54* gene has been shown to induce the synthesis of callose in response to pathogen infection. The *Pi54* gene is expressed constitutively at a basal level in both resistant as well as susceptible plants up to 48 hpi. Its expression is induced by *M. oryzae* infection, and a more than 2-fold higher induction was observed at 72 hpi in the resistant plant (Rai et al., 2011).

Pi54rh gene is an ortholog of *Pi54* gene cloned by allele mining approach from the wild variety *Oryza rhizomatis* (Das et al., 2012). The *Pi54rh* belongs to NBS-LRR family of disease resistance genes with a unique zinc finger domain. It is an extra cellularly localized protein and can play a role in signal transduction process. This is in contrast to other blast-resistance genes that are predicted to be intracellular NBS-LRR type resistance proteins. It has a pathogen inducible promoter and hence up-regulated after pathogen infection (Das et al., 2012). Structural relatedness of *Pi54rh* is clearly shown in phylogenetic tree. Both *Pita* and *Pi54rh* are put in a same clade and this shows their evolutionary relationship.

Pia and *Pish* are weaker effect *R* genes carried by the native Japanese cultivars. *Pia* consists of a cluster of four NBS-LRR genes, located on the short arm of rice chromosome 11. Out of four, only *Pia-3* is functional and is the strongest candidate gene of *Pia*. The *Pia* shows a narrow spectrum resistance to blast isolates (Kiyosawa, 1974; Hittalmani et al., 2000; Fukuta et al., 2007). On the other hand, *Pish* is a cluster of NBS-LRR genes, arranged as tandem repeats and are highly similar to each other. An *in silico* analysis revealed that the *Pish* locus is a hot spot for *Tos17* insertions. The *Tos17* inserts are most frequent in the functional genes within hot spot regions (Takahashi et al., 2010). Since both *Pia* and *Pish* are weaker genes and if present alone, they are not able to confer resistance. These *R* genes could have important roles if used in combination with other *R* genes, whether weaker or stronger *R* genes (Hittalmani et al., 2000; Liu et al., 2003).

Of the several known blast-resistance genes, *Pi40* has been identified in an *indica* introgression line that has inherited the resistance gene from an EE genome of wild species *O. australiensis* and advanced breeding lines derived from BC progenies were used to validate 9871.T7E and 9871.T7E2b as markers completely associated with the *Pi40(t)* gene (Jeung et al., 2007). Likewise, *Pi9* gene identified in *indica* rice lines has been derived from the BBCC genome of wild species *O. minuta*. Most of the resistance genes are race specific (Mackill & Bonman, 1992; Deng et al., 2006).

6.7   GENE DEPLOYMENT AGAINST PREVALENT PATHOTYPES

Complementary resistance spectra that exclude all the pathotypes of the pathogen are ideally suited for strategic resistant gene deployment. Since *Pi9* and *Pita-2* excluded all the pathotypes collected from a wide geographical region representing the different rice growing states of Eastern India, it was considered that a combination of *Pi-9* and *Pita-2* has potential for effective management of rice blast disease by excluding all virulences in this region. *Pi2* was also as effective as *Pi9* but had slightly variable pattern of virulence. With the aim of introducing blast-resistance genes *Pi2*, *Pi9*, and *Pita-2* a crossing program was initiated at Hazaribag using two popular varieties viz., Vandana and BPT5204 as recurrent parents and the monogenic lines viz., C101A51 (carrying *Pi-2*), IRBL9-W (carrying *Pi9*) and IRBLta2-Re (carrying *Pita-2*) as donor parents. Vandana is an upland rice cultivar of early duration popular in the plateau region of Jharkhand and adjoining table lands of Chhattisgarh and Orissa. It is moderately tolerant to leaf blast but susceptible to neck blast. BPT5204 is a long duration susceptible rice variety popular in several states of South and Eastern India. It was presumed that incorporation of blast resistance in a mega-variety like BPT5204 would increase the life-span of this variety which is preferred by the farmers due to its grain quality and high yields (this program was supported by the NAIP project on allele mining for blast resistance). Molecular markers closely linked to the major genes selected for incorporation were AP5930 for *Pi9* and *Pi2*, and *Pita3* for *Pita-2*. Parental analysis for Vandana, BPT5204, C101A51, and IRBL9-W was performed and the cycle of hybridization and selection with the foreground marker and a set of background markers carried on getting near isogenic lines of the recurrent parent with the target genes (Fig. 6.2). Around 30 HV SSR markers selected from chromosome 6 were used for background selection. Out of which, five HV SSRs were polymorphic between the recurrent parent and donor; and situated on either side of the target gene. Marker for *Pita-2* (*Pita3*) did not distinguish between the parents and hence efforts are on to identify and use other polymorphic markers. Near isogenic lines of Vandana and BPT5204 with *Pi2* and *Pi9* are under advanced field testing for agronomic traits.

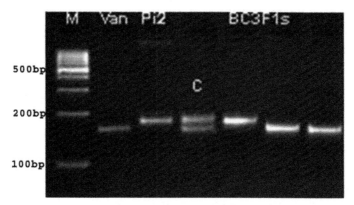

FIGURE 6.2   Gel picture showing confirmation of introgression of *Pi2* gene in the Vandana background at BC$_3$F$_1$ stage by foreground selection marker AP5930.

6.8   SEARCH FOR NOVEL ALLELES

Even though during the course of rice improvement many genes and their alleles from available land races, cultivars, elite rice lines, and wild rice species have been explored, still there is great potential to tap the rice germplasm for

the improvement of important traits of rice. Hence it is imperative to identify broad-spectrum blast-resistance genes for effective protection against
dynamic blast isolates of *M. oryzae*. Highly adaptive virulent isolates/races
of the pathogen often challenge the effectiveness of deployed *R* genes and
thus urge the need for the positive screening and identification of different
blast *R* genes in the germplasm collection (Wang et al., 2010). Identification and isolation of additional host resistance genes (*R*) and pathogen *Avr*
genes is now required to deepen understanding of molecular mechanisms
involved in the host–pathogen interaction (Kiyosawa et al., 1986; Valent,
1990; Inukai et al., 1994) and strategic deployment of resistance genes in
commercial cultivars.

The existence of genetic diversity has special significance in India, a
country characterized by highly varied agro-climates and diverse growing
conditions (Mahender et al., 2012). In the Northeast and Eastern parts of
India is epiphytotic to rice blast causing yield loss ranging from 40 to 46%
(Ngachan et al., 2011). Interestingly, only few of the genes identified/cloned
are effective against the different lineages prevalent in the region. Eastern
India is considered to be a rich pocket of rice genetic resources in the world
owing to the extremely diverse rice growing conditions as compared to other
parts of the country. Selection made unknowingly by various ethnic groups
inhabiting at different altitudes and climatic situations, practicing different
forms of cultivation might have contributed to some extent toward the diversity of rice crop in this region. It is roughly estimated that during the past,
more than 30,000 rice cultivars were grown in Northeast and Eastern parts of
India (Ngachan et al., 2011). Mahender et al. (2012) demonstrated the presence of six to seven genes in rice accessions from the Northeastern state of
India (Manipur), which was related to high level of resistance in the accessions. Molecular genetic markers are now widely used to characterize gene
bank collections that contain untapped resources of distinct alleles which
will remain hidden unless efforts are initiated to screen them for their potential use and function. Abundance of SNP polymorphisms has made it an
attractive tool for allele mining and marker assisted selection. SNPs can
be detected using allele-specific PCR primers and typed by the presence or
absence of PCR amplified products on standard agarose gels.

Presence of different blast-resistance genes in a collection of 32 germplasm from Northeast and Eastern India and another set of 47 germplasm
known to be resistant to blast at Hazaribag were examined by Imam et al.
(2013, 2014) using PCR-based SNP markers. The germplasm accessions
were screened for the presence of eight major blast-resistance (*R*) genes, *Piz*,
Piz-t, *Pik*, *Pik-p*, *Pik-h*, *Pita/Pita-2*, *Pib*, *Pi9* and one susceptible *Pita* gene

using a set of 10 SNPs, and gene-based STS markers. Genetic frequency of the major rice blast-resistance genes, *Piz*, *Piz-t*, *Pik*, *Pik-p*, *Pik-h*, *Pita/Pita-2*, *Pita*, *Pi9*, and *Pib*, ranged from 6 to 97% in the first set of 32 germplasm that exhibited different level of resistance in the uniform blast nursery. When the resistance pattern of the germplasm is examined vis a vis the presence of amplicon products of the major genes it was noted that though resistance is generally proportional to the frequency of the gene(s), certain genes (*Pi9*, *Pita-2*, *Piz-t*) were more effective than others in thwarting infection. They were also less frequently detected by marker amplification. Conversely, the less effective genes were more predominant but mostly ineffective against the different pathotypes/lineages that are normally present in the nursery, albeit with different frequencies. Multi-location evaluation of a set of isogenic lines carrying 24 major blast-resistance genes had earlier indicated that *Pi9*, *Piz-t*, and *Pita-2* were resistant at most locations, indicating their broader spectrum of resistance to pathotypes prevalent in Eastern India (Variar et al., 2009). Among the multi-genes near the waxy gene locus on chromosome 6 (Ballini et al., 2008) *Pi9* was more effective than *Pi2* (*Piz-t*). *Piz*, *Piz-t*, and *Pi40(t)* were commonly detected in germplasm originating from Northeast India earlier but *Pi9* was not detected among them (Mahender et al., 2012). Since the set of rice germplasm used by Imam et al. (2013) covered a wider geographical region (Chhattisgarh, Jharkhand, Sikkim, and Assam), the chances improved for detection of rare alleles like *Pi9* which was identified in two germplasm Kalchati and Bachi thima. While *Pi9* gene conferred complete resistance in Kalchati and Bachi thima, *Piz-t*, along with other genes, contributed to complete resistance in three germplasm (Nepali dhan, Chirakey-C, and Phaudel), and moderate resistance in others. Thirteen accessions (41%) showed positive bands for the *Piz* gene with z56592 SNP marker but its presence independently did not contribute to complete resistance in any of the germplasm evaluated. Both *Piz* and *Piz-t* genes have been used for conferring blast resistance to Japanese cultivars, because their importance was emphasized by Hayashi et al. (2004) in rice breeding in Japan. The *Piz-t* and *Piz* genes, co-segregated with z56592 and zt56591 markers, were flanked on one side by z4794 and on the other side by z60510 and z5765 markers (Hayashi et al., 2006). Two blast-resistance genes, *Pita* and *Pita-2*, have been located at the *Pita* locus near the centromere of chromosome 12 and are tightly linked to each other (Wang et al., 2002). These two genes were interacting in terms of their resistance specificity. *Pita-2* has a broader resistance spectrum than *Pita* (Rybka et al., 1997; Bryan et al., 2000). Dudhraj and Nepali dhan were the only two accessions positive of *Pita/Pita-2* genes using the SNP marker *Pita3* and dominant marker

YL155/YL87. These two accessions showed resistance pattern in UBN. The isogenic lines of *Pita* and *Pita-2* gene, evaluated earlier, were resistant to blast in multi-environment testing and this confirmed that the germplasm having *Pita/Pita-2* gene is likely be effective in rice breeding program. The 32 accessions were also screened for *Pita* susceptible gene with dominant marker YL183/YL87 and found that 29 accessions gave amplicons of 1,043 bp. The *Pita/Pita-2* markers used were based on DNA polymorphism of nucleotides between the resistant *indica Pita* allele and the susceptible *japonica Pita* allele (Jia et al., 2002, 2003, 2004a, 2004b). Identification of *Pita/Pita-2* genes and its validation with differential isolates of *M. oryzae* revealed that the Indian rice germplasm are diverse and potential source of blast resistant lines which can be exploited in rice blast breeding programs (Shikari et al., 2013). Members of the *Pik* multi-gene family and *Pib* were the most frequently detected genes in this study but neither the germplasm possessing them nor the isogenic lines in the previous evaluation (Variar et al., 2009) exhibited resistance.

Genotyping of the germplasm with allele-specific markers helped to identify alleles of nine major blast-resistance genes *Piz*, *Piz-t*, *Pik*, *Pik-p*, *Pik-h*, *Pita/Pita-2*, *Pib*, and *Pi9* and one recessive *Pita* gene in 32 Northeast and Eastern Indian rice germplasm. Another set of 47 rice accessions with consistent resistant reaction to leaf blast in a uniform blast nursery (UBN) at the Central Rainfed Upland Rice Research Station, Hazaribag, India, were subjected to molecular screening for the presence of eight major blast-resistance (*R*) genes, *Piz*, *Piz-t*, *Pik*, *Pik-p*, *Pita*, *Pita-2*, *Pi2*, and *Pib*, using a set of seven SNP markers. Isogenic lines in the background of LTH possessing these genes were used as controls to confirm the results. The genes were selected on the basis of differential reaction of rice accessions to the East Indian isolates of *M. oryzae* and availability of allele-specific markers. An allele-specific PCR marker assays genotype by analyzing the presence or absence of a PCR-amplification product. The UBN analysis revealed the blast disease pattern of these germplasm and showed that germplasm containing *Pita/Pita-2*, *Pi9* are also effective against the prevalent pathotypes of *M. oryzae*. Interestingly, several germplasm had multiple disease resistance genes. Members of the *Pik* multi-gene family and *Pib* were the most frequently detected genes in the rice accessions evaluated in this study. Though the accessions possessing them exhibited resistance in the nursery, the isogenic lines carrying these genes did not exhibit resistance when they were phenotyped in the blast nursery (Variar et al., 2009) or pathotyped against East Indian isolates (Alam et al., 2013). Results of this work showed that all of the 47 Indian rice accessions, which were consistently resistant to blast in the UBN across years, possessed one or more blast-resistance genes. Accessions carrying blast-resistance genes *Piz*, *Piz-t*, *Pi2*, and *Pita-2* can be of immediate use for varietal improvement as these genes were effective against the prevalent pathotypes (Alam et al., 2013). Being adapted to the locations where these germplasm originated, and having co-evolved with the local population of the blast fungus, use of these germplasm may have a competitive edge over other exotic blast-resistance donors like IRBLz-Fu (*Piz*), IRBLzt-T (*Piz-t*), C101A51 (*Pi2*), and IRBLta2-Re (*Pita-2*). Because this core set of accessions is consistently resistant to blast in outdoor blast nurseries and has the added advantage of local adaptation, knowledge of their genetic constitution with respect to the detected *R* genes, and availability of robust markers would be useful for rice breeding programs for resistance to *M. oryzae*.

6.9 DEFENSE RESPONSE GENES

Rice and *M. oryzae* interaction follows Flor's "gene-for-gene" hypothesis. The fungus suppresses plant innate immunity and promotes pathogenesis by injecting effector proteins in the cells (Hogenhout et al., 2009). The fungal effector protein acts in two ways. First, if the fungal effectors are recognized by cognate rice R proteins, effector-triggered immunity (ETI) gets activated and culminates in a HR that halts fungal growth within 48 hours. Second, if the fungal effectors are not recognized, a limited pathogen-associated molecular patterns (PAMP)-triggered immunity (PTI) response sets in upon the recognition of fungal PAMPs (Kaku et al., 2006; Shimizu et al., 2010) and during this intermittent period fungal hyphae continue to spread in plant tissues, resulting in disease symptoms in the initial few hours after infection (Kankanala et al., 2007). Thus, the early defense response immediately after *M. oryzae* invasion is critical to the final outcome of the level of rice blast resistance. Recent findings have shown that plants can rapidly develop effective ETI to inhibit pathogen growth when PTI overcome by newly evolved pathogen effectors (Xiang et al., 2008). In most cases, the defense responses which are accompanied by a HR in the infected cells trigger the rapid production of reactive oxygen species (ROS) (Hammond-Kosack & Jones, 1996). These defense responses are also associated with accumulation of SA, which functions in the induction of numerous pathogenesis-related (PR) genes, and the establishment of systemic acquired resistance (SAR) (Ryals et al., 1996; Dempsey et al., 1999; Durrant & Dong, 2004). It has been shown that SA is required in *R* gene-mediated defense signaling pathways.

Defense response genes are those genes which functions downstream of *R*- or HPRR-initiated defense signaling pathways. Out of 12 characterized genes contributing to QTL resistance, nine of them belong to the defense-responsive gene class. These genes are the components of defense signaling pathways leading to *R* gene-mediated race-specific resistance and/or initial resistance (Kou & Wang, 2010). These defense-responsive genes are either positive or negative regulators in rice–*M. oryzae* interaction.

6.10 FUTURE PERSPECTIVES

The potential for investigating rice–*M. oryzae* interaction for host derived specificity study using various genomics, proteomics and bioinformatics approaches is very clear nowadays, which can be exploited to uncover the mechanism underlying the rice–*M. oryzae* interaction. Harnessing this information effectively and learning lessons in how systematically to analyze genes in a high throughput manner and its mechanism of action will be the key challenges for the future.

KEYWORDS

- *Avr* genes
- gene interaction
- *R* genes
- resistance
- rice
- rice blast pathogen

REFERENCES

Ahn, S. N.; Kim, Y. K.; Hong, H. C.; Han, S. S.; Kwon, S. J.; Choi, H. C.; Moon, H. P.; McCouch, S. Molecular Mapping of a New Gene for Resistance to Rice Blast. *Euphytica.* **2000,** *116,* 17–22.

Alam, S.; Imam, J.; Maiti, D.; Prasad, C.; Sharma, T. R.; Mandal, N. P.; Variar, M. In *Sustainable Rice Production and Livelihood Security: Challenges and Opportunities,* Diversity of *Magnaporthe oryzae* and Resistance Spectrum of Major Genes to Rice Blast in Eastern India, ARRW; CRRI: Cuttack, 2013.

Amante-Bordeos, A.; Sitch, L. A.; Nelson, R.; Damacio, R. D.; Oliva, N. P.; Aswidinnoor, H.; Leung, H. Transfer of Bacterial Blight and Blast Resistance from the Tetraploid Wild Rice *Oryza minuta* to Cultivated Rice *Oryza sativa*. *Theor. Appl. Genet.* **1992**, *84*, 345–354.

Ashikawa, I.; Hayashi, N.; Yamane, H.; Kanamori, H.; Wu, J.; Matsumoto, T.; Ono, K.; Yano, M. Two Adjacent Nucleotide-Binding Site-Leucine-Rich Repeat Class Genes are Required to Confer *Pikm*-Specific Rice Blast Resistance. *Genetics*. **2008**, *180*, 2267–2276.

Ballini, E.; Morel, J. B.; Droc, G.; Price, A.; Courtois, B.; Notteghem, J. L.; Tharreau, D. A Genome-Wide Meta-Analysis of Rice Blast Resistance Genes and Quantitative Trait Loci Provides New Insights into Partial and Complete Resistance. *Mol. Plant Microbe Interact.* **2008**, *21*, 859–868.

Bohnert, H. U.; Fudal, I.; Dioh, W.; Tharreau, D.; Notteghem, J. L.; Lebrun, M. H. A Putative Polyketide Synthase/Peptide Synthetase from *Magnaporthe grisea* Signals Pathogen Attack to Resistant Rice. *Plant Cell*. **2004**, *16*, 2499–2513.

Bryan, G. T.; Wu, K. S.; Farrall, L.; Jia, Y.; Hershey, H. P.; McAdams, S. A.; Faulk, K. N.; Donaldson, G. K.; Tarchini, R.; Valent, B. A Single Amino Acid Difference Distinguishes Resistant and Susceptible Alleles of the Rice Blast Resistance Gene Pi-Ta. *Plant Cell*. **2000**, *12*, 2033–2046.

Chadha, S.; Gopalakrishna, T. Genetic Diversity of Indian Isolates of Rice Blast Pathogen (*Magnaporthe grisea*) Using Molecular Markers. *Curr. Sci.* **2005**, *88*, 1466–1469.

Chen, D. H.; Nelson, R. J.; Wang, G. L.; MacKill D. J.; Ronald, P. C. Use of DNA Markers in Introgression and Isolation of Genes Associated with Durable Resistance to Rice Blast. In *Advances in DNA-Based Markers in Plants*; Vasil, I. K., Phillips, R., Eds.; Kluwer Academic Publishers: Dordrecht, The Netherlands. 2000; pp 17–27.

Chen, D. H.; Dela-Vina, M.; Inukaim T.; Mackillm D. J.; Ronald, P. C.; Nelson, R. J. Molecular Mapping of the Blast-Resistance Gene, *Pi44(t)*, in a Line Derived from a Durably Resistant Rice Cultivar. *Theor. Appl. Genet.* **1999**, *98*, 1046–1053.

Chen, D. H.; Zeigler, R. S.; Ahn, S. W.; Nelson, R. J. Phenotypic Characterization of Rice Blast Resistance Gene *Pi2(t)*. *Plant Dis.* **1996**, *80*, 52–56.

Chen, J.; Shi, Y.; Liu, W.; Chai, R.; Fu, Y.; Zhuang, J.; Wu, J. A *Pid3* Allele from Rice Cultivar Gumei2 Confers Resistance to *Magnaporthe oryzae*. *J. Genet. Genomics*. **2011**, *38*, 209–216.

Chen, S.; Wang, L.; Que, Z.; Pan, R.; Pan, Q. Genetic and Physical Mapping of *Pi37 (t)*, a New Gene Conferring Resistance to Rice Blast in the Famous Cultivar St. No. 1. *Theor. Appl. Genet.* **2005**, *111*, 1563–1570.

Chen, X.; Shang, J.; Lei, C.; Xu, J.; Li, S.; Zhu, L. Genetic and Molecular Analysis of Blast Resistance in a Universal Blast Resistant Variety, Digu. In *Advances in Genetics, Genomics and Control of Rice Blast*; Wang, G. L., Valent, B., Eds.; Springer of Congress + Business Media B.V: USA, 2009; pp 149–159.

Chen, X. W.; Shang, J.; Chen, D.; Lei, C.; Zou, Y.; Zhai, W.; Liu, G.; Xu, J.; Ling, Z.; Cao, G.; Ma, B.; Wang, Y.; Zhao, X.; Li, S.; Zhu, L. A B-Lectin Receptor Kinase Gene Conferring Rice Blast Resistance. *Plant J.* **2006**, *46*, 794–804.

Constanzo, S.; Yulin, J. Sequence Variation at the Rice Blast Resistance Gene *Pi-Km* Locus: Implications for the Development of Allele Specific Markers. *Plant Sci.* **2010**, *178*, 523–530.

Couch, B. C.; Kohn, L. M. A Multilocus Genealogy Concordant with Host Preference Indicates Segregation of a New Species *Magnaporthe oryzae* from *M. grisea*. *Mycologia*. **2002**, *94*, 683–693.

Das, A.; Soubam, D.; Singh, P. K.; Thakur, S.; Singh, N. K.; Sharma, T, R. A Novel Blast Resistance Gene, *Pi54rh* Cloned from Wild Species of Rice, *Oryza rhizomatis* Confers Broad Spectrum Resistance to *Magnaporthe oryzae. Funct. Integr. Genomics.* **2012,** *12,* 215–228.

Dempsey, D.; Shah, J.; Klessig, D, F. Salicylic Acid and Disease Resistance in Plants. *Crit. Rev. Plant Sci.* **1999,** *18,* 547–575.

Deng, Y.; Zhu, X.; Shen, Y.; He, Z. Genetic Characterization and Fine Mapping of the Blast Resistance Locus *Pigm(T)* Tightly Linked to *Pi2* and *Pi9* in a Broad-Spectrum Resistant Chinese Variety. *Theor. Appl. Genetics.* **2006,** *113,* 705–713.

Durrant, W. E.; Dong X. Systemic Acquired Resistance. *Annu. Rev. Phytopathol.* **2004,** *42,* 185–209.

Ezuka, A.; Yunoki, T.; Sakurai, Y.; Shinoda, H.; Toriyama, K. Studies on the Varietal Resistance of Rice to Blast. I. Tests for Genotype of "True Resistance" (In Japanese with English Summary), *Bulletin Chugoku National Agriculture Experiment Stn.* E4, 1969a, pp 1–31.

Farman, M. L.; Leong, S. A. Chromosome Walking to the *AVR1-CO39* Avirulence Gene of *Magnaporthe grisea*: Discrepancy between the Physical and Genetic Maps. *Genetics.* **1998,** *50,* 1049–58.

Feng, S.; Wang, L.; Ma, J.; Lin, F.; Pan, Q. Genetic and Physical Mapping of *Avr Pi7*, a Novel Avirulence Gene of *Magnaporthe oryzae* Using Physical Position Ready Markers. *Chin. Sci. Bull.* **2007,** *52,* 903–911.

Flor, H. H. Current Status of the Gene-For-Gene Concept. *Annu. Rev. Phytopathol.* **1971,** *9,* 275–296.

Fukuoka, S.; Saka, N.; Koga, H.; Ono, K.; Shimizu, T.; Ebana, K.; Hayashi, N.; Takahashi, A.; Hirochika, H.; Okuno, K.; Yano, M. Loss of Function of a Proline-Containing Protein Confers Durable Disease Resistance in Rice. *Science.* **2009,** *325,* 998–1001.

Fukuta, Y.; Kobayashi, N.; Noda, T.; Hayashi, N.; Vera Cruz. C. M. In *a Research Network for Blast Disease to Build a Stable Rice Production System*, Proceedings of 28 Inter. Rice Research Conference, Pest, Disease, and Weed Management, Vietnam, Nov 8–12, 2010.

Fukuta, Y.; Ebron, L. A.; Kobayashi, N. Genetic and Breeding Analysis of Blast Resistance in Elite *Indica*-Type Rice (*Oryza sativa* L.) Bred in International Rice Research Institute. *Jap. Agr. Res.* **2007,** *41,* 101–114.

George, M. L. C.; Nelson, R. J.; Zeigler, R. S.; Leung, H. Rapid Population Analysis Of *Magnaporthe grisea* by Using Rep-PCR and Endogeneous Repetitive DNA Sequences. *Am. Phytopathol. Soc.* **1998,** *88,* 223–229.

Gouda, P. K.; Saikumar, S.; Varma, C. M. K.; Nagesh, K.; Thippeswamy, S.; Shenoy, V.; Ramesha, M. S.; Shashidhar, H. E. Marker-Assisted Breeding of *Pi-1* and *Piz-5* Genes Imparting Resistance to Rice Restorer Line PRR78, a Restorer Line for Pusa RH10 a Hybrid in Basmati. *Plant Breed.* **2012,** *132,* 61–69.

Hammond-Kosack, K. E.; Jones, J. D. Resistance Gene-Dependent Plant Defense Responses. *Plant Cell.* **1996,** *8,* 1773–1791.

Hayashi, K.; Yoshida, H.; Ashikawa, I. Development of PCR-Based Allele Specific and Indel Marker Sets for Nine Rice Blast Resistance Genes. *Theor. Appl. Genet.* **2006,** *113,* 251–260.

Hayashi, K.; *Yoshida, H.* Refunctionalization of the Ancient Rice Blast Disease Resistance Gene *Pit* by the Recruitment of a Retrotransposon as a *Promoter. Plant J.* **2009,** *57,* 413–425.

Hayashi, K.; Hashimoto, N.; Daigen, M.; Ashikawa, I. Development of PCR-Based SNP Markers for Rice Blast Resistance Genes at the *Piz* Locus. *Theor. Appl. Genet.* **2004,** *108,* 212–220.

Hayashi, N.; Inoue, H.; Kato, T.; Funao, T.; Shirota, M.; Shimizu, T.; Kanamori, H.; Yamane, H.; Hayano-Saito, Y.; Matsumoto, T. Durable Panicle Blast-Resistance Gene Pb1 Encodes an Atypical CC-NBS- LRR Protein and Was Generated by Acquiring a Promoter through Local Genome Duplication. *Plant J.* **2010,** *64,* 498–510.

Hittalmani, S.; Parco, A.; Mew, T. V.; Zeigler, R. S. Fine Mapping and DNA Marker-Assisted Pyramiding of the Three Major Genes for Blast Resistance in Rice. *Theor. Appl. Genet.* **2000,** *100,* 1121–1128.

Hogenhout S. A.; Vander Hoorn, R. A.; Terauchi, R.; Kamoun, S. Emerging Concepts in Effector Biology of Plant-Associated Organisms. *Mol. Plant Microbe Interact.* **2009,** *22,* 115–122.

Howard, R. J.; Valent, B. Breaking and Entering: Host Penetration by the Fungal Rice Blast Pathogen *Magnaporthe grisea. Annu. Rev. Microbiol.* **1996,** *50,* 491–512.

Huang, J.; Si, W.; Deng, Q.; Li, P.; Yang, S. Rapid Evolution of Avirulence Genes in Rice Blast Fungus *Magnaporthe oryzae. BMC Genet.* **2014,** *15,* 45–55.

Imam, J.; Alam, S.; Variar, M.; Shukla, P. Identification of Rice Blast Resistance Gene *Pi9* from Indian Rice Land Races with STS Marker and its Verification by Virulence Analysis. *PNAS. India B.* **2013,** *83,* 499–504.

Imam, J.; Mahto, D.; Mandal, N. P.; Maiti, D.; Shukla, P.; Variar, M. Molecular Analysis of Indian Rice Germplasm Accessions with Resistance to Blast Pathogen. *J. Crop Improv.* **2014,** *28,* 729–739.

Inukai, T.; Nelson, R. J.; Zeigler, R. S.; Sarkarung, S.; Mackill, D. J.; Bonman, J. M.; Takamure, I.; Kinoshita, T. Allelism of Blast Resistance Genes in Near-Isogenic Lines of Rice. *Phytopathology.* **1994,** *84,* 1278–1283.

Inukai, T.; Viet, D. L.; Imbe, T.; Ziegler, R. S.; Kinoshita, T.; Nelson, R. J. Identification of Four Resistant Gene in Vietnamese *Indica* Cultivar Tetep. *Rice Genet. Newsl.* **1995,** *12,* 237–238.

Inukai, T.; Zeigler, R. S.; Sarkarung, S.; Bronson, M.; Dung, L. V.; Kinoshita, T.; Nelson, R. J. Development of Pre-Isogenic Lines for Rice Blast-Resistance by Marker-Aided Selection from a Recombinant Inbred Population. *Theor. Appl. Genet.* **1996,** *93,* 560–567.

Jeon, J. S.; Chen, D.; Yi, G. H.; Wang, G. L.; Ronald, P. C. Genetic and Physical Mapping of *Pi5(t),* a Locus Associated with Broad Spectrum Resistance to Rice Blast. *Mol. Genet. Genomics.* **2003,** *269,* 280–289.

Jeung, J. U.; Kim, B. R.; Cho, Y. C.; Han, S. S.; Moon, H. P.; Lee, Y. T.; Jena, K. K. A Novel Gene, *Pi40(t),* Linked to the DNA Markers Derived from NBS-LRR Motifs Confers Broad Spectrum of Blast Resistance in Rice. *Theor. Appl. Genet.* **2007,** *115,* 1163–1177.

Jia, Y.; McAdams, S. A.; Bryan, G. T.; Hershey, H. P.; Valent, B. Direct Interaction of Resistance Gene and Avirulence Gene Products Confers Rice Blast Resistance. *EMBO J.* **2000,** *19,* 4004–4014.

Jia, Y.; Bryan, G. T.; Farrall, L.; Valent, B. Natural Variation at the *Pi-ta* Rice Blast Resistance Locus. *Phytopathology.* **2003,** *93,* 1452–1459.

Jia, Y.; Redus, M.; Wang, Z.; Rutger, J. N. Development of a SNLP Marker from the *Pi-ta* Blast Resistance Gene by Triprimer PCR. *Euphytica.* **2004a,** *138,* 97–105.

Jia, Y.; Wang, Z.; Singh, P. Development of Dominant Rice Blast *Pi-ta* Resistance Gene Markers. *Crop Sci.* **2002,** *42,* 2145–2149.

Jia, Y.; Wang, Z.; Fjellstrom, R. G.; Moldenhauer, K. A. K.; Azam, M. A.; Correll, J.; Lee, F. N.; Xia, Y.; Rutger, J. N. Rice *Pi-ta* Gene Confers Resistance to the Major Pathotypes of the Rice Blast Fungus in the US. *Phytopathology.* **2004b,** *94,* 296–301.

Jiang, H.; Feng, Y.; Bao, L.; Gao. G.; Ziang, Q.; Xiao, J.; Xu, C. Improving Blast Resistance of Jin 23B and its Hybrid Rice by Marker-Assisted Gene Pyramiding. *Mol. Breed.* **2012,** *30,* 1679–1688.

Jian-li, W.; Rong-yao, C.; Ye-yang, F.; De-Bao, L.; Kang-le, Z.; Leung, H.; Jie-yun, Z. Clustering of Major Genes Conferring Blast Resistance in a Durable Resistance Rice Cultivar Gumei 2. *Rice Sci.* **2004,** *11,* 161–164.

Kaku, H.; Nishizawa, Y.; Ishll-Minami, N.; Akimoto-Tomlyama, C.; Dohmae, N.; Taklo, K.; Minami, E.; Shlbuya, N. Plant Cells Recognize Chitin Fragments for Defense Signaling through a Plasma Membrane Receptor. *PNAS. U.S.A,* **2006,** *103,* 11086–11091.

Kang, S.; Sweigard, J. A.; Valent, B. The PWL Host Specificity Gene Family in the Rice Blast Fungus *Magnaporthe grisea. Mol. Plant Microbe Interact.* **1995,** *8,* 939–948.

Kankanala, P.; Czymmek, K.; Valent, B. Roles for Rice Membrane Dynamics and Plasmodesmata during Biotrophic Invasion by the Blast Fungus. *Plant Cell.* **2007,** *19,* 706–724.

Khang, C. H.; Berruyer R.; Giraldo M. C.; Kankanala P.; Park, S. Y. Translocation of *Magnaporthe oryzae* Effectors into Rice Cells and their Subsequent Cell-to-Cell Movement. *Plant Cell.* **2010,** *22,* 1388–1403.

Khang, C. H.; Park, S. Y.; Lee, Y. H.; Valent, B.; Kang, S. Genome Organization And Evolution of the *AVR-Pita* Avirulence Gene Family in the *Magnaporthe grisea* Species Complex. *Mol. Plant Microbe Interact.* **2008,** *21,* 658–670.

Kiyosawa, S. The Inheritance of Blast Resistance Transferred from Some *Indica* Varieties of Rice. *Bull. Nat. Inst. Agric. Sci.* **1972,** *23,* 69–95.

Kiyosawa, S. Studies on Genetics and Breeding of Blast Resistance in Rice. *Bull. Nat. Inst. Agric. Sci.* **1974,** D1, 1–58.

Kiyosawa, S.; Mackill, D. J.; Bonman, J. M.; Tanaka, Y.; Ling, Z. Z. An Attempt of Classification of World's Rice Varieties Based on Reaction Pattern to Blast Fungus Strains. *Bull. Nat. Inst. Agrobiol. Res.* **1986,** *2,* 13–39.

Kobayashi, N.; Ebron, L. A.; Fujita, D.; Fukuta, Y. Identification of Blast Resistance Genes in IRRI-Bred Rice Varieties by Segregation Analysis Based on a Differential System. *JIRCAS Work Report.* **2009,** *63,* 69–86.

Kobayashi, N.; Telebanco Yanoria, M. J.; Tsunematsu, H.; Kato, H.; Imbe, T.; Fukuta, Y. Development of New Sets of International Standard Differential Varieties for Blast Resistance in Rice (*Oryza sativa* L.). *JIRCAS Work Report.* **2007,** *41,* 31–37.

Kou, Y.; Wang, S. Broad-Spectrum and Durability: Understanding of Quantitative Disease Resistance. *Curr. Opin. Plant Biol.* **2010,** *13,* 181–185.

Lee, S. K.; Song, M. Y.; Seo, Y. S. Rice *Pi5*-Mediated Resistance to *Magnaporthe oryzae* Requires the Presence of two Coiled-Coil-Nucleotide-Binding-Leucine-Rich Repeat Genes. *Genetics.* **2009,** *181,* 1627–1638.

Lei, C.; Hao, K.; Yang, Y.; Ma, J.; Wang, S.; Wang, J.; Cheng, Z.; Zhao, S.; Zhang, X.; Guo, X.; Wang, C.; Wan, J. Identification and Fine Mapping of Two Blast Resistance Genes in Rice Cultivar. *Crop J.* **2013,** *1,* 2–14.

Leong S. A.; Farman M.; Smith J.; Budde A.; Tosa Y.; Nitta, N. Molecular Genetic Approach to the Study of Cultivar Specificity in the Rice Blast Fungus. In *Rice Blast Disease;* Zeigler, R. S., Leong, S. A., Teng, P. S., Eds.; CAB International: Wallingford, 1994; pp 51–64.

Levy, M.; Romao, J.; Marcheitti, M. A.; Hamer, J. E. DNA Fingerprinting with Dispersed Sequence Resolve Pathotype Diversity in the Rice Blast Fungus. *Plant Cell.* **1991,** *3,* 95–102.

Li, W.; Wang, B.; Wu, J.; Lu, G.; Hu, Y.; Zhang, X.; Zhang, Z.; Zhao, Q.; Feng, Q.; Zhang, H.; Wang, Z.; Wang, G.; Han, B.; Wang, Z.; Zhou, B. The *Magnaporthe oryzae* Avirulence

Gene *Avr-Piz-t* Encodes a Predicted Secreted Protein that Triggers the Immunity in Rice Mediated by the Blast Resistance Gene Piz-T. *Mol. Plant Microbe Interact.* **2009**, *22*, 411–420.

Lin, F.; Chen, S.; Que Z.; Wang L.; Liu, X. The Blast Resistance Gene *Pi37* Encodes an NBS-LRR Protein and is a Member of a Resistance Gene Cluster on Rice Chromosome 1. *Genetics.* **2007**, *177*, 1871–1880.

Liu, J.; Liu, X.; Dai, L.; Wang, G. L. Recent Progress in Elucidating the Structure, Function and Evolution of Disease Resistance Genes in Plants. *J. Genet. Genomics.* **2007**, *34*, 765–776.

Liu, S. P.; Li, X.; Wang, C. Y.; Li, X. H.; He, Y. Q. Improvement of Resistance to Rice Blast in Zhenshan 97 by Molecular Marker-Aided Selection. *Acta botanica. Sin.* **2003**, *45*, 1346–1350.

Liu, B.; Zhang, S.; Zhu, X.; Yang, Q.; Wu, S.; Mei, M.; Mouleon, R.; Leach, J.; Mew, M.; Leung, H. Candidate Defense Genes as Predictors of Quantitative Blast Resistance in Rice. *Mole. Plant Microbe Interact.* **2004**, *17*, 1146–1152.

Liu, G.; Lu, G.; Zeng, L.; Wang, G. L. Two Broad-Spectrum Blast Resistance Genes, *Pi9(t)* and *Pi2(T)* are Physically Linked on Rice Chromosome 6. *Mol. Genet. Genomics.* **2002**, *267*, 472–480.

Liu, X.; Lin, F.; Wang, L.; Pan, Q. The in Silico Map-Based Cloning of *Pi36*, a Rice Coiled-Coil Nucleotide-Binding Site Leucine-Rich Repeat Gene that Confers Race-Specific Resistance to the Blast Fungus. *Genetics.* **2007**, *176*, 2541–2549.

Ma, Jun-Hong.; Wang L.; Feng, S.; Lin, F.; Xiao, Y.; Pan, Q. Identification and Fine Mapping of *Avrpi15*, a Novel Avirulence Gene of *Magnaporthe grisea. Theor. Appl. Genet.* **2006**, *113*, 875–883.

Mackill, D. J.; Bonman, J. M. Inheritance of Blast Resistance in Near-Isogenic Lines of Rice. *Phytopathology.* **1992**, *82*, 746–749.

Mahender, A.; Swain, D. M.; Gitishree, D.; Subudhi, H. N.; Rao, G. J. N. Molecular Analysis of Native Manipur Rice Accessions for Resistance against Blast. *Afr. J. Biotechnol.* **2012**, *11*, 1321–1329.

McCouch, S. R.; Nelson, R. J.; Tohme, J.; Zeigler, R. S. Mapping of Blast Resistance Genes in Rice. In *Rice Blast Disease*; Zeigler, R. S., Leong, S. A., Teng, P. S., Eds.; CAB International: Wallingford, 1994; pp 167–186.

Mekwatanakarn, P.; Kositratana, W.; Phromraksa, T.; Zeigler, R. S. Sexually Fertile *Magnaporthe grisea* Rice Pathogens in Thailand. *Plant Dis.* **1999**, *83*, 939–943.

Moldenhauer, K. A. K.; Lee, F. N.; Norman, R. J.; Helms, R. S.; Well, R. H.; Dilday, R. H.; Rohman, P. C.; Marchetti, M. A. Registration of 'Katy' Rice. *Crop Sci.* **1990**, *30*, 747–748.

Nagai, K.; Fujimake, H.; Yokoo, M. Breeding of a Rice Variety Tordie 1 with Multi-Racial Resistance to Leaf Blast. *Jpn. J. Breed.* **1970**, *20*, 7–1 4.

Naqvi, N. I.; Chattoo, B. B. Molecular Genetic Analysis and Sequence Characterized Amplified Region-Assisted Selection of Blast Resistance in Rice. *Rice Genet.* **1996**, *3*, 570–576.

Ngachan, S. V.; Mohanty, A. K.; Pattanayak, A. Status Paper on Rice in North East India– Rice in North East India. Rice Knowledge Management Portal, 2011, 82. http://www.rkmp.

Okuyama, Y.; Kanzaki, H.; Abe, A. A Multi-Faceted Genomics Approach Allows the Isolation of Rice Pia Blast Resistance Gene Consisting of Two Adjacent NBS-LRR Protein Genes. *Plant J.* **2011**, *66,* 467–479.

Orbach, M. J.; Farral, L.; Sweigard, J. A.; Chumle, F. G.; Valent, B. A Telomeric Avirulence Gene Determines Efficacy for the Rice Blast Resistance Gene *Pi-ta. Plant Cell.* **2000**, *12*, 2019–2032.

Park, C.; Chen, S.; Shirsekar, G.; Zhou, B.; Khang, C. The *Magnaporthe oryzae* Effector *AVR-Piz-T* Targets the Ring E3 Ubiquitin Ligase APIP6 to Suppress Pathogen-Associated Molecular Pattern-Triggered Immunity in Rice. *Plant Cell.* **2012**, *24*, 4748–4762.

Peyyala, R.; Farman, M. L. *Magnaporthe oryzae* Isolates Causing Gray Leaf Spot of Perennial Ryegrass Possess a Functional Copy of the *AVR-Co39* Avirulence Gene. *Mol. Plant Pathol.* **2006**, *7*, 157–165.

Q, L.; Xu, X.; Shang, J.; Jiang, G.; Pang, Z.; Zhou, Z.; Wang, J.; Liu, Y.; Li, T.; Li, X.; Xu, J.; Cheng, Z.; Zhao, X.; Li, S.; Zhu, L. Functional Analysis of *Pid3-A4*, an Ortholog of Rice Blast Resistance Gene *Pid3* Revealed by Allele Mining in Common Wild Rice. *Phytopahology.* **2013**, *103*, 594–599.

Qu, S.; Liu, G.; Zhou, B.; Bellizzi, M.; Zeng, L.; Dai, L.; Han, B.; Wang, G. L. The Broad Spectrum Blast Resistance Gene *Pi9* Encodes a Nucleotide-Binding Site Leucine-Rich Repeat Protein and is a Member of a Multigene Family in Rice. *Genetics.* **2006**, *172*, 1901–1914.

Rai, A. K.; Kumar, S. P.; Gupta, S. K.; Gautam, N.; Singh, N. K.; Sharma, T. R. Functional Complementation of Rice Blast Resistance Gene *Pi-Kh* (*Pi54*) Conferring Resistance to Diverse Strains of *Magnaporthe oryzae. J. Plant Biochem. Biotechnol.* **2011**, *20*, 55–65.

Ram, T.; Majumder, N. D.; Mishra, B.; Ansari, M. M.; Padmavathi, G. Introgression of Broad-Spectrum Blast Resistance Gene(S) into Cultivated Rice (*Oryza sativa* sp *Indica*) From Wild Rice *O. rufipogon. Curr. Sci.* **2007**, *92*, 225–230.

Rathour, R.; Singh, B. M.; Plaha, P. Virulence Structure of the *Magnaporthe grisea* Rice Population from the Northwestern Himalayas. *Phytoparasitica.* **2006**, *34*, 281–291.

Ray, S.; Gupta, D. K.; Mehto, A. K.; Singh, N. K.; Sharma, T. R. *Identification and Cloning of Avr: Pikh (Pi54) Gene from Magnaporthe oryzae.* Abstract, 3838, International Rice Congress, Hanoi: Vietnam, 2011.

Ross, W. J.; Correll, J. C.; Cartwright, R. D. The Use of Rapds and MGR 586 DNA Fingerprinting to Characterize Rice Blast Isolates of *Pyricularia Grisea* from a Single Location in Arkansas. *Phytopathology.* **1995**, *85*, 1199.

RoyChowdhury, M.; Jia, Y.; Jackson, A.; Jia, M. H.; Fjellstrom, R.; Cartwright, R. D. Analysis of Rice Blast Resistance Gene *Pi-Z* in Rice Germplasm Using Pathogenictiy Assays and DNA Markers. *Euphytica.* **2012**, *184*, 35–46.

Ryals, J. A.; Neuenschwander, U. H.; Willits, M. G.; Molina, A.; Steiner H. Y.; Hunt, M. D. Systemic Acquired Resistance. *Plant Cell.* **1996**, *8*, 1809–1819.

Rybka, K.; Miyamoto, M.; Ando, I.; Saito, A.; Kawasaki, S. High Resolution Mapping of the *Indica*-Derived Rice Blast Resistance Genes II, *Pi-ta2* and *Pi-ta* and a Consideration of their Origin. *Mol. Plant Microbe Interact.* **1997**, *10*, 517–524.

Saleh, D.; Milazzo, J.; Adreit, H.; Fournier, E.; Tharreau, D. South East Asia is the Center of Origin, Diversity and Dispersion of the Rice Blast Fungus, *Magnaporthe oryzae. New Phytol.* **2014**, *201*, 1440–1456.

Sallaud, C.; Lorieux, M.; Roumen, E.; Tharreau, D.; Berruyer, R.; Svestasrani, P.; Garsmeur, O.; Ghesquiere, A.; Notteghem J. L. Identification of Five New Blast Resistance Genes in the Highly Blast-Resistant Rice Variety IR64 Using a QTL Mapping Strategy. *Theor. Appl. Genet.* **2003**, *106*, 794–803.

Shang, J.; Tao, Y.; Chen, X.; Zou, Y.; Lei, C.; Wang, J.; Li, X.; Zhao, X.; Zhang, M.; Lu, Z. Identification of a New Rice Blast Resistance Gene, *Pid3*, by Genome-Wide Comparison of Paired NBS-LRR Genes and their Pseudogene Alleles between the Two Sequenced Rice Genomes. *Genetics.* **2009,** *182,* 1303–1311.

Sharma, T. R.; Rai, A. K.; Gupta S. K.; Vijayan, J.; Devanna, B. N.; Ray, S. Rice Blast Management through Host–Plant Resistance: Retrospect and Prospects. *Agr. Res.* **2012,** *1,* 37–52.

Sharma, T. R.; Madhav, M. S.; Singh, B. K.; Shanker, P.; Jana, T. K.; Dalal, V.; Pandit, A.; Singh, A.; Gaikwad, K.; Upreti, H. C.; Singh, N. K. High-Resolution Mapping, Cloning and Molecular Characterization of the *Pi-Kh* Gene of Rice, Which Confers Resistance to *Magnaporthe grisea. Mol. Genet. Genomics.* **2005,** *274,* 569–578.

Sharma, T. R.; Rai, A. K.; Gupta, S. K.; Singh, N. K. Broad Spectrum Blast Resistance Gene *Pikh* Designated as *Pi54. J. Plant Biochem. Biotechnol.* **2010,** *19,* 987–989.

Shikari, A. B.; Khanna, A.; Krishnan, S. G.; Singh, U. D.; Rathour, R.; Tonapi, V.; Sharma, T. R.; Nagarajan, M.; Prabhu, K. V.; Singh, A. K. Molecular Analysis and Phenotypic Validation of Blast Resistance Genes *Pita* and *Pita2* in Landraces of Rice (*Oryza sativa* L.). *Indian J. Genet.* **2013,** *73,* 131–141.

Shimizu, T.; Nakano, T.; Takamizawa, D.; Desaki, Y.; Ishii-Minami, N.; Nishizawa, Y.; Minami, E.; Okada, K.; Yamane, H.; Kaku, H. Two LysM Receptor Molecules, Cebip and Oscerk1, Cooperatively Regulate Chitin Elicitor Signaling in Rice. *Plant J.* **2010,** *64,* 204–214.

Sweigard, J. A.; Carroll, A. M.; Kang, S.; Farrall, L.; Chumley, F. G.; Valent, B. Identification, Cloning and Characterization of *PWL2,* a Gene for Host Species Specificity in the Rice Blast Fungus. *Plant Cell.* **1995,** *7,* 1221–1233.

Tabien, R. E.; Li, Z.; Paterson, A. H.; Marchetti, M. A.; Stansel, J. W.; Pinson, S. R. M.; Park, W. D. Mapping of Four Major Rice Blast Resistance Genes from 'Lemont' and 'Teqing' and Evaluation of their Combinatorial Effect for Field Resistance. *Theor. Appl. Genet.* **2000,** *101,* 1215–1225.

Takahashi, A.; Hayashi, N.; Miyao, A.; Hirochika, H. Unique Features of the Rice Blast Resistance *Pish* Locus Revealed by Large Scale Retrotransposon-Tagging. *BMC Plant Biol.* **2010,** *10,* 175.

Tang, J.; Zhu, X.; Wang, Y.; Liu, L.; Xu, B.; Li, F.; Fang, J.; Chu, C. Semi-Dominant Mutations in the CC-NB- LRR-Type R Gene, NLS1, Lead to Constitutive Activation of Defense Responses in Rice. *Plant J.* **2011,** *66,* 996–1007.

Tanksley, S. D.; McCouch, S. R. Seeds Banks and Molecular Maps: Unlocking Genetic Potential from the Wild. *Science.* **1997,** *277,* 1063–1066.

Tharreau, D.; Fudal, I.; Andriantsilialona, D.; Santoso-Utami, D.; Fournier, E.; Lebrun, M. H.; Notteghem, J. L. World Population Structure and Migration of the Rice Blast Fungus, *Magnaporthe oryzae.* In *Advances in Genetics, Genomics and Control of Rice Blast Disease*; Wang, G., Valent, B., Eds.; Springer: New York, 2009; pp 209–215.

Tsunematsu, H.; Yanoria, M. J. T.; Ebron, L. A.; Hayashi, N.; Ando, I.; Kato, H.; Imbe, T.; Khush, G. S. Development of Monogenic Lines of Rice for Rice Blast Resistance. *Breed. Sci.* **2000,** *50,* 229–234.

Valent, B. Rice Blast as a Model System for Plant Pathology. *Phytopathology.* **1990,** *80,* 33–36.

Valent, B.; Khang, C. H. Recent Advances in Rice Blast Effector Research. *Curr. Opin. Plant Biol.* **2010,** *13,* 434–441.

Variar, M.; Vera Cruz, C. M.; Carrillo, M. G.; Bhatt, J. C.; Sangar, R. B. S. Rice Blast In India and Strategies to Develop Durably Resistant Cultivars. In *Advances in Genetics, Genomics and Control of Rice Blast Disease*; Liang Wang, G., Valent, B., Eds.; Springer Publication: New York, 2009; pp 359–374.

Veneault-Fourrey, C.; Barooah, M.; Egan, M.; Wakely, G.; Talbot, N. J. Cell Cycle-Regulated Autophagic Cell Death is Necessary for Plant Infection by the Rice Blast Fungus. *Science.* **2006**, *312*, 580–583.

Wang, C.; Hirano, K.; Kawasaki, S. Cloning of *Pita-2* in the Centromeric Region of Chr 12 with HEGS: High Efficiency Genome Scanning. Third International Rice Blast Conference, 25, 2002.

Wang, G. L.; Mackill, D. J.; Bonman, J. M.; McCouch, S. R.; Champoux, M. C.; Nelson, R. J. RFLP Mapping of Genes Conferring Complete and Partial Resistance to Blast in a Durably Resistance Rice Cultivar. *Genetics.* **1994**, *136*, 1421–1434.

Wang, X.; Fjellstrom, R. G.; Jia, Y.; Yan, W.; Jia, M. H.; Scheffer, B. E.; Wu, D.; Shu, Q.; McClung, A. M. Characterization of Pita Blast Resistance Gene in an International Rice Core Collection. *Plant Breed.* **2010**, *129*, 491–501.

Wang, Z.; Jia, Y.; Rutger, J. N.; Xia, Y. Rapid Survey for Presence of a Blast Resistance Gene *Pi-Ta* in Rice Cultivars Using the Dominant DNA Markers Derived from Portions of the *Pi-Ta* Gene. *Plant Breed.* **2007**, *126*, 36–42.

Wang, Z. X.; Yano, M.; Yamanouchi, U.; Iwamoto, M.; Monna, L.; Hayasaka, H.; Katayose, Y.; Sasaki, T. The *Pib* Gene for Rice Blast Resistance Belongs to the Nucleotide-Binding and Leucine-Rich Repeat Class of Plant Disease Resistance Genes. *Plant J.* **1999**, *19*, 55–64.

Widawsky, D. A.; O'Toole, J. C. *Prioritizing the Rice Biotechnology Research Agenda for Eastern India*; Rockefeller Foundation: New York, 1990; pp 86.

Wilson, R. A.; Talbot, N. J. Under pressure: Investigating the Biology of Plant Infection by *Magnaporthe oryzae*. *Nature Rev. Microbiol.* **2009**, *7*, 185–195.

Wu, J. L.; Fan, Y. Y.; Li, D. B.; Zheng, K. L.; Leung, H.; Zhuang, J. Y. Genetic Control of Rice Blast Resistance in the Durably Resistant Cultivar Gumei 2 against Multiple Isolates. *Theor. Appl. Genet.* **2005**, *111*, 50–56.

Xiang, T.; Zhong, N.; Zou, Y.; Wu, Y.; Zhang, J.; Xing, W.; Li, Y.; Tang, X.; Zhu, L.; Chai, J.; Zhou, J. M. *Pseudomonas syringae* Effector AvrPto Blocks Innate Immunity by Targeting Receptor Kinases. *Curr. Biol.* **2008**, *18*, 74–80.

Xu, X.; Hayashi, N.; Wang, C.; Kato, T.; Fujimura, H.; Kawasa, S. Efficient Authentic Fine Mapping of the Rice Blast Resistance Gene *Pik-h* in the *Pik* Cluster, Using New *Pik-H*-Differentiating Isolates. *Mol. Breed.* **2008**, *22*, 289–299.

Yuan, B.; Zhai, C.; Wang, W.; Zeng, X.; Xu, X.; Hu, H.; Lin, F.; Wang, L.; Pan, Q. The Pik-p Resistance to *Magnaporthe oryzae* in Rice is Mediated by a Pair of Closely Linked CC-NBS- LRR Genes. *Theor. Appl. Genet.* **2011**, *122*, 1017–1028.

Yoshida, K.; Saitoh, H.; Fujisawa, S.; Kanzaki, H.; Matsumura, H.; Yoshida, K.; Tosa, Y.; Chuma, I.; Takano, Y.; Win, J.; Kamoun, S.; Terauchia, R. Association Genetics Reveals Three Novel Avirulence Genes from the Rice Blast Fungal Pathogen *Magnaporthe oryzae*. *Plant Cell.* **2009**, *21*, 1573–1591.

Zeng, L. R.; Miguel, E. V. S.; Tong, Z.; Wang, G. L. Ubiquitination-Mediated Protein Degradation and Modification: An Emerging Theme in Plant-Microbe Interactions. *Cell Res.* **2006,** *16*, 413–426.

Zhai, C.; Lin, F.; Dong, Z. Q.; He, X. Y.; Yuan, B.; Zeng, X. S.; Wang, L.; Pan, Q. H. The Isolation and Characterization of Pik, a Rice Blast Resistance Gene Which Emerged after Rice Domestication. *New Phytol.* **2011,** *189,* 321–334.

Zhang, S.; Xu, J. R. Effectors and Effector Delivery in *Magnaporthe oryzae. PLoS Pathog.* **2014,** *10*, 1–4.

Zhou, B.; Wang, G. L. Functional and Evolutionary Analysis of the *Pi2/9* Locus in Rice. In *Advances in Genetics, Genomics and Control of Rice Blast Disease*; Wang G. L., Valent, B., Eds.; Springer: New York, 2009; pp 127–135.

Zhou, B.; Qu, S.; Liu, G.; Dolan, M.; Sakai, H.; Lu, G.; Bellizzi, M.; Wang, G. L. The Eight Amino-Acid Differences within Three Leucine-Rich Repeats between *Pi2* And *Piz-T* Resistance Proteins Determine the Resistance Specificity to *Magnaporthe grisea. Mol. Plant Microbe Interact.* **2006,** *19*, 1216–1228.

Zhou, E.; Jia, Y.; Singh, P.; Correll, J. C.; Lee, F. N. Instability of the *Magnaporthe oryzae* Avirulence Gene *AVR-Pita* Alters Virulence. *Fungal Genet. Biol.* **2007, 44,** 1024–1034.

CHAPTER 7

HOST SPECIFICITY IN *PHYTOPHTHORA*: A CONUNDRUM OR A KEY FOR CONTROL?

SANJOY GUHA ROY*

Department of Botany, West Bengal State University, Barasat, Kolkata 700126, India

Corresponding author. E-mail: s_guharoy@yahoo.com

CONTENTS

ABSTRACT

The Oomycete, *Phytophthora* causes devastating diseases in almost all ecological niches. It is a hemibiotroph, and has a 'two speed genome' which underpins a rapid evolution of the vast repertoire of virulence (effector) genes present. This brings about an ability to rapidly co-evolve and thereby adapt to resistant hosts has making it one of the most devastating of phytopathogens. Added to that is its variation in host ranges from a single host to hundreds of genera for certain species. Each having different host–pathosystem interaction and dynamics making complete control apparently impossible. Yet there seems to be a method in this diversity, which can be exploited. The effectors are often the key in determining host specificity and their interaction are essential for successful infection or vice versa. An understanding of how these effectors interact and function is perhaps a key out of this conundrum as they then can be targeted. This article discusses the above aspects.

7.1   INTRODUCTION

The sixth kingdom oomycetes, includes some of the most devastating pathogens on both cultivated crops and wild plants. Within them, *Phytophthora* is a genus of plant pathogenic filamentous oomycetes containing more than one hundred species. Virtually all of them are plant pathogens causing many well-known and important plant diseases worldwide, such as potato late blight, sudden oak death (SOD), and forest dieback caused by *Phytophthora infestans*, *Phytophthora ramorum*, and *Phytophthora cinnamomi*, respectively. Notwithstanding the fact that it causes numerous other diseases in almost all ecological niches (Erwin & Ribeiro, 1996, Lamour, 2013, Guha Roy & Grunwald, 2014). In the genus *Phytophthora* some closely related species have a broad host range, while others are very host specific. Pathogenicity and host adaptation are, therefore, essential traits to understand its biology and to come up with durable, efficient management. However, this becomes a challenge when one considers that most of species are pathogenic, have different host–pathogen dynamics, dearth of comparative genomics data of infection in different hosts and extremely broad host range of about 255 plant genera from 90 families (Cline et al., 2008) in some species like *P. nicotianae*. This is compounded when interspecific hybridization which is becoming increasingly evident as a common event in *Phytophthora* evolution extends the host range further and yet the consequences for its ecological fitness and distribution are not well understood. (Bertier et al., 2013).

Yet, perhaps the key to control lies in this very seemingly host complexity which can be unraveled with comparative proteomics and genomics of host–pathogen interaction to dissect the basis for difference in virulence strategies of the strains against the host range. The key here being the nature, differences and timing of the effectors secreted by the *Phytophthora* spp. *vis-a-vis* it's hosts. Host specificity of *Phytophthora* spp. is presumably based on differences in early infection events namely that of effector classes. The differential pathogenic response against a broad range of hosts is the key to strategizing control measures for the pathogen. Such responses to a large extent depend on the diversity, spatial and temporal regulation of effector classes, differences in early interaction events and on selection pressure of the pathogen populations which determine the virulence potential. Identifying effector elements responsible for interfering with pathogen-associated molecular patterns (PAMP) and effector triggered immunity (ETI) pathway and bringing about differential specificity in its host range will be the key to control as these can then be targeted.

7.2   THE CONUNDRUM

The development of modern agriculture has been shaped by oomycete plant pathogens. Three major epidemics spread over Europe in the middle nineteenth century. Potato, citrus, and grapevine productions were devastated by *Phytophthora infestans*, a complex of *P. citrophthora* and *P. nicotianae*, and *Plasmopara viticola*, respectively, (Erwin & Ribeiro, 1996) and recently that of forest tree pathogen, *Phytophthora ramorum*, the SOD causal agent (Werres et al., 2001). The potential risk due to this introduced pathogen contributed to the release of its complete genome only 5 years after its formal description (Tyler et al., 2006), and *P. ramorum* is now considered as one of the most devastating oomycetes (Kamoun et al., 2015). The human and economical losses were so important that they definitively impacted human history. This leads to the emergence of plant pathology as a formal science, the exploitation of empiric observations to favor the use of resistant plants (Laviola et al., 1990), and to elaborate preparations directed against pathogens, such as the Bordeaux mixture (Rivière et al., 2011). And accordingly, breeding for resistant cultivars and chemical control by fungicides became the cornerstones of nearly all crop protection strategies for more than a century.

The initial studies of twentieth century lead to the description initially of ~60 *Phytophthora* species which greatly differed in their biology, reproductive strategy and pathogenicity (Erwin & Ribeiro, 1996). Additionally, now more than 70 new species have been identified after 2000, and tens of provisional putative new species are awaiting a formal description in different laboratories around the world (Martin et al., 2014; Guha Roy & Grunwald, 2014). Considering that there are 200–600 extant *Phytophthora* species (Brasier, 2009), a large number of species, therefore, remain to be discovered.

Notwithstanding these facts, the ~4,400 host–pathogen associations identified with *Phytophthora* spp. worldwide are also rapidly evolving (Scott et al., 2013). These changes concern not only the newly identified species, but also some of the earliest *Phytophthora* species to be described, like *P. infestans* (Cooke et al., 2012) or *P. nicotianae* (Panabières et al., 2016).

7.3 QUESTIONS THAT NEED TO BE ANSWERED

So the questions that need to be answered are; do host-govern specificity in these different *Phytophthora*-phytopathosystems or are there different characteristics and mechanisms for selective pathogenesis within *Phytophthora* of different host origin? Are these mechanisms different or have they evolved from the basic repertoire and have differentiated due to different selection pressures including climate changes? If so, then, did host specific *Phytophthora* spp. evolve from broad host range species? Did host-specific *Phytophthora* spp. co-evolve with their host plants or did sympatric speciation occur at a much later stage in the evolution of the pathogen? The *Phytophthora* genus provides a fascinating range in complexities of these host–pathogen associations. Host ranges of *Phytophthora* species can vary from one extreme of being very diverse to that of a single host. Some species, like *P. cinnamomi* and *P. nicotianae* may attack hundreds of plants with *P. nicotianae* having the broadest range while others, like *P. infestans* have a narrow host range or *P. sojae,* which infects a single host. Host ranges of new invasive forest (*P. ramorum*) or other *Phytophthora* species are being determined but the range continues to expand with passage of time (Grünwald et al., 2008, Schwingle & Blanchette, 2008). Also, molecular validation and renaming of species are now often increasingly changing host boundaries for each species (e.g., *P. palmivora* MF4 [= *P. capsici*]–*P. tropicalis*). Most crops of agricultural importance and natural ecosystems were shown to be preferentially associated to a *Phytophthora* species, like potato and tomato (*P. infestans*), soybean (*P. sojae*), tobacco (*P. nicotianae*), or Australian jarrah trees (*P. cinnamomi*). Yet, some plants were hosts of several *Phytophthora* species. In these cases, a given species may be prominent and induce more severe symptoms than another, leading to the definition of primary and secondary pathogens.

7.4   HOST–PATHOGEN INTERACTIONS AND HOST SPECIFICITY

This outcome of host–pathogen interactions is determined by a fine-tuned molecular interplay between the two partners. Some species are soilborne (*P. sojae*, *P. nicotianae*) other are foliar (*P. infestans*) and understandably their mechanisms for host infection will differ to take into account this fact. The infection cycle of *Phytophthora* spp. is initiated by the attraction of swimming zoospores to plant roots. In most cases, penetration of the root epidermis is mediated by appressorium-like structures (Tyler, 2007; Attard et al., 2008), but the direct penetration of hyphae between root cells has been reported for *P. sojae* (Enkerli et al., 1997). Following penetration, bulbous hyphae invade the roots intercellularly (Benhamou & Côté, 1992; Widmer et al., 1998; Le Berre et al., 2008). During the interaction between soybean and *P. sojae*, this stage of infection involves a short, difficult-to-observe biotrophic phase (Hanchey & Wheeler, 1971) that seems to be associated with the differentiation of specialized feeding structures, called haustoria (Enkerli et al., 1997; Perfect & Green, 2001; Tyler, 2007).

However, differences also exist between those having similar mode of dissemination. *P. nicotianae* zoospores do not have plant species-specific root preferences in contrast to *P. sojae* zoospores which are attracted specifically to roots exuding the isoflavones diadzein and genistein. (Attard et al., 2008). But this specific chemotaxis toward host isoflavones is of limited importance in *Phytophthora sojae* and *Phytophthora vignae*, while, specific chemotaxis of *Phytophthora pisi* and *Phytophthora niederhauserii* indicated an adaptation to their pathogenicity on the host and lack of pathogenicity on non-host plants (Hosseini et al., 2014). Differences exist even between colonization of *Arabidopsis thaliana* roots by *P. nicotianae* and *P. capsici* both of which are successfully and are able to complete their disease cycle in this *a. thaliana* host. The two oomycetes caused similar symptoms on the plants, but symptoms develop later following the infection in case of *P. nicotianae* than in *P. capsici*. Differential responses have also been seen in activation of signaling pathways in response to infection in leaves and roots. In experiments with *A. thaliana* it has been seen that the salicylate- and jasmonate-dependent signaling pathways are concertedly activated when *P. nicotianae* penetrates the roots, but are down-regulated during invasive growth, when ethylene (ET)-mediated signaling predominates. Defense responses in *a. thaliana* roots are triggered immediately on contact with *P. nicotianae* but the pattern of early defense mechanism activation differs between roots and leaves (Attard et al., 2010). During leaf infections, switches from biotrophy to necrotrophy are frequently accompanied by a shift of plant defenses from salicylic acid (SA)- to jasmonic acid (JA)-mediated responses and is in contrast with the reported antagonistic action of the signaling pathways involving SA and JA/ET in leaves (Glazebrook, 2005). Similarly, it was found that beech root responses to *Phytophthora citrocola* differed from leaf responses, and showed that most of the genes activated in roots had no known function or no matches with database sequences for genes activated in aerial parts of plants Schlink (2009).

7.5   HOST ADAPTATION/HOST SPECIFICITY RELATED TO POPULATION BIOLOGY AND EVOLUTION OF THE PATHOGEN

Host adaptation also has a population biology perspective. While host specificity at the genus/species or at the cultivar level allows to define the host range and physiological races of the pathogen, the quantitative assessment of the disease induced in susceptible hosts is a major, but completely different component of pathogenicity. Related to which is the extent of pathogenic variation present in "old" and "new" pathogen populations which was responsible for the loss on *hitherto* non hosts or marginal/resistant cultivars hosts. In addition, rapid shifts among pathogen populations may generate strains that overcome fungicides and/or resistant varieties, and thus challenge disease management programs.

Host specificity is not only of pathological, but also of evolutionary significance, because the possibility for infecting more than one host determines to a large extent the availability of "green bridges" during the pathogen's life cycle. These are critical in maximizing survival opportunities in species with very low saprophytic abilities, such as *P. infestans*, and probably condition the extent of gene flow between isolates (Andrivon et al., 2004). Host specificity may also have led to a speciation event as between *P. infestans* and *P. mirabilis*, two species giving rise to fertile hybrids (Goodwin & Fry, 1994), morphologically indistinguishable from one another (Galindo & Hohl, 1985), but with mutually exclusive host ranges. This separation of host ranges explains the reproductive isolation of *P. infestans* and *P. mirabilis* in nature. Other similar speciation patterns have also been described in South America (Adler et al., 2002) which involve sympatric wild and/or cultivated hosts and which points to a selective advantage to host specialization in habitats where a number of potential hosts are present (Lapchin, 2002). As a result, current research (Lassiter et al., 2015) has proven that the *P. infestans* pathogen is closely related to four other *Phytophthora* species in the 1c clade including *P. phaseoli*, *P. ipomoeae*, *P. mirabilis*, and *P. andina* all of which are important pathogens of other wild and domesticated hosts and that *P. andina* is an interspecific hybrid between *P. infestans* and an unknown *Phytophthora* species. The formation of hybrids is perhaps the ultimate survival strategy: Reaching of new hosts through interspecific hybridizations as they would possess an unprecedented repertoire of virulence determinants inherited from both parents (Panabieries, 2015) and many examples of natural interspecific hybrids abound in this genus: *Phytophthora × pelgrandis* from different hosts (Man In't Veld et al., 1998; Man in 'T Veld et al., 2012; Faedda et al., 2013; Szigethy et al., 2013); *P. alni* subs *P. uniformis* and *P. alni* subs *P. multiformis* (Ioos et al., 2006).

However, this general trend toward specialization (i.e., restriction of host range) is sometimes reverted, as shown by the discovery in the Netherlands of isolates overcoming the resistance of *solanum nigrum*, until then regarded as a non-host for *P. infestans* (Flier et al., 2003a). Another such example could perhaps be *Phytophthora nicotianae* isolated from potato. *P. nicotianae* has been sporadically reported to cause foliar blight and tuber rot of potato over the past 75 years, but was generally considered of minor incidence (Taylor et al., 2015), but is now being increasingly reported as an important component of the tuber rot and foliar disease complex in US (Taylor et al., 2008), (Taylor et al., 2012). The concern here is that the *P. nicotianae* isolates recovered from potato are significantly more aggressive on this plant compared to *P. nicotianae* isolates recovered from other hosts (Taylor et al., 2012), suggesting potential host specialization and increasing of host range. This phenomenon leads one to recall the origins of *Phytophthora* species attacking legumes in Australia (Irwin et al., 1997). It is becoming increasingly evident that interspecific hybridization is a common event in *Phytophthora* evolution. Yet, the fundamental processes underlying interspecific hybridization and the consequences for its ecological fitness and distribution are not well understood. It has been hypothesized that interspecific hybridization and polyploidy are two linked phenomena in *Phytophthora*, and that these processes might play an important and ongoing role in the evolution of this genus (Bertier et al., 2013).

7.6 THE KEY: GENOME STRUCTURE AND PATHOGENICITY FACTORS (EFFECTOR) SPECIALIZATION LEADING TO HOST SPECIFICITY/DIVERSIFICATION AND SPECIATION

As a rule, *Phytophthora* diseases were more or less efficiently managed through cultural practices, fungicide applications, and the use of resistant varieties when available. In addition, rapid shifts among pathogen populations may generate strains that overcome fungicides and/or resistant varieties, and thus challenge disease management programs. The oomycetal world suffered extensive modifications while entering the twenty-first century both in terms of emergent species and advances in science, which resulted in resources, such as whole genome sequences of *Phytophthora* species. The first *Phytophthora* genomes, *P. ramorum* and *P. sojae,* became available in 2004, followed shortly by *P. infestans* in 2006 (Tyler et al., 2006; Haas et al., 2009) and *P. capsici* (Lamour et al., 2012) with the latest being *Phytophthora fragariae* var. *fragariae* (Gao et al., 2015). These genome sequences of *Phytophthora* will enable translational plant disease management and accelerate research (Grunwald, 2012) and have changed our understanding of host defenses and infection processes.

In general, the success of oomycetes as plant pathogens depend on their ability to suppress or evade host-defense responses and to gain nutrition and proliferate. During infection, oomycete pathogens secrete a variety of extracellular proteins such as cellulose binding elicitor lectin (CBEL) (Gaulin et al., 2006) and cell wall degrading enzymes that contribute to adhesion to the plant surface and plant cell wall degradation, respectively, and therefore to pathogenicity (Kamoun, 2006). In addition, *Phytophthora* species secrete effector proteins to modulate biochemical, morphological, and physiological processes of their hosts. These proteins can be divided into two broad categories, apoplastic, and cytoplasmic effectors with different target sites in the plant. Apoplastic effectors accumulate in the plant intracellular space and include necrosis-inducing proteins (NIPs) (Qutob et al., 2002), elicitins that are small cysteine-rich proteins (Kamoun, 2006) and different enzyme inhibitors such as serine protease inhibitor (EPI) (Tian et al., 2005) and glucanase inhibitor (GIP) (Denance et al., 2013). Cytoplasmic effectors are translocated into the plant cytoplasm and include two expanded gene families in *Phytophthora,* known as RXLR effectors (Birch et al., 2006) and Crinklers (CRNs) (Torto et al., 2003). The RXLR effectors share the conserved RXLR amino acid motif (arginine, any amino acid, leucine, arginine), the domain required for delivery inside plant cells, followed by diverse, rapidly evolving carboxy-terminal domains that are responsible for the virulence-related function of the effectors (Birch et al., 2008). CRNs are NIPs that have a conserved FLAK motif for translocation, and are targeted to the host nucleus upon delivery (Schornack et al., 2010). Differences in gene family expansion and diversity, in particular dynamic repertoires of effector genes, are probably responsible for different traits among *Phytophthora* species, such as altered host specificity. Interestingly, unlike the RXLR effectors, CRNs are present in the genome and transcriptome of all examined plant pathogenic oomycete species including *Pythium ultimum*, *Albugo candida*, and *A. euteiches* indicating that the CRNs form an ancient effector family that arose early in oomycete evolution (Schornack et al., 2010).

Genome structure analysis of these three *Phytophthora* species revealed that the conserved genes are present in regions where gene density is high and repeat content is relatively low (the core genome), whereas non-conserved genes are located in regions with low gene density and high repeat content (the plastic genome). The core genome contains genes involved in cellular processes such as DNA replication, transcription and protein translation, whereas genes involved in plant infection, such as fast-evolving effectors, are predominantly located in the gene-sparse or plastic region, which is highly dynamic (Mollahossein, 2015). This probably plays a crucial part in the rapid adaptability of these pathogens to host plants and derives their evolutionary potential (Haas et al., 2009).

More than 1000 effectors described for *Phytophthora* species have the potential to manipulate host metabolism (Vleeshouwers et al., 2006). Despite the broad range of compatible *Phytophthora*–host interactions that cause diseases worldwide, the biological functions of elicitins as virulence factors during susceptible infections are barely discussed in literature. The molecular mechanism of elicitins, a conserved protein secreted by almost all *Phytophthora* species, was deciphered and it was demonstrated that blocking elicitins caused loss of pathogen virulence. As a consequence, elicitins could be a target in plant–*Phytophthora* interactions to prevent infection. Le Berre et al. (2008) demonstrated for the first time the importance of elicitins, particularly of α-plurivorin, for pathogen penetration and its involvement in plant-defense suppression. This is in concert with similar data showing that a strain of *P. cinnamomi* silenced for the β-cinnamomin gene, which is involved in β-cinnamomin elicitin synthesis, was unable to invade root tissue actively and cause disease symptoms (Horta et al., 2010). These results give strong evidence that α-plurivorin is directly involved in manipulation of plant defenses by a broad down-regulation of defense-related genes, independently of the signaling pathways. Furthermore, up-regulation of WRKY, PR1, and ACO after blocking α-plurivorin suggests that the α-plurivorin can be also correlated with suppression of either PTI or ETI, therefore, acting as an effector triggering susceptibility (ETS).

Remarkably, the data of Le Berre et al. (2008) demonstrated that, even considering the presence of hundreds of effector genes in the *P. plurivora* genome, the blocking of α-plurivorin function compromises *P. plurivora* pathogenicity, thus suggesting its essential role for virulence. Because elicitins are highly conserved proteins (with high similarity (Yu et al., 1995) and are almost ubiquitously secreted by all *Phytophthora* species (Takemoto et al., 2005), it will be of interest to investigate their role as virulence factors in other *Phytophthora*-susceptible plant interactions. More importantly, the results found in this work open new perspectives toward the use of elicitins as specific targets for protecting plants against *Phytophthora* infection. It opens a new horizon to the plant-pathology field, since one can disturb this complex system between plants and pathogens, giving advantage to the plants. Most of the plants can defend themselves against pathogens; however, successful pathogens secrete effectors to mock plant-recognition of infection, including *P. plurivora* (Schlink, 2010). Any disturbance of the mode of action of effectors could activate plant-defense responses. In fact, a very punctual disturbance of the system, such as blocking the acidic elicitin α-plurivorin of *P. plurivora* among hundreds of other effectors, resulted in loss of virulence and simultaneously activation of plant defense. Scientists have also figured out that silencing of one single effector can compromise pathogenicity. One example is given by Yu et al. (2012), who proved that silencing the RxLR effector Avh241, from among the about 627 RxLR in total (Tyler et al., 2006) resulted in loss of virulence of *P. sojae* to soybeans. These growing evidences led Kale (2012) to state that "effector blocking technologies could be developed and utilized in a variety of important crop species against a broad spectrum of plant pathogens."

More recently, Researchers at Oxford University and The Sainsbury Laboratory, Norwich, (Dong et al., 2014) looked in unprecedented detail at how *Phytophthora infestans*, a pathogen that continues to blight potatoes and tomatoes today, evolved to target other plants. The study, used *Phytophthora infestans* and sister species *Phytophthora mirabilis*, (a pathogen that split from *P. infestans* around 1300 years ago) to target the *mirabilis jalapa* plant, commonly known as the four o'clock flower to show definitively for the first time that there is a direct molecular mechanism underpinning the change in host specialization allowing pathogens over time to switch from targeting one species to another through changes at the molecular level. They found that each pathogen species secretes specialized effectors to shut down the defenses of their target hosts. When a plant becomes infected, proteases help plants to attack the invading pathogens and trigger immune responses. *P. infestans* secretes protease inhibitor EPIC effectors that disable proteases in potatoes and tomatoes. These are highly specialized to block specific proteases in the host plant, fitting like a key into a "lock." The effectors secreted by *P. infestans* are less effective against proteases in other plants such as the four o'clock, as they do not fit well into the "locks." The researchers found that *P. mirabilis* evolved effectors that disable the defenses of the four o'clock plant but are no longer effective against potatoes or tomatoes. The EPIC effectors secreted by *P. infestans* have evolved to fit the structure of potato proteases just as *P. mirabilis* has evolved effectors that fit four o'clock proteases. Amino acid polymorphisms in both the inhibitors and their target proteases underpin this biochemical specialization. These results link effector specialization to diversification and speciation of this plant pathogen. Thus, the host specialization that led to evolutionary divergence depending on reciprocal single–amino acid changes that tailor the pathogen effector to a specific host protease, which is being disabled. Thus, small changes can open the door for a pathogen to jump to another species of host and, itself, diversify into another species of pathogen.

Dr. Renier van der Hoorn, co-author of the study from Oxford University's Department of Plant Sciences says that "If we could breed plants with proteases that can detect these stealthy EPIC effectors, we could prevent them from 'sneaking in' and thus make more resistant plants. Within the next decade, we plan to exploit the specialized nature of these effectors to develop proteases that are resistant to their action or can even trap them and destroy the pathogen. Potato and tomato plants with such proteases would be resistant to the blight pathogens, and combined with other resistant traits could provide another 'wall' of defence against the pathogens."

Similarly, studies on the extremely broad host pathogen *P. parasitica*, which is phylogenetically related to *P. infestans*, and overlaps its host range, including potato (Taylor et al., 2008), therefore, are expected to advance our knowledge on mechanisms underlying general pathogenicity and those governing host specificity. Additionally, comparative analyses on these two species varying in genome size (83 Mb for *P. parasitica* versus 240 Mb for *P. infestans*) will help in the understanding of the evolution of pathogenicity and host range among *Phytophthora* spp. Toward this end a through the sequencing of the genome of a cosmopolite isolate and the subsequent sequencing of isolates of diverse, narrow host range and geographical origins, the international *"Phytophthora parasitica* genome initiative" project is enabling the characterization of genes that determine host range. Currently, an in-depth analysis of 14 sequenced genomes of *P. nicotianae* has been completed along with the characterization of the repertoire of effector proteins. As a broad host range pathogen, *P. parasitica* provides a unique opportunity for intra- and inter-specific, comparative analyses, looking at the extent of these families, their organization, their role in plant recognition and infection, and their evolution among strains and species that display broad or restricted host ranges. The identification of conserved and accessory sets of effectors, as well as other pathogenicity genes, will give clues to evaluate the evolutionary pressure of exposure to different host-defense responses to the diversification of effectors and their role in adaptation to host plants. (Kamoun et al., 2015) which in turn will allow us to target the pathogen.

7.7 CONCLUSION

In conclusion, plants can be attacked by a vast range of pathogen classes, causing substantial agricultural losses. The mind boggling host range of some *Phytophthora* species makes control difficult, but the time has come for a paradigm shift in our approach by targeting pathogen effectors as they are secreted during infection playing a key role in disease biology and hence can be targeted allowing a translational output. What makes this focus on effectors more important is that effector specialization leads to diversification and speciation of this plant pathogen. But at the same time effector-induced adaptation to new hosts is an understudied topic and more studies are needed to investigate how *Phytophthora* effector proteins evolve the ability to specialize on new hosts.

KEYWORDS

- **pathogens**
- **comparative genomics**
- **effector**
- **molecular mechanism**
- **pathogenicity**

REFERENCES

Adler, N.; Chacon, G.; Forbes, G.; Flier, W. *Phytophthora infestans* Sensu Lato in South America-Population Sub Structuring through Host Specificity. In *Late Blight-Managing the Global Threat*; Lizarraga, C., Ed.; GILB-CIP: Lima, Peru, 2002; pp 13–17.

Andrivon, D.; Corbière, R.; Lebreton, L.; Pilet, F.; Montarry, J.; Pellé, R.; Ellissèche, D. Host Adaptation in *Phytophthora infestans*: A Review from a Population Biology Perspective. *Plant Breed. Seed Sci.* **2004**, *50*, 15–27.

Attard, A.; Gourgues, M.; Galiana, E.; Panabieres, F.; Ponchet, M.; Keller, H. Strategies of Attack and Defense in Plant–Oomycete Interactions, Accentuated for *Phytophthora parasitica* Dastur (Syn. P. Nicotianae Breda De Haan). *J. Plant Physiol.* **2008**, *165*, 83–94.

Attard, A.; Gourgues, M.; Callemeyn-Torre, N.; Keller, H. The Immediate Activation of Defense Responses in *Arabidopsis* Roots is not Sufficient to Prevent *Phytophthora parasitica* Infection. *New Phytol.* **2010**, *187*, 449–460. Doi: 10.1111/J.1469–8137.2010.03272.X.

Benhamou, N. B.; Côté, F. Ultrastructure and Cytochemistry of Pectin and Cellulose Degradation in Tobacco Roots Infected by *Phytophthora parasitica* var. *nicotianae*. *Phytopathol.* **1992**, *82*, 468–542.

Bertier, L.; Leus, L.; D'hondt, L.; De Cock, Arthur, W. A. M.; Höfte, M. Host Adaptation and Speciation Through Hybridization and Polyploidy in Phytophthora. *Plos ONE*. **2013**, *8*(12), E85385. Doi: 10.1371/Journal.Pone.00853 85.

Birch, P. R. J.; Boevink, P. C.; Gilroy, E. M.; Hein, I.; Pritchard, L.; Whisson, S. C. Oomycete RXLR Effectors: Delivery, Functional Redundancy and Durable Disease Resistance. *Curr. Opin. Plant Biol.* **2008**, *11*(4), 373–379.

Birch, P. R. J.; Rehmany, A. P.; Pritchard, L.; Kamoun, S.; Beynon, J. L. Trafficking Arms: Oomycete Effectors Enter Host Plant Cells. *Trends Microbiol.* **2006**, *14*(1), 8–11.

Brasier, C. *Phytophthora* Biodiversity: How Many *Phytophthora* Species are There? In *EM Goheen, SJ. Frankel (Tech Coords): Phytophthoras in Forests and Natural Ecosystems.* Proceedings of 4th meeting of IUFRO Working Party, Monterey, CA, July 02, 2009, Aug 26–31, 2007; Gen. Tech. Rep. PSW-GTR-221, Albany, CA, 2009, 101–115.

Cline, E. T.; Farr, D. F.; Rossman, A. Y. Synopsis of *Phytophthora* with Accurate Scientific Names, Host Range, and Geographic Distribution. *Plant Health Prog.* **2008**, Doi:10.1094/PHP-2008-0318–01-RS.

Cooke, D. E.; Cano, L. M.; Raffaele, S.; Bain, R. A.; Cooke, L. R.; Etherington, G. J.; Deahl, K. L.; Farrer, R. A.; Gilroy, E. M.; Goss, E. M.; Grunwald, N. J.; Hein, I.; Maclean, D.; Mcnicol, J. W.; Randall, E.; Oliva, R. F.; Pel, M. A.; Shaw, D. S.; Squires, J. N.; Taylor, M. C.; Vleeshouwers, V. G.; Birch, P. R.; Lees, A. K.; Kamoun, S. Genome Analyses of an Aggressive and Invasive Lineage of the Irish Potato Famine Pathogen. *PLoS Pathog.* **2012**, *8*, E1002940.

Denance, N.; Sanchez-Vallet, A.; Goffner, D.; Molina, A. Disease Resistance or Growth: The Role of Plant Hormones in Balancing Immune Responses and Fitness Costs. *Front. Plant Sci.* **2013**, *4*, 155.

Dong, S.; Stam, R.; Cano, L. M.; Song, J.; Sklenar, J.; Yoshida, K.; Bozkurt, T. O.; Oliva, R.; Liu, Z.; Tian, M.; Win, J.; Banfield, M. J.; Jones, A. M.; Van Der Hoorn, R. A. L.; Kamoun, S. Effector Specialization in a Lineage of the Irish Potato Famine Pathogen. *Science.* **2014**, *343*, 552–555.

Enkerli, K. Hahn, M.; Mims, C. W. Ultrastructure of Compatible and Incompatible Interactions of Soybean Roots Infected with the Plant Pathogenic Oomycete *Phytophthora sojae.* *Can. J. Bot.* **1997**, *75*, 1493–1508.

Erwin, D. C.; Ribeiro, O. K. *Phytophthora Diseases Worldwide*. American Phytopathological Society: St. Paul, MN, 1996.

Faedda, R.; Cacciola, S. O.; Pane, A.; Szygethy, A.; Bakonyi, J.; Man In 't Veld, W. A.; Martini, P.; Schena, L.; Magnano di San Lio, G. *Phytophthora* × *Pelgrandis* Causes Root and Collar Rot of *Lavandula stoechas* in Italy. *Plant Dis.* **2013**, *97*, 1091–1096.

Flier,W. G.; Bosch, G. B. M. Vanden; Turkensteen L. J. Epidemiological Importance of *Solanum sisymbriifolium, S. nigrum* and *S. dulcamara* as Alternative Hosts for *Phytophthora infestans*. *Plant Pathol.* **2003a**, *52*, 595–603.

Galindo, A. J.; Hohl, H. R. *Phytophthora mirabilis*, a New Species of *Phytophthora sydowia*. **1985**, 38, 87–96.

Gao, R.; Cheng, Y.; Wang, Y.; Wang, Y.; Guo, L.; Zhang, G. Genome Sequence of *Phytophthora Fragariae* var. *fragariae,* A Quarantine Plant-Pathogenic Fungus. *Genome. Announc.* **2015**, *3*(2), e00034–15. Doi:10.1128/Genomea.00034–15.

Gaulin, E.; Drame, N.; Lafitte, C.; Torto-Alalibo, T.; Martinez, Y.; Ameline-Torregrosa, C.; Khatib, M.; Mazarguil, H.; Villalba-Mateos, F.; Kamoun, S.; Mazars, C.; Dumas, B.; Bottin, A.; Esquerre-Tugaye, M. T.; Rickauer, M. Cellulose Binding Domains of a Phytophthora Cell Wall Protein are Novel Pathogen-Associated Molecular Patterns. *Plant Cell.* **2006**, 18(7), 1766–1777.

Glazebrook, J.Contrasting Mechanisms of Defense against Biotrophic and Necrotrophic Pathogens. *Annu. Rev. Phytopathol.* **2005**, *43*, 205–227.

Goodwin, S. B.; Fry, W. E. Genetic Analyses of Interspecific Hybrids between *Phytophthora infestans* and *Phytophthora mirabilis*. *Exp. Mycol.* **1994**, *18*, 20–32.

Grünwald, N. J. Genome Sequences of *Phytophthora* Enable Translational Plant Disease Management and Accelerate Research. *Can.J. Plant Pathol.* **2012**, *34*(1), 13–19.

Grünwald, N. J.; Goss, E. M.; Press, C. M. *Phytophthora mamorum*: A Pathogen with a Remarkably Wide Host-Range Causing Sudden Oak Death on Oaks and Ramorum Blight on Woody Ornamentals. *Mol. Plant Pathol.* **2008**, *9*, 729–740.

Guha Roy, S.; Grünwald, N. The Plant Destroyer Genus *Phytophthora* in the 21st Century. *Rev. Plant Pathol.* **2014**, *6*, 387–412.

Haas, B. J.; Kamoun, S.; Zody, M. C.; Jiang, R. H. Y.; Handsaker, R. E.; Cano, L. M. et al. Genome Sequence and Analysis of the Irish Potato Famine Pathogen *Phytophthora infestans*. *Nature.* **2009**, *461*, 393–398.

Hanchey, P.; Wheeler, H. Pathological Changes in Ultrastructure: Tobacco Roots Infected with *Phytophthora parasitica* var. *nicotianae*. *Phytopathol.* **1971**, *61*, 33–39.

Horta, M.; Caetano, P. Medeira, C.; Maia, I. Cravador, A. Involvement of the Betacinnamomin Elicitin in Infection and Colonisation of Cork Oak Roots by *Phytophthora cinnamomi*. *Eur. J. Plant Pathol.* **2010**, *127*, 427–436.

Hosseini, S.; Heyman, F.; Olsson, U.; Broberg, A.; Jensen Funck, D.; Karlsson, M. Zoospore Chemotaxis of Closely Related Legume-Root Infecting *Phytophthora* Species Towards Host Isoflavones. *Plant Pathol.* **2014**, *63*(3), 708–7014.

Ioos, R.; Andrieux, A.; Marcais, B.; Frey, P. Genetic Characterization of the Natural Hybrid Species *Phytophthora Alni* as Inferred from Nuclear and Mitochondrial DNA Analyses. Fungal Genet. Biol. **2006**, *43,* 511–529.

Irwin, J. A. G.; Crawford, A. R.; Drenth, A.The Origins of *Phytophthora* Species Attacking Legumes in Australia. *Adv. Bot. Res.* **1997**, *24*, 431–456.

Kale, S. D.; Oomycete and Fungal Effector Entry, a Microbial Trojan Horse. *New Phytol.* **2012**, *193*, 874–881.

Kamoun, S. A. Catalogue of the Effector Secretome of Plant Pathogenic Oomycetes. *Annu. Rev. Phytopathol.* **2006,** 44, 41–60.

Kamoun, S.; Furzer, O.; Jones, J. D. G.; Judelson, H. S.; Ali, G. S.; Dalio, R. J. D.; Guha Roy, S.; Schena, L.; Zambounis, A.; Panabières, F.; Cahill, D.; Ruocco, M.; Figueiredo, A.; Chen, X. R.; Hulvey, J.; Stam, R.; Lamour, K.; Gijzen, M.; Tyler, B. M.; Grünwald, N. J.; Mukhtar, M. S.; Tomé, D. F. A.; Tör, M.; Van Den Ackerveken, G.; Mcdowell, J.; Daayf, F.; Fry, W. E.; Lindqvist-Kreuze, H.; Meijer, H. J. G.; Petre, B.; Ristaino, J.; Yoshida, K.; Birch, P. R. J.; Govers, F. The Top 10 Oomycete Pathogens in Molecular Plant Pathology. *Mol. Plant Pathol.* **2015,** *16*, 413–434. Doi: 10.1111/Mpp.12190.

Lamour, K. *Phytophthora: Global Perspective*; CABI Plant Protection Series. CABI: UK, 2013.

Lamour, K. H.; Mudge, J.; Gobena, D.; Hurtado-Gonzales, O. P.; Schmutz, J.; Kuo, A.; Miller, N. A.; Rice, B. J.; Raffaele, S.; Cano, L. M.; Bharti, A. K.; Donahoo, R. S.; Finley, S.; Huitema, E.; Hulvey, J.; Platt, D.; Salamov, A.; Savidor, A.; Sharma, R.; Stam, R.; Storey, D.; Thines, M.; Win, J.; Haas, B. J.; Dinwiddie, D. L.; Jenkins, J.; Knight, J. R.; Affourtit, J. P.; Han, C. S.; Chertkov, O.; Lindquist, E. A.; Detter, C.; Grigoriev, I. V.; Kamoun, S.; Kingsmore, S. F. Genome Sequencing and Mapping Reveal Loss of Heterozygosity as a Mechanism for Rapid Adaptation in the Vegetable Pathogen *Phytophthora Capsici. Mol. Plant-Microb. Int. MPMI,* **2012,** *25*(10), 1350–1360. Doi: 10.1094/MPMI-02–12–0028-R.

Lapchin, L. Host-Parasitoid Association and Diffuse Coevolution: When to be a Generalist? *Am. Nat.* **2002,** *160*, 245–254.

Lassiter, E. S.; Russ, C.; Nusbaum, C.; Zeng, Q.; Saville, A. C.; Olarte, R. A.; Carbone, I.; Chia-Hui Hu, Seguin-Orlando, A.; Samaniego, J. A.; Thorne, J. L.; Ristaino J. B. Mitochondrial Genome Sequences Reveal Evolutionary Relationships of the *Phytophthora* 1c Clade Species. *Curr. Genet,* Mar 10, **2015.** [epub ahead of print].

Laviola, C.; Somma, V.; Evola, C. Present Status of *Phytophthora* Species in the Mediterranean Area, Especially in Relation to Citrus. *OEPP/EPPO Bull.* **1990,** *20*, 1–9.

Le Berre, J. Y.; Engler, G.; Panabieres, F. Exploration of the Late Stages of the Tomato–*Phytophthora parasitica* Interactions through Histological Analysis and Generation of Expressed Sequence Tags. *New Phytol.* **2008,** *177*, 480–492.

Martin, F. N.; Blair, J. E.; Coffey, M. D. A Combined Mitochondrial and Nuclear Multilocus Phylogeny of the Genus Phytophthora. *Fungal Genet. Biol.* **2014,** *66*, 19–32.

Mollahossein, S. H. Host-Pathogen Interactions in Root Infecting Oomycete Species Doctoral Thesis Faculty of Natural Resources and Agricultural Sciences Department of Forest Mycology and Plant Pathology Uppsala. Swedish University of Agricultural Sciences Uppsala. *Acta Uni. Agric. Suec.* **2015,** *9*.

Panabières, F.; Ali, G. S.; Allagui, M. B.; Dalio, R. J. D.; Gudmestad, N. C.; Kuhn, M. L.; Guha Roy, S.; Schena, L.; Zampounis, A. *Phytophthora nicotianae* Diseases Worldwide: Old Wine in New Bottles. *Phytopathol. Mediterr. (In Press),* **2016.**

Perfect, S. E.; Green, J. R. Infection Structures of Biotrophic and Hemibiotrophic Fungal Plant Pathogens. *Mol. Plant Pathol.* **2001,** *2*, 101–108.

Qutob, D.; Kamoun, S.; Gijzen, M. Expression of a *Phytophthora sojae* Necrosis inducing Protein Occurs during Transition from Biotrophy to Necrotrophy. *Plant J.* **2002,** 32(3), 361–373.

Rivière, M. P.; Ponchet, M.; Galiana, E. The Millardetian Conjunction in the Modern World. In *Pesticides in the Modern World-Pesticides Use and Management*; Stoytcheva, M., Ed.; Intech: Rijeka, Croatia - European Union, 2011; pp 369–390. ISBN: 978–953–307–459–7.

Schlink, K. Identification and Characterization of Differentially Expressed Genes from *Fagus sylvatica* Roots after Infection with *Phytophthora citricola*. *Plant Cell Rep.* **2009,** *28*(5), 873–882.

Schlink, K. Down-Regulation of Defense Genes and Resource Allocation into Infected Roots as Factors for Compatibility between *Fagus sylvatica* and *Phytophthora citricola*. *Funct. Integr. Genomics.***2010,** *10*, 253–264.

Schornack, S.; Van Damme, M.; Bozkurt, T. O.; Cano, L. M.; Smoker, M.; Thines, M.; Gaulin, E.; Kamoun, S.; Huitema, E. Ancient Class of Translocated Oomycete Effectors Targets the Host Nucleus. Proceedings of the National Academy of Sciences of the United States of America, 2010; 107(40), 17421–17426.

Schwingle, B.; Blanchette, R. A. Host Range Investigations of New, Undescribed, and Common *Phytophthora* sp. Isolated from Ornamental Nurseries in Minnesota. *Plant Dis.* **2008,** *92*, 642–647.

Szigethy, A.; Nagy, Z. A.; Vettraino, A. M.; Jozsa, A.; Cacciola, S. O.; Faedda, R.; Bakoni, J. First Report of *Phytophthora* × *Pelgrandis* Causing Root Rot and Lower Stem Necrosis of Common Box, Lavender and Port-Oxford Cedar in Hungary. *Plant Dis.* **2013,** *97*, 152.

Takemoto, D.; Hardham, A.; Jones, D. A. Differences in Cell Death Induction by *Phytophthora* Elicitins are Determined by Signal Components Downstream of MAP Kinase in Different Species Of *Nicotiana* and Cultivars of *Brassica mapa* and *Raphanus sativus*. *Plant Physiol.* **2005,** *138*, 1491–1504.

Taylor, R. J.; Pasche, J. S.; Gudmestad, N. C. Etiology of a Tuber Rot and Foliar Blight of Potato Caused by *Phytophthora nicotianae*. *Plant Dis.* **2015,** *99*, 474–481.

Taylor, R. J.; Pasche, J. S.; Gallup, C. A.; Shew, H. D.; Gudmestad, N. C. A Foliar Blight and Tuber Rot of Potato Caused by *Phytophthora nicotianae*: New Occurrences and Characterization of Isolates. *Plant Dis.* **2008,** *92*, 492–503.

Taylor, R. J.; Pasche, J. S.; Shew, H. D.; Klanno, K. R.; Gudmestad, N. C. Tuber Rot of Potato Caused by *Phytophthora nicotianae*: Isolate Aggressiveness and Cultivar Susceptibility. *Plant Dis.* **2012,** *96*, 693–704.

Tian, M.Y.; Benedetti, B.; Kamoun, S. A Second Kazal-Like Protease Inhibitor from *Phytophthora infestans* Inhibits and Interacts with the Apoplastic Pathogenesis-Related Protease P69B of Tomato. *Plant Physiol.* **2005,** *138*(3), 1785–1793.

Torto, T.A.; Li, S.; Styer, A.; Huitema, E.; Testa, A.; Gow, N. A. R.; van West, P.; Kamoun, S. EST Mining and Functional Expression Assays Identify Extracellular Effector Proteins from the Plant Pathogen *Phytophthora*. *Genome Res.* **2003,** *13*, 1675–1685.

Tyler, B. M. Genomics of Fungal Plant Pathogens. *Encyclopedia of Plant and Crop Science*. Taylor and Francis, 2006; pp 1–5. Online resource: http://www.informaworld. com/10.1081/E-EPCS-120019942)

Tyler, B. M. *Phytophthora sojae*: Root Rot Pathogen of Soybean and Model Oomycete. *Mol. Plant Patholo.* **2007,** *8*, 1–8.

Tyler, B. M.; Tripathy, S.; Zhang, X.; Dehal, P.; Jiang, R. H.; Aerts, A.; Arredondo, F. D.; Baxter, L.; Bensasson, D.; Beynon, J. L.; Chapman, J.; Damasceno, C. M.; Dorrance, A. E.; Dou, D.; Dickerman, A. W.; Dubchak, I. L.; Garbelotto, M.; Gijzen, M.; Gordon, S. G.; Govers, F.; Grunwald, N. J.; Huang, W.; Ivors, K. L.; Jones, R. W.; Kamoun, S.; Krampis, K.; Lamour, K. H.; Lee, M. K.; Mcdonald, W. H.; Medina, M.; Meijer, H. J.; Nordberg, E. K.; Maclean, D. J.; Ospina-Giraldo, M. D.; Morris, P. F.; Phuntumart, V.; Putnam, N. H.; Rash, S.; Rose, J. K.; Sakihama, Y.; Salamov, A. A.; Savidor, A.; Scheuring, C. F.; Smith, B. M.; Sobral, B. W.; Terry, A.; Torto-Alalibo, T. A.; Win, J.; Xu, Z.; Zhang, H.; Grigoriev,

I. V.; Rokhsar, D. S.; Boore, J. L. *Phytophthora* Genome Sequences Uncover Evolutionary Origins and Mechanisms of Pathogenesis. *Science.* **2006,** *313,* 1261–1266.

Tyler, B. M.; Tripathy, S.; Zhang, X.; Dehal, P.; Jiang, R. H. Y.; Aerts, A., et al. *Phytophthora* Genome Sequences Uncover Evolutionary Origins and Mechanisms of Pathogenesis. *Science,* **2006,** *313,* 1261–1266.

Vleeshouwers, V. G.; Driesprong, J. D.; Kamphuis, L. G.; Torto-Alalibo, T.; Van't Slot, K. A.; Govers, F.; Visser, R. G.; Jacobsen, E.; Kamoun, S. Agroinfection-Based High Throughput Screening Reveals Specific Recognition of INF Elicitins in *Solanum. Mol. Plant Pathol.* **2006,** *7,* 499–510.

Werres, S.; Marwitz, R.; Man In't Weld, W. A.; De Cock, A. W. A. M.; Bonants, P. J.; De Weert, M.; Themaénn, K.; Ilieva, E.; Baayen, R. P. *Phytophthora ramorum* sp. *Nov*: A New Pathogen on *Rhododendron* and *Viburnum. Mycol. Res.* **2001,** *105,* 1155–1165.

Widmer, T. L.; Graham, J. H.; Mitchell, D. J. Histological Comparison of Fibrous Root Infection of Disease-Tolerant and Susceptible Citrus Hosts by *Phytophthora nicotianae* and *P. palmivora. Phytopathology.* **1998,** *88,* 389–395.

Yu, L. Elicitins from *Phytophthora* and Basic Resistance in Tobacco. Proceedings of the National Academy of Sciences USA. 1995, 92, 4088–4094.

Yu, X.; Tang, J.; Wang, Q.; Ye, W.; Tao. K.; Duan, S.; Lu, C.; Yang, X.; Dong, S.; Zheng, X.; Wang, Y. The Rxlr Effector Avh241 from *Phytophthora sojae* Requires Plasma Membrane Localization to Induce Plant Cell Death. *New Phytol.* **2012,** *196*(1), 247–60.

CHAPTER 8

EVOLUTION OF HOST SELECTIVITY, HOST RESISTANCE FACTORS AND GENES RESPONSIBLE FOR DISEASE DEVELOPMENT BY *STREPTOMYCES SCABIES*

JAI S. PATEL[1], GAGAN KUMAR[2], ANKITA SARKAR[2], RAM S. UPADHYAY[2], HARIKESH B. SINGH[2], and BIRINCHI K. SARMA[2*]

[1]Department of Botany, Institute of Science, Banaras Hindu University, Varanasi 221005, India

[2]Department of Mycology and Plant Pathology, Institute of Agricultural Sciences, Banaras Hindu University, Varanasi 221 005, India

[*]Corresponding author. E-mail: birinchi_ks@yahoo.com

CONTENTS

ABSTRACT

Common scab of potato is a severe disease affecting tubers. This disease caused by soil-borne filamentous bacteria related to the genus *Streptomyces*. Generally streptomycetes were saprophytic in nature but a few species were modulated themselves to cause disease in underground parts of several plants. The causal agent of the potato common scab is the bacteria *Streptomyces scabies* which has worldwide occurrence. The pathogen produces certain phytotoxins like thaxtomin, which is one of the major virulence (*vir*) factors responsible for the common scab disease. A number of genes are responsible for production of this toxin and are clustered in a particular region with certain other *vir* factors in the genome of *S. scabies* commonly referred to pathogenicity associated island (PAI). The mobilizing and transferring abilities of the PAI are considered responsible for emergence of new pathogenic strains of *Streptomyces*. Synthesis of certain aromatic amino acids and phytohormones shows inhibitory effects on production of the toxin thaxtomin. This chapter deals with factors responsible for pathogenesis, host selectivity, non-host resistance, and evolution of new pathogenic strains of *S. scabies*.

8.1   INTRODUCTION

Host specificity is defined as the ability of microbial pathogens to grow, colonize, and infect their respective hosts (Kirzinger & Stavrinides, 2012). Bacterial pathogens express variety of host specificity mechanisms by modifying their genome such as through point mutation, duplication, horizontal gene transfer (HGT), etc. Some of the bacterial pathogens have very wide host range and recent studies on symbiotic bacteria showed that a single regulatory gene may determine their host specificity (Mandel et al., 2009). Differential expression of the single regulatory gene extends successful colonization by the bacteria on different host species. Gene regulation may be a prominent reason for host specificity in bacterial pathogens leading to emergence of new pathogenic strains leading to new diseases. The changes in the genome may thus lead to evolution in disease systems even in some of the symbiotic strains. Minute changes in the host–pathogen interactions may even be responsible for alteration in the host range of the bacterial strains and the degree of disease severity (Killiny & Alneida, 2011). Two things are observed to be very critical regarding the genetic and molecular mechanisms of host specificity in bacterial pathogens. They are (a) how bacterial pathogens distinguish their hosts by deploying molecular mechanisms and (b) how molecular signals or patterns deployed for host specificity are detected by the hosts. Understanding on these two areas can enhance our knowledge regarding pathogenesis and provide means to improve disease management strategies under natural conditions. Host specificity of bacterial pathogens is determined by the molecular interactions takes place between the hosts and pathogens (Pan et al., 2014). Further, host specificity in the pathogenic strains of bacteria is very high and well described. A number of pathovars are reported in different species of plant pathogenic bacteria. The pathovar group specificity of bacterial races toward their host cultivars supports the Flor's concept of "gene for gene" (Flor 1956). However, the probability of host specificity controlling by a single unique gene is very less (Hajri et al., 2009).

Certain diseases of crops emerge abruptly without any prior indication of severity of the danger. Such pathogens sometimes even travel across the borders or oceans before the researchers recognize them as infectious pathogens (Vinatzer et al., 2014). Bacterial association with other organisms is much diversified and it ranges from biofilm formation to mutualistic to pathogenic associations. Synthesis and secretion of specific proteins are very crucial in such aforesaid associations. Some of these secreted proteins are pathogenic and enter into the host cells and change their physiology, enhance colonization by producing toxins and other effector proteins resulting in disease development (Tseng et al., 2009). The genes involved in making such pathogenic proteins for their specific host mainly are categorized into two group *viz.*, avirulence (*avr*) and virulence (*vir*) genes (Vivian & Arnold, 2000; Vivian & Gibbon, 1997; Vivian et al., 2001). *Avr* gene products of the pathogenic bacteria act as negative acting epitope and enhance incompatibility between the host and the pathogenic strain while *vir* gene product acts as positive acting epitope, which enhances compatible interactions leading to disease development. Incompatible interactions developed by *avr* gene products result into hypersensitive response (HR) and programmed cell death in plant cells at the site of pathogen infection (Alfano & Collmer, 2004). The host selectivity is regulated by positive acting epitope translated from their host-specific *vir* (*hsv*) genes (Ezra et al., 2000; Waney et al., 1991). During the pathogenicity basal plant defense is inhibited, pathogen-specific HR defense is also inhibited and production of lesion and respective disease symptoms takes place (Alfano & Collmer, 2004).

Bacterial *avr* factors are diversified in structure and function (Leach et al., 2001). Their structural difference for example repetitive motifs of *avr* factors might also be responsible for host specificity (Herbers et al., 1992).

Post translational changes in *avr* gene products like acylation and methylation also strengthen the structural complexity simultaneously and enhance the pathogenic activities (Nimchuk et al., 2000; Shen et al., 2002). Effector proteins work at the site of pathogen race and host cultivar interaction and therefore may also get influenced by the host species and tissue specific determinants (Ryan et al., 2011). Type three secretion system (TTSS) effectors are also an important host specificity determinant in *Pseudomonas syringae* and *Xanthomonas campestris* groups. The allelic variation in effectors presents in all pathovars is also important determinants of host specificity (Alfano & Collmer, 2004). Development of resistant varieties against bacterial pathogens is a very tough task because of presence of limited *R* genes and the breeding program is time taking especially in crops like potato. Strategies involving genetic engineering are also not easy because a flurry of responses such as elicitation of plant defense, suppression of bacterial *vir* factors, production of antimicrobial of plant for effective management of a disease are needed to be generated in the host (Melchers & Stuiver, 2000). A significant level of protection is achieved against different diseases so far but broad range resistance still remains a challenge (Rivero et al., 2012).

Streptomyces scabies is a Gram-positive bacterium and its genome contains higher percentage of guanine (G) and cytosine (C) (> 71%). The bacteria on culture medium develop an aerial fragment bearing chain of spores, which give the culture a fluffy appearance. The spores are grey in color and appear like a corkscrew. The virulent species of the bacteria can be identified on the basis of spore chain formation. However, the numbers of spores are about 20 or more per chain and 0.5 to 1.0 μm roundish and smooth. *S. scabies* is able to grow on medium enriched with the sugar raffinose and unable to grow on xanthine enriched media. However, it produces melanin on media containing tyrosine. The pathogenic strains of the bacteria affect the potato tuber and produces corky brown lesions like spots of a few millimeters in size. The size of legions can vary on the basis of environmental conditions. Normally, there is no reduction in potato yield caused by this disease and tubers are edible even after the disease, but quality of the tubers goes down significantly and economic importance of the crop decreases (Lerat et al., 2012; St-Onge et al., 2010). Plenty of soil bacteria and species of streptomycetes are known for production of number of secondary metabolites. The species related to *Streptomyces* are generally saprophytic in nature. However, very few are pathogenic in nature and *S. scabies* (or *scabiei*) is one of them causing common scab disease in potato. *S. scabies* symptoms appear like corky patches or deep-pitted lesions on the surface of roots or tubers of potato, radish, etc. (Goyer & Beaulieu, 1997). Most often this disease causes economic loss for the farmers (Hill & Lazarovits, 2005). Thaxtomin, a phytotoxin, production makes *S. scabies* able to cause disease and produce symptoms on the surface of tubers. Thaxtomins are the secondary metabolites, produce after cyclization of two amino acids trypto-phan and a phenylalanine molecule. Thaxtomin A is the major form of this toxin (Lawrence et al., 1990). There are certain conditions, which induce thaxtomin production. In laboratory conditions oat bran broth medium was found best for induction of thaxtomin production (King & Lawrence, 1996; Lawrence et al., 1990). However, the medium was rich in carbohydrate and complex of plant derived materials; therefore, this is tedious to know main compound responsible for induction of thaxtomin. Pathogenic property of the genus *Streptomyces* is a very rare phenotype. The genus *Streptomyces* has especial and composite morphology because spores get germinated to produce mycelium and converted into aerial hyphae which later break up into spore chains. *Streptomyces* has three well-identified, pathogenic, and genetically different species *viz., S. acidiscabies*, *S. turgidiscabies*, and *S. scabies*. All three species cause common scab symptoms in tap root crops and potato (Loria et al., 1997). The common mechanism of pathogenicity is very similar between these three species in relation to selection of host range and symptom development. All the three species of *Streptomyces* are known to produce a family of thaxtomins (Bukhalid et al., 1998), which are impor-tant components for disease development (Healy et al., 2002). These three species are morphologically and genetically different and could be separated on the basis of DNA–DNA interactions and ribosomal sequences (Bukhalid et al., 2002; Healy & Lambert, 1991; Miyajima et al., 1998). The pathogenic species of *Streptomyces* do not show a high level of host specificity under *in vitro* controlled conditions. A report showed that virulent strains of *S. scabies* were able to cause disease on the seedlings of 14 crop plants including both monocot and dicot species (Leiner et al., 1996) and presence of this patho-genic species of *Streptomyces* even alters the shoot growth in 11 of crop species tested. The toxin thaxtomin produced by this pathogenic species targets the universal component of plant cell wall that is cellulose. However, some researchers reported that its host range is limited to a few crops only under field conditions (Goyer & Beaulieu, 1997) and a few other crops are recognized as resistant or moderately sensitive to common scab in natural conditions (Hiltunen et al., 2005). Earlier reports suggested that infection mechanism of common scab of potato caused by *S. scabies* is governed by a complex plant–microbe interaction process. Some level of host selectivity of this pathogen is due to presence of the nitrated dipeptide phytotoxin, thax-tomin having ability to inhibit biosynthesis of cellulose in the grooving plant tissue, also stimulates Ca^{++} spiking which results in cell death. It secretes a necrogenic toxin protein, nec1 which makes this pathogen able to cause disease in diverse agricultural crop plant species. The position of thaxtomin coding gene and *nec1* is on highly mobilizable locus of its genome known as pathogenicity associated islands (PAI) along with other *vir* genes including the genes for biosynthetic pathway of cytokinins.

Control measures used for the management of common scab of potato are problematic and there is no single reliable control treatment availability. A few researchers showed the role of foliar application of certain substituted phenoxy, benzoic, and picolinic acids on the growth of potato plants for suppression of the common scab disease (McIntosh et al., 1982; McIntosh et al., 1988). One of the most effective measures to control the phytopathogen is application of 2,4-dichlorophenoxyacetic acid (2,4-D) (Tegg et al., 2012; Thompson et al., 2014). A few researchers also have also reported that some bioagents can also suppress the common scab of potato (Singhai et al., 2011a).

8.2 PATHOGENESIS

8.2.1 SUITABLE ENVIRONMENT FOR PATHOGENICITY: ADAPTION IN SOIL ENVIRONMENT

The environmental condition of soil is very rough and competitive for microorganisms. These nutrient poor environmental conditions also provide house and food to various organisms, which release antimicrobial compounds. The organisms survive in soils by tolerating the extreme temperature and moisture content especially in upper layers of the soil profile. *Streptomyces* species also grows vigorously in such type of soil environment. The survivability of *Streptomyces* species is dependent upon their degradative capacity by producing variety of catabolic enzymes, which degrade biological polymers including the cellulose and chitin like fractious compounds (Hodgson, 2000; Williamson et al., 2000). Their degradative capacity plays an important role in nutrient recycling as well (Chamberlain & Crawford, 2000). *Streptomyces* species also produces various antimicrobial compounds, which are very crucial in survival of these species by competing with other microbes present in soil. However, *Streptomyces* species have adapted themselves in the soil environment by hydrolyzing the polymers derived from plants and animals for the production of biologically active molecules beyond their capacity. *Streptomyces* are Gram-positive bacteria having a complex developmental program within a few prokaryotes. They initiate their life from the hydrophobic, uninucleate, desiccation-resistant spores. Due to hydrophobic nature, these non-motile spores easily flow in moving water by free rides on motile eukaryotic organisms like soil arthropods and nematodes. These spores further germinate on nutrient dependent manner and grow as branched and multinucleated mycelium with sporadic cross wall. The web of mycelium releases catabolic enzymes, nutrients from fractious substrates present in soil, and provide required food for colony growth. The outgrowth of aerial mycelium relies on the production of antibiotics like sapB and other similar proteins (Kodani et al., 2005; Kodani et al., 2004), which have wetting agent like property for the emergence of aerial hyphae from the aqueous colony. The food for aerial hyphal growth is provided by degradation of vegetative mycelium. The secondary metabolite production and secretion having antimicrobial compounds is organized with the aging of substrate mycelium (Horinouchi, 2002). The pathogenic *Streptomyces* species use this ability for pathogenesis (Loria et al., 2006).

8.2.2 ROLE OF HGT IN PATHOGENICITY

In prokaryotic microbes HGT is a very important technique for evolution of pathogenic races. PAIs are bunches of pathogenic or *vir* genes, which are connected with multiple transfer and recombination and provide pathogenic phenotype (Loria et al., 2006). The PAI has the ability to mobilize and transfer themselves into non-pathogenic congeners, which provide an opportunity to generate new pathogenic species of *Streptomyces* (Lerat et al., 2009) (Fig. 8.1).

As we know that the genes frequently arise from phylogenetically diverse genomes and G+C content changes in genes present in PAI. These PAIs are like mosaic structure and known as field of persistent action of gene accomplishment, gene loss, and gene decay. *Streptomyces* species are one of the most fortunate soil-borne saprophytes. On the contrary their saprophytic nature and enormous multifariousness in soil has left under-appreciation as compared to their ability to work in various ecological niches. A number of species have complex symbiotic relationships in between the plants and insects. These systems explore their relationships, which is also one of the reasons for co-evolution in streptomycetes and eukaryotes.

The *nec1* gene does not exist in nucleotide and protein databases and it has very low G+C content (54%) while *Streptomyces* genomic DNA contains high G+C content (72%). This result reveals that *nec1* gene has

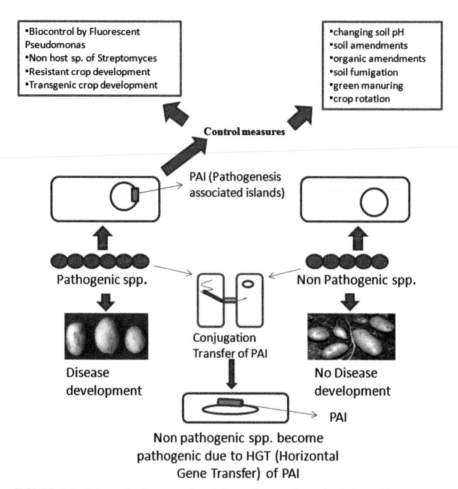

FIGURE 8.1   Schematic diagram of evolution of non-pathogenic strains of *Streptomyces* into pathogenic *Streptomyces* strains and their management tools.

come into *Streptomyces* genome from another genus by HGT. The *nec1* gene is conserved mostly in the pathogenic *Streptomyces* species (Bukhalid et al., 2002; Bukhalid & Loria, 1997). The N-terminal secretion signal of protein nec1 releases in early log phase in culture media indicates that this protein directly comes in the contact with the host plant. In *Arabidopsis*, *nec1* affects the pathogenicity of *S. turgidiscabies* on the entire plant (Bukhalid et al., 2002). An integration of PAI in SCO1326 region having 85% nucleotide homology with *bacA* suggests an in-frame insertion event in *S. turgidisca-bies*. Wild type and transconjugant strains of *S. coelicolor* are not different at the level of bacitracin resistance. Although the undecaprenyl pyrophosphate phosphatase (UPP) genes are conserved in bacteria, they are not described as targets for HGT. However, SCO1326 is instantly layed by low G+C content in wild type *S. coelicolor*, describing that SCO1326 has been the target for HGT. Recently, released draft genome of the emergent plant pathogen *S. acidiscabies* 84–104 demonstrated that it encodes a 100-kb PAI that has similarities with other plant pathogenic streptomycetes (Huguet-Tapia & Loria, 2012). They hypothesized that acquisition of the conserved PAI and the remnants of a conserved integrase/recombinase at its 3' end has helped the emergence of this streptomycete as a plant pathogen.

8.2.2.1   ROLE OF THE PAI IN EVOLUTION OF PATHOGENICITY

Among hundreds of known *Streptomyces* species only a few have the pathogenic ability without specified host selectivity. The pathogenic phenotype may improve the quality of recipient strains by maintaining PAI in the recipient host genomes often and effective transfer of *S. turgidiscabies* PAI to another saprophytic *Streptomyces* species *in vitro* gives the concept of mobilization of the pathogenicity factors in *Streptomyces* species, which occurs easily in nature. However, the presence of only a few pathogenic strains in nature reveals the genetic barrier of PAI instability, which forestalls the expression of pathogenicity genes present in recipient strains (Loria et al., 2006).

8.2.3   TOXIN RESPONSIBLE FOR PATHOGENICITY: THAXTOMIN AS A PATHOGENICITY DETERMINANT

Toxin production by *Streptomyces* species is believed to play a critical role in pathogenesis (Table 8.1). A toxin known as thaxtomin and many members of family tyhaxtomin have the ability to induce necrosis on excised potato tubers and there is appearance of scab like symptoms on immature potato tubers (Lawrence et al., 1990). The ability of thaxtomin to produce symptom on tuber suggests its role in pathogenesis of *S. scabies*. Disruption and complementation of genes for thaxtomin production confirmed the requirement of this toxin for pathogenicity (Healy et al., 2000). This is now well established that thaxtomin biosynthesis pathway has a critical role in the evolution of pathogenicity strains of *Streptomyces*. The gene *nec1*, which was discovered during screening of genetically modified strains, designed for identification of thaxtomin production genes. Bukhalid and Loria (1997)

TABLE 8.1   List of Toxins from *Streptomyces* spp. and Their Producers.

S. N.	Pathogenic Streptomyces	Toxin	Reference
1.	*S. scabies*	Thaxtomins, concanamycin A and B, COR-like metabolite(s)	King et al. (1989, 1992); Natsume et al. (1996, 1998, 2001); Bignell et al. (2010)
2.	*S. acidiscabies*	Thaxtomins	Bukhalid et al. (1998)
3.	*S. europaeiscabies*	Thaxtomins	Loria et al. (2006)
4.	*S. stelliscabiei*	Thaxtomins	Loria et al. (2006)
5.	*S. turgidiscabies*	Thaxtomin A	Bukhalid et al. (1998)
6.	*S. ipomoeae*	Thaxtomins	King et al. (1994); Guan et al. (2012)
7.	*S. cheloniumii*	FD-891	Natsume et al. (2005)
8.	*Streptomyces* spp. GK18	Borrelidin	Cao et al. (2012)

created a cosmid library of *S. scabies*, which was expressed in a non-pathogenic strain of *Streptomyces lividans* and screened for their ability to produce necrosis on potato tuber slices. The screening results confirmed the role of the cosmid library in inducing necrosis. Further, deletion mutation analysis identified a 666 bp in the gene *nec1* which facilitated non-pathogenic *S. lividans* to colonize and necrotize potato tuber slices and deletion of *nec1* showed no effect on production of thaxtomin by *S. turgidiscabies*. Similarly, the polysaccharide suberin is a major constituent of potato skin and reported to affect secondary metabolism leading to overproduction of BlDK proteins (Lauzier et al., 2008). This polysaccharide facilitates secondary metabolism and differentiation in different *Streptomyces* species (Lerat et al., 2012). A study demonstrated that presence of suberin in the culture medium for *S. scabies* induced the production of a number of glycosyl hydrolases. It was also suggested that production of suberin-degrading esterase by *S. scabies* may also be one of the reasons for pathogenicity of *S. scabies* (Komeil et al., 2013; Komeil et al., 2014). Another report shows the induction of thaxtomin production by *S. scabies*, this is due to induction of genes involved in biosynthesis of these secondary metabolites are known to be induced by cellobiose (Fig. 8.2), a plant disaccharide. However, in a minimal medium containing cellobiose carbon source, growth of *S. scabies* was found and there is only a trace amount of thaxtomins is produced. Bacterial growth doubles in the presence of both cellobiose and suberin and triggers the production of thaxtomin A (Fig. 8.3), which can be correlated with the upregulation of genes involved in thaxtomins synthesis by 342 folds. Addition of cellobiose and suberin singly does not affect the above process. Stimulation by suberin takes place at the onset of secondary metabolism, which is a pre requirement for the production of molecules such as thaxtomin A, while cellobiose induces biosynthesis of this secondary metabolite (Lerat et al., 2009; Lerat et al., 2010).

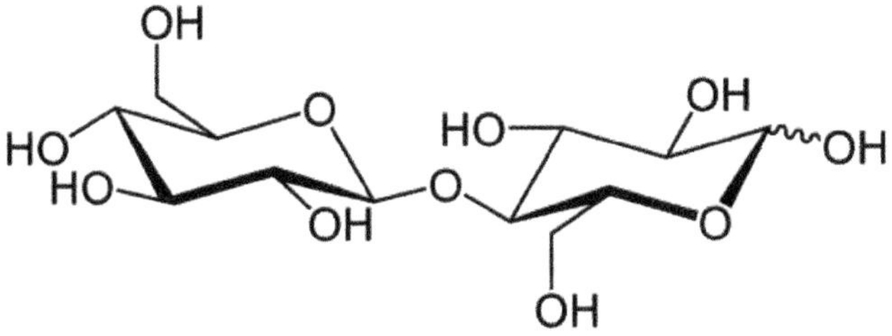

FIGURE 8.2   Cellobiose of plant cell wall that induces thaxtomin production.

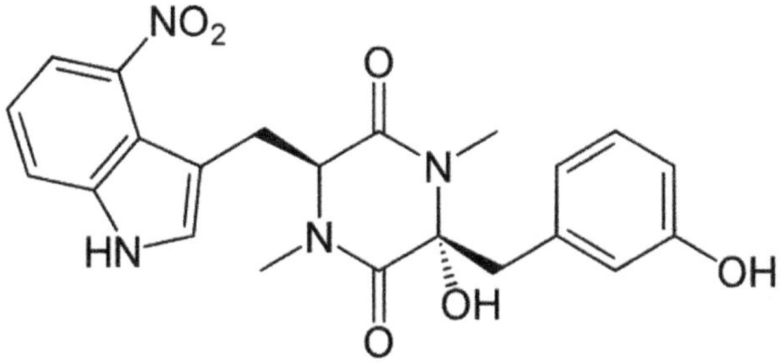

FIGURE 8.3   Chemical structure of thaxtomin A.

Bacteria produce an iron chelating compound known as siderophore that helps in iron uptake, can act as important *vir* determinants for plant pathogens (Poomthongdee et al., 2015). Siderophore production is the most common way for bacteria to grow in iron deficient conditions (Guerinot, 1994). It is important to note that some saprophytic *Streptomyces* species also produce siderophores (Barona-Gomez et al., 2006, 2004; Bickel et al., 1960; Imbert et al., 1995; Schupp, 1988). Another report demonstrated production of siderophore other than desferrioxamines and enterobactin, a type of catecholate siderophore which is typically produced by members of the family *Enterobacteriaceae* is produced by *Streptomyces* sp. Tu6125 (Fiedler et al., 2001). Another partially characterized siderophore griseobactin, containing catechol, threonine, and arginine is found to be produced by *S. griseus* and *Streptomyces* sp. ATCC 700974. Coelichelin, a novel trishydroxamate siderophore, is produced by *S. coelicolor* (Lautru et al., 2005; Patzer & Braun, 2010). Whole genome sequencing of pathogenic *S. scabies* showed the presence of pyochelin biosynthetic gene cluster (PBGC) in these bacteria (Seipke et al., 2011). Iron acquisition by the bacterial soft rot pathogen with help of siderophore and production of pectate lyase *vir* proteins is coupled in *Erwinia chrysanthemi* (Enard et al., 1988; Franza et al., 2002). Iron acquisition is also an important factor responsible for the pathogenicity of apple fire blight pathogen, *E. amylovora* (Dellagi et al., 1998; Seipke et al., 2011).

Some other phytotoxic secondary metabolites produced by *Streptomyces* were identified as concanamycins, FD-891, and borrelidin. A few other evidences reported, which may be responsible for the pathogenicity of *Streptomyces*. The phytotoxic compound concanamycins, polyketide macrolides, was reported first in *S. diastatochromogenes* (Kinashi et al., 1984). This toxin is characterized by presence of a methyl enol with 18-membered tetraenic macrolide ring ether and a b-hydroxyhemiacetyl side chain; it functions as vacuolar-type ATPase inhibitors and has antifungal, anti-neoplastic activity but not antibacterial activity (Kinashi et al., 1984; Seki-Asano et al., 1994). Whole genome sequence of *S. scabies* strain 87–22 tails the presence of a biosynthetic gene cluster, which are highly similar to biosynthetic gene cluster for coronafacic acid (CFA) biosynthesis, present in Gram-negative plant pathogens *Pseudomonas syringae* and *Pectobacterium atrosepticum* (Bignell et al., 2010). CFA, polyketide component of coronatine (COR), is a non-host specific phytotoxin (Gross & Loper, 2009). Biosynthetic gene cluster similar to CFA is reported in *S. scabies* 87–22 having 15 genes. Out of 15, nine are homologous to CFA biosynthetic gene clusters of *P. syringae* pv. tomato and *P. atrosepticum* (Bignell et al., 2010). However, there is no detection of thaxtomin A and gene coding this toxin from this strain of *Streptomyces*. A recent report identified a new pathogenic strain of *Streptomyces*, isolated from a scab lesion on a potato grown in Iran (Cao et al., 2012). The strain (GK18) induced deep-pitted lesions on potato tubers which was different than the raised lesions caused by *S. scabies* and other thaxtomin-producing species. This strain causes severe stunting of potato plants grown in pots. While, this strain is found to produce another 18-membered polyketide macrolide borrelidin toxin is identified first in *Streptomyces rochei* (Berger et al., 1949; Bignell et al., 2013).

Previously a few reports suggested that common scab-causing species have an ability to produce melanin. However, molecular analysis and characterization suggested that the species are not related to single species but a range of species. Some confirmation test like complementation tests assay (Lorang et al., 1995), DNA–DNA hybridization assay (Healy & Lambert, 1991; Paradis et al., 1994) also host range analysis (Lerat et al., 2009), and protein profile analysis (Paradis et al., 1994) suggested that there could be additional melanin producing species that cause common scab (Dees et al., 2012; Goyer & Beaulieu, 1996).

8.2.3.1 ROLE OF NITRIC OXIDE SIGNALING (NOS) IN THAXTOMIN PRODUCTION

A study investigated the NOS in the thaxtomin production by *Streptomyces* spp. (Johnson et al., 2008). They have demonstrated that NOS provides the active nitrogen species used in nitration. Cellulose biosynthesis inhibited by thaxtomin, which is an essential criterion for pathogenicity of most of the disease causing species of *Streptomyces* (Fry & Loria, 2002; Scheible et al., 2003). NOS plays role in nitration of thaxtomin and this is dependent on the NO produced by certain species of *Streptomyces* in the medium known for thaxtomin induction. Johnson et al. (2008) used *S. turgidiscabies* Car8, to determine whether NOS-dependent NO production occurs in thaxtomin-producing streptomycetes, they have added 1 mM CPTIO in oat bran broth (OBB, the thaxtomin inducer media) and grown the mutant (Car8nos) and wild type Car8 starin. After this study they have shown presence of NO but not directly as demonstrate the presence of NO in the medium by Akaike and Maeda (1996).

This is well known that thaxtomin production is stimulated by cello-biose (Wach et al., 2007), which was proved by detection of transcriptional upregulation of thaxtomin biosynthetic genes by the AraC-family regulator, TxtR (Johnson et al., 2007; Joshi et al., 2007), which proposed a hypothesis about the role of cellobiose in the biosynthesis of NO and its correlation with induction of thaxtomin production. Johnson et al. (2008) reported 3.9-fold increase in the rate of NO release by Car8 in OBAC (oat, bran, agar, cellobiose) compared to agar without cellobiose proves the induction of NO production by cellobiose. It is observed that cellulose synthesis occurs in expanding plant tissues like elongation zone of roots. Cellobiose and other cello-oligosaccharides are produced during cellulose synthesis by auxin induced cellulases enzymes of plant (Ohmiya et al., 2000; Shani et al., 2006). This is known that scab-causing streptomycetes infect expanding plant tissues only. A study demonstrated that only susceptible host tissues produce cello-oligosaccharides and this production stimulates the thaxtomin production (Johnson et al., 2007). This result indicates the role of NOS mediated NO production and its involvement in thaxtomin mediated host–pathogen interaction. Accumulation of thaxtomin and NO production correlation suggest that the NO generation out-passes thaxtomin production significantly. As the result showed the amount of NO produced is equal to NO incorporated in thaxtomin and free NO released. However, NO released from mycelium of streptomycetes travels through reactive internal environment of cell. Although, NO appears as byproduct of thaxtomin production in the pathogenic species of *Streptomyces*, NO plays an additional role in the interaction between plant and pathogenic *Streptomyces* spp. This might also be possible that NO serves as an intracellular signaling molecule, which serves as signaling molecule, coordinates the infection process in this filamentous pathogen. This is expected that pathogenic *Streptomyces* would encounter reactive oxygen species (ROSs) in early infection process. NO may provide protection against organic peroxyl radicals in the lipid membrane, which prevents peroxidation chain reactions (Moller et al., 2005; Rubbo et al., 2002).

There is the involvement of an operon containing the non-ribosomal peptide synthetase (NRPS) modules in the biosynthesis of thaxtomin A, B, and C and production of the dipeptide backbone by *S. turgidiscabies* (Healy et al., 2000). This operon also regulates a P450 mono-oxygenase, txtC, which adds two hydroxyl groups (Healy et al., 2002). Interestingly the gene for NOS lies in the upstream about 5 kb of the operon for thaxtomin, which allows its independent transcriptional regulation. NOS produces NO and NO donors have an ability to complement thaxtomin production in a NOS deletion strain (Wach et al., 2007). There is no direct effect of NOS on nitration of thaxtomin. However, this is demonstrated that reactive nitrogen species obtained from NOS is the source for nitration (Kers et al., 2004). It may happen that there is involvement of any other enzyme for site-specific nitration of either dipeptide backbone or the tryptophan. Presence of an unusual P450 mono-oxygenase encodes a directly upstream element of the *S. turgidiscabies* NOS (stNOS). However, people are still searching for proteins involved in the specific nitration for thaxtomin biosynthesis in *S. acidiscabies* and *S. scabies*. Another study showed that NOS has the ability to nitrate small amounts of tryptophan *in vitro* (Buddha et al., 2004) in *Deinococcus radiodurans*. Thaxtomin production includes only a few phytopathogenic members of streptomycetes. The streptomycetes, which involve in thaxtomin production, are NOS producing bacteria and NO producing genes are found adjacent to the secondary metabolite producing genes; most often these pathways are found clustered in prokaryotes. There is possibility that NO production present in most of the scab-causing *Streptomyces* and also have ability to produce thaxtomin.

8.2.4   MOLECULAR MECHANISMS OF PATHOGENICITY

The *fas* operon of *S. turgidiscabies* is collinear with *fas* operon of *Rhodococcus fascians,* an exception in case of *fas6,* which was shown to have undergone a rearrangement on the *S. turgidiscabies* chromosome. These genes have 67–84% similarity with amino acid sequence of *R. fascians fas* operon (Kers et al., 2005). In comparison to the *fas* operon in *S. turgidiscabies,* the *fas* operon in *R. fascians* is homologous and collinearly arranged and suggests a common origin of this operon. The *fas* operon is found conserved in both *S. scabies* and *S. acidiscabies,* which shows its differential nature to *nec1* and thaxtomin biosynthetic genes. A report showed the presence of a *fas* operon on the PAI of *S. turgidiscabies,* however, no report showing production of plant gall by these pathogenic streptomycetes. One report suggests the production of root gall by *Streptomyces* species, however, genetic basis of the phenotypes produced by these pathogens is not known. Reports are available regarding the ability of *S. turgidiscabies* Car8 that has a *fas* operon on the PAI for induction of leafy gall and primary symptoms similar to the symptoms induced by *R. fascians* (Yoshida & Kobayashi, 1991). In another report researchers mapped the integration site of PAI in *S. coelicolor* M145 by using comparative restriction fragment length polymorphism analyses of *S. coelicolor* wild type and transconjugant genomic DNA (Kers et al., 2005). The SCO1326 is integrated with PAI in an 876 bp open reading frame (ORF) *bacA,* which encodes an enzyme UPP. Involvement of this enzyme is presumed as an ortholog reported in *E. coli,* helps in peptidoglycan biosynthesis and resistance against bacitracin. The insertion points for PAI were sequenced in three transconjugants and recombination in SCO1326 at 11 bp regions on 8 bp palindromic sequences. A homologue of 876 bp *bacA* having 85% nucleotide similarity to SCO1326 was found in *S. turgidiscabies* on 5'-terminus of the PAI. PAI integration in the SCO1326 results the insertion in the ORF and creates a hybrid copy of ORF. Another report showed similar result as in *S. turgidiscabies* with *S. diastatochromogenes,* where trans-conjugates were analyzed. There is no difference in the level bacitracin resistance between transconjugant and wild type strains of *S. coelicolor.* Although, the genes for UPP are conserved among the bacteria, no previous study showed that these are the target sites for HGT. SCO1326 has the target for HGT, predicted by the presence of immediately flanked by low G+C DNA in wild type *S. coelicolor*.

8.3   SAPONINASE IN A BACTERIAL PATHOGEN

DNA sequencing in *nec1* gene region reveals the discovery of a saponi-nase gene homolog, which encodes a well characterized enzyme of fungal plant pathogens (Kers et al., 2005; Huguet-Tapia et al., 2011). This protein contains a predicted N-terminal secretion signal and is the member of glycosyl hydrolase family 10. Saponins preformed determinants of resis-tance to fungal attack and are plant glycosides. Formation of complexes is the primary mode of with membrane sterols present in eukaryotes, results in loss of membrane integrity (Osbourn, 1996). However, a homolog of saponinase discovered in *S. turgidiscabies* was found conserved in *S. acidicabies* and *S. scabies*, which suggested the role of this enzyme in pathogenicity (Kers et al., 2005). Characterization of several sections of the DNA sequence found in the region of PAIs and the *vir* determinants in *S. turgidiscabies* suggested the presence of PAI (Groisman & Ochman, 1996). Pathogenicity and *vir* gene conservation in the polyphyletic species reveal a typical G+C content in these regions along with the presence of multiple transposons and other insertion sequences. The clustering of these genes responsible for pathoge-nicity and *vir* supported the existence of PAI in scab-causing *Streptomyces* spp. and was the first report of a PAI in plant pathogenic Gram-positive bacteria (Kers et al., 2005). Emergence of newer plant pathogenic species of *Sterptomyces* suggests mobilization of PAI in some strains. Several reports showed that the PAI from *S. turgidiscabies* was transferred to other species of *Streptomyces* by conjugation (Miyajima et al., 1998; Takeuchi et al., 1996). A marker for antibiotic resistance was tagged with *nec1* to detect transfer of PAI from the donor *S. turgidiscabies* Car811 to the recipients *S. coelicolor* M145 and *S. diastatochromogenes*, which confirmed the mobiliz-able property of the PAI (Kers et al., 2005).

The mobile gene element is known as integrative conjugative element (ICE) and has the ability for lateral gene transfer (LGT) in prokaryotes. Transfer and its integration in a chromosome at specific location are mediated by conjugation from donor to recipient (Burrus & Waldor, 2004; Wozniak & Woldor, 2010). A site-specific recombinase helps in site-specific recom-bination. A few researchers reported that LGT plays an important role in the evolution of plant pathogenic streptomycetes (Loria et al., 2006; Miyajima et al., 1998). *S. turgidiscabies* has been best characterized for LGT of PAIs

(Huguet-Tapia et al., 2014; Kers et al., 2005). The PAIs have the ability to mobilize themselves from *S. turgidiscabies* and integrate in the chromosome of *S. coelicolor*, *S. diastochromogenes*, and *S. lividans* during conjugation (Huguet-Tapia et al., 2014; Kers et al., 2005).

8.4 HOST SELECTIVITY

Universal dominance of the pathogenic species *S. scabies* is known world-wide and many other species are also known to cause disease and show same type of symptoms on potato tuber. Researchers are working on the mechanisms of host dwelling of the pathogenic species by using molecular tools like PCR, RT-PCR, microarray, etc. Identification of the causal agent of common scab of potato was done in nineteenth century (Thaxter, 1892). In India this bacterium was first reported in Khasi Hills of Meghalaya (McRae, 1929). A recent review on this disease showed its presence worldwide in countries where potatoes are cultivated. Number of species has been reported for the disease in which *S. scabies* is a well studied species; many other species of *Streptomyces* are also recognized causing common scab. The discovery and isolation of actinomycetes from infected potato tuber open new opportunities for the search of control measures of the disease (Bouchek-Mechiche et al., 2000; Miyajima et al., 1998; Park et al., 2003; Wanner, 2006, 2007).

Studies on the structure of the pathogenic toxin thaxtomin revealed it as a dipeptide composed of phenyl alanine and nitrated tryptophan. This structure of the toxin suggests its synthesis to be non-ribosomal. Unreveling of the structure of the toxin had generated enthusiasm among the researchers to develop a knowledge regarding its biosynthesis and developed strategies to clone this gene in various microorganisms. Some researchers used degenerate oligo-nucleotide primers to amplify the conserved portions of the acyl-adenylation modules of NRPS genes by using genomic DNA isolated from toxin producing species of *Streptomyces*. Some others used another technique Southern hybridization and identified a conserved domain of NRPS acyladenylation. Sequencing of flanking DNA sequences of NRPS module opened the knowledge about the peptide synthesis genes (*txtA*, *txtB*) that encoded proteins consistent with the biosynthesis of the *N*-methylated cyclic dipeptide backbone of taxtomins. Downstream sequencing of *txtA* and *txtB* in *S. acidiscabies* resulted in the discovery of a P450 mono-xygenase, *txtC*. Genetic analysis confirmed the role of these NRPS domains in thaxtomin production and requirement of *txtC* for postcyclization hydroxylation of thaxtomin A (Healy et al., 2000).

8.5 CONTROL MEASURES FOR COMMON SCAB

Numerous efforts had been made to develop a suitable control measure for limiting this particular disease and the efforts included application of chemicals (Hooker, 1990), changing soil pH (Davis et al., 1974; Pavlista, 1992; Waterer, 2002), soil amendments (Mishra & Srivastava, 2005), organic amendments (Lazarovits et al., 2001), soil fumigation (Hooker, 1990), green manuring (Mishra & Srivastava, 2005), crop rotation (Hooker, 1990), excess irrigation during tuber formation (Loria et al., 1997), potato seed bacterization (Tanii et al., 1990), and agrochemicals (Neeno-Eckwall et al., 2001). However, so far no full proof management practices for common scab of potato are available. Moreover, any single approach such as cultural practices or fungitoxicants or host plant resistance or bio-agents, is not sufficient for management of this worldwide pathogen.

8.5.1 BIOCONTROL

In this approach eco-friendly microbes are used to control different diseases. Biocontrol agents such as *Pseudomonas* and *Bacillus* spp. are used to manage this pathogen (Table 8.2). In recent years, scientists are making efforts to manage the pathogen more effectively through integrated disease management practices by selection and application of useful combination of appropriate techniques to suppress the disease to a tolerable level. Application of biocontrol agents singly or in the combination of microorganisms with chemical control measures is realized to have potential in managing the pathogen. Use of plant growth-promoting rhizobacteria (PGPR) for biological management of the pathogen especially through antagonistic *Pseudomonas* spp. has emerged as a strong management tool for several soil-borne pathogens including bacteria (Kishore et al., 2005; Jain et al., 2012; Sarma et al., 2002; Singh et al., 2013). Soil amendments with specific biocontrol agents change the physicochemical and microbiological environment which suppresses the growth of soil invading pathogens (Funk & Lumsden, 1999; Sahni et al., 2008a, 2008b). Another report suggested the use of two suppressive strains of *Streptomyces* (*S. diastatochromogenes* strain Pon SS II and *S. scabies* strain Pon R) for the control of common scab in field condition (Liu et al., 1995). The suppressive strains did not affect the tuber yield and were re-isolated from the tubers grown in inoculated soil. In another experiment, suppressive species of the bacterial strain *Bacillus* sp. sunhua was isolated from the soil of potato cultivating area for the control of potato scab. The strain was found to produce an antibiotic which was stable at a broad range of pH. Another recent report showed the role of four pseudomonad strains as potential biocontrol agents particularly when amended with vermicompost reduced common scab symptoms in the potato tubers and also increased the tuber yield (Singhai et al., 2011a, 2011b) also suggested the involvement of secondary metabolites as possible contributory factors in resistance against *S. scabies* in the potato cultivars other than their genetic makeup. They found that the presence of phenolic acids in the peels of the potato cultivars may determine the basis of resistance in different cultivars of potato. A recent study showed the role of rhizospheric bacteria against six pathogens of potato including *S. scabies*. They used semi selective media to isolate number of rhizospheric bacteria and checked their effectiveness against six potato phytopathogens (Turnbull et al., 2012). Results of this study revealed that *Pseudomonas* spp. and *Lysobacter capsici* had the greatest antagonistic activity on media used in laboratory against six potato phytopathogens, and significantly decreased disease in plants grown in pathogen-infested soil also. Four other isolates namely *Asticcaculis*, *Methylobactrium*, *Paenibacillus*, *Pseudomonas umsongensis* increase growth of potato significantly in nodal explants in plant tissue culture.

TABLE 8.2   List of Biocontrol Agents for Common Scab and Their Mechanism of Action.

S. N.	Name of biocontrol agent	Mechanism of control	Reference
1.	*Pseudomonas* sp. LBUM223	Phenazine production	Arseneault et al. (2013)
2.	*Streptomyces melanosporofaciens* strain-EF-76	Antagonism	Beauséjour et al. (2003)
3.	*Pseudomonas*	Phenazine production	Mavrodi et al. (2006)
4.	*Bacillus* sp.	Antagonism	Han et al. (2005)
5.	*S. griseoviridis* strain-K61 (Mycostop), *Streptomyces* strain-346	Antagonism	Hiltunen et al. (2009), Liu et al. (1995)
6.	*Pseudomonas* spp.	Antimicrobial compounds	Paulin et al. (2009)
7.	*Pseudomonas* species and vermicompost	Antimicrobial compounds	Singhai et al. (2011a)
8.	*Pseudomonas* spp.	Repression of pathogenesis related genes	St-Onge et al. (2010)
9.	*Lysobacter capsici*	antagonistic activity	Turnbull et al. (2012)

8.5.2 NON-PATHOGENIC STREPTOMYCES

Potato is highly used as a staple food all over the world having fourth position on the basis of its use in food (Loria et al., 1997). No adequate control measures are there for the control of common scab disease, which decreases quality of potato by scabby wart like pitted lesions on the skin of potato tuber. Chater et al. (2010) reported that non-pathogenic *Streptomyces* spp. are mostly soil saprophytes degrading plant and fungal materials for their use as carbon sources and also reported the production of a number of extracellular enzymes including cellulases, pectinases, proteinases, and chitinases. Janssen (2006) showed that only a few *Streptomyces* spp., most of which are closely related to *Streptomyces scabies*, comprise aproximitely 1% of the 10^6–10^7 colony forming units (CFU) of *Streptomyces* per gram of typical agricultural soils and are apparently in endemic soil microbes.

A number of streptomycete strains had been utilized as biocontrol agents (Table 8.3). *Streptomyces griseoviridis* formulation is being used commercially (Verdera Oy, Espoo, Finland). Some non-pathogenic strains of *Streptomyces* species suppressed pathogenic *Streptomyces* species (Hiltunen et al., 2009; Ryan & Kinkel, 1997). However, the suppression results were not reproducible in multiyear field level experiments (Ryan & Kinkel, 1997).

TABLE 8.3   List of Pathogenic and Non-Pathogenic Species of *Streptomyces*.

S. N.	Pathogenic Streptomyces	Non-pathogenic Streptomyces
1.	S. scabiei	S. venezuelae
2.	S. acidiscabies	S. roseosporus
3.	S. europaeiscabies	S. fradiae
4	S. luridiscabiei	S. lincolnensis
5.	S. niveiscabiei	S. fradiae
6.	S. puniciscabiei	S. alboniger
7.	S. reticuliscabiei	S. griseus
8.	S. stelliscabiei	S. rimosus
9.	S. turgidiscabies	S. aureofaciens
10.	S. ipomoeae	S. clavuligerus
11.	S. aureofaciens	S. griseoviridis
12.	S. puniciscabiei	S. griseoruber
13.		S. violaceusniger
14.		S. albidoflavus
15.		S. atroolivaceus

Doumbou et al. (2001) reported non-pathogenic isolates of *Streptomyces* species from potato common scab lesions, and larger proportions of non-pathogenic isolates of *Streptomyces* species were associated with control of common scab disease (Wanner, 2006, 2007). Successful field application of *Streptomyces* strains to suppress common scab of potato was described by several earlier workers (Jiang et al., 2012; Neeno-Eckwall, 2001; Ryan et al., 2011). Non-pathogenic *Streptomyces* species had been reported to function as biocontrol agents in many other cases as well (Jobin et al., 2005; Mazzola & Manica, 2012).

8.6   FUTURE PROSPECTS

Mostly *Streptomyces* species are not plant pathogenic in nature but sometimes their behavior similar to fungal-like strategy is very prominent for plant root attachment and infection. Further work is needed to know the genetic and molecular aspects of fungal-like strategy of *Streptomyces* species, which are still unknown. The Gram-positive bacteria are unable to inject their protein through type III secretion system due to lack of T3SS. Knowledge of T3SS in *Streptomyces* species may reveal one of the various infection strategies. HGT, genome reduction, and recombination are some important processes which are required in genetic repertoires and their associations for the adaptation in particular niches. Further determination of novel repertoire for pathogenecity by using bioinformatical tools in *S. scabies* genome is also essential for characterization of functional PAIs. Characterization of diversity in PAIs of pathogenic *Streptomyces* can explore the island evolution and involvement of vital genes to survive in nature. A much needed research work in area of control of *S. scabies* infection through thaxtomin production by using transgenic approach is also required. Transgenic expression of plant thaxtomin inhibiting proteins may control the loss due to *S. scabies* infection in plants. To develop transgenic approach identification of host cell targets for thaxtomin A and the nec1 protein is most important step. Furthermore, the research is also required in the area to know the compounds of host defense which interact with *vir* repertoires. It can be improved by using genome sequencing of the host targets and their interactions for enhanced disease resistance. Information derived from genome-scale sequencing of pathogens is being used to develop improved methods for disease diagnosis and analysis of population dynamics, and to reveal *vir* strategies that can be manipulated for disease control.

KEYWORDS

- horizontal gene transfer (HGT)
- host specificity
- pathogenesis associated islands (PAI)
- pathogenicity
- potato scab
- thaxtomin

REFERENCES

Akaike, T.; Maeda, H. Quantitation of Nitric Oxide Using 2-Phenyl-4,4,5,5-Tetramethylimid-azoline-1-Oxyl 3-Oxide. *Meth. Enzymol.* **1996,** *268,* 211–221.

Alfano, J. R.; Collmer, A. Type III Secretion System Effector Proteins: Double Agents in Bacterial Disease and Plant Defense. *Annu. Rev. Phytopathol.* **2004,** *42,* 385–414.

Arseneault, T.; Goyer, C.; Martin, F. Phenazine Production by *Pseudomonas* sp. LBUM223 Contributes to the Biological Control of Potato Common Scab. *Biol. Control.* **2013,** *103*(10), 995.

Barona-Gomez, F.; Lautru, S.; Francou, F. X.; Leblond, P.; Pernodet J. L.; Challis, G. L. Multiple Biosynthetic and Uptake Systems Mediate Siderophore-Dependent Iron Acquisition in *Streptomyces coelicolor* A3 (2) and *Streptomyces ambofaciens* ATCC 23877. *Microbiology.* **2006,** *152,* 3355–3366.

Barona-Gomez, F.; Wong, U.; Giannakopulos, A. E.; Derrick, P. J.; Challis, G. L. Identification of a Cluster of Genes that Directs Desferrioxamine Biosynthesis in *Streptomyces coelicolor* M145. *J. Am. Chem. Soc.* **2004,** *126,* 16282–16283.

Beauséjour, J.; Clermont, N.; Beaulieu, C. Effect of *Streptomyces melanosporofaciens* Strain EF-76 and of Chitosan on Common Scab of Potato. *Plant. Soil.* **2003,** *256,* 463–468.

Berger, J.; Jampolsky, L. M.; Goldberg; Borrelidin, M. W. A New Antibiotic with Antiborrelia Activity and Penicillin Enhancement Properties. *Arch. Biochem.* **1949,** *22,* 476–478.

Bickel, H.; Bosshardt, R.; Gaumann, E.; Reusser, P.; Vischer, E.; Voser, W.; Wettstein, A.; Zahner H. Metabolic Products of *Actinomycetaceae. Helv. Chim. Acta.* **1960,** *43,* 2118–2128.

Bignell, D. R. D; Fyans, J. K.; Cheng, Z. Phytotoxins Produced by Plant Pathogenic *Streptomyces. J. Appl. Microbiol.* **2013,** *116,* 223–235.

Bignell, D. R.; Seipke, R. F.; Huguet-Tapia, J. C.; Chambers, A. H.; Parry, R.; Loria, R. *Streptomyces scabies* 87-22 Contains a Coronafacic Acid-Like Biosynthetic Cluster that Contributes to Plant-Microbe Interactions. *Mol. Plant-Microbe Interact.* **2010,** *23,* 161–175.

Bouchek-Mechiche, K.; Gardan L.; Normand P.; Jouan B. DNA Relatedness among Strains of *Streptomyces* Pathogenic to Potato in France: Description of Three New Species, *S. europeiscabiei* sp. nov. and *S. stelliscabiei* sp. nov. Associated with Common Scab, and

S. reticuliscabiei sp. nov. Associated with Netted Scab. *Int. J. Syst. Evol. Microbiol.* **2000,** *50,* 91–99.

Buddha, M. R.; Tao, T.; Parry, R. J.; Crane, B. R. Regioselective Nitration of Tryptophan by a Complex Between Bacterial Nitric-Oxide Synthase and Tryptophanyl-Trna Synthetase. *J. Biol. Chem.* **1992,** *279,* 49567–49570.

Bukhalid, R. A.; Chung S. Y.; Loria R. *nec1,* a Gene Conferring a Necrogenic Phenotype, is Conserved in Plant-Pathogenic *Streptomyces* Species and Linked to a Transposase Pseudo-gene. *Mol. Plant-Microbe Interact.* **1998,** *11,* 960–967.

Bukhalid, R. A.; Takeuchi T.; Labeda, D.; Loria R. Horizontal Transfer of the Plant Virulence Gene, *nec1,* and Flanking Sequences among Genetically Distinct *Streptomyces* Strains in the Diastatochromogenes Cluster. *Appl. Environ. Microbiol.* **2002,** *68,* 738–44.

Bukhalid, R. B.; Loria, R. Cloning and Expression of a Gene from *Streptomyces scabies* Encoding a Putative Pathogenicity Factor. *J. Bacteriol.* **1997,** *179,* 7776–7783.

Burrus, V.; Waldor, M. K. Shaping Bacterial Genomes with Integrative and Conjugative Elements. Genome Plasticity and the Evolution of Microbial Genomes. *Res. Microbiol.* **2004,** *155,* 376–386.

Cao, Z.; Khodakaramian, G.; Arakawa, K.; Kinashi, H. Isolation of Borrelidin as a Phytotoxic Compound from a Potato Pathogenic *Streptomyces* Strain. *Biosci. Biotechnol. Biochem.* **2012,** *76,* 353–357.

Chamberlain, K.; Crawford, D. L. Thatch Biodegradation and Antifungal Activities of Two Lignocellulolytic *Streptomyces* Strains in Laboratory Cultures and in Golf Green Turfgrass. *Can. J. Microbiol.* **2000,** *46,* 550–58.

Chater, K. F.; Biro, S.; Lee, K. J.; Palmer, T.; Schrempf H. The Complex Extracellular Biology of *Streptomyces. FEMS Microbiol. Rev.* **2010,** *34,* 171–198.

Davis, J. R.; Garner, J. G.; Callihan, R. H. Effects of Gypsum, Sulphur, Terrachalor and Terrachalor-X for Scab Control. *Am. J. Potato Res.* **1974,** *51,* 36–43.

Dees, M. W.; Somervuo, P.; Lysøe, E.; Aittamaa, M.; Valkonen, J. P. T. Species' Identification and Microarray-Based Comparative Genome Analysis of *Streptomyces* Species Isolated from Potato Scab Lesions in Norway. *Mol. Plant Pathol.* **2012,** *13*(2), 174–186.

Dellagi, A.; Brisset, M. N.; Paulin, J. P.; Expert, D. Dual Role of Desferrioxamine in *Erwinia amylovora* Pathogenicity. *Mol. Plant-Microbe Interact.* **1998,** *11,* 734–742.

Doumbou, C. L.; Akimov, V.; Cote, M.; Charest, P. M.; Beaulieu, C. Taxonomic Study on Nonpathogenic *Streptomycetes* Isolated from Common Scab Lesions on Potato Tubers. *Syst. Appl. Microbiol.* **2001,** *24,* 451–456.

Enard, C.; Diolez A.; Expert, D. Systemic Virulence of *Erwinia chrysanthemi* 3937 Requires a Functional Iron Assimilation System. *J. Bacteriol.* **1988,** *170,* 2419–2426.

Ezra, D.; Barash, I.; Valinsky, L.; Manulis, S. The Dual Function in Virulence and Host Range Restriction of a Gene Isolated from the $pPATH_{Ehg}$ Plasmid of *Erwinia herbicola* pv. *gypsophilae. Mol. Plant-Microbe Interact.* **2000,** *13,* 683–692.

Fiedler, H. P.; Krastel, P.; Muller, J.; Gebhardt, K.; Zeeck, A. Enterobactin: The Characteristic Catecholate Siderophore of *Enterobacteriaceae* is Produced by *Streptomyces* species.(1). *FEMS Microbiol. Lett.* **2001,** *196,* 147–151.

Flor, H. H. The Complementary Genetic Systems in Flax and Flax Rust. *Adv. Genet.* **1956,** *8,* 29–54.

Franza, T.; Michaud-Soret, I.; Piquerel, P.; Expert, D. Coupling of Iron Assimilation and Pecti-nolysis in *Erwinia chrysanthemi* 3937. *Mol. Plant-Microbe Interact.* **2002,** *15,* 1181–1191.

Fry, B.A.; Loria, R. Thaxtomin A: Evidence for a Plant Cell Wall Target. *Physiol. Mol. Plant Pathol.* **2002**, *60*, 1–8.

Funk Jensen, D.; Lumsden, R. D. Biological Control of Soil Borne Pathogens. In *Integrated Pest and Disease Management in Greenhouse Crops;* Albajes, R., M. L., Gullino, J. C., Van Lenteren, Y. Elad., Eds.; Kluwer Academic Publishers: Dordrecht, The Netherlands, 1999, pp 319–337.

Goyer, C.; Beaulieu, C. Host Range of *Streptomycete* Strains Causing Common Scab. *Plant Dis.* **1997**, *81*, 901–904.

Goyer, C.; Otrysko, B.; Beaulieu, C. Taxonomic Studies on *Streptomycetes* Causing Potato Common Scab: A Review. *Can. J. Plant Pathol.* **1996**, *18*, 107–113.

Groisman, E. A.; Ochman, H. Pathogenicity Islands: Bacterial Evolution in Quantum Leaps. *Cell.* **1996**, *87*, 791–794.

Gross, H.; Loper, J. E. Genomics of Secondary Metabolite Production by *Pseudomonas* sp. *Nat. Prod. Rep.* **2009**, *26*, 1408–1446.

Guan, D.; Grau, B. L.; Clark, C. A.; Taylor, C. M.; Loria, R.; Pettis, G. S. Evidence that Thaxtomin C is a Pathogenicity Determinant of *Streptomyces ipomoeae*, the Causative Agent of *Streptomyces* Soil Rot Disease of Sweet Potato. *Mol. Plant-Microbe Interact.* **2012**, *25*, 393–401.

Guerinot, M. L. Microbial Iron Transport. *Annu. Rev. Microbiol.* **1994**, *48*, 743–772.

Hajri, A.; Brin, C.; Hunault, G.; Lardeux, F.; Lemaire, C.; Manceau, C.; Boureau, T.; Poussier, S. A. Repertoire for Repertoire Hypothesis: Repertoires of Type Three Effectors are Candidate Determinants of Host Specificity in *Xanthomonas*. *PLoS ONE.* **2009**, *4*, e6632.

Han, J. S.; Cheng, J. H.; Yoon, T. M.; Song, J.; Rajkarhikar, A.; Kim, W. G.; Yoo, I. D.; Yang, Y. Y.; Suh, J. W. Biological Control Agent of Common Scab Disease by Antagonistic Strain *Bacillus* sp. Sunhua. *J. Appl. Microbiol.* **2005**, *99*, 213–221.

Healy, F. G.; Krasnoff, S. B.; Wach, M.; Gibson, D. M.; Loria, R. Involvement of a Cytochrome P450 Monooxygenase in Thaxtomin-A Biosynthesis by *Streptomyces acidiscabies*. *J. Bacteriol.* **2002**, *184*, 2019–2029.

Healy, F. G.; Lambert D. H. Relationships among *Streptomyces* sp. Causing Potato Scab. *Int. J. Syst. Bacteriol.* **1991**, *41*, 479–482.

Healy, F. G.; Wach, M.; Krasnoff, S. B.; Gibson, D. M.; Loria, R. The txtAB Genes of the Plant Pathogen *Streptomyces acidiscabies* Encode a Peptide Synthetase Required for Phytotoxin Thaxtomin A Production and Pathogenicity. *Mol. Microbiol.* **2000**, *38*, 794–804.

Herbers, K.; Conrads-Strauch, J.; Bonas, U. Race-Specificity of Plant Resistance to Bacterial Spot Disease Determined by Repetitive Motifs in a Bacterial Avirulence Protein. *Nature.* **1992**, *356*, 172–173.

Hill, J.; Lazarovits, G. A. Mail Survey of Growers to Estimate Potato Common Scab Prevalence and Economic Loss in Canada. *Can. J. Plant Pathol.* **2005**, *27*, 46–52.

Hiltunen, L. H.; Ojanpera, T.; Kortemaa, H.; Richter, E.; Lehtonen, M. J.; Valkonen J. P. T. Interactions and Biocontrol of Pathogenic *Streptomyces* Strains Co-Occurring in Potato Scab Lesions. *J. Appl. Microbiol.* **2009**, *106*, 199–212.

Hiltunen, L. H.; Weckman, A.; Ylhäinen, A.; Rita, H.; Richter, E.; Valkonen, J. P. T. Responses of Potato Cultivars to the Common Scab Pathogens, *Streptomyces scabies* and *S. turgidiscabies*. *Ann. Appl. Biol.* **2005**, *146*, 395–403.

Hodgson, D. A. Primary Metabolism and its Control in *Streptomycetes:* A Most Unusual Group of Bacteria. *Adv. Microb. Physiol.* **2000**, *42*, 247–238.

Hooker, W. J. Common Scab. In *Compendium of Potato Diseases;* Hooker, W. J., Ed., The American Phytopathological Society: St. Paul, MN, 1990, pp 33–34.

Horinouchi, S. A. Microbial Hormone, A-factor, as a Master Switch for Morphological Differentiation and Secondary Metabolism in *Streptomyces griseus*. *Front. Biosci.* **2002**, *7*, 45–57.

Huguet-Tapia, J. C.; Badger, J. H.; Loria, R.; Pettis, G. S. *Streptomyces turgidiscabies* Car8 Contains a Modular Pathogenicity Island that Shares Virulence Genes with Other Actinobacterial *Plant Pathog. Plasmid.* **2011**, *65*, 118–124.

Huguet-Tapia, J. C.; Bignell, D. R. D.; Loria, R. Characterization of the Integration and Modular Excision of the Integrative Conjugative Element PAISt in *Streptomyces turgidiscabies* Car8. *PLoS ONE.* **2014**, *9*, e99345.

Huguet-Tapia, J. C.; Loria, R. Draft Genome Sequence of *Streptomyces acidiscabies* 84-104, an Emergent Plant Pathogen. *J, Bacteriol.* **2012**, *194*, 1847.

Imbert, M.; Bechet, M.; Blondeau, R. Comparison of the Main Siderophores Produced by Some Species of *Streptomyces*. *Curr. Microbiol.* **1995**, *31*, 129–133.

Jain, A.; Singh, S.; Sarma, B. K.; Singh, H. B. Microbial Consortium–Mediated Reprogramming of Defence Network in Pea to Enhance Tolerance against *Sclerotinia sclerotiorum*. *J. Appl. Microbiol.* **2012**, *112*(3), 537–550.

Janssen, P. H. Identifying the Dominant Soil Bacterial Taxa in Libraries of 16S rRNA and 16S rRNA Genes. *Appl. Environ. Microbiol.* **2006**, *72*, 1719–1728.

Jiang, H. H.; Meng, Q. X.; Hanson, L. E.; Hao, J. J. First Report of *Streptomyces stelliscabiei* Causing Potato Common Scab in Michigan. *Plant Dis.* **2012**, *96*(6), 904.

Jobin, G.; Couture, G.; Goyer, C.; Brzezinski, R.; Beaulieu, C. Streptomycete Spores Entrapped in Chitosan Beads as a Novel Biocontrol Tool against Common Scab of Potato. *Appl. Microbiol. Biotechnol.* **2005**, *68*, 104–110.

Johnson, E. G.; Joshi, M. V.; Gibson, D. M.; Loria, R. Cello-Oligosaccharides Released from Host Plants Induce Pathogenicity in Scab-Causing *Streptomyces* Species. *Physiol. Mol. Pathol.* **2007**, *71*(1–3), 18–25.

Johnson, E. G.; Sparks, J. P.; Dzikovski, B.; Crane, B. R.; Gibson, D. M.; Loria R. Plant–Pathogenic *Streptomyces* Species Produce Nitric Oxide Synthase-Derived Nitric Oxide in Response to Host Signals. *Chem. Biol.* **2008**, *15*, 43–50.

Joshi, M.; Rong, X.; Moll, S. Kers, J.; Franco, C.; Loria, R. *Streptomyces turgidiscabies* Secretes a Novel Virulence Protein, Nec1, which Facilitates Infection. *Mol. Plant–Microbe Interact.* **2007**, *20*, 599–608.

Kers, J. A.; Cameron, K. D.; Joshi, M. V.; Bukhalid, R. A.; Morello, J. E.; Wach, M. J.; Gibson, D. M.; Loria, R. A Large, Mobile Pathogenicity Island Confers Plant Pathogenicity on *Streptomyces* Species. *Mol. Microbiol.* **2005**, *55*, 1025–1033.

Kers, J. A.; Wach, M. J.; Krasnoff, S. B.; Widom, J.; Cameron, K. D.; Bukhalid, R. A; Gibson, D. M.; Crane, B. R.; Loria, R. Nitration of a Peptide Phytotoxin by Bacterial Nitric Oxide Synthase. *Nature.* **2004**, *429*, 79–82.

Killiny, N.; Almeida, R. P. P. Gene Regulation Mediates Host Specificity of a Bacterial Pathogen. *Environ. Microbiol.* **2011**, *3*, 791–797.

Kinashi, H.; Someno, K.; Sakaguchi, K. Isolation and Characterization of Concanamycins A, B and C. *J. Antibiot.* **1984**, *37*, 1333–1343.

King, R. R.; Lawrence, C. H. Characterization of New Thaxtomina Analogues Generated *in vitro* by *Streptomyces scabies*. *J. Agr. Food Chem.* **1996**, *44*, 1108–1110.

King, R. R.; Lawrence, C. H.; Calhoun, L. A.; Ristaino, J. B. Isolation and Characterization of Thaxtomin-Type Phytotoxins Associated with *Streptomyces ipomoeae*. *J. Agr. Food Chem.* **1994**, *42*, 1791–1794.

King, R. R.; Lawrence, C. H.; Calhoun, L. A. Chemistry of Phytotoxins Associated with *Streptomyces Scabies*, the Causal Organism of Potato Common Scab. *J. Agric. Food Chem.* **1992**, *40*, 834–837

King, R. R.; Lawrence, C. H.; Clark, M. C.; Calhoun, L. A. Isolation and Characterization of Phytotoxins Associated with *Streptomyces scabies*. *J. Chem. Soc. Chem. Commun.* **1989**, *13*, 849–850.

Kirzinger, M. W.; Stavrinides, J. Host Specificity Determinants as a Genetic Continuum. *Trends Microbiol.* **2012**, *20*, 88–93.

Kishore, G. K.; Pande S.; Podile A. R. Biological Control of Collar Rot Disease with Broad Spectrum Antifungal Bacteria Associated with Groundnut. *Can. J. Microbiol.* **2005**, *51*, 123–132.

Kodani, S.; Hudson, M. E.; Durrant, M. C.; Buttner, M. J.; Nodwell, J. R; Willey, J. M. The *SapB* Morphogen is a Lantibiotic-Like Peptide Derived from the Product of the Developmental Gene *ramS* in *Streptomyces coelicolor*. *Proc. Natl. Acad. Sci. U.S.A.* **2004**, *101*, 11448–11453.

Kodani, S.; Lodato M. A.; Durrant, M. C.; Picart, F.; Willey, J. M. SapT, a Lanthionine Containing Peptide Involved in Aerial Hyphae Formation in the *Streptomycetes*. *Mol. Microbiol.* **2005**, *58*, 1368–1380.

Komeil, D.; Padilla-Reynaud, R.; Lerat, S.; Simao-Beaunoir, A.; Beaulieu, C. Comparative Secretome Analysis of *Streptomyces scabiei* During Growth in the Presence or Absence of Potato Suberin. *Proteome Sci.* **2014**, *12*, 35.

Komeil, D.; Simao-Beaunoir, A. M.; Beaulieu, C. Detection of Potential Suberinase-Encoding Genes in *Streptomcyes scabiei* Strains and Other *Actinobacteria*. *Can. J. Microbiol.* **2013**, *59*, 294–303.

Lautru, S.; Deeth, R. J.; Bailey, L. M.; Challis, G. L. Discovery of a New Peptide Natural Product by *Streptomyces coelicolor* Genome Mining. *Nat. Chem. Biol.* **2005**, *1*, 265–269.

Lauzier, A.; Simao-Beaunoir, A. M.; Bourassa, S.; Poirier, G. G.; Talbot, B.; Beaulieu, C. Effect of Potato Suberin on *Streptomyces scabies* Proteome. *Mol. Plant Pathol.* **2008**, *9*, 753–762.

Lawrence, C. H.; Clark, M. C.; King, R. R. Induction of Common Scab Symptoms in Aseptically Cultured Potato Tubers by the Vivotoxin, Thaxtomin. *Phytopathol.* **1990**, *80*, 606–608.

Lazarovits, G.; Tenuta, M.; Conn, K. L. Organic Amendments as a Disease Control Strategy for Soil Born Diseases of High Value Agricultural Crops. *Aus. Plant Pathol.* **2001**, *30*, 111–117.

Leach, J. E.; Vera Cruz, C. M.; Bai, J.; Leung, H. Pathogen Fitness Penalty as a Predictor of Durability of Disease Resistance Genes. *Annu. Rev. Phytopathol.* **2001**, *39*, 187–224.

Leiner, R. H.; Fry, B. A.; Carling, D. E.; Loria, R. Probable Involvement of Thaxtomina in Pathogenicity of *Streptomyces scabies* on Seedlings. *Phytopathol.* **1996**, *86*, 709–713.

Lerat, S.; Forest, M.; Lauzier, A.; Grondin, G.; Lacelle, S.; Beaulieu, C. Potato Suberin Induces Differentiation and Secondary Metabolism in the Genus *Streptomyces*. *Microbes Environ.* **2012**, *27*, 36–42.

Lerat, S.; Simao-Beaunoir, A. M.; Beaulieu, C. Genetic and Physiological Determinants of *Streptomyces scabies* Pathogenicity. *Mol. Plant Pathol.* **2009**, *10*, 579–585.

Lerat, S.; Simao-Beaunoir, A. M.; Wu, R.; Beaudoin, N.; Beaulieu, C. Involvement of the Plant Polymer Suberin and the Disaccharide Cellobiose in Triggering ThaxtominA Biosynthesis, a Phytotoxin Produced by the Pathogenic Agent *Streptomyces scabies*. *Phytopathol.* **2010**, *100*(1), 91–96.

Liu, D.; Anderson, N. A.; Kinkel, L. L. Biological Control of Potato Scab in the Field with Antagonistic *Streptomyces scabies*. *Phytopathol.* **1995**, *85*, 827–831.

Lorang, J. M.; Liu, D.; Anderson, N. A.; Schottel, J. L. Identification of Potato Scab Inducing and Suppressive Species of *Streptomyces*. *Phytopathology*. **1995**, *85*, 261–268.

Loria, R.; Bukhalid, R. A.; Fry, B. A.; King, R. R. Plant Pathogenicity in the Genus *Streptomyces*. *Plant Dis.* **1997**, *81*, 836–846.

Loria, R.; Kers, J.; Joshi, M. Evolution of Plant Pathogenicity in *Streptomyces*. *Annu. Rev. Phytopathol.* **2006**, *44*, 469–487.

Mandel, M. J.; Wollenberg, M. S.; Stabb, E. V.; Visick, K. L.; Ruby, E. G. A Single Regulatory Gene is Sufficient to Alter Bacterial Host Range. *Nature*. **2009**, *458*, 215–218.

Mavrodi, D. V.; Blankenfeldt, W.; Thomashow, L. S. Phenazine Compounds in Fluorescent *Pseudomonas* sp. Biosynthesis and Regulation. *Annu. Rev. Phytopathol.* **2006**, *44*, 417–445.

Mazzola, M.; Manici, L. M. Apple Replant Disease: Role of Microbial Ecology in Cause and Control. *Annu. Rev. Phytopathol.* **2012**, *50*, 45–65.

McRae, W. India: New Plant Diseases Reported During the Year 1928. *Int. Bull. Plant Protect.* **1929**, *3*(2), 21–22.

McIntosh, A. H.; Bateman, G. L.; Chamberlain, K. Substituted Benzoic and Picolinic Acids as Foliar Sprays against Potato Common Scab. *Ann. Appl. Biol.* **1988**, *112*(3), 397–401.

McIntosh, A. H.; Burrell, M. M.; Hawkins, J. H. Field Trials of Foliar Sprays of 3, 5-Dichlorophenoxyacetic Acid (3, 5-D) against Common Scab on Potatoes. *Potato Res.* **1982**, *25*(4), 347–350.

Melchers, L. S.; Stuiver, M. H. Novel Genes for Disease-Resistance Breeding. *Curr. Opin. Plant Biol.* **2000**, *3*, 147–152.

Mishra, K. K.; Srivastava, J. S. Soil Amendments to Control Common Scab of Potato. *Potato Res.* **2005**, *47*, 101–109.

Miyajima, K.; Tanaka, F.; Takeuchi, T.; Kuninaga, S. *Streptomyces turgidiscabies* sp. nov. *Int. J. Syst. Bacteriol.* **1998**, *48*, 495–502.

Moller, M.; Botti, H.; Batthyany, C.; Rubbo, H.; Radi, R.; Denicola, A. Direct Measurement of Nitric Oxide and Oxygen Partitioning into Liposomes and Low Density Lipoprotein. *J. Biol. Chem.* **2005**, *280*, 8850–8854.

Natsume, M.; Komiya, M.; Koyanagi, F.; Tashiro, N.; Kawaide, H.; Abe, H. Phytotoxin Produced by *Streptomyces* sp. Causing Potato Russet Scab. *J. Gen. Plant Pathol.* **2005**, *71*, 364–369.

Natsume, M.; Ryu, R.; Abe, H. Production of Phytotoxins, Concanamycins A and B by *Streptomyces* sp. *Ann. Phytopathol. Soc. Japan.* **1996**, *62*, 411–413.

Natsume, M.; Taki, M.; Tashiro, N.; Abe, H. Phytotoxin Production and Aerial Mycelium Formation by *Streptomyces scabies* and *S. acidiscabies* in vitro. *J. Gen. Plant Pathol.* **2001**, *67*, 299–302.

Natsume, M.; Yamada, A.; Tashiro, N.; Abe, H. Differential Production of the Phytotoxins Thaxtomin A and Concanamycins A and B by Potato Common Scab-Causing *Streptomyces* sp. *Ann. Phytopathol. Soc. Japan.* **1998**, *64*, 202–204.

Neeno-Eckwall, E. C.; Kinkel, L. L.; Schottel, J. L. Competition and Antibiosis in the Biological Control of Potato Scab. *Can. J. Microbiol.* **2001**, *47*, 332–340.

Nimchuk, N.; Marois, E.; Kjemtrup, S.; Leister, R. T.; Katagiri, F.; Dangl, J. L. Eukaryotic Fatty Acylation Drives Plasma Membrane Targeting and Enhances Function of Several Type III Effector Proteins from *Pseudomonas syringae*. *Cell.* **2000**, *101*, 353–363.

Ohmiya, Y.; Samejima, M.; Shiroishi, M.; Amano, Y.; Kanda, T.; Sakai, F.; Hayashi, T. Evidence that Endo-1,4-B-Glucanases Act on Cellulose in Suspension-Cultured Poplar Cells. *Plant J.* **2000**, *24*, 147–158.

Osbourn, A. Saponins and Plant Defence – A Soap Story. *Trends Plant Sci.* **1996**, *1*, 4–9.

Pan, X.; Yang Y.; Zhang, J. Molecular Basis of Host Specificity in Human Pathogenic Bacteria. *Emerg. Microb. Infect.* **2014**, *3*(3), e23.

Paradis, E.; Goyer, C.; Hodge, N. C.; Hogue, R.; Stall, R. E.; Beaulieu, C. Fatty-Acid and Protein Profiles of *Streptomyces scabies* Strains Isolated in Eastern Canada. *Int. J. Syst. Bacteriol.* **1994**, *44*, 561–564.

Park, D. H.; Kim, J. S.; Kwon, S. W.; Wilson, C.; Yu, Y. M.; Hur, J. H.; Lim C. K. (*Streptomyces luridiscabiei* sp. nov., *Streptomyces puniciscabiei* sp. nov. and *Streptomyces niveiscabiei* sp. nov., which Cause Potato Common Scab Disease in Korea. *Int. J. Syst. Evol. Microbiol.* **2003**, *53*, 2049–2054.

Patzer, S. I.; Braun, V. Gene Cluster Involved in the Biosynthesis of Griseobactin, a Catechol-Peptide Siderophore of *Streptomyces* sp. ATCC700974. *J. Bacteriol.* **2010**, *192*, 426–435.

Paulin, M. M.; Novinscak, A.; St-Arnaud, M.; Goyer, C.; DeCoste, N. J.; Privé, J. P.; Owen, J.; Filion, M. Transcriptional Activity of Antifungal Metabolite-Encoding Genes *phlD* and *hcnBC* in *Pseudomonas* sp. Using qRT-PCR. *FEMS Microbiol. Ecol.* **2009**, *68*, 212–222.

Pavlista, A. D. Common Scab: Control of Common Scab with Sulfur and Ammonium Sulfate. *Spudman.* **1992**, *11*, 13–15.

Poomthongdee, N.; Duangmal, K.; Pathom-aree, W. Acidophilic *Actinomycetes* from Rhizosphere Soil: Diversity and Properties Beneficial to Plants. *J. Antibiot.* **2015,** *68*(2), 106–114.

Rivero, M.; Furman. N.; Mencacci, N.; Picca, P.; Toum, L.; Lentz, E.; Bravo-Almonacid F.; Mentaberry, A. Stacking of Antimicrobial Genes in Potato Transgenic Plants Confers Increased Resistance to Bacterial and Fungal Pathogens. *J. Biotechnol.* **2012**, *157*, 334–343.

Rubbo, H.; Botti, H.; Batthyany, C.; Trostchansky, A.; Denicola, A.; Radi, R. Antioxidant and Diffusion Properties of Nitric Oxide in Low-Density Lipoprotein. *Meth. Enzymol.* **2002**, *359*, 200–209.

Ryan, A. D.; Kinkel, L. L. Inoculum Density and Population Dynamics of Suppressive and Pathogenic *Streptomyces* Strains and Their Relationship to Biological Control of Potato Scab. *Biocontrol.* **1997**, *10*, 180–186.

Ryan, R. P.; Vorhölter, F. J.; Potnis, N.; Jones, J. B.; Van Sluys, M. A.; Bogdanove, A. J.; Dow. J. M. Pathogenomics of *Xanthomonas*: Understanding Bacterium–Plant Interactions. *Nature Rev. Microbiol.* **2011**, *9*, 344–355.

Sahni, S.; Sarma B. K.; Singh, D. P.; Singh, H. B.; Singh, K. P. Vermicompost Enhances Performance of Plant Growth-Promoting Rhizobacteria in *Cicer arietinum* Rhizosphere against *Sclerotium rolfsii*. *Crop Protection.* **2008a**, *27*, 369–376.

Sahni, S.; Sarma, B. K.; Singh, K. P. Management of *Sclerotium rolfsii* with Integration of Non-Conventional Chemicals, Vermicompost and *Pseudomonas syringae*. *World J. Microbiol. Biotechnol.* **2008b**, *24*, 517–522.

Sarma, B. K.; Singh, D. P.; Mehta, S.; Singh, H. B.; Singh, U. P. Plant Growth Promoting Rhizobacteria-mediated Alterations in Phenolic Profile of Chickpea (*Cicer arietinum*) Infected by *Sclerotium rolfsii*. *J. Phytopathol.* **2002**, *150*, 277–282.

Scheible, W. R.; Fry, B.; Kochevenko, A.; Schindelasch, D.; Zimmerli, L.; Somerville, S.; Loria, R.; Somerville, C. R. An Arabidopsis Mutant Resistant to Thaxtomin A, a Cellulose Synthesis Inhibitor from *Streptomyces species*. *Plant Cell.* **2003**, *15*, 1781–1794.

Schupp, T.; Toupet, C.; Divers, M. Cloning and Expression of Two Genes of *Streptomyces pilosus* Involved in the Biosynthesis of the Siderophore Desferrioxamine B. *Gene.* **1988,** *64,* 179–188.

Seipke, R. F.; Song, L.; Bicz, J.; Laskaris, P.; Yaxley, A. M.; Challis, G. L.; Loria, R. The Plant Pathogen *Streptomyces scabies* 87-22 has a Functional Pyochelin Biosynthetic Pathway that is Regulated by TetR- and AfsR-Family Proteins. *Microbiology.* **2011,** *157,* 2681–2693.

Seki-Asano, M.; Okazaki, T.; Yamagishi, M.; Sakai, N.; Hanada, K.; Mizoue, K. Isolation and Characterization of New 18-Membered Macrolides FD-891 and FD-892. *J. Antibiot.* **1994,** *47,* 1226–1233.

Shani, Z.; Dekel, M.; Roiz, L.; Horowitz, M.; Kolosovski, N.; Lapidot, S.; Alkan, S.; Koltai, H.; Tsabary, G.; Goren, R.; Shoseyov, O. Expression of Endo-1,4-Bglucanase (Cel1) in *Arabidopsis thaliana* is Associated with Plant Growth, Xylem Development and Cell Wall Thickening. *Plant Cell Rep.* **2006,** *25,* 1067–1074.

Shen, Y.; Sharma, P.; da Silva, F. G.; Ronald, P. The *Xanthomonas oryzae* pv. *oryzae raxP* and *raxQ* Genes Encode an ATP Sulphurylase and Adenosine-5'-Phosphosulphate Kinase that are Required for AvrXa21 Avirulence Activity. *Mol. Microbiol.* **2002,** *44,* 37–48.

Singh, A.; Jain, A.; Sarma, B. K.; Upadhyay, R. S.; Singh, H. B. Rhizosphere Microbes Facilitate Redox Homeostasis in *Cicer arietinum* against Biotic Stress. *Ann. Appl. Biol.* **2013,** *163,* 33–46.

Singhai, P. K.; Sarma, B. K.; Srivastava, J. S. Biological Management of Common Scab of Potato through *Pseudomonas* Species and Vermicompost. *Biol. Control.* **2011a,** *57,* 150–157.

Singhai, P. K.; Sarma, B. K.; Srivastava, J. S. Phenolic Acid Content in Potato Peel Determines Natural Infection of Common Scab Caused by *Streptomyces* sp. *World J. Microbiol. Biotechnol.* **2011b,** *27,* 1559–1567.

St-Onge, R.; Goyer, C.; Filion, M. *Pseudomonas* sp. can Inhibit *Streptomyces scabies* Growth and Repress the Expression of Genes Involved in Pathogenesis. *J. Bacteriol. Parasitol.* **2010,** *1,* 101.

Takeuchi, T.; Sawada, H.; Tanaka, F.; Matsuda. I. Phylogenetic Analysis of *Streptomyces* sp. Causing Potato Scab Based on 16S rRNA Sequences. *Int. J. Syst. Bacteriol.* **1996,** *46,* 476–79.

Tanii, A.; Takeuchi, T.; Horita, H. Biological Control of Scab, Black Scurf and Soft Rot of Potato by Seed Tuber Bacterization. In *Biological Control of Soil–Borne Plant Pathogens;* Hornby, D., Ed.; C.A.B. International; Wallingford, England, 1990; pp 143–164.

Tegg, R. S.; Corkrey, R.; Wilson, C. R. Relationship Between the Application of Foliar Chemicals to Reduce Common Scab Disease of Potato and Correlation with Thaxtomin A Toxicity. *Plant Dis.* **2012,** *96*(1), 97–103.

Thaxter, R. Potato Scab. *Annual Report of the Connecticut Agricultural Experiment Station for 1891.* 1892, pp 153–160.

Thompson, H. K.; Tegg, R. S.; Corkrey, R.; Wilson, C. R. Foliar Treatments of 2,4-Dichlorophenoxyacetic Acid for Control of Common Scab in Potato have Beneficial Effects on Powdery Scab Controle. *Scientific World J.* **2014,** *947167,* 1–5.

Tseng, T. T.; Tyler, B. M.; Setubal, J. C. Protein Secretion Systems in Bacterial-Host Associations, and Their Description in the Gene Ontology. *BMC Microbiol.* **2009,** *9*(Suppl 1), S2.

Turnbull, A. L.; Liu, Y.; Lazarovits, G. Isolation of Bacteria from the Rhizosphere and Rhizoplane of Potato (*Solanum tuberosum*) Grown in Two Distinct Soils Using Semi Selective Media and Characterization of Their Biological Properties. *Am. J. Potato Res.* **2012,** *89,* 294–305.

Vinatzer, B. A.; Monteil, C. L.; Clarke, C. R. Harnessing Population Genomics to Understand How Bacterial Pathogens Emerge, Adapt to Crop Hosts, and Disseminate. *Annu. Rev. Phytopathol.* **2014**, *52,* 19–43.

Vivian, A.; Gibbon, M. J. Avirulence Genes in Plant Pathogenic Bacteria: Signals or Weapons? *Microbiology.* **1997**, *143,* 693–704.

Vivian, A.; Arnold, D. L. Bacterial Effector Genes and Their Role in Host-Pathogen Interactions. *J. Plant Pathol.* **2000**, *82,* 163–178.

Vivian, A.; Murillo, J.; Jackson, R. W. The Roles of Plasmids in Phytopathogenic Bacteria: Mobile Arsenals? *Microbiology.* **2001**, *147,* 763–780.

Wach, M. J.; Krasnoff, S. B.; Loria, R.; Gibson, D. M. Effect of Carbohydrates on the Production of Thaxtomin A by *Streptomyces acidiscabies. Arch. Microbiol.* **2007**, *188,* 81–88.

Waney, V. R.; Kingsley, M. T.; Gabriel, D. W. *Xanthomonas campestris* pv. *translucens* Genes Determining Host-Specific Virulence and General Virulence on Cereals Identified by Tn*5-gusA* insertion mutagenesis. *Mol. Plant-Microbe Interact.* **1991**, *4,* 623–627.

Wanner, L. A. A Survey of Genetic Variation in *Streptomyces* Isolates Causing Potato Common Scab in the United States. *Phytopathol.* **2006**, *96,* 1363–1371.

Wanner, L. A. A New Strain of *Streptomyces* Causing Common Scab in Potato. *Plant Dis.* **2007**, *91,* 352–359.

Waterer, D. Impact of High Soil pH on Potato Fields and Grade Losses to Common Scab. *Can. J. Plant Sci.* **2002**, *82,* 583–586.

Williamson, N.; Brian, P.; Wellington, E. M. Molecular Detection of Bacterial and *Streptomycete* Chitinases in the Environment. *Antony Van Leuwenhoek.* **2000**, *78,* 315–321.

Wozniak, R. A. F.; Waldor, M. K. Integrative and Conjugative Elements: Mosaic Mobile Genetic Elements Enabling Dynamic Lateral Gene Flow. *Nature Rev. Microbiol.* **2010**, *8,* 552–563.

Yoshida, M.; Kobayashi, K. Taxonomic Characterization of the Actinomycete Causing Root Tumor of Melon. *Ann. Phytopathol. Soc. Japan.* **1991**, *57,* 540–548.

CHAPTER 9

EVOLUTION, ADAPTATION, AND HOST SELECTION BY PLANT VIRUSES: CURRENT UNDERSTANDING AND FUTURE PERSPECTIVES

ABHISHEK SHARMA*, PREM L. KASHYAP, SUMIT I. KAUR and SANTOSH S. KANG

Department of Vegetable Science, Punjab Agricultural University, Ludhiana 141001, Punjab, India

Corresponding author. E-mail: abhishek@pau.edu

CONTENTS

ABSTRACT

Plant viruses are the most abundant and genetically diverse biological entities. They are infectious agents with small genome sizes and require living
host cells to complete their life cycles. So far, the origin and evolution of
plant viruses is mystery, but several experimental and molecular evidences
indicate that plant viruses have been evolved as a result of symbiogenesis
or through competition. Different selection pressures in different hosts will
result in fitness trade-offs across hosts, leading to specialization. Newly
emerged viruses have their origin in host species or populations in which they
are well established and their emergence get influenced by various factors
viz. alternation in the host plant and/or virus ecology, changes in genetic
composition of the host and virus population, and in the case of vectored
viruses, changes in the ecology or genetic composition of the vector. Like
other living entities, viruses substantially resemble the parent during their
replication, but can change to give rise to new type strains. This inherent
variation enables viruses to adapt to new and changing situations. However,
the complete picture of virus–virus and virus–host interactions is much more
complex, since the viral populations constantly evolves and adapt to their
new hosts and vectors and ultimately affects ecosystem dynamics. Considering these points, this chapter provides deep insights into the current understanding and future perspectives of evolution, adaptation, and host selection
by plant viruses.

9.1 INTRODUCTION

Plant viruses are obligate parasites and depend on the host cell for their
survival and replication. They have a great potential for genetic variation,
rapid evolution, and adaptation. Broadly, the factors influencing the virus
evolutionary process can be divided into macro and micro levels. Generally,
at the macro level, the evolutionary process is influenced by selection pressures arising from major agricultural changes that cause significant alterations in the biology of viruses and vectors. Various factors influencing the
evolution of viruses at the vegetation interface in nature have been described
in Table 9.1. Briefly, the major factors involve expansion in natural host range,
greater adaptation to infect hosts; introduction of new, more efficient virus-
vector species, emergence of more efficient virus-vector biotypes or variants
of existing vector species; and overcoming of host defenses (Jones, 2009).
At micro level, changes to virus genomes result when selection pressure is

TABLE 9.1 Lists of Ecological And Genetic Factors Driving Evolution, Adaptation, And Host Selection by Plant Viruses.

Factor(s)	Effect/ Consequences	Reference(s)
Agricultural diversification through crop breeding, new crop introductions, cultivation of susceptible genotypes and agricultural intensification	Increased encounter rates facilitate virus movement from native plants to introduced crops by facilitating direct contact between newly introduced crops and native vegetation	Thresh (1982), Morales and Anderson (2001), Anderson et al. (2004), Morales (2006)
Genetic drift due to geographical demarcation	Random extinction of virus variants within an isolated site alters virus population balance	Fargette et al. (2006), Gibbs et al. (2008a)
Genome integration	Virus sequences embedded in host genome cause episomal infection and if present in native vegetation such viruses might emerge to infect susceptible introduced crops	Harper et al. (2002), Jones (2009)
Host-range expansion	Selection of variant of indigenous virus that infects new host and helping virus to spread from native plants to introduced crop	Fargette et al. (2006), Gibbs et al. (2008a)
Injudicious use of chemical control measures	Increased potential for viruliferous vectors to migrate from native plants to introduced crops by developing pesticide resistance in vector insects and killing beneficial insects	Anderson et al. (2004), Morales (2006)
Introduction of new and more efficient virus-vector species	New vector introduction enhanced prevalence and distribution of indigenous virus in introduced crop	Anderson et al. (2004), Morales (2006)
Modular evolution, recombination, pseudo recombination and reassortment	Host range expansion and appearance of new virus diseases due to genetic exchange of genomic segments associated with host switches	Fargette et al. (2006), Jones (2009)
Mutation	Virus variation due to alterations in nucleotide sequences led to the production of variant of indigenous virus that can invade new host	Jeger et al. (2006)
Overcoming host defense by gene silencing	New variant of indigenous virus invades new host successfully and increased incidence and distribution of indigenous virus in introduced crop	Seal et al. (2006b)

TABLE 9.1 (*Continued*)

Factor(s)	Effect/ Consequences	Reference(s)
Population bottlenecks and "founder effects" associated with vector transmission and virus spread	Increased potential for virus spread from native plants to introduced crop due to change in virus population size and greater diversity between virus populations	Fargette et al. (2006), Sacristán and Garcia-Arenal (2008), Jones (2009)
Satellite viruses and nucleic acids	Enhanced virus concentration and virulence led to enhance virus adaptation capabilities	Fargette et al. (2006), Seal et al. (2006b), Gibbs et al. (2008a)
Selection of variants within virus populations	Production of variant of indigenous virus that can invade new host and enhanced potential for virus spread and their adaptation from native plants to introduced crop	Jeger et al. (2006), Gibbs et al. (2008a)
Symbiosis between different viruses	Production of variant of indigenous virus that can invade new host due to co-infection	Fargette et al. (2006), Jeger et al. (2006)
Wider adaptation to infect introduced host	Pathogenesis and virulence of indigenous virus enhanced incidence and distribution of indigenous virus in introduced crop	Jeger et al. (2006), Sacristán and Garcia-Arenal (2008)

applied to viral genetic material. The genetic exchange of genomic segments that occurs during recombination, pseudo-recombination, reassortment, and modular evolution has often been associated with host switches, host-range expansion, and appearance of new virus types (Gibbs et al., 2008). There are five basic mechanisms determining the genetic structure and evolution of virus populations including mutation, recombination, natural selection, genetic drift, and migration (Domingo-Calap & Sanjuán, 2011). In fact, the viral population produced during pathogenesis is composed of a group of complex variants recognized as viral quasispecies (Lauring & Andino, 2010), which act as a genetic reservoir and are selected due to the pressure of encountering virus and the host. Such type of selection results in specific footprints such as genome composition, mutation bias, amino acid usage, synonymous codon usage, and dinucleotide usage in the viral genome (Cheng et al., 2014). In addition, other factors that drive plant virus emergence include selection from existing variants within virus populations, new mutations, synergism, genetic drift arising from geographical isolation, population bottlenecks, and "founder effects" associated with vector transmission or virus spread within individual plants, presence of satellite viruses, and genome integration (Cooper & Jones, 2006). Moreover, the requirement of optimum conditions for propagation and various plant defense mechanisms compel the viruses to evolve parallel with the host system. Thus, exploring the complexities of evolution, adaptation, and host selection by plant virus is one of biggest challenge, since the viral populations constantly evolve and adapt to their new hosts, vectors and ultimately affects ecosystem dynamics. Briefly, in this chapter, attempts have been made to provide updates on current understanding and future perspectives of evolution, adaptation, and host selection by plant viruses.

9.2 EVOLUTION OF PLANT VIRUSES

Like fungal and bacterial pathogens, plant viruses substantially resemble the parent during their replication but can change to give rise to new strain types. This inherent variation allows viruses to adapt to new and changing situations. During long course of evolution, new viruses arise, and there must have been a time at which the archetypical virus arose. Several studies provide evidences coming from the fossil record of macroscopic and microscopic remnants for the origin and evolutionary pathways of higher organisms. Viruses do not form conventional fossils, though insects preserved in amber and seeds preserved in archaeological sites provide some clues regarding their origin and evolution (Hull, 2014). Due to the fact that the genome of viruses underlies mutation and genetic recombination, viruses probably evolve according to a form of natural selection, very analogous to that governing other living things. Currently, three main hypotheses have been articulated: (a) the regressive, or reduction, hypothesis asserts that viruses are remnants of cellular organisms; (b) the progressive, or escape, hypothesis states that viruses arose from genetic elements that gained the ability to move between cells and (c) the virus-first hypothesis states that viruses predate or coevolved with their current cellular hosts. The first hypothesis is "regressive evolution," which explains that viruses descend from free-living and more complex parasites. According to this theory, ancestral viruses developed a growing dependence on host-cell intracellular machinery through evolutionary time, while retaining the ability to auto-replicate, like mitochondria that have their own genetic information and replicate on their own (Desjardins et al., 2005). The second hypothesis is "independent" or "parallel" evolution of viruses and other organisms, which assumes that viruses appeared at the same time as the most primitive organisms (Margulis & Sagan, 1997). According to this hypothesis, viruses originated through a progressive process. Mobile genetic elements, pieces of genetic material capable of moving within a genome, gained the ability to exit one cell and enter another. Finally, there is the theory of "cell origin," which states that viruses reflect their origin from cell DNA or mRNA, which acquired the ability to auto-replicate, create extracellular virions, exist, and function independently. All the three, regressive, progressive, and cell origin hypotheses assume that cells existed before viruses whereas a recently proposed hypothesis of "a pre-cellular origin of viruses" or "the virus first hypothesis," assumes that viruses may have been the first replicating entities (Forterre, 2006; Koonin et al., 2009). Under such a situation, a "virus world" developed within networks of inorganic compartment that contained a diverse network of virus-like genetic elements, which were the ancestors of both cellular and viral genes. RNA viruses evolved first followed by retroid elements and then DNA viruses (Fig. 9.1). The main evidence for this hypothesis is that there are key proteins known as "virus hallmark genes" involved in viral replication and morphogenesis that are not present in other living cells. These include the double β-barrel capsid protein of isometric virus particles (Krupovič & Bamford, 2008) and super family 3 helicases, both found in RNA and DNA viruses of eukaryotes, phage and archaeal viruses, RNA-dependent RNA polymerase and reverse transcriptase (RT) (Koonin et al., 2006).

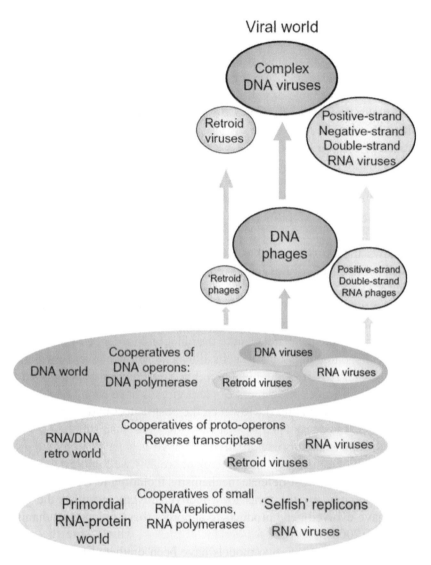

FIGURE 9.1 The origin and evolution of plant viruses (Source: Koonin & Dolja, 2006; © Koonin et al; licensee BioMed Central Ltd. 2006.) http://creativecommons.org/licenses/by/2.0)

Plant viruses have evolved as combinations of genes whose products interact with cellular components to produce progeny throughout the plants and, in most cases, to interact with a vector to be moved to other plants. These processes require several layers of precise interactions with the host. Some viral genes tend to be relatively conserved, whereas other genes appear to be completely unrelated to any gene in otherwise similar viruses or, in some cases, to any known gene. In general, the more conserved genes tend to be those whose products interact within the cell to replicate the viral genome. This process appears to be somewhat generic because many viruses have been found to be capable of multiplying in individual cells (protoplasts) but cannot move throughout the intact plant. The less conserved genes tend to be those that have evolved for interactions with the specific host for movement and to counter host-defense systems (Li & Ding, 2006).The ability of the virus to move from the initially infected cell throughout the plant appears to be one of the major selective forces for the evolution of plant viruses. Successful systemic infection of plant viruses results from replication in initially infected cells, followed by two distinct processes: cell-to-cell and long-distance movement (Lucas, 2006). Cell-to-cell movement is a process that allows the virus to pass to adjacent cells by successful interactions between virus-encoded movement proteins and host factors (Lucas, 2006; Harries & Ding, 2011). Long-distance movement is a multistep process that allows the virus to enter the sieve element from an adjacent cell, followed by passive movement of virus through the phloem to a distal region of the plant by exiting into a cell adjacent to the phloem. Further cell-to-cell movement from the phloem-associated cells allows the virus to invade most of the cells at a distal region of the plant. Viral proteins and host factors that are involved in cell-to-cell movement of plant viruses have been widely examined (Waigmann et al., 2004). However, the host factors that are involved in long-distance transport of plant viruses and the mechanisms of long-distance movement such as factors that are involved in virus entry into phloem tissue and virus exit at a distal region of the plant are less well understood. Additionally, plants have host-defense mechanisms including RNA silencing that must be overcome by the virus for effective movement within the plant. Viruses have evolved gene products to suppress these defense mechanisms (Li & Ding, 2006).

In case of plant virus, two models have been predicted for their evolution on the basis of their relationship with host. The initial development of virus-like genetic elements became intracellular parasites post evolution of bacteria and archaea. Selection against excessively aggressive parasites that would kill off the host ensembles of genetic elements would lead to early evolution of temperate virus-like agents and primitive defense mechanisms, possibly based on the RNA interference principle (Wang et al., 2004). Then, when the eukaryotic cell had evolved, new forms of viruses originated from extensive recombination of genes from various bacteriophages, archaeal viruses, plasmids, and the evolving eukaryotic genomes. The model also predicates viral involvement in the emergence of eukaryotic cells under which archaeobacterial symbiosis was the starting point of eukaryogenesis. At this time, it is suggested that there was a major diversification of picorna-like viruses that originated from prokaryotic viruses, which acquired some bacterial elements and genes (Koonin et al., 2008).

Usually, relationship between plant viruses and their hosts is thought to be antagonistic, but now there are a number of evidences that show that viruses apparently are not transmitted from plant to plant, but are passed from generation to generation through seed (Fukuhara et al., 2006). It is interesting to note that these viruses belong to fungal virus group (Boccardo & Candresse, 2005; Hacker et al., 2005), hence, it seems possible that they originated as fungal viruses that became trapped in plants during a plant–fungus endophytic interaction (Roossinck, 1997). Later on, Márquez et al. (2007) found that the mutualistic endophytic fungus *Curvularia protuberata* requires a virus to confer thermal tolerance to plants. Virus–host symbiogenesis can also result in virus speciation through the acquisition of new genes by virus from the host (Mayo & Jolly, 1991). Through several experimental evolution studies, Allison et al. (1996) reported that the movement proteins of plant viruses have been acquired from their hosts, as these proteins show an extraordinary level of diversity in viruses whose replicase genes are closely related (Melcher, 2000), and plants express proteins that are functionally very similar to plant virus movement proteins (Xoconostle-Cázares et al., 1999). Similarly, mixed infections of plant viruses are common and well documented. Plant virus synergy, is a well-known example of plant virus symbiosis (Rochow & Ross, 1955). In this phenomenon, the disease symptoms are enhanced by the mixed infection. The most well-documented examples of synergy involve potyviruses, and result in enhanced replication of the partner virus, while the potyvirus replication is not affected (Pruss et al., 1997). Scheets (1998) also reported fully mutualistic relationships between *Wheat streak mosaic virus* and *Maize chlorotic mottle virus*, where concentration of both viruses increases in the synergistic fashion. Synergy occurs among DNA viruses as well as RNA viruses of plants, and has been documented for the geminiviruses (Fondong et al., 2000; Pita et al., 2001), and between an RNA virus and a DNA virus (Hii et al., 2002). Support of satellite RNAs or viruses is another form of symbiosis that is found in plant viruses (Roossinck, 2005). Satellites do not provide any essential functions for the helper virus, but they can alter the symptom phenotype by either attenuating or exacerbating symptoms (Simon et al., 2004). In general, these elements have an antagonistic affect on the replication of the helper virus. A number of plant viruses exhibit another form of symbiosis called interdependence. These symbiotic interactions are generally obligate in nature. For instance, they function during vector transmission in umbravirus/luteovirus symbiosis (Gibbs, 1995) and the rice tungro disease viruses (Hull, 2002), and establishment of systemic infection, in case of *Pepper veinal mottle virus* and *Potato virus Y* (Marchoux et al., 1993) or with *Pepper mottle virus* and *Cucumber mosaic virus* (CMV) (Guerini & Murphy, 1999).

Besides this, modular evolution of plant virus have been reported by Botstein (1980), where viruses in symbiotic relationships in mixed infections have been frequently re assorted and recombined to form new strain types. This provides plant viruses an enormous level of flexibility and a capacity for very rapid evolutionary changes. A number of recently described examples of new RNA virus species which likely formed through symbiogenesis include *tobravirus* strains I6 and N5, which resulted from the reassortment of elements from *Tobacco rattle virus* and *Pea early browning virus* (Robinson et al., 1987), *Bean distortion mosaic virus*, the result of both reassortment and recombination between *Peanut stunt virus* and *CMV* (White et al., 1995), *Bean leafroll virus* and *Sugarcane yellow leaf virus*, the products of recombination events between luteo-like and polero-like viruses (Moonan et al., 2000; Domier et al., 2002), *Poinsettia latent virus*, which appears be a recombinant product of a polerovirus and a sobemovirus (Aus Dem Siepen et al., 2005). The phenomenon of modular evolution through symbiogenesis has been reported in case of plant DNA viruses. A great deal of the emergence of new geminiviruses has been attributed to reassortment and recombination (Rojas et al., 2005; Seal et al., 2006a). Hu et al. (2007) found that reassortment played an important role in the evolution of the extant *Banana bunchy top virus* group. Thus, in the broader sense, evolution of plant viruses can be seen as a form of mutualistic symbiosis or competition. Evolution of some important plant viruses and their associated molecules is discussed further to understand the role of genome type, host, and vector in evolution.

9.2.1 GEMINIVIRUSES AND THEIR SATELLITES

The *Geminiviridae* family encompasses diverse group of insect vector-transmitted plant viruses with circular single-stranded DNA (ssDNA) genome comprises of one or two components of 2.7–3.0 kb encapsidated in twinned icosahedral particles. The *Geminiviridae* are currently divided into six genera, *Begomovirus*, *Curtovirus*, *Eragrovirus*, *Mastrevirus*, *Topocuvirus*, and *Turncurtovirus* on the basis of their genome organizations, host ranges, and insect vectors (Varsani et al., 2014). They differ from other plant viruses in the fact that they are ssDNA viruses and multiply through rolling circle replication (RCR). They constitute one of the three recognized groups of episomal replicons that use RCR, the other being circular ssDNA bacterio-phages, and plasmids of bacteria or archaea (Hanley-Bowdoin et al., 2013). There are ample of evidences suggesting that modern geminiviruses may have evolved from ancient extra-chromosomal DNA replicons, present in prokaryotic or primitive eukaryotic ancestors of red algae (*Porphyra pulchra*) (Fig. 9.2). In such a situation, the ancestral geminivirus is envi-sioned as a circular extra-chromosomal ssDNA plasmid that replicated via a rolling circle mechanism involving a dsDNA replicative form. It is generally admitted that the Rep protein has been the sole common protein throughout this long evolutionary process. Phylogenetic and clustering analyses suggest that geminivirus Reps share an ancestor with Reps encoded on plasmids of phytoplasmas (Krupovic et al., 2009). There is significant homology between Wheat dwarf mastrevirus Rep protein and Rep-like sequences encoded by the red alga (*Porphyra pulchra*) plasmid (Nawaz-ul-Rehman & Fauquet, 2009). Thus, the conserved nature of the Rep protein-like genes among modern prokaryotic and eukaryotic DNA replicons and the prokary-otic features such as CP promoter, polycistronic mRNA, and replication capacity present in modern geminiviruses appears to be the initial point of evolution of ancestral geminivirus. However, still, it is unclear how a replicated plasmid in an algae evolved into an independent virus capable of movement between cells, tissues, and plant hosts and it is a mystery how this pro-virus acquired *de novo* genes such as AC2, AC3, AC4, AV1, and AV2 to constitute the genome of modern geminiviruses. Recently, based on the similarities among replication associated proteins and comparative homology-based structural modeling of viral capsid proteins, Krupovic et al. (2009) proposed a plasmid-to-virus transition hypothesis, where a phyto-plasma plasmid acquired a capsid-coding gene from a plant RNA virus to give rise to the ancestor of geminiviruses. Once the efficient movement and insect transmission ability via coat protein (CP) was obtained, geminivirus spread at an alarming rate. Furthermore, human interference resulted in an unbalanced ecosystem, which resulted in the invasion of geminiviruses onto introduced domesticated crops.

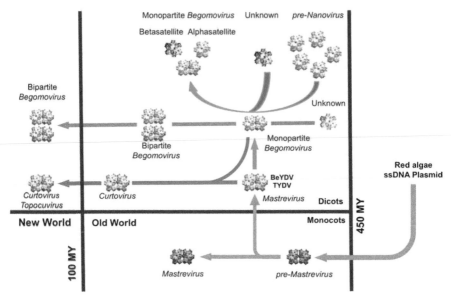

FIGURE 9.2   Putative path of origin and evolution of geminivirus and their satellites (Source: Muhammad Shah Nawaz-ul-Rehman, Claude M. Fauquet, "Evolution of geminiviruses and their satellites," in FEBS Letters, 2009 Jun 18;583(12):1825-32. With permission from John Wiley.)

Ge et al. (2007) reported that background mutation pressure is a key element in geminivirus evolution because recombination cannot create genetic variation *de novo*. Although, there are several other reports of the appearance of geminivirus strains with altered pathogenicity (Seal et al., 2006a) that provide indication of rapid genetic change due to recombination or reassortment among different viral genomes. Unlike RNA viruses, geminiviruses replicate their genomes inside the nucleus by using the host replication machinery, presumably involving DNA polymerase α- and δ-like activities (Gutierrez, 1999). Thus, these viruses were assumed to have higher replication fidelity and lower rates of mutation than RNA viruses (Rojas et al., 2005). However, the large number of species and the continued reports of new species, as well as the high degree of genetic diversity within a species (Fauquet et al., 2003; Bull et al., 2006; Patil et al., 2005), advocate that geminiviruses have a high mutation rate and that they generate highly diverse populations in a short time. It was reported that DNA methylation inhibits the replication of *Tomato gold mosaic virus* in tobacco protoplasts (Brough et al., 1992), suggesting that geminivirus DNA may not be methylated and that the normal mechanisms for mismatch repair probably do not operate during virus replication cycle (Inamdar et al., 1992). Thus, it is possible that the mechanisms of mismatch repair may function differently during the replication of geminivirus DNA and cellular DNA and that the lack of post-replication repair may be responsible for higher mis-incorporation in the Geminivirus progeny DNA (Seal et al., 2006a). Mutations in the helix 4 motif of the AL1 gene of two distantly related begomoviruses revert at 100% frequency, suggesting that nucleotide substitutions occur with high incidence and are under strong selective pressure during geminivirus infection (Seal et al., 2006a). Thus, in agreement with recent reports of high mutation rates for other ssDNA viruses infecting vertebrates and bacteria, nucleotide substitution events are likely to contribute to the diversity and rapid evolution of geminivirus ssDNA genomes (Shackelton et al., 2005).

9.2.2 REOVIRUSES

Reoviridae is a family of viruses that can affect the gastrointestinal system (such as *Rotavirus*) and respiratory tract. Viruses in the family Reoviridae have genomes consisting of segmented, double-stranded RNA (dsRNA). The name "Reo" is derived from *Respiratory enteric orphan viruses*. The term "orphan virus" means that a virus that is not associated with any known disease. The family Reoviridae is divided into nine genera, four of which can infect humans and animals, four other genera infect only plants and insects, and one infects fish. Reovirus group (*Fijivirus*, *Oryzavirus*, and *Phytoreovirus*) contains viruses infecting plants, invertebrates, and vertebrates (Quito-Avila et al., 2012). They constitute one of the three non-enveloped virus groups that are transmitted in a persistent replicative manner by plant hoppers (*Fijivirus* and *Oryzavirus* spp.) and leafhoppers (*Phytoreovirus* spp.) (Hogenhout et al., 2008; Evangelina et al., 2013). The following points mentioned by Nault and Ammar(1989) support the hypothesis that plant reoviruses originated in the leafhopper vectors.

Fijiviruses are morphologically similar to viruses such as *Leafhopper A virus* (LAV) that replicate only in the insect and to the *Peregrinus maidis virus* (Falk et al., 1988).

- All known plant reoviruses replicate in their hopper vectors.
- Plant reoviruses are not seed-borne, nor are they transmitted by mechanical means, except in special circumstances. Thus, they are entirely dependent on the hopper vectors for survival.
- The plant species infected by reoviruses are usually the prime food and breeding hosts of their hopper vectors.

- The plant reoviruses appear more closely adapted to their hopper hosts because (a) they replicate to higher titer in the insects; (b) several plant reoviruses are transmitted through insect eggs but none is transmitted through plant seed; (c) the percentage of virus-carrying insects in a given vector population is higher than the percentage of plants that can be infected by feeding single hoppers on them; and (d) some cause cytopathic effects in the hopper vectors, but in general these viruses have less severe pathogenic effects in the insects than in their plant hosts. In fact, most can be considered as causing latent infections.

9.2.3   RHABDOVIRUSES

Rhabdoviruses are taxonomically classified in the family *Rhabdoviridae*, order *Mononegavirales*. As a group, rhabdoviruses can infect plants, invertebrates, and vertebrates. Plant cyto- and nucleorhabdoviruses infect a wide variety of species across both monocot and dicot families, including agriculturally important crops such as lettuce, wheat, barley, rice, maize, potato, and tomato. Plant rhabdoviruses are transmitted by and replicate in hemipteran insects such as aphids (*Aphididae*), leafhoppers (*Cicadellidae*), or planthoppers (*Delphacidae*). Rhabdoviruses are enveloped negative strand RNA viruses that belong to the order *Mononegavirales* and include *Bornaviridae*, *Filoviridae*, and *Paramyxoviridae* families (Jackson et al., 2005). They are bullet or cone-shaped (from vertebrates and invertebrates) or bacilliform (from plants). It appeared that like the *Reoviridae*, rhabdoviruses have a very similar particle morphology and genome strategy and infect either vertebrates and invertebrates or invertebrates and plants. Thus, a common origin for this family among the insect vectors is indicated. The viruses are not seed-transmitted but are transmitted through the eggs of hopper vectors. The situation is somewhat more complex than with the plant reoviruses since different plant rhabdoviruses have hopper, aphid, piesmatidae, or mite vectors. Perhaps, as a rare event in the past, one rhabdovirus could transfer from a vector in one family to a vector in another where the vectors had a common host plant. It may be relevant to this idea that vesicular stomatitis virus, a rhabdovirus infecting vertebrates, replicates when injected into the leafhopper vector. In general, the maximum diversity in the Rhabdoviridae is seen among the insect-borne plant viruses (Cytorhabdovirus and Nucleorhabdovirus genera), followed by the Novirhabdovirus, isolated from fish and other aquatic animals (including invertebrates), the dimarhabdoviruses (with the possibility to replicate in vertebrate and invertebrate hosts), and finally by the lyssaviruses, which replicate exclusively in mammals. Rhabdoviral infection, as infections caused by other arboviruses, may have at least two results. As suggested earlier, insect infection is typically persistent, lasting for the life of the insect, and can result in transovarial transmission from infected females to their offspring. Persistent infections are characterized by an initial phase of rapid replication and substantial viral production (acute phase), followed by a decrease in the extent of viral replication to levels that are orders of magnitude lower, but continue for life (persistent phase). A hallmark of these persistent infections is little or no cell killing. In contrast, mammalian infection with these viruses is typically acute and cytolytic. Virus production is massive and occurs rapidly, but ceases with the death of the cells. Inferred major forces of rhabdovirus evolution include point mutations and purifying selection. There is no strong supportive evidence of homologous recombination in the family, or within the whole order *Mononegavirales*, and no suggestions for positive selection toward host adaptation have been reported (Kuzmin et al., 2009).

9.2.4 TOSPOVIRUSES

Tospoviruses (Genus *Tospovirus*, Family *Bunyaviridae*) are phytopathogens responsible for significant worldwide crop losses. They have a tripartite negative and ambisense RNA genome segments, termed S (Small), M (Medium) and L (Large) RNA. The vector-transmission is mediated by thrips in a circulative-propagative manner. Tospovirus group contains virus species that infect vertebrates, plants, and arthropods. These viruses are generally transmitted by mosquitoes, sandflies, or ticks. The tospoviruses are transmitted by thrips insects (Order Thysanoptera) in a circulative-propagative manner. Despite the existence of more than 5,000 thrips species, only 14 species are known as potential tospovirus vectors and most of them belong to the genera *Frankliniella* and *Thrips*. The *Frankliniella* genus is neotropical with all but seven species considered endemic to the new world, whereas the worldwide distributed genus *Thrips* has no species native to South America. Interestingly, the natural distribution of these vector species is somewhere reflected in the tospovirus phylogenetic relationships, with *Frankliniella*-transmitted tospoviruses clustering in an "American lineage"; and *Thrips*-transmitted tospoviruses clustering in a "Eurasian lineage". Another evolutionary lineage is formed by two tospoviruses isolated from peanut and transmitted by thrips from genus *Scirtothrips* (Pappu et al. 2009). It seems very probable that, like the Reoviridae and Rhabdoviridae families, the bunyaviruses originated in their insect hosts. The tenuiviruses almost certainly originated in insects. The same may be applicable for some marafiviruses (e.g., *MRFV*, OBDV, and BELV) that are transmitted in a circulative propagative manner, but the evidence is less compelling.

9.2.5   NODAVIRUSES

Nodavirus group consists of ssRNA viruses that infect animals and which have small (< 5 kb) bipartite genomes. Flock house virus (FHV) is an insect virus belonging to this viral group. It has no known relationship with plants in nature (Hao et al., 2014). However, Selling et al. (1990) showed that the virus replicated to low levels in the leaves of several plant species following mechanical inoculation with FHV RNA. No replication could be detected following inoculation with whole virus. However, inoculation of barley protoplasts with intact FHV resulted in the synthesis of small amounts of progeny virus particles, indicating that the virus could be uncoated in plant cells. FHV particles move systemically in *Nicotiana benthamiana* expressing the movement protein of TMV or RCNMV but show no symptoms in infected plants (Dasgupta et al., 2001). They also replicate in inoculated leaves of alfalfa, Arabidopsis, *Brassica*, cucumber, maize, and rice. These results suggest that the internal milieu of diverse organisms may be sufficiently similar to be able to support the replication of simple RNA viruses once inside a cell. This may have been a factor in ancient evolutionary processes.

9.2.6   CIRCO-AND NANO-VIRUSES

The vertebrate-infecting circoviruses and the plant infecting nanoviruses each have small circular ssDNA genomes with several features in common (Gibbs, 1999; Abraham et al., 2012). In an analysis of circovirus and nanovirus Rep protein sequences, Gibbs and Weiller (1999) noted similarities between the N-terminal regions, whereas the C-terminal region of calicivirus Rep is related to protein 2C of picorna-like viruses. They concluded that circoviruses evolved from nanoviruses by transference of nanovirus DNA to an animal followed by recombination.

9.3   PLANT VIRUS ADAPTATION

Viruses adapt to their hosts by evading defense mechanisms and taking over cellular metabolism for their own benefit. Alterations in cell metabolism as well as side effects of antiviral responses contribute to symptoms development and virulence. Sometimes, a virus may spill over from its usual host species into a novel one, where usually it will fail to successfully infect and further transmit to new host. However, in some cases, the virus transmits and persists after fixing beneficial mutations that allow for a better exploitation of the new host. One of the first consequences of organisms' adaptation to new environments is the manipulation of resources. Whether the relationship between a host and a parasite evolves toward a more or less virulent or benign situation depends on several genetic and ecological factors that may affect virus accumulation and transmission between hosts (Agudelo-Romero et al., 2008). Understanding host-range evolution in plant viruses is central for the control of viral diseases: changes in host range resulting in host switches or host-range expansions are at the root of virus emergence either in new host species or in host genotypes previously resistant to the virus. Theory predicts that different selection pressures in different hosts will result in fitness trade-offs across hosts, leading to specialization. Generally, plant viruses are host generalists and vector specialists (Power & Flecker, 2003). Differential adaptation of generalist viruses to their different plant and their genotypes has been investigated. Power and Mitchell (2004) strongly supported the concept of differential host adaptation by observing the prevalence of the PAV strain of *Barley yellow dwarf virus* (*BYDV*) in seven species of Poaceae. Later on, Cronin et al. (2010) confirmed this hypothesis by performing quantification of within-host multiplication and transmission assay. However, Malpica et al. (2006) showed that host selectivity is a successful strategy for generalist viruses while analyzing the occurrence and prevalence of five generalist viruses in 21 wild plant species. These results were in agreement with the hypothesis that supports specialization as advantageous. These studies thus contradicts that generalist viruses that exploit their different hosts with similarly high efficiencies may not be common. It is a well-known fact that if virus isolates from one host are serially passaged in another host, they become adapted to the new host and often involves phenotypic alterations in the original hosts suggesting fitness penalties (Yarwood, 1979). The generation of across-host trade-offs in plant viruses have been explained by two different mechanisms. Antagonistic pleiotropy is the first mechanism that results from opposite phenotypic effects of mutations across hosts (Fry, 1990; Bedhomme, 2012). Mutation accumulation is another mechanism that results from the accumulation by drift of mutations that are neutral in one host, but deleterious in another one (Kawecki, 1994). Although both mechanisms imply host-dependent fitness effects, natural selection is the only cause of trade-offs in the first mechanism, whereas genetic drift is the cause for the second.

9.3.1   TRADE-OFF ACROSS HOSTS AND ANTAGONISTIC PLEIOTROPY

The first example of a trade-off across hosts was reported by Matthews in 1949, when *Potato virus X (PVX)* from potato was inoculated in tobacco, it enhanced its virulence and infectivity in tobacco, but after 19 artificial inoculations, it lost virulence (Yarwood, 1979). Liang et al. (2002) also re-examined host adaptation *Hibiscus chlorotic spot virus* (HCSV)–*Chenopodium quinoa* system and provided evidence for across-host trade-offs. Serial passage in peas of *Plum pox virus* (PPV) isolates from peach resulted in adaptation to the new host by enhanced infectivity and within-host multiplication, while it decreased transmission efficiency in the original one (Wallis et al., 2007). Similarly, after 17 serial passages, the pathosystem *Tobacco etch potyvirus* (TEV)–*Arabidopsis thaliana* ecotype *Ler* showed that the viral genome has accumulated only five changes, three of which were non-synonymous. An amino acid substitution in the viral genome-linked protein (VPg) was responsible for the appearance of symptoms, whereas one substitution in the viral P3 protein the epistatically contributed to exacerbate severity. DNA microarray analyses show that the evolved and ancestral viruses affect the global patterns of host gene expression in radically different ways. A major difference is that genes involved in stress and pathogen response are not activated upon infection with the evolved virus, suggesting that selection has favored viral strategies to escape from host defenses (Agudelo-Romero et al., 2008). It is worth to mention here that *HCSV, PPV,* and *TEV* are specialists with narrow natural host ranges, and that adaptation to new hosts in passage experiments resulted in host-range expansion. Interestingly, when six isolates of CMV were passaged in their original three hosts, neither fitness nor virulence was improved in any host, suggesting that the fitness landscape of this generalist virus over its host range is at or near its maximum (Sacristán et al., 2003). Thus, the differences in fitness among hosts are determinants of the host range and of the consequences of host jumps: a virus may have a high fitness in its primary host(s) but a very low one in non-host species. Recently Moreno-Pérez et al. (2014) provided evidence for the existence of a relationship between the taxonomic proximity among hosts and the fitness of a specialist virus. They reported that narrow host-range viruses have been transmitted to a cultivated plant species from wild members of the same genera (e.g., *LMV* and *PepMV*) or family (e.g., *CSSV* and MSV). Accordingly, *PepMV* adaptation to tomato (*Solanum lycopersicum*) does not seem to be associated to a trade-off in the closely related *Solanum peruvianum* or *Solanum chilense*, whereas a trade-off is apparent in the more distant *Solanum muricatum*. These all studies suggest that generalist viruses did not show equal fitness across all potential hosts, but that their capacity of multiplication and transmission varied among hosts, with possible adaptive trade-offs (Betancourt et al., 2011; Power et al., 2011).

The mechanism of antagonistic pleiotropy in plant viruses have been evident from the studies of across-host effects of mutations introduced in viral genomes through manipulation of biologically active cDNA clones (Moury & Simon, 2011). For instance, Liang et al. (2002) showed that five different HCSV lineages adapted to *C. quinoa* had fixed eight amino acid mutations in the CP, three of which were sufficient, when introduced by site-directed mutagenesis in the original genotype, to cause its fitness loss in *Hibiscus*. Rico et al. (2006) also reported same mechanisms in case of *Pelargonium flower break virus* (PFBV). A single nucleotide substitution engineered in the CI cistron of *TuMV* that resulted in overcoming TuRB01 resistance in oilseed rape showed decreased competitive ability with the wild type in passage experiments in the susceptible host (Jenner et al., 2002). Double mutants at *PVY* VPg that were more efficient than single mutants at breaking *pvr23* resistance in pepper were out-competed by single mutants in susceptible host genotypes (Quenouille et al., 2013). Similarly, amino acid substitution N25I in the CP of *PVY* support rapid virus multiplication in tobacco, but in potato showed multiplication with decline rate (Moury & Simon, 2011). It was also shown that multiplication, in the susceptible rice genotype, of a rymv1-2-resistance-breaking mutant with the substitution R48I in the VPg was less than the multiplication of wild type virus (Sorho et al., 2005; Poulicard et al., 2010). However, multiplication of rymv1-2-resistance-breaking mutants (R48G and R48E) was several folds of lower magnitude than the wild type susceptible rice genotypes (Poulicard et al., 2010). On the other hand, the mutation H52Y, which results in overcoming both rymv1-2 and rymv1-3, showed no evidence of a fitness penalty in the susceptible host (Poulicard et al., 2012). Similar case of antagonistic pleiotropy was also observed for the *ToMV* mutants overcoming Tm1 resistance (Ishibashi et al., 2012). Hence, different mutations resulting resistance-breaking showed different fitness penalties in the susceptible host, that were positively correlated with the level of virus multiplication in the resistant host.

9.3.2 DISTRIBUTION OF EPISTASIS AND MUTATIONAL FITNESS EFFECTS

The evolutionary fate of a viral population in a given host also depends on the distribution of mutational fitness effects (DMFE). DMFE includes the fraction of all possible mutations that are beneficial, neutral, and deleterious. Carrasco et al. (2007) characterized DMFE for random nucleotide substitutions in the genome of *Tobacco etch virus* (TEV) in the primary host *Nicotiana tabacum* and highlighted that the most of mutations were lethal and on average responsible for fitness reduction. These results are in good agreement with the findings of Domingo-Calap et al. (2009). Recently, Lalić et al. (2011) measured the fitness of a subset of the *TEV* single-nucleotide substitution mutants generated by Carrasco et al. (2007) across a panel of eight susceptible hosts that differ in terms of taxonomic relatedness to tobacco (primary host). The results show that the host species wherein fitness is evaluated has a major effect on the DMFEs. The mean of the distribution moves toward smaller values (more deleterious effects) as the degree of genetic relatedness between the test host and the primary host decreases. They also observed the existence of a significant genotype-by-environment (G×E) interaction (Lalić et al., 2011). This may be explained by antagonistic pleiotropy as most mutations change the sign of their effect depending on the host; and by reduction in genetic variance for fitness among hosts (Lalić et al., 2011). The existence of significant G×E component has important implications for viral emergence and adaptation (Elena et al., 2014a). First, it introduces a degree of uncertainty in terms of phenotypic effect of a mutation in a given host, or its impact on an alternative host. Second, the likelihood of host specialization would be proportional to the extent by which the G×E component is generated by antagonistic pleiotropy. Third, the fact that a reduction in genetic variance for fitness also contributes significantly to generate a G×E component implies that genetic drift becomes important relative to natural selection during viral evolution, thus making the process of adaptation difficult for a new host.

The genome of RNA viruses carries multiple mutations, and interactions of these mutations determine their fitness (Sanjuán et al., 2010). In terms of quantitative genetics, the intensity and type of epistasis shaping the genome of plant RNA viruses has only been explored for pairs of random mutations introduced in *TEV* genome (Lalić & Elena, 2012a). The average epistasis for TEV was found to be negative. Elena et al. (2006) related the cause for this dominance of negative epistasis with the lack of genetic redundancy characteristic of RNA genomes, with overlapping genes and multifunctional proteins. However, to understand the genetics of viral evolution, it is essential to determine the significance of epistasis-by-host component ($G \times G \times E$). Recently Lalić and Elena (2012b) using *TEV* showed that the sign of epistasis among pairs of random mutations depends on the degree of genetic relatedness between the primary host and the alternative ones. These results indicate that selection more efficient promoting emergence as more distantly related the primary and the new hosts would be.

9.3.3 EVASION, SUPPRESSION AND PLANT DEFENSE TROUNCING

The most important factor to determine virus fitness in a given host is its ability to trounce plant defense system. Recently, Carr et al. (2010) described several resistance mechanisms adopted by plants against viral infections. García-Arenal and Fraile (2013) highlighted that the viral genotypes overcoming resistances are less fit in susceptible plants in comparison to those viruses which are unable to overcome the resistance. Consequently, such fitness cost represents a specific example of fitness trade-off across hosts that are generated as a result of antagonistic pleiotropy. This trade-off also provides explanation for higher durability of resistances against viruses than other cellular pathogens (Fraile & García-Arenal, 2010). In past few years, attention has been made to understand the factors that determine trouncing of host defense system (García-Arenal & McDonald, 2003; Janzac et al., 2009). The studies based on the analysis of posteriori of epidemiological data suggested that virus' evolutionary potential and the number of mutations required to overcome the resistance are the major factors to decide trouncing of host resistance.

RNA silencing suppression is a common feature of plant viruses. Overcoming plant defense system through RNA silencing have been evaluated by analyzing the durability of the resistance against *Turnip mosaic virus* (TuMV) conferred by transgenic expression in *Arabidopsis thaliana* of artificial microRNAs (amiRs) specifically designed to be complementary to viral cistron encoding for the RNA silencing suppressor protein of this virus (HC-Pro) (De la Iglesia et al., 2012). A series of experiments conducted on the evolution of TuMV lineages by serial passages in susceptible or partially resistant plants, suggested that the dynamics of overcoming resistance depend on the extent of protection conferred by the *amiR* (Lafforgue et al., 2011; Martínez et al., 2012). They also mentioned that such mechanism of virus evolution was rapid in partially resistant plants than highly susceptible plants. It is important to mention here that overcoming *amiR*-mediated resistance was always associated to the presence of mutations at any of the 21 positions of the *amiR* target within HC-Pro. The detection of ancestral genotype of *TuMV* in resistant genotypes (Martínez et al., 2012) suggesting replication of viral populations at higher rate in resistant plants. It will be very interesting to see application of such type of resistance in virus disease management in agricultural fields, since it will favor the reversion to the wild type viral genotype, if escape mutations would have a fitness cost in susceptible plants.

9.3.4   GENETIC DRIFT AND METAPOPULATION DYNAMICS

Spatial structure of a viral population within a host plant generally influence the relative contribution of genetic drift and natural selection to evolution and play role in virus adaptation to new hosts. Recently, García-Arenal and Fraile (2011) reported that the process of infection and colonization of plants by viruses is due to evolutionary consequences. The fact that plant viruses lack mechanisms that actively allow them to move across cell walls indicate symplastic mode of virus colonization in their host system. There might be a possibility that plant viruses do not follow mass-action law, which otherwise work well in case of bacteriophages (Elena et al., 2014a). The intercellular connections, tissues architecture, plant type, and smooth connectivity between distal parts by the vascular system creates favorable environment for the replication and evolution of virus population. Thus, virus population replicating within an infected host system behave as a meta-population. Such type of meta-population structures have been reported in both RNA and DNA viruses, and their hosts (Hall et al., 2001; Jridi et al., 2006; González-Jara et al., 2009). These spatial structures enhance the fitness of an emerging virus in the new host by imposing strong restrictive conditions to the expansion of new beneficial mutations. Furthermore, Zwart et al. (2012) found that spatial segregation reduces intracellular competition in *TEV* and suggested that the efficiency of natural selection to optimize the average meta-population fitness is independent of the magnitude of the beneficial effect and conferred by mutation. French and Stenger (2003) used population genetics to analyze the differences in viral fitness during systemic colonization by infecting plants with a mixture of two or more strains of a virus on the different plant parts. They found that effective population size (Ne) vary widely among viruses and hosts (Sacristán et al., 2003; Li & Roossinck, 2004; Monsion et al., 2008). It is interesting to note that the magnitude of Ne is always several folds smaller than the number of viral genomes in any virus population (García-Arenal & Fraile, 2011). Consequently, these findings clearly provide evidences that genetic drift may play an important role in the evolution of viral populations. Frank (2001) highlighted that under severe infection situation, complementation between genetic variants can reduce the rate of fixation of beneficial mutations in virus population. There might be another possibility that the effective ploidy level of the system is high in virus population system, which dilutes the contribution of each locus to the phenotype and relaxing the effect of selection pressure over them. Usually, a weaker selection implies more genetic diversity and the maintenance of deleterious variants in the viral population during longer periods of time. Fraile et al. (2008) while working on two different virus population systems showed that trans-complementation results in the maintenance of mutants at high frequency. In such circumstances, minimization of complementation and acceleration in the evolution of linked loci is possible due to the phenomenon of super-infection inhibition (Folimonova, 2012), which would be beneficial for the virus in long term.

9.3.4   MIXED VIRAL INFECTION

In nature, co-infection and complementation among genetic variants carrying beneficial or deleterious alleles determined the rate of evolution. Inter specific co-infections are a very common phenomenon in plant viruses (Salvaudon et al., 2013; Elena et al., 2014b). Malpica et al. (2006) found that the prevalence of certain viruses is dependent on the prevalence of other viruses, where some viral combinations appeared more frequently. Hammond et al. (1999) found that in an individual host, co-infection may have variable consequences, from the development of milder symptoms to their exacerbation. Several reports indicate that mixed viral infections influenced the host range, cellular tropism, viral accumulation, and rate of transmission (Moreno et al., 1997; Guerini & Murphy, 1999; García-Cano et al., 2006; Sánchez-Navarro et al., 2006; Wintermantel et al., 2008; Martín & Elena, 2009). In case of potyviruses and some ssRNA virus, it was noticed that during synergy of two ssDNA or two ssRNA viruses, the viral load of the non-potyvirus is increased, whereas that of the potyvirus remains unchanged.

Later on, this point was explained by the activity of the potyvirus HC-Pro protein as suppressor of RNA silencing (Dunoyer & Voinnet, 2005). It is interesting to note that the interactions between viruses do not always lead to the synergistic effect on symptom development but the result depends from the particular combination of viral species (Kokkinos & Clark, 2006). From these studies, it is clear that two isolates of the same virus may exclude each other from a cell, creating spatially segregating distributions of geno-types with minimal overlap. Contrarily to this, Dietrich and Maiss (2003) observed the phenomenon of exclusion in variants of PPV. They found that *Potyviruses* from different species did not exclude each other, and moreover infecting the same cells, although exclusion was limited to *PPV* variants. Thus, these findings conclude that potyvirus species still show significant sequence similarities, inter-specific co-infections between members of the same genus and hence open the possibility for the evolution and origin of new viral species due to recombination and reassortment.

9.4   HOST SELECTION BY PLANT VIRUSES

Host plant influences virus evolution by affecting genome stability, protein emergence, amino acid usage, synonymous codon usage, and dinucleo-tide usage. However, the impact of the host on viral evolution is varied for different viruses and their hosts. Host alternations play an important role in maintaining genome integrity and elimination of lethal mutants of viruses, especially arthropod-borne viruses, such as *Rice dwarf virus* (RDV) and *Tomato spotted wilt virus* (TSWV) (Sin et al., 2005; Pu et al., 2011). It is interesting to note that the distribution of mutational deletions in viral genome is not random but concentrated in one or more mutational hot spots. de Oliveira et al. (1991) observed that the major deletions within the *TSWV* genome was localized in the gene encoding the precursor of glycoproteins and generated by serial mechanical transmissions (de Oliveira et al., 1991). Similarly, the 3a gene of RNA3 is the hot spot for CMV deletion mutations generated during infection of various host plants (Takeshita et al., 2008). Besides this, host plant also played critical role in the making the genome more complex and emergence of new viral genes. For instance, the genome of closterovirus group encodes two types of genes *viz.*, conserved core genes regulating virus replication and species-specific genes (Martelli et al., 2012). The species-specific genes show dramatic variation in their locations, numbers, functions, and more over their products showed differentiation in terms of proteins present in closteroviruses. These results clearly indicate that species-specific genes originated in unique fashion in closteroviruses and underwent evolution recently. Molecular analysis showed that the three non-conserved genes (p33, p18, and p13) of *Citrus tristeza virus* (CTV) have different roles in infecting various citrus cultivars. All of these genes are dispensable for *CTV* replication in *Citrus macrophylla* and *C. aurantifolia* but are also required for other citrus plants, including sour orange (*C. aurantium*), lemon (*C. limon*), grapefruit (*C. paradise*), or calamondin (*C. mitis*) (Tatineni et al., 2011). These results revealed that these genes in *CTV* evolved during adaptation to the new citrus hosts. Interestingly, the movement proteins encoded by plant viruses are assumed to have evolved during the adaptation of ancient viruses to plant hosts as well (Lucas, 2006). Besides this, Barr and Fearns (2010) reported that viruses have also devised several mechanisms like evolving additional viral protein(s) to increase the fidelity of RNA-dependent RNA polymerase (RdRp), genome length checking and restoring damaged genome termini to maintain their genome integrity. Moreover, the need for adaptation to plants encourages viruses to quickly alter their genomes and such changes further leads to the emergence of new viral gene(s). In addition to this, plants are also able to influence the evolution of existing viral genes, especially those viral genes that encode proteins that directly interact with host proteins. This may be due to the fact that compatibility between viral proteins and host factors affects the efficiencies of viral replication, viral particle assembly, cell-to-cell movement, and other viral processes. Therefore, the interfaces of viral proteins that directly interact with host factors will be subjected to stronger selection pressure than other positions. For example, the plant eukaryotic translation initiation factors (eIF4E and eIF4G) interact directly with the potyviral genome-linked protein (VPg), which is essential for potyvirus multiplication. Nieto et al. (2011) reported that the eIF containing amino acid substitutions at the interaction interface will hamper the interaction between the viral VPg and eIFs, abolishing the replication of potyvirus and finally providing recessive resistance against potyviruses in the host plant. Diaz-Pendon et al. (2004) mentioned that majority of recessive resistance genes in plants against the potyviruses is eukaryotic translation initiation factors. Further, Truniger and Aranda (2009) also reported that potyviruses are able to induce amino acid changes in their VPgs to overcome host eIF4E-mediated resistance. These results suggest that the VPgs of potyviruses may have coevolved with the plant eIF4E (Charron et al., 2008). The point mutations introduced during viral replication can result in synonymous (nucleotide substitution) and non-synonymous (amino acid substitution) mutations for the synthesis of viral proteins. Hurst (2002) showed that the comparison of the number of non-synonymous substitutions per non-synonymous site (dN) to the number of synonymous substitution per synonymous site (dS) in a given gene provide an important indicator of the selective pressure acting on that gene. For example, in case of a Darwinian selection, a gene is under strong positive selection and therefore, the value of dN/dS should be higher than 1.0 and vice versa. Similarly, the nucleotide substitution model in the capsid genes of plant RNA viruses reflects the mode of transmission, suggesting the existence of host-specific selection pressures on certain viruses (Chare & Holmes, 2004). Thus, from these studies, it is clear that host plants affect the evolution of all viral proteins directly or indirectly during viral multiplication in their cells.

The viral mutational preference is one of the most important factors that determine the viral synonymous codon usage. Synonymous codon usage is influenced by several parameters, such as translation selection, mutation pressure, gene transfer, amino acid conservation, RNA stability, hypersaline adaptation, and growth conditions (Paul et al., 2008). However, the translational pressure due to tRNA availability, nucleotide acid abundance, and selection of CpG-suppressed clones in the host cell by the immune system, also affects viral synonymous codon usage and even determines the synonymous codon usage bias in some viral genes or particular viruses (Karlin et al., 1990, Lobo et al., 2009). The first comprehensive analysis of the synonymous codon usage of plant viruses was performed by Adams and Antoniw (2004). They analyzed synonymous codon usage bias of 385 plant viruses by using effective number of codons (ENC), generally used to measure the distance between codon usage of a gene and equal usage of synonymous codons (Wright, 1990). They found that the ENC values of these viruses were positively correlated with those of viral GC contents in the third codon position but not with the host type they infect. As a result, they reported that mutational bias, rather than translational selection, accounts for the observed variations in synonymous codon usage in plant viruses. Similarly, detailed analysis of the synonymous codon usage of begomoviruses was performed by Xu et al. (2008). They detected translational selection in the highly expressed genes of virus genomes, although mutation bias appears to be the major factor to determine the synonymous codon usage pattern of begomoviruses. It is interesting to note that Cheng et al. (2012) found a high degree of similarity of the synonymous codon usage between *CTV* and its citrus host. Additionally, the synonymous codon usage resemblance between woody plant-infecting closteroviruses and their woody hosts is higher than herbaceous plant-infecting closteroviruses and their herbaceous hosts (Cheng et al., 2012). These findings further confirm the influence of the host on synonymous codon usage in plant viruses. In another study, Cheng et al. (2012) also found that linear specific synonymous codon usage exists in viruses within the *Bunyaviridae*, *Tenuivirus*, and *Emaravirus*, although the synonymous codon usage of most of these viruses shows a high degree of resemblance, suggesting that the mutational accumulation is the major factor affecting synonymous codon usage. From these studies, it is clear that the synonymous codon usage of plant viruses within the same genus achieve maximum similarity due to mutational pressure. In case of some plant viruses, translational pressure from the host plant may influence the viral synonymous codon usage, especially in those plant viruses that coevolved with their hosts.

The dinucleotide frequency is the incidence of a given neighbor dinucleotide in a sequence (e.g., a gene or a genome). Under no selection pressure, all the nucleotides exist in randomized fashion and frequencies of the sixteen dinucleotide pairs should be similar. However, studies conducted by Cheng et al. (2014) confirmed the existence of selection pressure by reporting the ubiquitous extremities in the presentations of dinucleotide pairs in the genomes.

9.5   CONCLUSION AND FUTURE PERSPECTIVES

From the perspective of viral origin and evolution, it is clear that virus adaptation to plant genotype results in parallel evolution and the convergence of independent lineages upon a shared adaptive solution via the same genetic targets. The evolution through continual selection leads to increased pathogenicity and wider population divergence. In addition, host-range expansion during co-evolution occurs by stepwise evolution involving multiple mutations, and large host-shifts that appear to be constrained by mutational accessibility, pleiotropic costs, and ecological factors. Thus, it will be very important to examine the ecological roles of plant-associated viruses and their vectors in managed and unmanaged ecosystems and the reciprocal influence of ecosystem properties on the distribution and evolution of plant viruses and their vectors. As a result of new insights from the studies of molecular epidemiology of plant viruses, there is growing interest to understand and predict the dynamics of plant viruses and their evolution; hence, future research must consider virus influence not only in agriculture, but also in nature. So far, there are several plant viruses, undiscovered in nature, that will determine host fitness, which will require advanced techniques like next-generation sequencing for the analysis of metagenome and metabolome of existed virome in the wild and cultivated plant species. Current understanding of plant–virus dynamics is blooming for molecular to population-level interactions and in future, understanding of key issues such as the mechanisms determining host plant resistance, fitness trade-off and transmission across hosts are likely to be extended to wild plants and links between molecular mechanisms and ecological consequences will be explored. As the research focus extends from managed to natural systems, it is useful to consider differences along this gradient in species identities and population structures, and the need for studies of virus influence on plant demography. Overall, it is realized that there is an urgent need to integrate consideration of plant viruses into ecological research and theory, in which viruses have generally been overlooked for making clear picture regarding the evolution, adaption, and host selection by plant viruses.

KEYWORDS

- **adaptation**
- **evolution**
- **fitness**
- **origin**
- **virus**

REFERENCES

Abraham, A. D.; Varrelmann, M.; Vetten, H. J. Three Distinct Nanoviruses, One of Which Represents a New Species, Infect Faba Bean in Ethiopia. *Plant Dis*. **2012**, *96*, 1045–1053.

Adams, M. J.; Antoniw, J. F. Codon Usage Bias amongst Plant Viruses. *Arch. Virol*. **2004**, *149*, 113–135.

Agudelo-Romero, P.; De La Iglesia, F.; Elena, S. F. The Pleiotropic Cost of Host-Specialization in *Tobacco etch potyvirus*. Infect. *Genet. Evol*. **2008**, *8*, 806–814.

Allison, R. F.; Schneider, W. L.; Greene, A. E. Recombination in Plants Expressing Viral Transgenes. *Semin. Virol*. **1996**, *7*, 417–422.

Anderson, P. K.; Cunningham, A. A.; Patel, N. G.; Morales, F. J.; Epstein, P. R.; Daszak, P. Emerging Infectious Diseases of Plants: Pathogen Pollution, Climate Change and Agrotechnology Drivers. *Trends Ecol. Evol*. **2004**, *19*, 535–544.

Aus Dem Siepen, M.; Pohl, J. O.; Koo, B-J.; Wege, C.; Jeske, H. *Poinsettia latent virus* is not a Cryptic Virus, but a Natural Polerovirus-Sobemovirus Hybrid. *Virology*. **2005**, *336*, 240–250.

Barr, J. N.; Fearns, R. How RNA Viruses Maintain their Genome Integrity. *J. Gen. Virol.* **2010**, *91*, 1373–1387.

Bedhomme, S.; Lafforgue, G.; Elena, S. F. Multihost Experimental Evolution of a Plant RNA Virus Reveals Local Adaptation and Host-Specific Mutations. *Mol. Biol. Evol.* **2012**, *29* (5), 1481–1492.

Betancourt, M.; Escriu, F.; Frail, A.; García-Arenal, F. Virulence Evolution of a Generalist Plant Virus in a Heterogeneous Host System. *Evol. Appl.* **2013**, *6*, 875–890.

Betancourt, M.; Fraile, A.; García-Arenal, F. Cucumber Mosaic Virus Satellite RNAs that Induce Similar Symptoms in Melon Plants Show Large Differences in Fitness. *J. Gen. Virol.* **2011**, *92,* 1930–1938.

Boccardo, G.; Candresse, T. Complete Sequence of the RNA1 of an Isolate of *White Clover Cryptic Virus* 1, Type Species of the Genus *Alphacryptovirus. Arch. Virol.* **2005**, *150*, 399–402.

Botstein, D. A. Theory of Modular Evolution for Bacteriophages. *Ann. N. Y. Acad. Sci.* **1980**, *354*, 484–490.

Brough, C. L.; Gardiner, W. E.; Inamdar, N. M.; Zhang, X. Y.; Ehrlich, M.; Bisaro, D. M. DNA Methylation Inhibits Propagation of Tomato Golden Mosaic Virus DNA in Transfected Protoplasts. *Plant Mol. Biol.* **1992**, *18*, 703–712.

Bull, S. E.; Briddon, R. W.; Sserubombwe, W. S.; Ngugi, K.; Markham, P. G.; Stanley, J. Genetic Diversity and Phylogeography of Cassava Mosaic Viruses in Kenya. *J. Gen. Virol.* **2006**, *87*, 3053–3065.

Carr, J. P.; Lewsey, M. G.; Palukaitis, P. Signaling in Induced Resistance. *Adv. Virus Res.* **2010**, *76*, 57–121.

Carrasco, P.; De La Iglesia, F.; Elena, S. F. Distribution of Fitness and Virulence Effects Caused by Single-Nucleotide Substitutions in *Tobacco Etch Virus. J. Virol.* **2007**, *81*, 12979–12984.

Chare, E. R.; Holmes, E. C. Selection Pressures in the Capsid Genes of Plant RNA Viruses Reflect Mode of Transmission. *J. Gen. Virol.* **2004**, *85*, 3149–3157.

Charron, C.; Nicolaï, M.; Gallois, J. L.; Robaglia, C.; Moury, B.; Palloix, A.; Caranta, C. Natural Variation and Functional Analyses Provide Evidence for Co-Evolution between Plant Eif4e and Potyviral Vpg. *Plant J.* **2008**, *54*, 56–68.

Cheng, X-F.; Virk, N.; Wang, H-Z. Impact of the Host on Plant Virus Evolution. In *Plant Virus–Host Interaction: Molecular Approaches and Viral Evolution;* Gaur, R. K., Hohn, T., Sharma, P., Eds.; Academic Press: USA, 2014; pp 361–371.

Cheng, X-F.; Wu, X-Y.; Wang, H-Z.; Sun, Y-Q.; Qian, Y-S.; Luo, L. High Codon Adaptation in *Citrus Tristeza Virus* to its Citrus Host. *Virol. J.* **2012**, *9*, 113.

Cooper, I.; Jones, R. A. C. Wild Plants and Viruses: Under-Investigated Ecosystems. *Adv. Virus Res.* **2006**, *67*, 1–47.

Cronin, J. P.; Welsh, M. E.; Dekkers, M. G.; Abercrombie, S. T.; Mitchell, C. E. Host Physiological Phenotype Explains Pathogen Reservoir Potential. *Ecol. Lett.* **2010**, *13*, 1221–1232.

Dasgupta, R.; Garcia, B. H.; Goodman, R. M. Systemic Spread of an RNA Insect Virus in Plants Expressing Plant Viral Movement Protein Genes. *Proc. Natl. Acad. Sci. U.S.A.* **2001**, *98*, 4910–4915.

De La Iglesia, F.; Martínez, F.; Hillung, J.; Cuevas, J. M.; Gerrish, P. J.; Darós, J. A.; Elena, S. F. Luria-Delbrück Estimation of *Turnip Mosaic Virus* Mutation Rate *In Vivo. J. Virol.* **2012**, *86*, 3386–3388.

de Oliveira, R. R.; de Haan, P.; de Avila, A. C.; Kitajima, E. W.; Kormelink, R.; Goldbach, R.; Peters, D. Generation of Envelope and Defective Interfering RNA Mutants of Tomato Spotted Wilt Virus by Mechanical Passage. *J. Gen. Virol.* **1991,** *72,* 2375–2383.

Desjardins, C.; Eisen, J. A.; Nene, V. New Evolutionary Frontiers from Unusual Virus Genomes. *Genome Biol.* **2005,** *6,* 212–213.

Diaz-Pendon, J. A.; Truniger, V.; Nieto, C.; Garcia-Mas, J.; Bendahmane, A.; Aranda, M. A. Advances in Understanding Recessive Resistance to Plant Viruses. *Mol. Plant Pathol.* **2004,** *5,* 223–233.

Dietrich, C.; Maiss, E. Fluorescent Labelling Reveals Spatial Separation of Potyvirus Populations in Mixed Infected *Nicotiana benthamiana* Plants. *J. Gen. Virol.* **2003,** *84,* 2871–2876.

Domier, L. L.; McCoppin, N. K.; Larsen, R. C.; D'Arcy, C. J. Nucleotide Sequence Shows that *Bean Leafroll Virus* has a Luteovirus-Like Genome Organization. *J. Gen. Virol.* **2002,** *83,* 1791–1798.

Domingo-Calap, P.; Cuevas, J. M.; Sanjuán, R. The Fitness Effects of Random Mutations in Single-Stranded DNA and RNA Bacteriophages. *PLoS Genet.* **2009,** *5,* E1000742.

Domingo-Calap, P.; Sanjuán, R. Experimental Evolution of RNA versus DNA Viruses. *Evolution.* **2011,** *65,* 2987–2994.

Dunoyer, P.; Voinnet, O. The Complex Interplay between Plant Viruses and Host RNA-Silencing Pathways. *Curr. Opin. Plant Biol.* **2005,** *8,* 415–423.

Elena, S. F.; Carrasco, P.; Daròs, J. A.; Sanjuán, R. Mechanisms of Genetic Robustness in RNA Viruses. *EMBO Rep.* **2006,** *7,* 168–173.

Elena, S. F.; Fraile, A.; García-Arenal, F. Evolution and Emergence of Plant Viruses. *Adv. Virus Res.* **2014a,** *88,* 161–191.

Elena, S. F.; Bernet, G. P.; Carrasco, J. L. The Games Plant Viruses Play. *Curr. Opin. Virol.* **2014b,** *8,* 62–67.

Evangelina, B. Caro, A.; Guillermo, A. Maroniche; Analía, D. Dumón; Sagadín Mónica, B.; Mariana Del Vas; Truol, G. High Viral Load in the Planthopper Vector *Delphacodes kuscheli* (Hemiptera: *Delphacidae*) is Associated with Successful Transmission of Mal De Río Cuarto Virus. *Ann. Entomol. Soc. Am.* **2013,** *106*(1), 93–99.

Falk, B. W.; Kim, K. S.; Tsai, J. H. Electron Microscopic and Physicochemical Analysis of a Reo-Like Virus of the Planthopper *Peregrinus maidis*. *Intervirology.* **1988,** *29,* 195–206.

Fargette, D.; Konate, G.; Fauquet, C.; Muller, E.; Peterscmitt, M.; Thresh, J. M. Molecular Ecology and Emergence of Tropical Plant Viruses. *Ann. Rev. Phytopathol.* **2006,** *44,* 235–260.

Fauquet, C. M.; Bisaro, D. M.; Briddon, R. W.; Brown, J. K.; Harrison, B. D.; Rybicki, E. P.; Stenger, D. C.; Stanley, J. Revision of Taxonomic Criteria for Species Demarcation in the Family Geminiviridae, and an Updated List of Begomovirus Species. *Arch. Virol.* **2003,** *148,* 405–421.

Folimonova, S. Y. Superinfection Exclusion is an Active Virus-Controlled Function that Requires a Specific Viral Protein. *J. Virol.* **2012,** *86,* 5554–5561.

Fondong, V. N.; Pita, J. S.; Rey, M. E. C.; De Kochko, A.; Beachy, R. N.; Fauquet, C. M. Evidence of Synergism between *African Cassava Mosaic Virus* and a New Double-Recombinant Geminivirus Infecting Cassava in Cameroon. *J. Gen. Virol.* **2000,** *81,* 287–297.

Forterre, P. The Origin of Viruses and their Possible Roles in Major Evolutionary Transitions. *Virus Res.* **2006,** *117,* 5–16.

Fraile, A.; García-Arenal, F. The Coevolution of Plants and Viruses: Resistance and Pathogenicity. *Adv. Virus Res.* **2010,** *76,* 1–32.

Fraile, A.; Sacristán, S.; García-Arenal, F. A Quantitative Analysis of Complementation of Deleterious Mutants in Plant Virus Populations. *Span. J. Agric. Res.* **2008,** *6,* 195–200.

Frank, S. A. Multiplicity of Infection and the Evolution of Hybrid Incompatibilities in Segmented Viruses. *Heredity.* **2001,** *87,* 522–529.

French, R.; Stenger, D. C. Evolution of Wheat Streak Mosaic Virus: Dynamics of Population Growth within Plants May Explain Limited Variation. *Ann. Rev. Phytopathol.* **2003,** *41,* 199–214.

Fry, J. D. Trade-Offs in Fitness on Different Hosts: Evidence from a Selection Experiment with a Phytophagous Mite. *Am. Nat.* **1990,** *136,* 569–580.

Fukuhara, T.; Koga, R.; Aioki, N.; Yuki, C.; Yamamoto, N.; Oyama, N.; Udagawa, T.; Horiuchi, H.; Miyazaki, S.; Higashi, Y.; Takeshita, M.; Ikeda, K.; Arakawa, M.; Matsumoto, N.; Moriyama, H. The Wide Distribution of Endornaviruses, Large Double-Stranded RNA Replicons with Plasmid Like Properties. *Arch. Virol.* **2006,** *151,* 995–1002.

García-Arenal, F.; Fraile, A. Trade-Offs in Host-Range Evolution of Plant Viruses. *Plant Pathol.* **2013,** *62,* 2–9.

García-Arenal, F.; Mcdonald, B. A. An Analysis of the Durability of Resistance to Plant Viruses. *Phytopathology.* **2003,** *93,* 941–952.

García-Arenal, F.; Fraile, A. Population Dynamics and Genetics of Plant Infection by Viruses. In *Recent Advances in Plant Virology*; Caranta, C., Aranda, M. A., Tepfer, M., López-Moya, J. J., Eds.; Caister Academic Press: United Kingdom, 2011; pp 263–281.

García-Cano, E.; Resende, R. O.; Fernández-Muñoz, R.; Moriones, E. Synergistic Interaction between *Tomato Chlorosis Virus* and *Tomato Spotted Wilt Virus* Results in Breakdown of Resistancein Tomato. *Phytopathology.* **2006,** *96,* 1263–1269.

Ge, L. M.; Zhang, J. T.; Zhou, X. P.; Li, H. Y. Genetic Structure and Population Variability of *Tomato Yellow Leaf Curl* China Virus. *J. Virol.* **2007,** *81*(11), 5902–5907.

Gibbs, A. J.; Gibbs, M. J.; Ohshima, K.; Garcia-Arenal, F. More about Plant Virus Evolution: Past Present and Future. In *Origin and Evolution of Viruses;* 2nd Ed; Domingo, E., Parish, C., Holland, J., Eds.; Academic Press: London, UK, 2008; pp 229–250.

Gibbs, M. The Luteovirus Supergroup: Rampant Recombination and Persistent Partnerships. In *Molecular Basis of Virus Evolution*; Gibbs, A. J., Calisher, C. H., García-Arenal, F., Eds.; Cambridge University Press: Cambridge, 1995; pp 351–368.

Gibbs, M. Evidence that a Plant Virus Switched Hosts to Infect a Vertebrate and then Recombined with a Vertebrate-Infecting Virus. *Proc. Natl. Acad. Sci. U.S.A.* **1999,** *96*(14), 8022–8027.

Gibbs, M. J.; Weiller, G. F. Evidence that a Plant Virus Switched Hosts to Infect a Vertebrate and then Recombined with a Vertebrate Infecting Virus. *Proc. Natl. Acad. Sci. U.S.A.* **1999,** *96,* 8022–8027.

González-Jara, P.; Fraile, A.; Canto, T.; García-Arenal, F. The Multiplicity of Infection of a Plant Virus Varies during Colonization of its Eukaryotic Host. *J. Virol.* **2009,** *83,* 7487–7494.

Guerini, M. N.; Murphy, J. F. Resistance of *Capsicum Annuum* 'Avelar' to *Pepper Mottle Potyvirus* and Alleviation of this Resistance by Co-Infection with *Cucumber Mosaic Cucumovirus* are Associated with Virus Movement. *J. Gen. Virol.* **1999,** *80,* 2785–2792.

Guerini, M. N.; Murphy, J. F. Resistance Of *Capsicum Annuum* 'Avelar' to *Pepper Mottle Potyvirus* and Alleviation of this Resistance by Co-Infection with Cucumber Mosaic Cucumovirus are Associated with Virus Movement. *J. Gen. Virol.* **1999,** *80,* 2785–2792.

Gutierrez, C. Geminivirus DNA Replication. *Cell. Mol. Life Sci.* **1999,** *56,* 313–329.

Hacker, C. V.; Brasier, C. M.; Buck, K. W. A Double-Stranded RNA from a *Phytophthora* Species is Related to the Plant Endornaviruses and Contains a Putative UDP Glycosyltransferase Gene. *J. Gen. Virol.* **2005,** *86,* 1561–1570.

Hall, J. S.; French, R.; Hein, G. L.; Morris, T. J. Y.; Stenger, D. C. Three Distinct Mechanisms Facilitate Genetic Isolation of Sympatric *Wheat Streak Mosaic Virus* Lineages. *Virology.* **2001,** *282,* 230–236.

Hammond, J.; Lecoq, H.; Raccah, B. Epidemiological Risks from Mixed Virus Infections and Transgenic Plants Expressing Viral Genes. *Adv. Virus Res.* **1999,** *54,* 189–314.

Hanley-Bowdoin, L.; Bejarano, E. R.; Robertson, D.; Mansoor, S. Geminiviruses: Masters at Redirecting and Reprogramming Plant Processes. *Nature Rev. Microbiol.* **2013,** *11,* 777–788.

Hao, L.; Lindenbach, B.; Wang, X.; Dye, B.; Kushner, D.; He, Q.; Newton, M.; Ahlquist, P. Genome-Wide Analysis of Host Factors in Nodavirus RNA Replication. *PLoS One.* **2014,** *9*(4), E95799.

Harper, G.; Hull, R.; Lockhart, B.; Olszewski, N. Viral Sequences Integrated into Plant Genomes. *Annu. Rev. Phytopathol.* **2002,** *40,* 119–136.

Harries, P.; Ding, B. Cellular Factors in Plant Virus Movement: At the Leading Edge of Macromolecular Trafficking in Plants. *Virology.* **2011,** *411,* 237–243.

Hii, G.; Pennington, R.; Hartson, S.; Taylor, C. D.; Lartey, R.; Williams, A.; Lewis, D.; Melcher, U. Isolate-Specific Synergy in Disease Symptoms between Cauliflower Mosaic and Turnip Vein-Clearing Viruses. *Arch. Virol .* **2002,** *147,* 1371–1384.

Hogenhout, S. A.; Ammar, E. D.; Whitfield, A. E.; Redinbaugh, M. G. Insect Vector Interactions with Persistently Transmitted Viruses. *Annu. Rev. Phytopathol.* **2008,** *46,* 327–359.

Hu, J. M.; Fu, H. C.; Lin, C. H.; Su, H. J.; Yeh, H. H. Reassortment and Concerted Evolution *in Banana Bunchy Top Virus* Genomes. *J. Virol.* **2007,** *81*(4), 1746–1761.

Hull, R. *Matthews' Plant Virology*; Academic Press: San Diego, CA, 2002.

Hull, R. *Plant Virology*; Academic Press: New York, 2014; pp 1118.

Hurst, L. D. The Ka/Ks Ratio: Diagnosing the form of Sequence Evolution. *Trends Genet.* **2002,** *18,* 486–487.

Inamdar, N. M.; Zhang, X. Y.; Brough, C. L.; Gardiner, W. E.; Bisaro, D. M.; Ehrlich, M. Transfection of Heteroduplexes Containing Uracilguanine or Thymine-Guanine Mispairs into Plant Cells. *Plant Mol. Biol.* **1992,** *20,* 123–131.

Ishibashi, K.; Mawatari, N.; Miyashita, S.; Kishino, H.; Meshi, T.; Ishikawa, M. Coevolution and Hierarchical Interactions of Tomato Mosaic Virus and the Resistance Gene Tm-1. *PLoS Pathog.* **2012,** *8* (10), E1002975.

Jackson, A. O.; Dietzgen, R. G.; Goodin, M. M.; Bragg, J .N.; Deng, M. Biology of Plant Rhabdoviruses. *Annu. Rev. Phytopathol.* **2005,** *43,* 623–660.

Janzac, B.; Fabre, F.; Palloix, A.; Moury, B. Constraints on Evolution of Virus Avirulence Factors Predict the Durability of Corresponding Plant Resistances. *Mol. Plant Pathol.* **2009,** *10,* 599–610.

Jeger, M. J.; Seal, S. E.; Van Den Bosch, F.; Evolutionary Epidemiology of Plant Virus Disease. *Adv. Virus Res.* **2006,** *67,* 163–203.

Jenner, C. E.; Wang, X.; Ponz, F.; Walsh, J. A. A Fitness Cost for *Turnip Mosaic Virus* to Overcome Host Resistance. *Virus Res.* **2002,** *86,* 1–6.

Jones, R. A. C. Plant Virus Emergence and Evolution: Origins, New Encounter Scenarios, Factors Driving Emergence, Effects of Changing World Conditions, and Prospects for Control. *Virus Res.* **2009,** *141,* 113–130.

Jridi, C.; Martin, J. F.; Mareie-Jeanne, V.; Labonne, G.; Blanc, S. Distinct Viral Populations Differentiate and Evolve Independently in a Single Perennial Host Plant. *J. Virol.* **2006,** *80,* 2349–2357.

Karlin, S.; Blaisdell, B. E.; Schachtel, G. A.; Contrasts in Codon Usage of Latent versus Productive Genes of Epstein–Barr Virus: Data and Hypotheses. *J. Virol.* **1990,** *64,* 4264–4273.

Kawecki, T. J. Accumulation of Deleterious Mutations and the Evolutionary Cost of Being a Generalist. *Am. Nat.* **1994,** *144,* 833–838.

Kokkinos, C. D.; Clark, C. A. Interactions among Sweet Potato Chlorotic Stunt Virus and Different Potyviruses and Potyvirus Strains Infecting Sweet Potato in the United States. *Plant Dis.* **2006,** *90,* 1347–1352.

Koonin, E. V.; Senkevich, T. G.; Dolja, V. V. The Ancient Virus World and Evolution of Cells. *Biol. Direct.* **2006,** *1,* 29.

Koonin, E. V.; Senkevich, T. G.; Dolja, V. V. Compelling Reasons why Viruses are Relevant for the Origin of Cells. *Nat. Rev. Microbiol.* **2009,** *7,* 615–616.

Koonin, E. V.; Wolf, Y. I.; Nagasaki, K.; Dolja, V. V. The Big Bang of Picorna-Like Virus Evolution Antedates the Radiation of Eukaryotic Supergroups. *Nat. Rev. Microbiol.* **2008,** *6,* 925–939.

Krupovič, M.; Bamford, D. H. Virus Evolution: How Far does the Double B-Barrel Viral Lineage Extend? *Nat. Rev. Microbiol.* **2008,** *6,* 941–948.

Krupovic, M.; Ravantti, J. J.; Bamford, D. H. Geminiviruses: A Tale of a Plasmid Becoming a Virus. *BMC Evol. Biol.* **2009,** *9,* 112.

Kuzmin, I. V.; Novella, I. S.; Dietzgen, R. G.; Padhi, A.; Rupprecht, C. E. The Rhabdoviruses: Biodiversity, Phylogenetics, and Evolution. *Infect. Genet. Evol.* **2009,** *9,* 541–553.

Lafforgue, G.; Martínez, F.; Sardanyés, J.; De La Iglesia, F.; Niu, Q. W.; Lin, S. S.; Solé, R. V.; Chua, N. H.; Darós, J. A.; Elena, S. F. Tempo and Mode of Plant RNA Virus Escape From RNA Interference-Mediated Resistance. *J. Virol.* **2011,** *85,* 9686–9695.

Lalić, J.; Cuevas, J. M.; Elena, S. F. Effect of Host Species on the Distribution of Mutational Fitness Effects for an RNA Virus. *PLoS Genet.* **2011,** *7,* E1002378.

Lalić, J.; Elena, S. F. Epistasis between Mutations is Host-Dependent for an RNA Virus. *Biol. Lett.* **2012b,** *9,* 20120396.

Lalić, J.; Elena, S. F. Magnitude and Sign Epistasis among Deleterious Mutations in a Positive-Sense Plant RNA Virus. *Heredity.* **2012a,** *109,* 71–77.

Lauring, A. S.; Andino, R. Quasispecies Theory and the Behavior of RNA Viruses. *PLoS Pathog.* **2010,** *6,* E1001005.

Li, H. Y.; Roossinck, M. J. Genetic Bottlenecks Reduce Population Variation in an Experimental RNA Virus Population. *J. Virol.* **2004,** *78,* 10582–10587.

Li, F.; Ding, S. W. Virus Counter Defense: Diverse Strategies for Evading the RNA-silencing Immunity. *Annu. Rev. Microbiol.* **2006,** *60,* 503–531.

Liang, X. Z.; Lee, B. T. K.; Wong, S. M. Covariation in the Capsid Protein of *Hibiscus Chlorotic Ringspot Virus* Induced by Serial Passaging in a Host that Restricts Movement Leads to Avirulence in its Systemic Host. *J. Virol.* **2002,** *76,* 12320–12324.

Lobo, F. P.; Mota, B. E. F.; Pena, S. D. J.; Azevedo, V.; Macedo, A. M.; Tauch, A.; Machado, C. R.; Franco, G. R. Virus–Host Coevolution: Common Patterns of Nucleotide Motif Usage in *Flaviviridae* and their Hosts. *PLoS One.* **2009,** *4,* E6282.

Lucas, W. J. Plant Viral Movement Proteins: Agents for Cell-To-Cell Trafficking of Viral Genomes. *Virol.* **2006,** *344,* 169–184.

Malpica, J. M.; Sacristán, S.; Fraile, A.; García-Arenal, F. Association and Host Selectivity in Multi-Host Pathogens. *PLoS One.* **2006,** *1*, E41.

Marchoux, G.; Delecolle, B.; Selassie, K. G. Systemic Infection to Tobacco by *Pepper Veinal Mottle Potyvirus* (PVMV) Depends on the Presence of *Potato Virus Y* (PVY). *J. Phytopathol.* **1993,** *137*, 283–292.

Margulis, L.; Sagan, D. *Microcosmos: Four Billion Years of Evolution from our Microbial Ancestors*; University of California Press: Berkeley, USA, 1997.

Márquez, L. M.; Redman, R. S.; Rodriguez, R. J.; Roossinck, M. J. A Virus in a Fungus in a Plant-three Way Symbiosis Required for Thermal Tolerance. *Science.* **2007,** *315*, 513–515.

Martelli, G. P.; Agranovsky, A. A.; Bar-Joseph, M.; Boscia, D.; Candresse, T.; Coutts, R. H.; Dolja, V.; Hu, J.; Jelkmann, W.; Karasev, A. V.; Martin, R. R.; Minafra, A.; Namba, S.; Vetten, H. J. Closteroviridae. In *Virus Taxonomy*; Andrew, M. Q. K., Ed.; Academic Press, San Diego, 2012; pp 987–1001.

Martínez, F.; Lafforgue, G.; Morelli, M. J.; González-Candelas, F.; Chua, N. H.; Daròs, J. A.; Elena, S. F. Ultradeep Sequencing Analysis of Population Dynamics of Virus Escape Mutants in RNAi-Mediated Resistant Plants. *Mol. Biol. Evol.* **2012,** *29*, 3297–3307.

Mayo, M. A.; Jolly, C. A. The 5'-Terminal Sequence of Potato Leafroll Virus RNA: Evidence of Recombination between Virus and Host RNA. *J. Gen. Virol.* **1991,** *72*, 2591–2595.

Melcher, U. The '30K' Superfamily of Viral Movement Proteins. *J. Gen. Virol.* **2000,** *81*, 257–266.

Monsion, B.; Froissart, R.; Michalakis, Y.; Blanc, S. Large Bottleneck Size in Cauliflower Mosaic Virus Populations during Host Plant Colonization. *PLoS Pathog.* **2008,** *4*, E1000174.

Moonan, F.; Molina, J.; Mirkov, T. E. Sugarcane Yellow Leaf Virus: An Emerging Virus that has Evolved by Recombination between Luteoviral and Poleroviral Ancestors. *Virology.* **2000,** *269*, 156–171.

Morales, F. J. History and Current Distribution of Begomoviruses in Latin America. *Adv. Virus Res.* **2006,** *67*, 127–162.

Morales, F. J.; Anderson, P. K. The Emergence and Dissemination of Whitefly Transmitted Geminiviruses in Latin America. *Arch. Virol.* **2001,** *146*, 415–441.

Moreno, I.; Malpica, J. M.; Rodríguez-Cerezo, E.; García-Arenal, F. A Mutation in Tomato Aspermy Cucumovirus that Abolishes Cell-To-Cell Movement is Maintained to High Levels in the Viral RNA Population by Complementation. *J. Virol.* **1997,** *71*, 9157–9162.

Moreno-Pérez, M. G.; Pagán, I.; Aragón-Caballero, L.; Cáceres, F.; Fraile, A.; García-Arenal, F. Ecological and Genetic Determinants of Pepino Mosaic Virus Emergence. *J. Virol.* **2014,** *88*(6), 3359–3368.

Moury, B.; Simon, V. Dn/Ds-Based Methods Detect Positive Selection Linked to Trade-Offs between Different Fitness Traits in the Coat Protein of *Potato Virus Y. Mol. Biol. Evol.* **2011,** *28*, 2707–2717.

Nault, L. R.; Ammar, D. Leafhopper and Planthopper Transmission of Plant Viruses. *Annu. Rev. Entomol.* **1989,** *34,* 503–529.Nault, L. R.; Madden, L. V. Phylogenetic Relatedness of *Maize Chlorotic Dwarf Virus* Leafhopper Vectors. *Phytopathology.* **1988,** *78*, 1683–1687.

Nawaz-Ul-Rehman, M.; Fauquet, C. M. Evolution of Geminiviruses and their Satellites. *FEBS Lett.* **2009,** *583*, 1825–1832.

Nieto, C.; Rodríguez-Moreno, L.; Rodríguez-Hernández, A. M.; Aranda, M. A.; Truniger, V. *Nicotiana benthamiana* Resistance to Non-Adapted *Melon Necrotic Spot Virus* Results from a n Incompatible Interaction between Virus RNA and Translation Initiation Factor 4E. *Plant J.* **2011,** *66*, 492–501.

Pappu, H. R.; Jones, R. A.; Jain, R. K. Global Status of Tospovirus Epidemics in Diverse Cropping Systems: Successes Achieved and Challenges Ahead. *Virus Res.* **2009**, *141*(2), 219–236.

Patil, B. L.; Rajasubramaniam, S.; Bagchi, C.; Dasgupta, I. Both Indian Cassava Mosaic Virus and Sri Lankan Cassava Mosaic Virus are Found in India and Exhibit High Variability as Assessed by PCR-RFLP. *Arch. Virol.* **2005**, *150*, 389–397.

Paul, S.; Bag, S.; Das, S.; Harvill, E.; Dutta, C.; Molecular Signature of Hypersaline Adaptation: Insights from Genome and Proteome Composition of Halophilic Prokaryotes. *Genome Biol.* **2008**, *9*, R70.

Pita, J. S.; Fondong, V. N.; Sangaré, A.; Otim-Nape, G. W.; Ogwal, S.; Fauquet, C. M. Recombination, Pseudorecombination and Synergism of Geminiviruses are Determinant Keys to the Epidemic of Severe Cassava Mosaic Disease in Uganda. *J. Gen. Virol.* **2001**, *82*, 655–665.

Poulicard, N.; Pinel-Galzi, A.; Hébrard, E.; Fargette, D. Why Rice Yellow Mottle Virus, A Rapidly Evolving RNA Plant Virus, Is Not Efficient at Breaking Rymv1-2 Resistance. *Mol. Plant Pathol.* **2010**, *11*, 145–154.

Poulicard, N.; Pinel-Galzi, A.; Traore, O.; Vignols, F.; Ghesquiere, A.; Konate, G.; Hebrard, E.; Fargette, D. Historical Contingencies Modulate the Adaptability of *Rice Yellow Mottle Virus*. *PLoS Pathog.* **2012**, *8*, 2482–2490.

Power, A. G.; Borer, E. T.; Hosseini, P.; Mitchell, C. E.; Seabloom, E. W. The Community Ecology of *Barley/Cereal Yellow Dwarf Viruses* in Western US Grasslands. *Virus Res.* **2011**, *159*, 95–100.

Power, A. G.; Mitchell, C. E. Pathogen Spillover in Disease Epidemics. *Am. Nat.* **2004**, *164*, S79–S89.

Power, A. G.; Flecker, A. S. Virus Specificity In Disease Systems: Are Species Redundant? In *The Importance of Species*; Kareiva, P., Levin, S. A., Eds.; Princeton University Press: New Jersey, USA, 2003; pp 330–347.

Pruss, G.; Ge, X.; Shi, X. M.; Carrington, J. C.; Vance, V. B. Plant Viral Synergism: The Potyviral Genome Encodes a Broad-Range Pathogenicity Enhancer that Transactivates Replication of Heterologous Viruses. *Plant Cell.* **1997**, *9*, 859–868.

Pu, Y.; Kikuchi, A.; Moriyasu, Y.; Tomaru, M.; Jin, Y.; Suga, H.; Hagiwara, K.; Akita, F.; Shimizu, T.; Netsu, O.; Suzuki, N.; Uehara-Ichiki, T.; Sasaya, T.; Wei, T.; Li, Y.; Omura, T. Rice Dwarf Viruses with Dysfunctional Genomes Generated in Plants are Filtered out in Vector Insects: Implications for the Origin of the Virus. *J. Virol.* **2011**, *85*, 2975–2979.

Quenouille, J.; Montarry, J.; Palloix, A.; Moury, B. Farther, Slower, Stronger: How the Plant Genetic Background Protects a Major Resistance Gene from Breakdown. *Mol. Plant Pathol.* **2013**, *14*, 109–118.

Quito-Avila, D. F.; Lightle, D.; Lee, J.; Martin, R. R. Transmission Biology of Raspberry Latent Virus, the First Aphid-Borne Reovirus. *Phytopathol.* **2012**, *102*, 547–553.

Rico, P.; Herna'ndez, C. Infectivity of *in Vitro* Transcripts from a Full-Length cDNA Clone of Pelargonium Flower Break Virus in an Experimental and a Natural Host. *J. Plant Pathol.* **2006**, *88*, 103–106.

Robinson, D. J.; Hamilton, W. D. O.; Harrison, B. D.; Baulcombe, D. C. Two Anomalous Tobravirus Isolates: Evidence for RNA Recombination in Nature. *J. Gen. Virol.* **1987**, *68*, 2551–2561.

Rochow, W. F.; Ross, A. F. Virus Multiplication in Plants Doubly Infected with Potato Virus X and Y. *Virology.* **1955**, *1*, 10–27.

Rojas, M. R.; Hagen, C.; Lucas, W. J.; Gilbertson, R. L. Exploiting Chinks in the Plant's Armor: Evolution and Emergences of Geminiviruses. *Annu. Rev. Phytopathol.* **2005**, *43*, 361–394.

Roossinck, M. J. Mechanisms of Plant Virus Evolution. *Annu. Rev. Phytopathol.* **1997**, *35*, 191–209.

Roossinck, M. J. Symbiosis versus Competition in the Evolution of Plant RNA Viruses. *Nat. Rev. Microbiol.* **2005**, *3*(12), 917–924.

Sacristán, S.; García-Arenal, F. The Evolution of Virulence and Pathogenicity in Plant Pathogen Populations. *Mol. Plant Pathol.* **2008**, *9*, 369–384.

Sacristán, S.; Malpica, J. M.; Fraile, A.; García-Arenal, F. Estimation of Population Bottlenecks during Systemic Movement of Tobacco Mosaic Virus in Tobacco Plants. *J. Virol.* **2003**, *77*, 9906–9911.

Salvaudon, L.; De Moraes, C. M.; Mescher, M. C. Outcomes of Co-Infection by two Potyviruses: Implications for the Evolution of Manipulative Strategies. *Proc. Biol Sci.* **2013**, *280*, 20122959.

Sánchez-Navarro, J. A.; Herranz, M. C.; Pallás, V. Cell-to-Cell Movement of Alfalfa Mosaic Virus can be Mediated by the Movement Proteins of Ilar-, Bromo-, Cucumo-, Tobamo- And Comoviruses and does not Require Virion Formation. *Virology.* **2006**, *346*, 66–73.

Sanjuán, R.; Nebot, M. R.; Chirico, N.; Mansky, L. M.; Belshaw, R. Viral Mutation Rates. *J. Virol.* **2010**, *19*, 9733–9748.

Scheets, K. Maize Chlorotic Mottle Machlovirus and Wheat Streak Mosaic Rhymovirus Concentrations Increase in the Synergetic Disease Corn Lethal Virus. *Virology.* **1998**, *242*, 28–38.

Seal, S. E.; Vandenbosch, F.; Jeger, M. J. Factors Influencing Begomovirus Evolution and their Increasing Global Significance: Implications for Sustainable Control. *Crit. Rev. Plant. Sci.* **2006a**, *25*, 23–46.

Seal, S. E.; Jeger, M. J.; Van Den Bosch, F. Begomovirus Evolution and Disease Management. *Adv. Virus Res.* **2006b**, *67*, 297–316.

Selling, B. H.; Allison, R. F.; Kaesberg, P. Genomic RNA of an Insect Virus Directs Synthesis of Infectious Virions in Plants. *Proc. Natl. Acad. Sci. U. S. A.* **1990**, *87*, 434–438.

Shackelton, L. A.; Parrish, C. R.; Truyen, U.; Holmes, E. C. High Rate of Viral Evolution Associated with the Emergence of Carnivore Parvovirus. *Proc. Natl. Acad. Sci. U. S. A.* **2005**, *102*, 379–384.

Simon, A. E.; Roossinck, M. J.; Havelda, Z. Plant Virus Satellite and Defective Interfering RNAs: New Paradigms for a New Century. *Ann. Rev. Phytopathol.* **2004**, *42*, 415–437.

Sin, S. H.; Mcnulty, B. C.; Kennedy, G. G.; Moyer, J. W.; Viral Genetic Determinants for Thrips Transmission of Tomato Spotted Wilt Virus. *Proc. Natl. Acad. Sci. U. S. A.* **2005**, *102*, 5168–5173.

Sorho, F.; Pinel, A.; Traoré, O.; Bersoult, A.; Ghesquière, A.; Hébrard, E.; Konaté, G.; Séré, Y.; Fargette, D. Durability of Natural and Transgenic Resistances in Rice to *Rice Yellow Mottle Virus. Eur. J. Plant Pathol.* **2005**, *112*, 349–359.

Takeshita, M.; Matsuo, Y.; Yoshikawa, T.; Suzuki, M.; Furuya, N.; Tsuchiya, K.; Takanami, Y. Characterization of a Defective RNA Derived from RNA 3 of the Y Strain of Cucumber Mosaic Virus. *Arch. Virol.* **2008**, *153*, 579–583.

Tatineni, S.; Robertson, C. J.; Garnsey, S. M.; Dawson, W. O. A Plant Virus Evolved by Acquiring Multiple Nonconserved Genes to Extend its Host Range. *Proc. Natl. Acad. Sci. U. S. A.* **2011**, *108*(42), 17366–17371.

Thresh, J. M. Cropping Practices and Virus Spread. *Annu. Rev. Phytopathol.* **1982,** *20,* 193–218.

Truniger, V.; Aranda, M. A. Recessive Resistance to Plant Viruses. *Adv. Virus Res.* **2009,** *75,* 119–159.

Varsani, A.; Navas-Castillo, J.; Moriones, E.; Herna´ndez-Zepeda, C.; Idris, A.; Brown, J. K.; Zerbini, F. M.; Martin, D. P. Establishment of Three New Genera in the Family Gemini-viridae: Becurtovirus, Eragrovirus and Turncurtovirus. *Arch. Virol.* **2014,** *159,* 2193–2203.

Waigmann, E.; Ueki, S.; Trutnyeva, K.; Citovsky, V. The Ins and Outs of on destructive Cell-to-Cell and Systemic Movement of Plant Viruses. *Crit. Rev. Plant Sci.* **2004,** *23,* 195–250.

Wallis, CM.; Stone, A. L.; Sherman, D. J.; Damsteegt, V. D.; Gildow, F. E.; Schneider, W. L. Adaptation of *Plum Pox Virus* to a Herbaceous Host (*Pisum sativum*) Following Serial Passages. *J. Gen. Virol.* **2007,** *88,* 2839–2845.

Wang, M.-B.; Bian, X.-Y.; Wu, L.-M.; Liu, L.-X.; Smith, N. A.; Isenegger, D.; Wu, R. M.; Masuta, C.; Vance, V. B.; Watson, J. M.; Rezaian, A.; Dennis, E. S.; Waterhouse, P. M. On the Role of RNA Silencing in the Pathogenicity and Evolution of Viroids and Viral Satel-lites. *Proc. Natl. Acad. Sci. U. S. A.* **2004,** *101,* 3275–3280.

White, P. S.; Morales, F. J.; Roossinck, M. J. Interspecific Reassortment in the Evolution of a Cucumovirus. *Virology.* **1995,** *207,* 334–337.

Wintermantel, W. M.; Cortez, A. A.; Anchieta, A. G.; Gulati-Sakhuja, A.; Hladky, L. L. Co-Infection by Two Criniviruses Alters Accumulation of Each Virus in a Host Specific Manner and Influences Efficiency of Virus Transmission. *Phytopathology.* **2008,** *98,* 1340–1345.

Wright, F. The 'Effective Number of Codons' Used in a Gene. *Gene.* **1990,** *87,* 23–29.

Xoconostle-Cázares, B.; Xiang, Y.; Ruiz-Medrano, R.; Wang, H-L.; Monzer, J.; Yoo, B-C.; Mcfarland, K. C.; Franceschi, V. R.; Lucas, W. J. Plant Paralog to Viral Movement Protein that Potentiates Transport of mRNA into the Phloem. *Science.* **1999,** *283,* 94–98.

Xu, X. Z.; Liu, Q. P.; Fan, L. J.; Cui, X. F.; Zhou, X. P. Analysis of Synonymous Codon Usage and Evolution of Begomoviruses. *J. Zhejiang Uni. Sci. B.* **2008,** *9,* 667–674.

Yarwood, C. E. Host Passage Effects with Plant Viruses. *Adv. Virus Res.* **1979,** *24,* 169–191.

Zwart, M. P.; Darò, J. A.; Elena, S. F. Effects of Potyvirus Effective Population Size in Inoculated Leaves on Viral Accumulation and the Onset of Symptoms. *J. Virol.* **2012,** *86,* 9737–9747.

CHAPTER 10

HOST PREFERENCE BY EVOLVING INSECT VECTORS IN RELATION TO INFECTION OF PLANT VIRUSES

SOMNATH K. HOLKAR[1*], PRATIBHA KAUSHAL[2], and SANJEEV KUMAR[2]

[1]*Division of Crop Protection, Indian Institute of Sugarcane Research, Rai Bareli Road, P.O. Dilkusha, Lucknow 226002, Uttar Pradesh, India*

[2]*Division of Crop Improvement, Indian Institute of Sugarcane Research, Rai Bareli Road, P.O. Dilkusha, Lucknow 226002, Uttar Pradesh, India*

Corresponding author. E-mail: somnathbhu@gmail.com

CONTENTS

ABSTRACT

Host preference of evolving insect vectors has a direct correlation in transmitting plant viruses. The presence of more than 50 lakh species of insects on the earth has been estimated and, ~5000 virus species infecting plants, animals, bacteria, fungi have been known to exist. Among several insect orders, Insecta is the major group in transmitting plant viruses in general and homopterans in particular. Among arachnids, certain members of two families, *viz.*, Eryophididae and Tetranychidae are known to transmit plant viruses. Aphids are the largest group of insect in transmitting at least 300 plant viruses belonging to 16 genera. Aphids are having the broad host range and more than 20 aphid species are known to transmit plant viruses. More than 13 thrips species are known to naturally transmit more than 20 tospoviruses. Thrips prefer to select major hosts from families including *Asteraceae*, Cucurbitaceae, Leguminosae, and Solanaceae. Whiteflies are the second most important vector of more than 200 plant viruses. Four species of whiteflies, *viz.*, *Bemisia tabaci*, *Trialeurodes vaporariorum*, *T. abutilonea*, and *T. ricini* have been reported in transmitting plant viruses recognized as species in the genera *Begomovirus*, *Carlavirus*, *Crinivirus*, and *Ipomovirus*. Among these four species, *B. tabaci* is the major and the most efficient vector having a broader host range. Leafhoppers are the third largest group of insects in transmitting 82 plant viruses belonging to the genus *Curtovirus*, *Machlomovirus*, *Mastrevirus*, *Polerovirus*, and *Waikavirus*. All the insect species respond differently to different host species and their preference is affected by the nature of the host, their feeding behavior and certain chemicals or exudates secreted by plants that attract or distract insect vectors.

10.1 INTRODUCTION

Transmission of plant viruses is effected by several invertebrates, nematodes, fungi, mechanical means, seed, and pollen. Majority of the reported plant viruses are transmitted by invertebrates generally referred to as insect vectors. Insect vectors transmitting plant viruses belong to several orders including Coleoptera, Dermaptera, Diptera, Hemiptera, Lepidoptera, Orthoptera, and Thysanoptera. Most of the plant viruses transmitting insects are from the Hemiptera and Thysanoptera. The order Hemiptera is comprised of three suborders, *viz.*, Heteroptera-the true bugs, Auchenorrhyncha-the leaf and planthoppers and the most important Sternorrhyncha-the aphids, whiteflies, mealy bugs, and psyllids. The suborders, Auchenorrhyncha and

Sternorrhyncha together referred as Homopterans and comprising more than 55% of insect transmitted plant viruses (Richards & Davis, 1977; Nault, 1997). Likewise, Thysanoptera (family: Thripidae) are the vectors of more than twenty tospoviruses (family: *Bunyaviridae*, genus: *Tospovirus*), are the emerging threats to the crop diversity worldwide (Whitfield et al., 2005; Pappu et al., 2009; King et al., 2012).

Aphids (family: Aphididae; subfamily: Aphidinae) are the largest group of insect vectors of 3742 species, but only 300 species are known to transmit plant viruses in persistent, non-persistent, and semi-persistent manner (Table 10.1). Aphids are the most important vector of plant viruses causing economic loss to agricultural and horticultural crops worldwide (Watson & Roberts, 1939; Sylvester, 1956; Table 10.2). The major genera of aphids contributing in transmitting plant viruses are *Aphis*, *Myzus*, and *Macrosiphum*. Aphids are known to transmit viruses belonging to several genera including, *Alfamovirus*, *Babuvirus*, *Coremovirus*, *Carlavirus*, *Caulimovirus*, *Cucumovirus*, *Closterovirus*, *Enamovirus*, *Fabavirus*, *Luteovirus*, *Nanovirus*, *Polerovirus*, *Potyvirus*, *Sequivirus*, *Sobemovirus*, and *Rhabdoviruses* (Eastop, 1983). Among Homopterans, whiteflies are the second largest group of insect vectors comprising of 1300 reported species so far but only four species, *viz.*, *Bemisia tabaci*, *Trialeurodes abutiloneus*, *T. ricini* and *T. vaporariorum* have been known to transmit more than 200 plant viruses. The whitefly transmitting virus genera including, *Begomovirus*, *Carlavirus*, *Crinivirus*, *Ipomovirus*, and *Toradovirus* (Varma et al., 2001; Jones, 2003; Morales, 2007; Navas-Castillo et al., 2011). *Bemisia tabaci* is a polyphagous pest of horticultural and ornamental crops is transmitting most of the whitefly transmitted plant viruses in circulative and semi-persistent manner in tropical and subtropical regions of the world (Novas-Castillo et al., 2014; Table 10.2).

Leaf and planthoppers (order: Hemiptera; family: Cicadellidae and Delphacidae) are one of the most efficient vectors of plant viruses and phytoplasma, leafhoppers alone transmit more than 80 viruses, whereas, planthoppers transmit more than 20 plant viruses of various genera mentioned (Table 10.1). Leafhoppers are one of the largest families within the order Hemiptera that includes most of the sap sucking herbivores. The phylogenetic relationships of leafhoppers have been conclusively studied on the major lineages instead of the species belonging to a single genus (Zahniser & Dietrich, 2010). The phylogenetic results revealed that the preference of the natural hosts is solely based on the mouth part morphology. Most of the genera of leafhoppers, *viz.*, *Agalia*, *Aphrodes*, *Coelidia*, *Deltocephalus*, *Gypona*, *Lassus*, *Ideocerus*, *Macropsis*, and *Megophtalamus*

TABLE 10.1 Transmission of Plant Viruses and their Vectors Belonging to Several Families and Groups (up to March, 2015).

Group, family/ sub-family	Common name	Reported species (genera)	Number of vector species (genera) that transmit viruses	Number of viruses transmitted and genera which they belong	Reference
Insecta					
Aphididae S. F. Aphidinae	Aphids	3419–3742 >3740	300	300; Alfamo-, Babu-, Coremo-, Carla-, Caulimo-, Cucumo-, Clostero-, Enamo-, Faba-, Luteo-, Nano-, Polero-, Poty-, Sequi-, Soymo-, Rhabdoviruses	Eastop (1983); Harris & Maramorosch (2014); Remaudiere & Remaudiere (1997)
Aleyrodidae S. F. Aleyrodinae	White flies	1500	4	>200; Begomo-, Carla-, Crini-, Ipomoviruses	Jones (2003) Morales (2001, 2010); Varma et al. (2001)
Cicadellidae	Leaf hoppers	~20000 (2400)	≥200 (21)	82; Curto-, Machlomo-, Mastre-, Polero-, Waikaviruses	Conti (1985) Dietrich (2013) Nault & Madden (1988); Nielson (1979)
Delphacidae	Planthoppers	1100 (137)	27 (13)	24; Fiji-, Nucleorhabdo-, Oryza-, Tenuiviruses	Dietrich (2013)
Chrysomelidae	Leaf beetles	38000 (2500)	61	42; Bromo-, Carmo-, Como-, Machlomo-, Sobemo-, Tymoviruses	Nault (1997); Seeno & Wilcox (1982)
Coccinellidae	Lady bird beetle	6000	2	7; Sobemoviruses	Abo et al. (2004); Nault (1997); Vandenberg (2002)
Cucucrhionidae	Weevil	62000 (5800)	10	4; Bromo-, Como-, Sobemo-, Tobamo-, Tymoviruses	Slipinski et al. (2011)
Miridae	Bugs	6700	–	1; Sobemoviruses	

TABLE 10.1 (Continued)

Group, family/sub-family	Common name	Reported species (genera)	Number of vector species (genera) that transmit viruses	Number of viruses transmitted and genera which they belong	Reference
Thripidae/Thripinae	Thrips	>5000	14	>20; Carmo-, Ilar-, Machlomo-, Sobemo-, Tospoviruses	Ciuffo et al. (2010); Hassani-Mehraban et al. (2010); Jones (2005); Mandal et al. (2012); Ullman (1997)
Arachnida					
Eryophiddae	Mites	—	8	10; Allexi-, Clostero-, Emara-, Tritimoviruses	Agrios (2005); Satyanarayana et al. (1996, 1998); Young et al. (2012)
Tetranychidae	Spider mites	>900	2	3; Allexi-, Clostero-, Emara-, Tritimoviruses	
Nematoda					
Longidoridae	*Longidorus, Paralongidorus, and Xiphinema*	480 (6)	325	17; Nepoviruses	Lamberti & Roca (1987); Nayudu (2008)
Trichodoridae	*Trichodorus, Paratrichodorus*	100 (5)	14	14; Tobraviruses	Decraemer & Robbins (2007)
Fungi					
Olpidiaceae	*Olpidium*	—	3	12; Tombus-, Ophio-, Varicosaviruses	Brunt et al. (1996)
Synchytriaceae	*Synchytrium*	—	1	1; Potexviruses	Darozhkin & Chykava (1974)
Protista					
Plasmodiopho-raceae	*Polymyxa* and *Spongospora*	—	3	18; Beny-, Bymo-, Furo-, Peclu- and Pomoviruses	Adams et al. (2001); Dieryck et al. (2009)

TABLE 10.2 Aphids, Whiteflies and Thrips, their Natural Hosts, and Virus Species Transmitted by them.

Vector	Natural host	Virus species which is transmitted	Reference
Aphis fabae	*Euonymus europaeus* and *Vicia faba*	BtMV, BYV	Landis & Van der Werf (1997)
A. gossypii	*Cucurbita pepo* cv. Dixie, *Cucumis sativus* and many other cucurbits, *Lycopersicon esculentum*, *Spinacia oleracea* and *Musa sapientum*	CMV, ZYMV	Mauck et al. (2010); Kaper & Waterworth (1981); Madhubala et al. (2005)
	M. sapientum and *Citrus* sp.	CTV, BBrMV	Yokomi & Garnsey (1987); Dheepa & Paranjothi (2010)
	Lactuca sp., *Senecio vulgaris*, *Sonchus* sp., *Cicer arietinum* and *Pisum sativum*	LMV	Tomlinson (1970)
	Ipomoea sp.	SPFMV	Brunt et al. (1996)
	Arachis hypogaea, A. pintoi, Cassia bicapsularis, C. leptocarpa, C. occidentalis, C. tora, Glycine max, Phaseolus vulgaris, P. sativum and *Stylosanthes* sp.	PMV	Paguio & Kuhn (1974); Kuhn (1965); Behncken (1970)
A. craccivora	*C. sativus* and many other cucurbits, *L. esculentum, S. oleracea*	CMV	Kaper & Waterworth (1981)
	Prunus armeniaca, P. cerasifera, P. domestica, P. glandulosa, P. persica, P. insititia, P. spinosa and *P. salicina*	PPV	Capote et al. (2008)
	Ipomoea ssp.	SPFMV	Brunt et al. (1996)
	Arachis hypogaea, A. pintoi, Cassia bicapsularis, C. leptocarpa, C. occidentalis, C. tora, G. max, P. vulgaris, P. sativum and *Stylosanthes* sp.	PMV	Paguio & Kuhn (1974); Kuhn (1965); Behncken (1970)
A. spiraecola	Primary hosts-*Prunus armeniaca, P. persica, P. persica* var. nectarine, *P. domestica, P. salicina, P. insititia, P. cerasifera, P. glandulosa, P. avium, P. cerasus, P. amygdalus*. Secondary hosts-*P. spinosa, P. americana, P. bessey, P. mahaleb, P. mume, P. pumila, P. hortulana, P. davidana, P. tomentosa, P. nigra, P. maritime* and *P. laurocerasus*	PPV	Wallis et al. (2005)

TABLE 10.2 (Continued)

Vector	Natural host	Virus species which is transmitted	Reference
A. frangulae, A. nasturtii	Solanum tuberosum	PVY	Clinch et al. (1936)
A. glycines	Caryopteris incana, Cicer arietinum, L. esculentum, Apium graveolens, A. graveolens var. rapaceum, Lactuca sativa, Malva parviflora, Trifolium incarnatum, T. repens, Viburnum opulus, Medicago sativa, Nicotiana tabacum, Lupinus., P. vulgaris, Vigna unguiculata, V. radiata, Astragalus glycyphyllos, G. max, Lablab purpureus, Lens culinaris, Capsicum annuum, Philadelphus sp., Pisum sativum, Solanum tuberosum	AMV, SMV TRSV& TBMV	Davis et al. (2005)
	Cucurbita pepo cv. Dixie, Cucumis sativus and many other cucurbits, L. esculentum, Spinacia oleracea and Musa sapientum	CMV	Davis et al. (2005)
	Arachis hypogaea, A. pintoi, Cassia bicapsularis, C. leptocarpa, C. occidentalis, C. tora, Glycine max, P. vulgaris, P. sativum and Stylosanthes sp.	PMV	Davis et al. (2005); Kuhn (1965); Behncken (1970)
Lipaphis erysimi	Ipomoea spp.	SPFMV	Brunt et al. (1996)
Schizaphis graminum	Solanum tuberosum	PVY, CYDV & RPV	Yang et al. (2008)
Melanaphis sacchari	Cynodon dactylon, Miscanthus sinensis, Oryza sativa, Panicum colonum, P. maximum, Paspalum sanguinale Lamarck, Pennisetum sp., Saccharum officinarum, Setaria italic, Sorghum bicolor, S. halepense, S. verticilliflorum and Zea mays	MRLV, SCYLV SCMV & SrMV	Wilbrink (1922); Setokuchi (1988); Kawada (1995); Miao & Sunny (1987); Denmark (1988); White et al. (2001); Patil (1992); Agarwal (1985); Schenck (2000); Blackman & Eastop (1984) Bhargava et al. (1971)

TABLE 10.2 (Continued)

Vector	Natural host	Virus species which is transmitted	Reference
Myzus persicae	*Cicer arietinum, Lathyrus odoratus, Lens culinaris, Medicago iscose, Pisum sativum, Trifolium incarnatum, Vicia faba, Vicia sativa.* Primary host-*Prunus* spp. Other hosts- *C. arietinum, Lathyrus odoratus, Lens culinaris* Medik., *Medicago, Pisum sativum, Trifolium incarnatum, Vicia faba, V. sativa* and *Cucurbita pepo*	PEMV	Hodge & Powell (2010)
	Cucumis sativus and many other cucurbits, *Lycopersicon esculentum* and *Spinacia oleracea*	CMV	Mauck et al. (2010); Kaper & Waterworth (1981)
	S. tuberosum,	PVY	Radcliffe & Ragsdale (2002, 2008)
	S. tuberosum ssp. *S. tuberosum* sp. *Andigena* and *Lycopersicon esculentum*	PLRV	Braithwaite & Blake (1961)
	Lactuca ssp., *Senecio vulgaris, Sonchus* ssp., *C. arietinum* and *Pisum sativum*	LMV	Tomlinson (1970)
	Prunus armeniaca, P. cerasifera, P. domestica, P. glandulosa, P. persica, P. insititia, P. spinosa and *P. salicina*	PPV	Labonne et al. (1995)
	Ipomoea sp.	SPFMV	Brunt et al. (1996)
	Caryopteris incana, C. arietinum, L. esculentum, Apium graveolens, A. graveolens var. *rapaceum, Lactuca sativa, Malva parviflora, Trifolium incarnatum, T. repens, Viburnum opulus, Medicago sativa, Nicotiana tabacum, Lupinus., P. vulgaris, Vigna unguiculata, V. radiata, Astragalus glycyphyllos, G. max, Lablab purpureus, Lens culinaris, Capsicum annuum, Philadelphus* sp., *P. sativum* and *S. tuberosum*	AMV & TuYV	Weimer (1931); Brault et al. (2010)
	Capsella bursa-pastoris, Brassica nigra, B. campestris ssp. *pekinensis, Stellaria media, Trifolium hybridum, Alliaria officinalis, Calanthe* sp., *Hesperis matronalis, B. campestris* sp. *chinensis, B. campestris* ssp. *rapa* and *B. japonica*	TuMV	Pirone & Perry (2002)
	Arachis hypogaea, A. pintoi, Cassia bicapsularis, C. leptocarpa, C. occidentalis, C. tora, G. max, P. vulgaris, P. sativum and *Stylosanthes* sp.	PMV	Paguio & Kuhn (1974); Kuhn (1965); Behncken (1970)

TABLE 10.2 *(Continued)*

Vector	Natural host	Virus species which is transmitted	Reference
M. varians, M. humuli	Prunus armeniaca, P. cerasifera, P. domestica, P. glandulosa, P. persica, P. insititia, P. spinosa and P. salicina	PPV	
M. persicae, M. nicotianae	Nicotiana sp.	PVY	Kanavaki et al. (2006)
Macrosiphum euphorbiae	Lactuca sp., Senecio vulgaris, Sonchus ssp., C. arietinum and P. sativum	LMV	Tomlinson (1970)
Rhopalo-siphum padi	Triticum sp.	BYDV	Bosque-Perez & Eigenbrode (2011)
	S. tuberosum,	PVY	Radcliffe et al. (2008)
	A. hypogaea, A. pintoi, Cassia bicapsularis, C. leptocarpa, C. occidentalis, C. tora, G. max, P. vulgaris, P. sativum and Stylosanthes sp.	PMV	Kuhn (1965); Behncken (1970)
R. maidis	S. tuberosum	PVY	Radcliffe et al. (2008)
Brachycaudus helichrysi and B. cardui	Prunus armeniaca, P. cerasifera, P. domestica, P. glandulosa, P. persica, P. insititia, P. spinosa and P. salicina	PPV	Isac et al. (1998)
Hyperomyzus lactucae	A. hypogaea, Arachis pintoi, Cassia bicapsularis, C. leptocarpa, C. occidentalis, C. tora, G. max, P. vulgaris, P. sativum and Stylosanthes sp.	PMV	Kuhn (1965)
Sitobion avenae	S. tuberosum	BYDV-MAV PVY	Radcliffe et al. (2008)
Acyrthosi-phon pisum	C. arietinum, Lathyrus odoratus, Lens culinaris, Medicago iscose, P. sativum, Trifolium incarnatum, Vicia faba and Vicia sativa	PEMV	Hodge & Powell (2010)
	C. arietinum L., Lathyrus odoratus L., Lens culinaris Medik., Medicago sp., P. sativum L., Trifolium incarnatum L., Vicia faba L., and V. sativa L., Medicago sativa and T. pratense	CMV FBNYV BYMV	Hull (2002); Zitter & Provvidenti (1984); Hogenhout et al. (2008)

TABLE 10.2 *(Continued)*

Vector	Natural host	Virus species which is transmitted	Reference
Thrips			
Frankiniella occidentalis	*P. sativum, S. lycopersicum, C. annum, Fragaria × ananassa, Rosa* sp., *Prunus persica, P. armeniaca, P. domestica, Dianthus caryophyllus, Lathyrus odoratus, Cucumis sativus, Gladiolus* sp., *Beta vulgaris, Daucus carota, Vitis vinifera, Allium cepa, Carthamus tinctorius, Chrysanthemum* sp., *Gerbera* sp., and *Saint-paulia ionantha*	GRSV, INSV, PDV, PNRSV, TCSV TSWV, PFBV & CSNV	Hull et al. (2000); Jones (2005); Bezzara et al. (1999)
F. fusca	*Nicotinia tabacum, S. lycopersicum, Citrullus lanatus, C. annum, A. hypogaea, G. max, Gossypium* spp., *Vigna unguiculata* and *Zea mays*	TSWV & INSV	Jones (2005); Naidu et al. (2001)
F. intonsa	*Chrysanthemum* sp., *C. annum, S. lycopersicum, A. hypogea, Asparagus officinalis, G. max, Gossypium* sp., *Abelmoschus esculentus, Medicago sativa, P. sativum, Oryza sativa, P. vulgaris* and *Prunus persicae*	GRSV, TCSV & TSWV	Hull (2000); Jones (2005); Wijkamp et al. (1995)
F. schulzei	*S. lycopersicum, A. hypogea, G. max, Gossypium* sp., *Cajanus cajan, Lactuca sativa, Nicotiana tabacum, Annas comosus, Lens culinaris, Vigna mungo, V. unguiculata, Hyacinthus orientalis, Orchis* sp., and *cactus* sp.	GRSV, TCSV, TSWV & CSNV	Hull (2002); Jones (2005)
F. bispinosa	*Citrus sinensis, Fragaria x ananassa, Nicotiana tabacum, Raphanus raphanistrum, Rosa* sp., and *Triticum aestivum*	TSWV	Jones (2005); Webb et al. (1998)
F. zucchini	*Cucurbita pepo, L. esculentum* and *Citrullus lanatus*	ZLCV	Jones (2005)
Micro-cephalothrips abdominalis	*Citrus* sp., *Solanum lycopersicum, Oryza sativaorchis* sp., *Cucumis sativus, Helianthus annus* and *Chenopodium giganteum, Chrysanthemum* sp., *Tagetes erecta, Ageratum houstonianum, Bidens formosa, Tanacetum coccineum* and *Zinnia elegans*	PNRSV, TSV & PNRSV	Greber et al. (1991); Anon. (2002); Hull (2002)

TABLE 10.2 (Continued)

Vector	Natural host	Virus species which is transmitted	Reference
T. palmi	*C. annum, Nicotiana,* sp., *Cucumis sativus, Citrullus lanatus, Cucumis melo, Cucurbita* sp., *S. tuberosum, Gossypium* sp., *Vigna unguiculata, Pisum sativum, Phas eolusvulgaris, Glyci nemax, Helianthus annus* and *Sesamumi ndicum,* and *Actinidea chinensis*	TSWV, MYSV, GBNV, WBNV, WSMV &WSMoV	Akram et al. (2013); Singh & Krishnareddy (1995); Hull (2002); Persley et al. (2006)
T. setosus	*S. melongena, P. vulgaris, Chrysanthemum* sp., *Citrus* sp., *C. sativus, Nicotiana* sp., *Solanum lycopersicum, Dahlia* sp., *Ficus carica, Vitis vinifera, Impatiens* sp., lettuce, melon, *G. max,* sesame, strawberry, sweet pepper, pea, tea, *Mentha arvensis, Cucurbita moschata* and *Trifolium repens*	TSWV	Hull (2002); Miyazuki & Kudo (1988); Fujisawa et al. (1988)
T. tabaci	Primary hosts: *Gossypium* sp., *Cucumis sativus, Nicotiana* sp., *Allium sativum, Allium ampeloprasum* var. porrum, *Brassica oleracea* var. capitata, *Piper nigrum.* Secondary hosts: *Solanum tuberosum, Cucurbita* sp., *Solanum lycopersicum, Betavulgaris* var. saccharifera, *Cajanuscajan, Manihot esculenta, Citrusunshiu, Diospyruskaki* and *Asparagusofficinalis*	PNRSV, SoMV, TSV, TSWV & IYSV	Cortes et al. (1998); Hull (2002)
T. parvispinus	*C. lanatus, Carica papaya, S. tuberosum, C. annum, Nicotiana* sp., *Bidens pliosa, Coffea* sp. *Impatiens balsamina* and *Gardenia* sp.	TSV	Klose et al. (1996); Mound & Collins (2000) Anon (2002); Jones (2005)
Cerato- thripoids claratris	*L. esculentum, C. lanatus, Capsicum, Hoya australis* and *Arachis hypogaea*	CaCV	Premachandra et al. (2005); Jones (2005)
Scirtothrips dorsalis	*Acacia* species, *L. esculentum, Nicotiana tabacum, Citrus* sp., *Gossypium* sp., *Alliumcepa, Anacardium occidentale, Ricinus communis, C. annuum, Mangifera indica, Camellia sinensis* and *A. hypogea*	GBNV PYSV	Akram et al. (2013); Akram & Naimuddin, (2010, 2012); Sivaprasad et al. (2011); Gopal et al. (2010)

TABLE 10.2 *(Continued)*

Vector	Natural host	Virus species which is transmitted
Whitefly (adapted from Jones, 2005)		
Bemisia tabaci	*Abutilon spp., Hibiscus spp., Malva spp., Sida spp. Acalypha indica Hewittia sublobata, Jatropha multifida, J. gossypifolia, J. podagrica, J. tamnifolia, Laportea aestuans, Manihot esculenta, M. glaziovii, Zinnia elegans Ageratum sp., Asystasia gangetica, P. vulgaris, Malva parviflora, Sida sp. Phaseolus sp., Glycine max, Macroptilium lathyroides, Malvastrum coromandelianum, Abelmoschus esculentus, Brassica oleracae, Calopogonium mucunoides, Sechium edule, C. annuum, C. frutescens, S. lycopersicum, Castanospermum sp., Gossypium hirsutum, G. barbadense, Raphanus sativus, V. unguiculata, V. mungo, V. radiate, V. aconitifolia Croton bonplandianum, C. lobatus C. melo, C. pepo, C. maxima, C. moschata, Dicliptera sexangularis, Lablab purpurea, M. Eclipta prostrate, Eupatorium glehni, E. makinoi, Nicotiana tabacum, Euphorbia heterophylla, E. prunifolia, Alcea rosea, Lonicera japonica, Cajanus cajan, Indigifera sp.. Macrotyloma uniflorum, M. glaziovii, Ipomoea spp., but not I. batatas, Passiflora edulis, P. foetida, Hibiscus cannabinus, Leonurus sibiricus, Lupinus hartwegii, Macroptilium lathyroides, Macrotyloma spp. Malva spp., Calonyction acueatum, Merremia aegyptia, M. quinquefolia, Cajanus cajan, A. moschatus Euphorbia pulcherrima, E. heterophylla, S. tuberosum, S. chacoense, S. sisymbrifolium, Alternanthera tenella., Calopogonium mucunoides, Desmodium frutescens, Rhynchosia minima, Sida acuta,*	AbMV, AYMV, ACMV, AYVV, AgYVV-PK, AYVSLV, AYVTV, AYVV, AGMV, BcaMV, BDDV, BDMV, BGMV, BGYMV, BYVMV, CaLCV, CalGMV, ChMV, ChiLCuV, CdTV, CLCrV, CLCuAV, CLCuGV, CLCuKV, CLCuMV, CLCuRV, CYMV, CuLCrV, DiYMoV, DoYMV, ACMCV, EACMV (recombinant), EACMZV, EYVV, EYMV, EpYVV, EuMV, HLCrV, HYVMV; UK isolate not transmissible by B. tabaci, HgYMV, ICMV, ICLCV, JMV, KIGV, LeMV, LGMV, LLCuV, MGMV, MaMPRV, MaYMFV, MaYMV, MaMV, MCV, MIGV-PR, MYVV, MCLCV, MLCV, MeMV, MYMIV, MYMMV, MYMV, OkLCuIV, OkLCuV, OkMMV, OYVMV, PaLCV, PaMV, PLMV, PepGMV, PHYVV, PepLCBV, PepLCV, PepMTV, PepYLCV, PLCV, PDMV, PYMPV, PYMTV, PYMV, PYVV, RhGMV, RhMV, SiGMCRV, SiGMFV, SiGMHV, SiGMIV, SiMoV, SiYMV, SiYVV, SALCV, SACMIV, SCLV, SoyGMV, SLCCNV, SLCV, SLCCYV, SMLCV, SYMMoV, SLCMV, StaLCV, SPLCGV, SPLCV, TbASV, TbCSV, TbLCCV, TbLCIV, TbLCJV, TbLCKoV, TbLCYV, TbLCZWV, TbLRV, ToCMoV, ToCVV, ToCrV, ToCSV, TDLCV, TGMV, TGMoV, ToLCBV, ToLCBDV, ToLCBBV, ToLCGV, ToLCIV, ToLCIDV, ToLCKV, ToLCLV, ToLCMV, ToLCNDV, ToLCNV, ToLCPV, ToLCSV, ToLCSinV, ToLCSLV, ToLCTWV, ToLCTZV, ToLCVV, ToLCV, ToMBV, ToMHV, ToMoLCV,

TABLE 10.2 (Continued)

Vector	Natural host	Virus species which is transmitted
	S. rhombifolia, Pseuderanthemum spp., Alcea rosea, Bastardia, Corchorus aestuans, Hibiscus brasilensis, Physalis peruviana, Manihot esculenta, Carica papaya, Stachytarpheta sp., Oxalis corniculata, Parthenium hysterophorus, Solanum nigrum, Luffa acutangula Zinnia elegans, Boerhavia, erecta, Chenopodium murale, Cleome viscosa, Convolvulus spp., Conyza sumatrensis, Croton lobatus, Cuscuta sp., Cynanchum acutum, D. stramonium, Dittrichia viscosa, Eustoma grandiflorum, Malva parviflora, M. nicaeensis, Mercurialis ambigua, Wissadula spp., L. hirsutum, L. pimpinellifolium, S. tuberosum, S. quitoense, Triumffeta rhomboidiaceae, C. colocynthis, Wissadula amplissima, Lactuca sativa, Beta vulgaris, Chenopodium murale, Citrullus lanatus, Daucus carota, Lactuca sativa, Physalis wrightii., Citrullus vulgaris, Convolvulus arvensis, Ecballium elaterium, Malva parviflora, Sonchus asper, S. oleraceus, S. tenerrimus, Arachis hypogaea, Calopogonium mucunoides, Canavalia ensiformis, Psophocarpus tetragonolobus, Mucana pruriens, Stylothanthes gracile, Vicia faba, Voandzeia, subterranean Oxalis latifolia, O. martian	ToMoTV, ToMoV, ToRMV, ToSLCV, ToSRV, ToUV, ToYDV, TYLCV, TYLCMV; a recombinant between TYLCV and TYLCSV, ToYMMV, ToYMoV, ToYMoV, ToYVSV, TYVV, WmCSV, WmCMV, WGMV, ZiLCV
		Crinivirus
		CYSDV, LCV, LIYV, SPCSV, ToCV, CBSV, CVYV, CPMMV, PYVMV, SYLCV, SPMMV, and SPYDV
Trialeuroides vaporariorum	Callistephus chinensis, Cynara, Cardunculus, C. scolymus, Lactuca sativa, Nicotiana glauca, Petunia, Physalis ixocarpa, Picris echioides, Ranunculus spp. Catharanthus roseus, Lycopersicon sp., Polygonium spp., Rumex obtusifolium, Solanum nigrum, S. tuberosum, Tagetes spp. Datura stramonium,	TICV, PYDV, ToLCV, BPYV, and CCSV
T. abutilonea	Abutilon spp., Diodia virginiana, Ipomoea batatas, Datura stramonium, S. lycopersicum, S. nigrum	AbYV, DVCV, SPCSV, and ToCV

TABLE 10.2 (Continued)

Vector	Natural host	Virus species which is transmitted
T. ricini	*Boerhavia erecta, C. annuum, Chenopodium murale, Cleome viscose, Convolvulus* spp., *Conyza sumatrensis, Croton lobatus, Cuscuta* spp., *Cynanchum acutum, Datura stramonium, Dittrichia viscosa, Eustoma grandiflorum, Lycopersicon esculentum, Macroptilium* spp., *Malva parviflora, M. nicaeensis, Mercurialis ambigua, P. vulgaris, Physalis* spp., *Sida* spp., *S. nigrum, Wissadula* spp.	TYLCV

Note: AbMV: *Abutilon mosaic virus*; AbYV: *Abutilon yellows virus*; ACMV: *African cassava mosaic virus*; AGMV: *Asystasia golden mosaic virus*; AgYVV-PK: Ageratum yellow vein virus-Pakistan; AMV: *Alfalfa mosaic virus* x; AYMV: *Acalypha yellow mosaic virus*; AYVV: *Ageratum yellow vein virus*; AYVSLV: Ageratum yellow vein virus Sri Lanka; AYVTV: Ageratum yellow vein Taiwan virus; AYVV: *Ageratum yellow vein virus*; BBrMV: *Banana bract mosaic virus*; BcaMV: Bean calico mosaic virus; BDDV: Bean distortion dwarf virus; BDMV: Bean dwarf mosaic virus; BGMYV: *Bean golden mosaic yellow virus*; BPYV: Beet pseudo yellows virus; BtMV: *Beet mosaic virus*; BYDV-MAV: *Barley yellow dwarf virus-* MAV; BYMV: *Bean yellow mosaic virus*; BYV: *Beet yellows virus*; BYVMV: *Bhendi yellow vein mosaic virus*; CaCV: *Capsicum chlorosis virus*; CaLCV: Cabbage leaf curl virus; CalGMV: Calopogonium golden mosaic virus; CBSV: Cassava brown streak virus; CCSV: Cucumber chlorotic spot virus; hMV: Chayote mosaic virus; ChiLCuV: Chilli leaf curl virus; CdTV: Chino del tomate virus; CLCuMV: Cassava Cotton leaf curl Multan virus; CLCuRV: Cotton leaf curl Rajasthan virus; CuLCrV: Cucurbit leaf crumple virus; CLCrV: Cotton leaf crumple virus; CLCuAV: Cotton leaf curl Alabad virus; CLCuGV: Cotton leaf curl Gezira virus; CLCuKV: Cotton leaf curl Kokhran virus; CMV: *Cucumber mosaic virus*; CPMMV: Cowpea mild mottle virus; CSNV: *Chrysanthemum stem necrosis virus*; CTV: *Citrus tristeza virus*; CVYV: Cucumber vein yellowing virus; CYDV: *Cereal yellow dwarf virus*; CYMV: Cowpea yellow mosaic virus; CYSDV: Cucurbit yellowing stunting disorder virus; DVCV: Diodia vein chlorosis virus; DiYMoV: Dicliptera yellow mottle virus; DoYMV: Dolichos yellow mosaic virus; EACMV (recombinant): East African cassava mosaic virus; EACMZV: East African cassava mosaic Zanzibar virus; EYVV: Eclipta yellow vein virus; EYMV: Eggplant yellow mosaic virus; EpYVV: Eupatorium yellow vein virus; EuMV: Euphorbia mosaic virus; FBNYV: *Faba bean necrotic yellows virus*; GBNV: *Groundnut bud necrosis virus*; GRSV: *Groundnut ringspot virus*; HLCrV: Hollyhock leaf crumple virus; HYVMV-UK isolate not transmissible by B. tabaci: Honeysuckle yellow vein mosaic virus; HgYMV: Horsegram yellow mosaic virus; ICMV: Indian cassava mosaic virus; ICLCV: Ipomoea crinkle leaf curl virus; INSV: *Impatiens necrotic spot virus*; IYSV: *Iris yellow spot virus*; JMV: Jatropha mosaic virus; KIGV: Kigluaik phantom virus; LCV: Lettuce chlorosis virus; LeMV: Leonurus mosaic virus; LGMV: Limabean golden mosaic virus; LIYV: *Lettuce infectious yellows virus*; LLCuV: LMV: *Lettuce mosaic virus*; MGMV: Macroptilium golden mosaic virus; MaMPRV: Macroptilium mosaic Puerto Rico virus; MaYMFV: Macroptilium yellow mosaic Florida virus; MaYMV: Macroptilium Yellow Mosaic Virus; MaMV: Macroptilium Mosaic Virus; MCV: Maize chlorotic virus; MYVV: Malvastrum yellow vein virus; MCLCV: Melon chlorotic leaf curl virus; MLCV: *Melon leaf curl virus*; MeMV: Melilotus mosaic virus; MYMIV: *Mungbean yellow mosaic India virus*; MYMV: *Mungbean yellow mosaic virus*; MYSV: *Melon yellow spot virus*; OkLCuIV: Okra leaf curl India virus OkLCuV:Okra leaf curl virus; OkMMV: Okra mosaic Mexico virus; OYVMV: Okra yellow vein mosaic virus; PaLCV: *Papaya leaf curl virus*; PDV: *Prune dwarf virus*; PYVMV: Pumpkin yellow vein mosaic virus; PAMV:

TABLE 10.2 (Continued)

Potato aucuba mosaic virus; PEMV: *Pea enation mosaic virus*; PFBV: *Pelargonium flower break virus*; PLMV: Peach latent mosaic virus; PepGMV: Pepper golden mosaic virus; PHYVV: Pepper huasteco yellow vein virus; PepLCBV: Pepper leaf curl Bangladesh virus; PepLCV: Pepper leaf curl virus; PepMTV: Pepper mild tigre virus PepYLCV: Pepper yellow leaf curl virus; PLCV: *Pelargonium leaf curl virus*; PLRV: *Potato leafroll virus*; PMV: *Panicum mosaic virus*; PNRSV: *Prunus necrotic ringspot virus*; PPV: *Plum pox virus*; PVY: *Potato virus Y*; PYMPV: Potato yellow mosaic Panama virus; PYMTV: Potato yellow mosaic Trinidad virus; PYMV: Potato yellow mosaic virus; PYSV: *Peanut yellow spot virus*; PYVV: *Potato Yellow Vein Virus*; RhGMV: Rhynchosia golden mosaic virus; RhMV: Rhynchosia mosaic virus; SCYLV: *Sugarcane yellow leaf virus*; SiGMCRV: Sida golden mosaic Costa Rica virus; SiGMFV: Sida golden mosaic Florida virus; SiGMHV: Sida golden mosaic Honduras virus; SiGMIV: Sida golden mosaic Jamaica virus; SiMoV: Sida mottle virus; SiYMV: Sida yellow mosaic virus; SiYVV: Sida yellow vein virus; SALCV: Solanum apical leaf curling virus; SACMV: South African cassava mosaic virus; SCLV: Soybean crinkle leaf virus; SCMV: *Sugarcane mosaic virus*; SLCCNV: Squash leaf curl China virus; SLCV: Squash leaf curl virus; SLCCYV: Squash leaf curl Yunnan virus; SMLCV: Squash mild leaf curl virus; SMV: *Soybean mosaic virus*; SrMV: *Sorghum mosaic virus*; StaLCV: Stachytarpheta leaf curl virus; SYMMoV: Squash yellow mild mottle virus; SLCMV: Sri Lankan cassava mosaic virus; SPCSV: Sweet potato chlorotic stunt virus; SPCSV: Sweet potato chlorotic stunt virus; SPMMV: Sweet potato mild mottle virus; SPYDV: Sweet potato yellow dwarf virus; SPLCV: Sweet potato leaf curl virus; SPLCGV: Sweet potato leaf curl Georgia virus; SPFMV: *Sweet potato feathery mottle virus*; SPLCV: TbASV: Tobacco apical stunt virus; TbCSV: Tobacco curly shoot virus; TbLCCV: Tobacco leaf curl China virus; TbLCIV: Tobacco leaf curl India virus; TbLCJV: Tobacco leaf curl Japan virus; TbLCKoV: Tobacco leaf curl Kochi virus; TbLCYV: Tobacco leaf curl Yamaguchi virus; TbLCZWV: Tobacco leaf curl Zimbabwe virus; TbLRV: Tobacco leaf rugose virus; TBSV: *Tomato bushy stunt virus*; TCSV: *Tomato chlorotic spot virus*; TICV: Tomato infectious chlorosis virus; ToCMoV: Tomato chlorotic mottle virus; ToCV: *Tomato chlorosis virus*; ToCVV: Tomato chlorotic vein virus; ToCrV: Tomato crinkle virus; ToCV: Tomato chlorosis virus; TDLCV: Tomato dwarf leaf curl virus; TGMV: Tomato golden mosaic virus; TGMoV: Tomato golden mottle virus; ToLCBV: Tomato leaf curl Bangalore virus; ToLCBDV: Tomato leaf curl Bangladesh virus; ToLCBBV: Tomato leaf curl Barbados virus; ToLCGV: Tomato leaf curl Gujarat virus; ToLCIV: Tomato leaf curl Iran virus; ToLCIDV: Tomato leaf curl Indonesia virus; ToLCKV: Tomato leaf curl Karnataka virus; ToLCLV: Tomato leaf curl Laos virus; ToLCMV: Tomato leaf curl Malaysia virus; ToLCNDV: Tomato leaf curl New Delhi virus; ToLCNV: Tomato leaf curl Nicaragua virus; ToLCPV: Tomato leaf curl Philippines virus; ToLCSV: Tomato leaf curl Senegal virus; ToLCSinV: Tomato leaf curl Sinaloa virus; ToLCSLV: Tomato leaf curl Sri Lanka virus; ToLCTWV: Tomato leaf curl Taiwan virus; ToLCTZV: Tomato leaf curl Tanzania virus; ToLCVV: Tomato leaf curl Vietnam virus; ToLCV: Tomato leaf curl virus; ToMBV: Tomato mosaic Barbados virus ToMHV: Tomato Mosaic Havana Virus; ToMoLCV: Tomato mottle leaf curl virus; ToMoTV: Tomato mottle Taino virus; ToMoV: *Tomato mottle virus*; ToRMV: Tomato rugose mosaic virus; ToSLCV: Tomato severe leaf curl virus; ToSRV: Tomato severe rugose virus; ToUV: ToYDV: Tomato yellow dwarf virus; TRSV: *Tobacco ringspot virus*; TSV: *Tobacco streak virus*; TSWV: *Tomato spotted wilt virus*; TYLCSV: ToYMMV, ToYMoV: *Tomato yellow leaf curl virus*; TYLCV: Tomato yellow leaf curl virus; TYLCMV: Tomato yellow leaf curl Morondava virus is a recombinant between TYLCV and TYLCSV; ToYMMV, TYLCSV: *Tomato yellow mottle virus*; ToYVSV: Tomato yellow vein streak virus; TSWV: *Tomato spotted wilt virus*; TuMV: *Turnip mosaic virus*; TYV: Tobacco yellow vein virus; WBNV: *Watermelon bud necrosis virus*; WmCSV: Watermelon chlorotic stunt virus; WmCMV: Watermelon curly mottle virus; WGMV: *Wissadula golden mosaic virus*; WSMoV: *Watermelon silver mottle virus*; ZiLCV: Zinnia leaf curl virus; ZLCV: *Zucchini lethal chlorosis virus*; ZYMV: *Zucchini yellow mosaic virus* preferentially feeds on phloem, whereas, some of the species like *Cicadella* and *Typhlocyba* feeds on xylem and mesophyll cells, respectively (Hamilton, 1999; Hamilton & Zack, 1999; Zahniser & Dietrich, 2010).

Thrips are known to transmit tospoviruses (family: *Bunyaviridae*; genus: *Tospovirus*). Globally, more than 20 tospovirus species are known (Table 10.3). It has been reported that a total of 14 thrips species (family: Thripidae) associated with the transmission of tospoviruses (Ullman 1997; Jones 2005; Pappu et al., 2009; Ciuffo et al., 2010; Hassani-Mehraban et al., 2010). Besides, tospoviruses the most economical important group of plant viruses, thrips transmit certain members of *Carmovirus*, *Ilarvirus*, *Machlomovirus*, and *Sobemovirus* (Ciuffo et al., 2010; Hassani-Mehraban et al., 2010).

Beetles belonging to order Coleoptera, the members from several families including Chrysomelidiae, Coccinellidae, Curculionidae, and Meloidae are known to transmit plant viruses. A total of 75 viruses have been reported as beetle transmitted of which family Chrysomelidiae is one of the predominant and transmitting 61 species of plant viruses. The several genera of beetles as virus vector includes *Acalymma*, *Ceratoma*, *Diabrotica*, *Epitrix*, *Phaedon*, and *Phyllotreta* (Seeno & Wilcox, 1982; Fulton et al., 1987; Nault, 1997; Meier et al., 2008). Apart from the insect vectors, plant-viruses have also been transmitted by nematodes and fungi as mentioned in Table 10.1, however, the detailed description is beyond the scope of this chapter.

10.2 ADAPTATION OF PLANT VIRUSES WITH INSECTS

Plant viruses have been known to be associated with insect vectors for their possible spread from one plant to another and their survival. The effective transmission from plant-to-plant by the insect vector particularly depends upon the ability of plant virus to transfer from one vector to other host plant which is common to both. This has been known as *colonization hypothesis*. The adaptation of plant viruses with insect vectors possibly involves several processes including the evolutionary origins, through mutational changes in the genome particularly in coat protein/helper component and descent and direct colonization of a new vector group (Hull, 2002). Based on the fossil records, Homoptera has been diverted in to Auchenorhhyncha and Sternorhhyncha by the Upper Triassic, 180 million years ago (Evans, 1963; Hennig, 1981) and early herbivores by 70 million years ago (Gould & Shaw, 1983). Since long, adaptation of plant viruses to their vectors has been a subject of interest and, several studies in the

TABLE 10.3 Distribution and Natural Host Range of Thrips-Transmitted Tospoviruses.

Virus [Acronym]	Distribution	Vector	Natural hosts of virus and vector	Reference
Distinct tospoviruses approved by ICTV				
Groundnut bud necrosis virus [GBNV]	India, Bangladesh, Nepal, Vietnam, Thailand, Iran, China, Sri Lanka	*Thrips palmi, Frankliniella Schultzei, Scirtothrips dorsalis,*	Members of Solanaceae, Fabaceae, Euphorbiaceae, Cucurbitaceae, Asteraceae, Amaryllidae and Malvaceae	Reddy et al. (1968); Reddy et al. (1992); Meena et al. (2005); Akhter et al. (2012); Holkar et al. (2012)*; Jain et al. (2007); Akram & Naimuddin (2010); Thien et al. (2003); Sujitha et al. (2012); Pundhir et al. (2012); Pappu et al. (2009)
Groundnut ringspot virus (GRSV)	South Africa, Brazil, India, U.S.A., Argentina	*Frankliniella occidentalis, F. Schultzei*	*Capsicum annuum, Solanum americanum* and other *Solanum* spp., *Physalis angulata*	Wijkamp et al. (1995); De Bordon et al. (1999); Nagata et al. (2004); Webster et al. (2015)
		F. intosa, F. gemina		
Groundnut yellow spot virus [GYRSV]	India, Thailand, China	*Scirtothrips dorsalis*	*Fabaceae* and *Solanaceae*	Satyanarayana et al. (1996, 1998); Reddy et al. (1991); Yin et al. (2006)*; Gopal et al. (2010)
Impatiens necrotic spot virus [INSV]	U.S.A., The Netherlands, Italy, U.K., Bosnia, Iran, Israel, New Zealand, Japan, China, Yunnan, Australia, Austria	*F. intosa* *F. occidentalis* *F. fusca*	*Antherium* sp., *Anemone* sp., *Exacum* sp., *Bowardia* sp., *Gloxinia* sp., *Ranunculus* sp., *Antirrhinum* sp., *Fatsia* sp., *Limonium* sp., *Aphelendra* sp., *Chrysanthemum, Gazania* sp., *Gerbera* sp., *Gladiolus* sp., *Helichrysum*sp., *Hydrangea* sp., *Penstemon* sp., *Ruscus* sp., *Sinningia* sp., *Aeschynanthus* sp., *Alstromeria* sp., *Bromelia* sp., *Arabidopsis* sp., *Ardsia* sp., *Cardomine* sp., *Curcuma* sp., *Eustoma* sp., *Gentiana* sp., *Hippeastrum* sp., *Iris* sp., *Kalanchoe* sp., *Lobelia* sp., *Nemesia* sp., *Phalaenopsis* sp., *Primula* sp., *Saxiferaga* sp., *Schizanthus* sp., *Senecio* sp., *Zantedeshia* sp., *Cineraria* sp., *Agrostemma* sp., *Capsicum* sp., *Pericallis* sp., *Browallia* sp.,	Mertelik et al. (2002); Koike & Mayhew (2001); Vicchi et al. (1999); Naidu et al. (2001); Lebas et al. (2004); Sakurai et al. (2004); Gonzalez-Pacheco & Silva-Rosales (2014); Kuo et al. (2014); Crosslin & Hamlin (2014); Zang et al. (2010); Elliot et al. (2009)*; Naidu et al. (2007)*; Fineti-Sialer et al. (2006)*; Dietzgen et al. (2012); Prihodko et al. (2012)*.

TABLE 10.3 *(Continued)*

Virus [Acronym]	Distribution	Vector	Natural hosts of virus and vector	Reference
			Caceolaria sp., *Callistephus* sp., *Diascia* sp., *Eustoma* sp., *Platycodon* sp., *Portulaca* sp., *Primula* sp., *Schizanthus* sp., *Solenstemon* sp., *Stephanotis* sp., *Oncidium* sp., *Opuntia* sp., *Microdasys* sp., *Solanum municatum*, *Sinaciaoleracea*, *Capsicum annuum*, *Cucumissativus*, *Eustoma grandifolium*, *Begonia* sp., *GentionaScabra*, *Pericallis hybrid*, *Physalis*, *Antirrhinum* sp., *Spinach radicchio*, *Orchis* sp., *Viciafaba*, *Cynara cardunculus var: scolymus*, *Lactucasativa*, *Anthriscuscerefolium*, *Erucasativa*, *Gentiana* sp., *Macrophylla* sp. *Solanum tuberosum*, *Phalaenopsis* sp., *Hymenocallis littoralis (Spider lily), Rununculus* sp., *Primula* sp., *Obconica* sp., *Impateins* sp., *Walleriana* sp., *Gardenia Jasminoides, Monadadidyma, Solanum esculentum* and *Ruellia* sp.	
Iris yellow spot virus [IYSV]	Brazil, The Netherlands, Israel, India, U.S.A., Solvenia, Australia, Japan, Iran	*Thrips tabaci*	*Hippeastrum hybridum, Allium cepa, Allium tuberosum, Alstromeria, Lisianthus* sp., *Clivia minata, Bessera elegans, Portulaca, Cycas* sp., *Rosa* sp. and *Scidapsis* sp.	Cortes et al. (1998); Kritzman et al. (2001); Mohan & Moyer (2002); Gent et al. (2004); Coutts et al. (2003); Zen et al. (2003); Murai (2004); Hsu et al. (2010); Srinivasan et al. (2012); Iftikar et al. (2014); Gawande et al. (2015)
Melon yellow spot virus [MYSV]	China, India, Japan, Thailand	*T. palmi*	*Citrullus lanatus, Cucumis melo* and *Physalis minima*	Kato et al. (2000); Takeuchi et al. (2009); Liu & Rao (2010)*
Polygonum ringspot virus [PolRSV]	Italy	*Dictyothrips betae*	*Fallopia dumetorum*	Ciuffo et al. (2010); Margaria et al. (2014)

TABLE 10.3 *(Continued)*

Virus [Acronym]	Distribution	Vector	Natural hosts of virus and vector	Reference
Tomato chlorotic spot virus [TCSV]	Brazil, U.S.A., and South Africa	*F. occidentalis, F. Schultzei, F. intosa*	*Catharanthus roseus*, members of *Solanaceae, Cichorumendiva, Solanum gilo, Lactica sp., lisianthus sp., Solanum tuberosum, Capsicum annum, S. lycopersicum, Amaranthus* sp., *Schlumbergera truncate, Hoya wayetii, Datura ferox, D. stramonium, Coronopus didymus*, and *Portulaca oleracea*	Wijkamp et al. (1995); Rabelo et al. (2002); Nagata et al. (2004); Jones et al. (2005); Vanderspool et al. (2014)*
Tomato spotted wilt virus [TSWV]	Australia, Brazil, China, U.S.A., East and West Egypt, Europe, India, U.K., Portugal, Noth & South America, Africa and Australasia, Italy, Iran, France, Serbia, South Korea, Kenya, Venezuela, Hungary, Turkey, Saudi Arabia, Japan, Jordan, New Zealand, Dominican Republic, Bulgaria, Bosnia and Herzegovina, Spain	*T. tabaci, F. schultzei, T. palmi, T. setosus, F. intosa, F. bispinosa, F. fusca, F. gemina, F. cephalica*	*Allium cepa, Solanum lycopersicum, C. annum, Solanum melongena, Brassica oleracea, Emilia sonchifolia, Vigna unguiculata, Cineraria* sp., *Convolvulus arvensis, C. sativus, Phaseolus vulgaris, Lactuca* sp., *Leuzea carthamoides, Lobelia cardinalis, Carica papaya, Pisumsativum, Arachis hypogea, Ananas comostus, Solanum tuberosum, Glycine max*, spinach, *Nicotiana* sp., *Brassica oleracea, Emilia sonchifolia, Chrysanthemum* sp., *Anemone* sp., *Calendula* sp., *Callistephus* sp., *Cineria* sp., *Dahlia* sp., *Dieffenbachia* sp., *Gazania* sp., *Gerbera* sp., *Impatiens walleriana* and other *Impatiens* sp., *Hoya arnottiana, Rununculus* sp., *Sonchus oleraceus, Salvia* sp., *Zinnia* sp., *Tropaeolum* sp., *Antirrhinum* sp., *Aster* sp., *Beloperone* sp., *Catharanthus* sp., *Cellosia* sp., *Coleus* sp., *Dianthus* sp., *Fuchsia* sp., *Saintpaulia* sp., *Tagets* sp., *Zantedeschia* sp., *Allium* sp., *Chicorium intybus, Convolvulus, Conyza, Dracaena* sp., *Fuchsia* sp., *Glueliolus* sp., *Hieracium* sp., *Hydrangea* sp., *Jasminum* sp., *Nerium* sp., *Pelargonium* sp., *Penstemon* sp., *Sinningia* sp., *Solanum capsiastrum, Stellaria media, Pinellia tripartita, Primula* sp., *Tagetes patula, Capsicum* sp. *Nicotiana* sp., and *Ocimum basilicum*	Nagata et al. (2004); Persley et al. (2006); Parrella et al. (2003); Naidu et al. (2001); Avila et al. (2006); Ohnishi et al. (2006); Sundaraj et al. (2014); Lian et al. (2013); Hu et al. (2011); Renukadevi et al. (2015); Perez et al. (2014)*; Margaria et al. (2015); Almasi et al. (2014); Deligoz et al. (2014)*; Al-Saleh et al. (2014); Tomitaka et al. (2015). Timmerman-Vaughan et al. (2014); Trkulja et al. (2013); Salem et al. (2011)*

TABLE 10.3 *(Continued)*

Virus [Acronym]	Distribution	Vector	Natural hosts of virus and vector	Reference
Watermelon bud necrosis virus [WBNV]	Taiwan, Japan and India	*T. palmi*	*C. lanatus, C. sativus, C. melo, Luffa acutangula,* and *Chrysanthemum* sp.	Singh & Krishnareddy (1995); Jain et al. (1998); Mound et al. (1996); Mandal et al. (2003); Holkar et al. (2012)*
Watermelon silver mottle virus [WSMoV]	China, Japan, Russia, U.S.A., Taiwan, Thailand	*T. palmi*	*Capsicum annuum, Citrullus lanatus, Zantedeschia* sp. *C. melo, Solanum lycopersicum, Physalis minima, Luffa* spp. *Physalis heterophylla, Euphorbia heterophylla, Amaranthus* sp.	Yeh et al. (1995); Okuda et al. (2001); Chiemsombat & Manecchoat (2010)*; Rao et al. (2013)
Zucchini lethal chlorosis virus [ZLCV]	Brazil, Japan	*F. zucchini*	*Cucumis pepo, Solanum esculentum, Chrysanthemum* sp., *Capsicum annum, Datura stramonium* and *C. sativus*	Nakahara & Monteiro (1999); Boiteux et al. (1994)
Tentative tospovirus species (unclassified)				
Alstroemeria necrotic streak virus [ANSV]	Colombia	*F. occidentalis*	*Alstroemeria* sp.	Hassani-Mehraban et al. (2010).
Bean necrotic mosaic virus [BeNMV]	Brazil	N.D.	*Physalis pubescens, P. vulgaris*	de Oliveira et al. (2012)
Calla lily chlorotic spot virus [CCSV]	Taiwan, China	*T. palmi*	*Zantedeschiaaethiopica, C. lanatus, cucumber, Pumpkin, Benica sahispida, Nicotiana* sp. *Hymeno callislitteralis*	Chen et al. (2005); Liu et al. (2012)
Capsicum chlorosis virus [CaCV]	Australia, China, India, Indonesia, Queensland, Taiwan, Thailand and U.S.A.	*F. occidentalis, Ceratothripoidesclaratris*	members of Solanaceae, Apocynaceae, Amaryllidaceae, Orchidaceae, Amaranthaceae, Araceae, Gesneriaceae, Begoniaceae	Persley et al. (2006); Premchandra et al. (2005) Kunkalikar et al. (2010); Chen et al. (2009); Melzer et al. (2014); Zheng et al. (2011); Knierim et al. (2006)
Chrysanthemum stem necrosis virus [CSNV]	Belgium, Brazil, U.K., Solvenia, Japan	*F. occidentalis, F. Schultzei*	*Chrysanthemum x morifolium Ramat, Lisianthus, Chrysanthemum* sp.	Nagata et al. (1994); Mumford et al. (2003); Nagata et al. (2004); Dullemans et al. (2015); De Jonghe et al. (2013)*

TABLE 10.3 (Continued)

Virus [Acronym]	Distribution	Vector	Natural hosts of virus and vector	Reference
Gloxinia ringspot virus [GloxRSV]	Taiwan	N.D.	*Gloxinia* sp.	Chang & Chen (2003)*
Hippeastrum chlorotic ringspot virus [HCRSV]	China	N.D.	*Hymenocallislittoralis,*	Xu et al. (2014); Dong et al. (2013)
Melon yellow spot virus [MYSV]	China, Thailand, Japan, Ecuador, India	*T. palmi*	*C. lanatus, Momordicacharantia, C. sativus, C. melo, M. charantia, B. hispida, Luffa* sp. *Physalisheterophylla*	Kato et al. (2000); Seepiban et al. (2008)*; Takeuchi et al. (2009); Okuda et al. (2006); Laxmi Devi et al. (2010)*
Peanut chlorotic fanspot virus [PCFV]	Taiwan	*S. dorsalis*	Fabaceae	Chu et al. (2001)
Physalis severe mottle virus [PSMV]	The Netherlands	N.D.	*Physalis* sp.	Cortez et al. (2001)
Soybean vein necrosis virus [SVNaV]	U.S.A.	*Neohydatothrips variabilis* (USDA)	*Glycine max*	Zhou & Tzanetakis (2013)
Tomato necrosis virus TD8 [TNeV]	Thailand	N.D.	*S. lycopersicum*	Chiemsombat et al. (2008)
Tomato necrotic ringspot virus [TNRSV]	Thailand	N.D.	*S. lycopersicum, Capsicum annum, Citrullus lanatus*	Seepiban et al. (2011)
Tomato yellow fruit/ ring virus [TYRSV]	Iran, Kenya	*T. tabaci*	*S. lycopersicum, S. tuberosum, Gazania* sp., *Chrysanthemum* sp., *Alstroemeria* sp., *cineraria* sp., *anemone* sp. *Glycine max*	Hassani-Mehraban et al. (2011)*; Rasoulpour & Iza panah (2007); Birithia et al. (2016)
Tomato zonate spot virus [TZSV]	China	N.D.	*S. tuberosum, Iris tectorum, S. lycopersicum*	Huang & Liu (2014)*; Liu et al. (2015). Dong et al. (2008)

*Reference obtained from the NCBI GenBank and not included in the reference list at the end of this chapter.

past revealed the concept behind the association of virus and vectors. Zettler (1967) put forth the evidence that the aphids of the sub-family Aphidinae were the efficient vectors of *Bean common mosaic virus* (BCMV) than the aphids from sub-families Callaphidinae or Chaitophorinae. Basically, these differences were nothing but the variations in the probing behavior of the aphids. Subsequently, Francki et al. (1985) suggested that no consistent differences were recorded in comparison of leafhoppers and whiteflies transmitted begomoviruses in their host plants in the form of aggregation in virions or cytopathological changes but in view of this further evidences suggested the differences were due to the genome compositions of these two distinct groups of plant viruses. This was further supported by Harrison (1985) that the bipartite condition of the whitefly-transmitted begomoviruses is derived from monopartite form with consistent variation in the genomes of whitefly and leafhopper transmitted plant viruses. Later, the origin and evolution of Auchenorrhyncha-transmitted plant viruses have been reviewed by Nault (1986) and described the association of reo and rhabdoviruses with their vectors.

10.3   HOST PREFERENCE BY EVOLVING INSECTS

Vectors of plant viruses perform series of events during recognition of the host plant in general and identification of feeding sites in particular (Fereres & Moreno, 2009). The behavioral events for homopterans including, aphids, whiteflies, leafhoppers, and planthoppers have been described (Powell et al., 2006; Fereres & Moreno, 2009). These events including insect behavior before landing on the host surface, contact with the host surface, selection of the hosts after landing, probing into superficial tissues followed by finding of the exact location and insertion of the stylets for feeding and finally salivation and food ingestion. Before going to select a host by the insects they get attracted toward specific colors, for example, the aphids are attracted toward intensive yellow colored targets and land on yellow–green surface (Todd et al., 1990). Aphids, after landing on the host surface walk and with the help of movement of antennae they judge the different types of odours and surface textures as chemo-and mechano-receptors are present on the antennal region (Park & Hardie, 2004). Aphids in general prefer to stay and probe on the smooth textured surface because too much hairiness and undulated structure of epicuticular wax creates difficulty in walking (Doring et al., 2004, 2007). Aphids prefer to settle on abaxial leaf surface due to positive geotaxis and negative phototaxis nature but the first probe may occur on either of the leaf surfaces (Klingauf, 1987; Fig. 10.1).

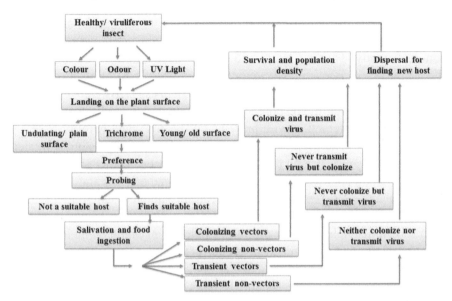

FIGURE 10.1 Schematic representation of the host-plant selection behavior of aphids and transmission of plant viruses in general. (From Klingauf, F. A. Feeding, Adaptation and Excretion. In Aphids: Their Biology, Natural Enemies, and Control; Minks, A. K., Harrewijn, P., Ed.; Elsevier: Amsterdam, 1987; pp 225–253. Used with permission from Cambridge University Press.)

Irwin et al. (2007) classified the aphids based on the landing, stay, colonization, and ability to transmit the virus as: (a) transient non-vectors, (b) transient vectors, (c) colonizing non-vectors, and (d) colonizing vectors. Transient non-vectors have the ability to land and probe on the crop but they neither colonize nor transmit virus, whereas, transient vectors have ability to land, probe, and transmit the virus without colonizing crop. Colonizing non-vectors are the one that can land, stay, and colonize the crop but do not transmit virus, colonizing vectors on the other hand can stay, colonize crop and transmit the virus. With the advancement in development of electronic devices in order to determine probing and feeding behavior of homopterans, the first electronic device has been developed by Mclean and Kinsley in 1964 since then the device was modified and greatly improved (Tjallingh, 1988; Bucks & Benett, 1992). With the help of this instrument insect activities, particularly, probing, feeding and positions of stylets can easily be monitored (Tjallingii, 1985; Fig. 10.1). At the time of puncturing cell membrane by insect a distinctive electrical penetration graph (EPG) signal is recorded in the form of potential drops (pd), this EPG graph indicates different phases like E1 and E2. The phase E1 represents phloem salivation by aphids and whiteflies transmit virus, whereas, E2 indicates phloem ingestion including sub phases II-1, II-2, and II-3. The II-2 and III-3 are sub phases associated with the inoculation and acquisition of viruses particularly those transmitted non-persistently. This has been recorded in inoculation of *Potato virus-Y* (PVY) by *Myzus persicae, Cucumber mosaic virus* (CMV) by *A. gossypii* and *Cauliflower mosaic virus* (CaMV) by *Brevicoryne brassicae* (Moreno et al., 2005). Moreover, the response of phloem wound toward salivary secretions by aphids interacting with proteins has been studied with the help of EPG technique (Tjallingii, 2006). Subsequently, spatially explicit model were described by Sisterson (2008) indicated that the vectors prefer to feed on healthy plant and that enhance the rate of spread of virus and feeding on infected plants decreases the rate of spread of pathogens. Similarly, recent studies on the use of EPG technique to determine vector behavior, it has been shown that aphids prefer prolonged phloem ingestion on their suitable host and delay ingestion on non-host plants (Nalam et al., 2012; ten-Broeke et al., 2013). EPG technique has demerits of restricted capacity, it requires limited plants (~8) per set up and is labor intensive (ten Broeke et al., 2013). To overcome the demerits of the EPG technique, an automated video tracking software was introduced in early 1990s to record feeding behavior of homopterans. Later, it has been re-modified, further validated and extensively utilized for recording precise behavior of *M. persicae* on two natural *Arabidopsis* accessions, Co-2; identified as resistant, and Sanna-2; identified as susceptible (Kloth et al., 2015). This technique could be another option for identifying the natural sources of resistance against homopterans feeding on large number of agricultural and horticultural crops.

Whiteflies (*Bemisia tabaci*) during their flight, get attracted towards green reflected light (Isaacs et al., 1999). The female *B. tabaci* shows definite color preference and get attracted toward yellow color and this phototactic behavior relates their migration (Berlinger, 1986). Whiteflies after landing on the plant surface keeps on walking until they reach to edge of the shaded leaf and move toward the abaxial surface and start probing, this is not only due to the gravity but also the negative phototaxis nature of whiteflies. Whiteflies prefer to settle and feed on the shaded leaves irrespective of the lower or upper surface and for oviposition they prefer young leaves (Coombe, 1982). As like aphids, leaf hairs decide the host selection by whiteflies because it create a physical barrier (Duffey, 1986). Whiteflies are able to differentiate between non-host and host plants for feeding by probing apoplast of mesophyll cells below the epidermis instead of probing phloem but the stylets of *Trialeurodes vaporariorum* and *Bemisia tabaci* end in the phloem vessels (Ekbom & Xu, 1990).

Recently, the stylet penetration behavior of four species of leafhoppers including *Cicadulina arachidis*, *C. dabrowskii*, *C. mbila*, and *C. storeyi* were studied with the help of EPG technique by Oluwafemi and Jakai (2013). The results indicate that these vectors prefer to feed on healthy seedlings rather than virus infected crops and this behavior of leafhoppers possibly helped spread of *Maize streak virus* (MSV). Moreover, these four vectors differed significantly to stylet penetration and recorded that *C. mbila* and *C. storeyi* spent more time without probing, required least time for phloem feeding and probing. The results indicated that these two vectors transmitted the MSV efficiently than the other two vectors.

The host selection, feeding, probing, ingestion, and transmission of plant viruses differ from insect to insect and host to hosts of same or different families. The variable response is due to several factors including, host morphology, healthy/infected hosts, certain volatiles or chemicals that attract or arrest insect-vectors from feeding for example, potato plants infected with *Potato leaf roll virus* (PLRV) attract and arrest *Myzus persicae* for feeding (Eigenbrode et al., 2002). Labiate oils from the plant roosmary (*Rosamarinus officinalis*) belonging to family Laminaceae have been shown the deterrent and repellent properties against the onion thrips (*Thrips tabaci* Lindeman) in host selection, settling, oviposition by females (Koschier & Sedy, 2003).

10.4   VIRUSES TRANSMITTED BY APHIDS

10.4.1   GENUS: *ALFAMOVIRUS* (BROMOVIRIDAE)

Alfalfa mosaic virus (AMV), the type species of the genus *Alfamovirus*. Alfamoviruses contain tripartite RNA (RNA 1, RNA 2, and RNA 3) genome of positive strand polarity. RNAs 1 and 2 encode proteins involved in viral RNA replication, whereas, RNA 3 encodes a protein involved in cell-to-cell movement and the coat protein (CP), which is translated from a sub-genomic messenger, RNA 4. The four viral RNAs are separately encapsidated into bacilliform particles (Bol, 1999). In 1971, it was reported that a mixture of the three genomic RNAs of AMV, the type species of the alfamoviruses, was not infectious to plants unless a few molecules of CP were added per RNA molecule to the inoculum (Bol et al., 1971). Alfamoviruses are transmitted in a non-persistent manner by at least 14 aphid species and it shares many properties at the molecular level with members of *Ilarvirus*. Virions are generally bacilliform, 18×30 to 57 nm in size. The genus has a very broad host range (King et al., 2012).

10.4.2 GENUS: *BABUVIRUS* (NANOVIRIDAE)

Babuviruses (type species: *Banana bunchy top virus*; BBTV), babuviruses naturally infect *Musa* and *Amomum* spp. and are transmitted by *Pentalonia nigronervosa* (banana aphid). BBTV has a genome consisting of six different single-stranded (ss) DNA components, known as DNA-R, -S, -C, -M, -N, and -U3. These components have been studied from all the isolates and are considered to be an integral part of the *Babuvirus* genome. A polymerase chain reaction (PCR) assay was used to study BBTV transmission efficiency and to determine the minimum acquisition-access period, minimum inoc-ulation-access period, retention time, transovarial ability of transmission in this vector. Results revealed that BBTV was acquired by banana aphids within 4 h and was transmitted within 15 min. after feeding and recorded 65% transmission efficiency by banana aphids (Hu et al., 2008).

10.4.3 GENUS: *CAULIMOVIRUS* (CAULIMOVIRIDAE)

Caulimovirus (type species: *Cauliflower mosaic virus*; CaMV) contains dsDNA as genetic material of 7.8–8.2 kb in size. Virus having the plus and minus DNA strands possesses discontinuity. The minus-strand DNA possesses a single discontinuity, whereas, the plus-strand DNA having two or three discontinuities. The *Caulimovirus* genome contains six single open reading frames (ORFs) and large and small intergenic regions (King et al., 2012).

10.4.4 GENUS: *CLOSTEROVIRUS* (CLOSTEROVIRIDAE)

Closterovirus belongs to family *Closteroviridae* with the type species: *Beet yellows virus* (BYV), members of the genus composed of particle length more than 1200 nm, and monopartite RNA. Size is ranged from 14 to 20 kb and is naturally transmitted by aphids. Wide range of insect vectors are involved in the transmission of closteroviruses. BYV is transmitted by not less than 20 aphid species (*Myzus persicae* and *Aphis fabae* are the major natural vectors), *Citrus tristeza virus* (CTV) is transmitted by seven species of which *Toxoptera citricida* and *Aphis gossypii* are the most efficient one, and a number of other viruses including *Carnation necrotic fleck virus* (CNFV), *Watercress yellow spot virus* (WYLV), *Burdock yellows virus* (BdYV), and *Mint virus* 1 (MV-1) are transmitted by single aphid species, or do not have known vectors of *Grapevine leafroll associated virus* (GLRaV-2). Some species can be sap transmitted by inoculation (Martelli & Candresse, 2014).

10.4.5 GENUS: *CUCUMOVIRUS* (BROMOVIRIDAE)

Cucumovirus belongs to the family *Bromoviridae*, the type species is *Cucumber mosaic virus* (CMV) is a multicomponent virus having single-stranded positive sense RNA. RNAs 1 and 2 encode viral genome replication, whereas, the RNA 3 is associated with the movement protein and coat protein. The strains of cucumoviruses have been mainly classified in to two sub-groups, that is, sub-groups I and II on the basis of serological properties and nucleotide sequence homology (Palukaitis et al., 1992; Madhubala et al., 2005; Oraby et al., 2008). Based on phylogenetic analysis, the subgroup I has been further divided in to two groups (IA and IB) by Roossinck (1999). The RNA 2 produces a 2b protein that is associated with suppression of RNA interference (RNAi). Virions are uniform sized icosahedral. Besides the *Cucumber mosaic virus*, other cucumovirus species have narrower host ranges and *Peanut stunt virus* (PSV) is confined to legumes and solanaceous hosts, whereas, *Tomato aspermy virus* (TAV) largely infects members of compositeae and solanaceous hosts. The major cucumovirus species including, *Cucumber mosaic virus* subgroup I strain, *Cucumber mosaic virus* subgroup II strain, PSV, TAV (King et al., 2012). Cucumoviruses are having the very broad host range of ~1000 species belonging to agricultural and horticultural crops (Roossinck, 1999). Members of Cucumovirus are transmitted in a non-persistent manner by >75 aphid species in >30 genera (Palukaitis et al., 1992).

10.4.6 GENUS: *POLEROVIRUS* (LUTEOVIRIDAE)

Polerovirus belongs to family *Luteoviridae* (Type species: *Potato leaf roll virus*). Poleroviruses have non-enveloped, monopartite and isometric virus particles of 24–26 nm diameter encapsidated in a ssRNA (+) molecule of 5.3–5.9 kb (Mayo & Ziegler-Graff, 1996) with a small viral genome-linked protein (VPg) at the 5' end and having no poly (A) tail or tRNA-like structure at the 3' end (van der Wilk et al., 1997). Poleroviruses are restricted to the phloem and solely depend on their aphid vectors for transmission but the mechanical inoculation through biolistic inoculation procedure *Beet western yellows virus* (BWYV) and *Potato leafroll virus* (PLRV) has been proved in several host plants (Hoffmann et al., 2001). The genus *Polerovirus* is comprised about 17 species (King et al., 2012). Polerovirus species infecting several crops, *viz.*, beet, carrot, cereals, chickpea, cucurbits, potato, tobacco, turnip, and sugarcane. Polerovirus infecting sugarcane including, *Sugarcane yellow leaf virus* (SCYLV) causing Yellow leaf disease (YLD) in sugarcane. SCYLV is known to be transmitted by *Melanaphis sacchari*, *Rhopalosiphum* maidis, and *R. rufiabdominalis*. In India, YLD is a serious threat to the sugarcane crop and it adversely affects some of the ruling varieties, *viz.*, Co 6304, CoC 86062, CoV 92102, CoV 94102, CoV 06356, CoA 92081, CoA 05323, CoC 92061, Co 86032, and Co 94012 (Viswanathan et al., 2002; Viswanathan & Rao, 2011). In India, the disease incidence of YLD has been recorded from different states during 2012–2013 under All India Co-ordinated Research Project on Sugarcane (AICRP-S), *viz.*, traces to 1–4% in Madhya Pradesh (Hoshangabad, Betul, and Bankhedi), 1–5% in Maharashtra (Pune and Kolhapur), 10 to 60% in Andhra Pradesh (Vijayanagaram, Visakhapatnam, Karimnagar, Nizamabad, East Godavari, and Chittoor), 2–11% in Karnataka (Haveri, Bagalkot, Dharwad, and Belgaon), 15–30% in Tamil Nadu (Pondicherry, Villupuram, Seithiathope, Mundiyapakkam) and traces in Haryana (Shahabad and Rohtak), Uttarakhand (Sitarganj, Kiccha, Gadarpur, Kashipur, and Haridwar), Gujarat (Bardoli), Kerala (Iramallikk, Thiruvandoor and Vallamkulam). The symptoms of this disease are visualized at 6–8 months stage of the crop. In U.S.A., up to 15% yield loss has been reported, whereas, as high as 50% yield loss in Brazil has been recorded due to YLD (Grisham et al., 2001; Vega et al., 1997). In India, preliminary work on detection and diagnosis using RT-PCR and qRT-PCR has been done (Viswanathan et al., 2009) including characterization of the complete genome (Chinnaraja et al., 2013) and impact of SCYLV infection on the physiological efficiency and growth parameters under tropical conditions has been well studied by Viswanathan et al. (2014).

10.4.7 GENUS: *POTYVIRUS* (POTYVIRIDAE)

Potyvirus belongs to family *Potyviridae* and it was named with the type species *Potato virus Y* (PVY) having worldwide distribution. The genus *Potyvirus* is one of the largest genera in this family and among the plant viruses. The members of potyviruses infect large numbers of monocots and dicot plants. Potyviruses are transmitted by aphids in a non-persistent manner. Virions are flexuous filaments rods of 680–900 nm length and 11–13 nm wide, for example, the electron microscopic image of *Sugarcane mosaic virus* (SCMV) is presented in Figure 10.2 (b). The virions contain positive sense ssRNA, about 10 kb in size. Genome is composed of terminal untranslated regions and a translated large polyprotein of ORF overlapped with another ORF. Potyviruses have been transmitted by mechanical inoculations, certain species are not at all transmitted by aphids, whereas, few are known to be seed borne. The genus is comprised >140 distinct species and 30 tentative species (King et al., 2012). In India, at least 40 potyvirus species have been known to infect members of *Solanaceae*, Leguminosae, and *Cucurbitaceae* (Sharma et al., 2014).

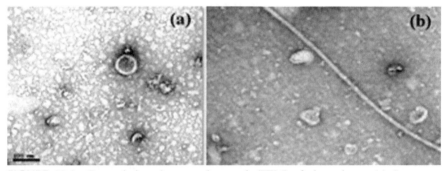

FIGURE 10.2   Transmission electron micrograph (TEM) of plant viruses (a) *Sugarcane mosaic virus* (SCMV) infecting sugarcane (*Sachharum officinarum*) (b) *Groundnut bud necrosis virus* (GBNV) infecting chrysanthemum (*Chrysanthemum indicum*) and in India.

10.5   ECONOMICALLY IMPORTANT VIRUSES TRANSMITTED BY WHITEFLIES

10.5.1   GENUS: *BEGOMOVIRUS* (GEMINIVIRIDAE)

Begomovirus belongs to the family *Geminiviridae* (type member: *Bean golden mosaic virus*; BGMV), the largest genus among the four genera of this family. Currently there are ~200 virus species of begomoviruses reported from various plant samples. The members of begomoviruses contain ssDNA as genetic material which is encapsidated in twinned quasi-icosahedral (geminate) virions. Besides begomoviruses, this family comprised other genera, viz., *Mastrevirus*, *Curtovirus*, and *Topucovirus* which differ in genome organization, natural hosts, and mode of transmission (Fauquet et al., 2008). Based on the genome organization, begomoviruses are classified as bipartite and monopartite, the bipartite containing ssDNA of two segments, DNA-A and DNA-β with similar size of 2.5 to 2.7 kb. Whereas, monopartite bego-moviruses contain ssDNA of DNA-A alone and are known to associated with betasatellites and are solely dependent on the helper component for its repli-cation, encapsidation, and transmission. Similarly, some of the members are referred as alfa and betasatellite complexes (Nawaz-ul-Rahman & Fauquet, 2009). Members of begomoviruses particularly phloem limited are trans-mitted by *B. tabaci* in a persistent manner (Jones, 2003). In India, more than 30 begomoviruses are known to infect several dicotyledonous plants and this is reviewed by Borah and Dasgupta (2012).

10.5.2   GENUS: *CRINIVIRUS* (CLOSTEROVIRIDAE)

Crinivirus belongs to the family *Closteroviridae*, members of this family possess linear ssRNA with the genome size of up to 20 kb. are among the largest RNA viruses infecting plants. Genome is encapsidated in very long and flexuous particles and it comprised two molecules of RNA-1 and RNA-2. The RNA-1 encodes for proteins associated with replication whereas, RNA-2 encodes for proteins associated for the encapsidation, movement, and vector transmission (Karasev, 2000; Martelli et al., 2002; Livieratos et al., 2004). Besides crinivi-ruses, this family is composed of *Closterovirus* and *Ampelovirus* that are trans-mitted by aphids (Martelli et al., 2002). *Crinivirus* is the only genus of this family transmitted by whiteflies (*B. tabaci, Trialeurodes vaporariorum* and *T. abutiloneus*) in semi-persistent manner (Wisler et al., 1998).

10.5.3   GENUS: *IPOMOVIRUS* (POTYVIRIDAE)

Ipomovirus is the only genus of the family *Potyviridae* transmitted by white-flies (*Bemisia tabaci*) in semi-persistent manner, whereas, the other genera including poty- and machluraviruses are transmitted by aphids, rymo- and tritimoviruses are transmitted by mites, and bymoviruses are knowm to be transmitted by plasmodiophorales (Jones, 2003; Berger et al., 2005; Carstens et al., 2010). Ipomoviruses transmitted by whiteflies include four species, *viz., Cassava brown streak virus* (CBSV), *Cucumber vein yellowing virus* (CVYV), *Squash vein yellowing virus* (SqVYV) and *Sweet potato mild mottle virus* (SPMMV) (King et al., 2012).

10.6   VIRUSES TRANSMITTED BY THRIPS

10.6.1   GENUS: *TOSPOVIRUS* (BUNYAVIRIDAE)

Tospoviruses (type member: *Tomato spotted wilt virus*; TSWV) are the major constraints for 15 monocotyledonous and 69 dicotyledonous crop families worldwide (Parrella et al., 2003). More than 20 tospoviruses have been identified globally and known to transmit by through least 14 thrips species (Ullman, 1997; Jones, 2005; Pappu et al., 2009; Ciuffo et al., 2010; Hassani-Mehraban et al., 2010). Thrips (order: Thysanoptera and family: *Thripidae*) transmit tospoviruses in a propagative manner (Ullman, 1997). Larval and adult thrips are known to feed actively on virus infected plants to acquire and transmit virus. Late instar larvae and adults only transmit virus after the latent period. Likewise, adults acquire virus but do not transmit it due to insufficient multiplication in midgut. Transovarial transmission is not known to occur in thrips (Wijkamp et al., 1996a, b; Ullman, 1997; Whitfield et al., 2005; Persley et al., 2006). In India, six tospoviruses are known to occur including, *Capsicum chlorosis virus* (CaCV) during 2006 in chilli (Krishnareddy et al., 2008), *Groundnut bud necrosis virus* (GBNV) Fig. 10. 2 (b) would be here during 1968 in groundnut (Reddy et al., 1992), *Groundnut yellow spot virus* (GYSV) during 1991 in groundnut (Satyana-rayana et al., 1998), *Iris yellow spot virus* (IYSV) during 2002 and 2003 in onion (Ravi et al., 2006) and *Tomato spotted wilt virus* (TSWV) during 2013 in chrysanthemum (Renukadevi et al., 2015) and *Watermelon bud necrosis virus* (WBNV) during 1991 and 1992 (Jain et al., 1998). WBNV and GBNV cause bud necrosis disease and are transmitted by the same vector, *Thrips palmi* (Vijayalakshmi, 1994; Rebijith et al., 2012). The detailed descriptions about different thrips species and their mount have been extensively reviewed by Jones (2005) and Riley et al. (2011), whereas, the distribution of tospoviruses transmitting thrips species has been mentioned in Table 10.3. Tospoviruses are enveloped isometric in shapes having tripartite genome of small (S), medium (M), and large (L) ssRNA. The L RNA encoding replicase associated protein in negative sense, whereas, the M RNA encoding movement protein (NSm) and glycoprotein (Gn/Gc) and S RNA encoding nonstructural (NSs) and nucleocapsid protein (N), M and S RNA each contain two genes in an ambisense orientation, whereas, the L RNA harbors single ORF in viral complementary strand and the L protein is RNA dependent RNA polymerase (King et al., 2012) (Fig. 10.3).

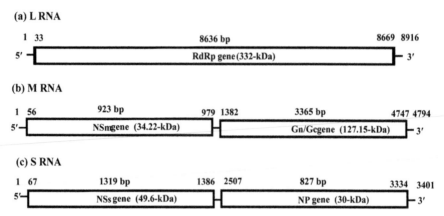

FIGURE 10.3 Schematic representation of genome organization of *watermelon bud necrosis virus* (WBNV), a distinct tospovirus species transmitted by thrips (*Thrips palmi*) in India.

10.7 CONCLUSION

Host preference by insect vectors mainly depends on their feeding behavior, which is generally governed by the texture of leaf and stem surfaces followed by the certain plant exudates or chemicals. In the process of host selection insects practice various phenomena like, adherence to host surface, probing, stylets penetration, salivation, and finally ingestion. During these practices of host selection naturally insects acquire the viruses and are transmitted to healthy plants. It is presumed that the pest of one crop may transmit the associated virus with the host and insect. Insect transmitted viruses are reported from several new hosts day by day due to the shift/preference of insects to new hosts may be due to non-availability of the main natural hosts for their survival. There is no doubt that certain biological and molecular interaction between virus and vector decides the possible spread and multiplication of viruses. The triangle of interaction of virus–vector–hosts and *vice versa* needs to establish to determine the etiology, and proper understanding of new or reported virus species infecting new natural hosts. Insect transmission studies with the help of certain automated video tracking tools will definitely determine the feeding or non-feeding behavior to different experimental, agricultural, and horticultural crops. This non-feeding behavior of various insects to economically important crops may serve as one of the option for identification of source of natural resistance in order to develop insect resistant varieties.

KEYWORDS

- aphid
- host preference
- plant viruses
- thrips
- whitefly

REFERENCES

Abo, M. E.; Alegbejo, M. D.; Sy, A. A.; Sere, Y. Retention and Transmission of *Rice yellow mottle virus* (RYMV) by Beetle Vectors in Cote Divoire. *Agron. Afr.* **2004,** *16* (1), 71–75.

Adams, M. J.; Antoniw, J. F.; Mullins, J. G. L. Plant Virus Transmission by Plasmodiophorid Fungi is Associated with Distinctive Transmembrane Regions of Virus-Encoded Proteins. *Arch. Virol.* **2001,** *146,* 1139–1153.

Agarwal, B. K. Notes on Some Aphids (Homoptera: Aphididae) Affecting Economically Important Plants on Bhutan. *Indian Agric.* **1985,** *27,* 261–262.

Agrios, G. N. *Plant Pathology*; 5th Ed., Elsevier Acadmic Press: U. S. A., 2005.

Akhter, M. S.; Holkar, S. K.; Akanda, A. M.; Mandal, B.; Jain, R. K. First Report of *Groundnut bud necrosis virus* in Tomato in Bangladesh. *Plant Dis.* **2012,** *96*(6), 917.

Akram, M.; Naimuddin. First Report of *Groundnut bud necrosis virus* Infecting Pea (*Pisum sativum*) in India. *New Dis. Rep.* **2010,** *21,* 10.

Akram, M.; Pratap, A.; Malviya, N.; Yadav, P. First Report of *Groundnut bud necrosis virus* Infecting Wild Species of Vigna, Based on NP Gene Sequence Characteristics. *Phyto-pathol. Mediterranea.* **2013,** *52*(3), 532–540.

Akram, Md.; Naimuddin. Characterization of *Groundnut bud necrosis virus* Infecting French Bean. *J. Food Legumes.* **2012,** *25*(1), 54–57.

Almasi, A.; Csillery, G.; Csomor, Z.; Nemes, K.; Palkovics, L.; Salanki, K.; Tobias, I. Phylogenetic Analysis of *Tomato spotted wilt virus* (TSWV) Nss Protein Demonstrates the Isolated Emergence of Resistance-Breaking Strains in Pepper. *Virus Genes.* **2014,** *50,* 71–78.

Al-Saleh, M. A.; Al-Shahwan, I. M.; Amer, M. A.; Shakeel, M. T.; Ahmad, M. H.; Kamran, A.; Katis, N. I. First Report of *Tomato spotted wilt virus* in Lettuce Crops in Saudi Arabia. *Plant Dis.* **2014,** *98*(11), 1591–1591.

Anon. Plant Protection Compendium, CAB International, Wallingford, UK, 2002.

Avila, Y.; Stavisky, J.; Hague, S.; Funderburk, J.; Reitz, S.; Momol, T. Evaluation of *Frankliniella bispinosa* (Thysanoptera: Thripidae) as a Vector of the *Tomato spotted wilt virus* in Pepper. *Fla. Entomol.* **2006,** *89,* 204–207.

Behncken, G. M. Some Properties of a Virus from *Galinsoga parviflora*. *Aust. J. Biol. Sci.* **1970,** *23*(2), 497–502.

Berger, P. H.; Adams, M. J.; Barnett, O. W.; Brunt, A. A.; Hammond, J.; Hill, J. H.; Jordan, R. L.; Kashiwazaki, S.; Rybicki, E.; Spence, N.; Stenger, D. C.; Ohki, S. T.; Uyeda, I.; Van Zaayen, A.; Valkonen, J.; Vetten, H. J. Family Potyviridae. In *Virus Taxonomy: Eighth Report of the International Committee on Taxonomy of Viruses;* Fauquet, C. M., Mayo, M. A., Maniloff, J., Desselberger, U., Ball, L. A., Eds.; Academicpress: San Diego, 2005; pp 819–841.

Berlinger, M. J. Host Plant Resistance To *Bemisia tabaci. Agric. Ecosyst. Environ.* **1986,** *17,* 69–82.

Bhargava, K. S.; Joshi, R. D.; Rizvi, M. A. Some Observations on the Insect Transmission of *Sugarcane mosaic virus. Sugarcane Pathol. Newslett.* **1971,** *7,* 20–22.

Blackman, R. L.; Eastop, V. F. *Aphids on the World's Crops: An Identification and Information Guide;* Wiley: New York, 1984.

Boiteux, L. S.; De Avila, A. C.; Dutra, W. P. Natural Infection of Melon by a *Tospovirus* in Brazil. *Plant Dis.* **1994,** *78,* 102.

Bol, J. F. *Alfalfa mosaic virus* and Ilarviruses: Involvement of Coat Protein in Multiple Steps of the Replication Cycle. *J. Gen. Virol.* **1999,** *80*(5), 1089–1102.

Bol, J. F.; Van Vloten-Doting, L.; Jaspars, E. M. J. A Functional Equivalence of Top Component a RNA in the Initiation of Infection by *Alfalfa mosaic virus. Virol.* **1971,** *46,* 73–85.

Borah, B. K.; Dasgupta, I. Begomovirus Research in India: A Critical Appraisal and the Way Ahead. *J. Biosci.* **2012,** *37*(4), 791–806.

Bosque-Perez, N. A.; Eigenbrode, S. D. The Influence of Virus-Induced Changes in Plants on Aphid Vectors: Insights from *Luteovirus* Pathosystems. *Virus Res.* **2011,** *159*(2), 201–205.

Braithwaite, B. M.; Blake, C. D. Tomato Yellow Top Virus: Its Distribution, Characteristics and Transmission by the Aphid *Macrosiphum euphorbiae* (Thom.). *Crop Pasture Sci.* **1961,** *12*(6), 1100–1107.

Brault, V.; Uzest, M.; Monsion, B.; Jacquot, E.; Blanc, S. Aphids as Transport Devices for Plant Viruses. *Comptes. Rendus. Biologies.* **2010,** *333*(6), 524–538.

Brunt, A. A.; Crabtree, K.; Dallwitz, M. J.; Gibbs, A. J.; Watson, L. *Viruses of Plants, Descriptions and Lists from the VIDE Database;* CAB International: Wallingford, UK, 1996.

Capote, N.; Perez-Panades, J.; Monzo, C.; Carbonell, E.; Urbaneja, A.; Scorza, R.; Ravelonandro, M.; Cambra, M. Assessment of the Diversity and Dynamics of *Plum pox virus* and Aphid Populations in Transgenic European Plums under Mediterranean Conditions. *Trans. Res.* **2008,** *17*(3), 367–377.

Carmo-Sousa, M.; Moreno, A.; Garzo, E.; Fereres, A. A Non-Persistently Transmitted-Virus Induces a Pull–Push Strategy in its Aphid Vector to Optimize Transmission and Spread. *Virus Res.* **2014,** *186,* 38–46.

Carstens, E. B. Ratification Vote on Taxonomic Proposals to the International Committee on Taxonomy of Viruses. *Arch. Virol.* **2010,** *155,* 133–146.

Chinnaraja, C.; Viswanathan, R.; Karuppaiah, R.; Bagyalakshmi, K.; Malathi, P.; Parameswari, B. Complete Genome Characterization of *Sugarcane yellow leaf virus* from India: Evidence for RNA Recombination. *Eur. J. Plant Pathol.* **2013,** *135,* 335–349.

Ciuffo, M.; Mautino, G. C.; Bosco, L.; Turina, M.; Tavella, L. Identification Of *Dictyothrips betae* as the Vector of *Polygonum* L. R. *Ring spot virus. Ann. Appl. Biol.* **2010,** *157,* 299–307.

Conti, M. Transmission of Plant Viruses by Leafhoppers and Planthoppers, In *The Leafhoppers And Planthoppers;* Nault, L. R., Rodriguez, J. G., Eds.; Wiley: New York, 1985; pp 289–307.

Coombe, P. E. Visual Behaviour of the Greenhouse Whitefly, *Trialeurodes vaporariorum*. *Physiol. Entomol.* **1982,** *7,* 243–251.

Cortes, I.; Livieratos, I. C.; Derks, A.; Peters, D.; Kormelink, R. Molecular and Serological Characterization of *Iris yellow spot virus*, a New and Distinct *Tospovirus* Species. *Phytopathol.* **1998,** *88*(12), 1276–1282.

Cortez, I.; Saaijer, J.; Wongjkaew, K. S.; Pereira, A. M.; Goldbach, R.; Peters, D.; Kormelink, R. Identification and Characterization of a Novel *Tospovirus* Species Using a New RT-PCR Approach. *Arch. Virol.* **2001,** *146*(2), 265–278.

Davis, J. A.; Radcliffe, E. B.; Ragsdale, D. W. Soybean Aphid, *Aphis glycines* Matsumura, a New Vector of *Potato virus Y* in Potato. *Am. J. Potato Res.* **2005,** *82*(3), 197–201.

De Bordon, C. M.; Gracia, O.; De Santis, L. Survey of Thysanoptera Occurring on Vegetable Crops as Potential *Tospovirus* Vectors in Mendoza, Argentina. *Rev. Soc. Entomol. Argent.* **1999,** *58,* 59–66.

De Oliveira, A. S.; Melo, F. L.; Inoue-Nagata, A. K.; Nagata, T.; Kitajima, E. W.; Resende, R. O. Characterization of *Bean necrotic mosaic virus*: A Member of a Novel Evolutionary Lineage within the Genus *Tospovirus*. *PLoS ONE.* **2012,** *7*(6), E38634.

Denmark, H. A. Sugarcane Aphids in Florida. Florida Department of Agriculture and Consumer Services, Division of Plant Industry. *Entomol. Circular.* 1988, 302.

Dheepa, R.; Paranjothi, S. Transmission of *Cucumber mosaic virus* (CMV) Infecting Banana by Aphid and Mechanical Methods. *Emirates J. Food Agric.* **2010,** *22*(2), 117–129.

Dietrich, C. H. Overview of the Phylogeny, Taxonomy and Diversity of the Leafhopper (Hemiptera: Auchenorrhyncha: Cicadomorpha: Membracoidea: Cicadellidae) Vectors of Plant Pathogens. In *Proceedings of the International Symposium on Insect Vectors and Insect-Borne Diseases;* Special Publication of TARI: Taichung, Taiwan, ROC, 2013; Vol. 173, pp 47–70.

Dong, J. H.; Cheng, X. F.; Yin, Y. Y.; Fang, Q.; Ding, M.; Li, T. T.; Zhang, Z. K. Characterization of *Tomato zonate spot virus*, a New *Tospovirus* in China. *Arch. Virol.* **2008,** 153(5), 855–864.

Dong, J. H.; Yin, Y. Y.; Fang, Q.; Mcbeath, J. H.; Zhang, Z. K. A New *Tospovirus* Causing Chlorotic Ringspot on *Hippeastrum sp.* in China. *Virus Genes.* **2013,** *46*(3), 567–570.

Döring, T. F.; Kirchner, S. M.; Kuhne, S.; Saucke, H. Response of Alate Aphids to Green Targets on Coloured Backgrounds. *Entomol. Exp. Appl.* **2004,** *113*(1), 53–61.

Döring, T. F.; Chittka, L. Visual Ecology of Aphids- A Critical Review on the Role of Colours in Host Finding. *Arthropod Plant Interact.* **2007,** *1*(1), 3–16.

Duffey, S. S. Plant Grandular Trichomes: Their Partial Role in Defence against Insects. In *Insects and Plant Surface;* Jupiner, B. E, Southwood, T. R. E., Eds.; Edward Arnold: London, 1986; pp 173–183.

Eastop, V. F. The Biology of the Principal Aphid Virus Vectors. In *Plant Virus Epidemiology*; Plumb, R. T., Thresh, J. M., Eds.; Blackwell Scientific Publications: Oxford, 1983; pp 115–132.

Eigenbrode, S. D.; Ding, H.; Shiel, P.; Berger, P. H. Volatiles from Potato Plants Infected with *Potato leafroll virus* Attract and Arrest the Virus Vector, *Myzus persicae* (Homoptera: Aphididae). *Proc. Biol. Sci.* **2002,** *269*(1490), 455–460.

Fauquet, C. M.; Briddon, R. W.; Brown, J. K.; Moriones, E.; Stanley, J.; Zerbini, M.; Zhou, X.. Geminivirus Strain Demarcation and Nomenclature. *Arch. Virol.* **2008,** *153*(4), 783–821.

Fereres, A.; Moreno, A. Behavioural Aspects Influencing Plant Virus Transmission by Homopteran Insects. *Virus Res.* **2009,** *141*(2), 158–168.

Francki, R. I. B.; Milne, R. G.; Hatta, T. *Atlas of Plant Viruses;* CRC Press: Boca Raton, FL, USA., 1985.

Fujisawa, I.; Tanaka, K.; Ishii, M. Tomato Spotted Wilt Virus Transmissibility by Three Species of Thrips, *Thrips setosus*, *Thrips tabaci* and *Thrips palmi. Ann. Phytopathol. Soc. Japan.* **1988,** *54,* 392.

Fulton, J. P.; Gergerich, R. C.; Scott, H. A. Beetle Transmission of Plant Viruses. *Ann. Rev. Phytopathol.* **1987,** *25*(1), 111–123.

Gopal, K.; Krishna Reddy, M.; Reddy, D. V. R.; Muniyappa, V. Transmission of Peanut Yellow Spot Virus (PYSV) by Thrips, *Scirtothrips dorsalis* Hood in Groundnut. *Arch. Phytopathol. Plant Prot.* **2010,** *43*(5), 421–429.

Greber, R. S.; Klose, M. J.; Milne, J. R.; Teakle, D. S. Transmission of Prunus Necrotic Ringspot Virus Using Plum Pollen and Thrips. *Ann. Appl. Biol.* **1991,** *118*(3), 589–593.

Grigoras, I.; Gronenborn, B.; Vetten, H. J.; First Report of a Nanovirus Disease of Pea in Germany. *Plant Dis.* **2010,** *94,* 642.

Grigoras, I.; Timchenko, T.; Katul, L.; Grande-Perez, A.; Vetten, H. J.; Gronenborn, B. Reconstitution of Authentic Nanovirus from Multiple Cloned Dnas. *J. Virol.* **2009,** *83,* 10778–10787.

Grisham, M. P.; Pan, Y. B.; Legendre, B. L.; Godshall, M. A.; Eggleston, G. Effect of Sugarcane Yellow Leaf Syndrome on Sugarcane Yield and Juice Quality. *Proc. Intern. Soc. Sugarcane Technol.* **2001,** *24,* 434–438.

Gutierrez, S.; Michalakis, Y.; Blanc, S. Virus Population Bottlenecks during Within-Host Progression and Host-To-Host Transmission. *Curr. Opin. Virol.* **2012,** *2*(5), 546–555.

Hamilton, K. A.; Zack, R. S. Systematics and Range Fragmentation of the Nearctic Genus *Errhomus* (Rhynchota: Homoptera: Cicadellidae). *Ann. Entomol. Soc. Am.* **1999,** *92*(3), 312–354.

Hamilton, K. G. A. The Ground-Dwelling Leafhoppers Sagmatiini and Myerslopiidae (Rhynchota: Homoptera: Membracoidea). *Invertebr. Taxon.* **1999,** *13,* 207–235.

Harris, K. F.; Maramorosch, K. Eds. *Aphids as Virus Vectors*; Elsevier: USA, 2014.

Harrison, B. D. Advances in Geminivirus Research. *Ann. Rev. Phytopathol.* **1985,** *23,* 55–82.

Harrison, B. D.; Swanson, M. M.; Fargette, D. Begomovirus Coat Protein: Serology, Variation and Functions. *Physiol. Mol. Plant Pathol.* **2002,** *60*(5), 257–271.

Hassani-Mehraban, A.; Botermans, M.; Verhoeven, J. T. J.; Meekes, E.; Saaijer, J.; Peters, D.; Kormelink, R. A Distinct Tospovirus Causing Necrotic Streak on *Alstroemeria sp.* in Colombia. *Arch. Virol.* **2010,** *155*(3), 423–428.

Hodge, S.; Powell, G. Conditional Facilitation of an Aphid Vector, *Acyrthosiphonpisum*, by the Plant Pathogen, Pea Enation Mosaic Virus. *J. Insect Sci.* **2010,** *10*(1), 155.

Hoffmann, K.; Verbeek, M.; Romano, A.; Dullemans, A. M.; Van Den Heuvel, J. F. J. M.; Van Der Wilk, F. Mechanical Transmission of Poleroviruses. *J. Virol. Meth.* **2001,** *91*(2), 197–201.

Hogenhout, S. A.; Ammar, E. D.; Whitfield, A. E.; Redinbaugh, M. G. Insect Vector Interactions with Persistently Transmitted Viruses. *Ann. Rev. Phytopathol.* **2008,** *46,* 327–359.

Hsu, C. L.; Hoepting, C. A.; Fuchs, M.; Shelton, A. M.; Nault. B. A. Temporal Dynamics of Iris Yellow Spot Virus and its Vector, *Thrips tabaci* (Thysanoptera: *Thripidae*) in Seeded and Transplanted Onion Fields. *Environ. Entomol.* **2010,** *39,* 266–277.

Hu, Z. Z.; Feng, Z. K.; Zhang, Z. J.; Liu, Y. B.; Tao, X. R. Complete Genome Sequence of a Tomato Spotted Wilt Virus Isolate From China and Comparison to other TSWV Isolates of Different Geographic Origin. *Arch. Virol.* **2011,** *156*(10), 1905–1908.

Hull, R. *Mathew's Plant Virology;* Academic Press: San Diego, USA, 2002.

Hull, R.; Harper, G.; Lockhart, B. Viral Sequences Integrated into Plant Genomes. *Trends Plant Sci.* **2000,** *5*(9), 362–365.

Irwin, M. E.; Kampmeier, G. E.; Weisser, W. W. Aphid Movement: Process and Consequences. In *Aphids as Crop Pests*; CABI: Cambridge, USA, 2007; pp 153.

Isac, M.; Preda, S.; Marcu, M. Aphid Species-Vectors of Plum Pox Virus. *Acta. Virol.* **1998,** *2*(4), 233–234.

Jain, R. K.; Pappu, H. R.; Pappu, S. S.; Krishna Reddy, M.; Vani, A. Watermelon Bud Necrosis Tospovirus is a Distinct Virus Species Belonging to Serogroup IV. *Arch. Virol.* **1998,** *143*(8), 1637–1644.

Jaspars, E. M. J. Interaction of Alfalfa Mosaic Virus Nucleic Acid and Protein. In *Molecular Plant Virology;* Davies, J. W., Ed.; CRC Press: Boca Raton, 1985; pp 155–225.

Jones, D. R. Plant Viruses Transmitted By Thrips. *Eur. J. Plant Pathol.* **2005,** *113*(2), 119–157.

Jones, D. R. Plant Viruses Transmitted By Whiteflies. *Eur. J. Plant Pathol.* **2003,** *109*(3), 195–219.

Jones, R. A. Plant Virus Emergence and Evolution: Origins, New Encounter Scenarios, Factors Driving Emergence, Effects of Changing World Conditions, and Prospects for Control. *Virus Res.* **2009,** *141*(2), 113–130.

Jousselin, E.; Cruaud, A.; Genson, G.; Chevenet, F.; Foottit, R. G.; Cœur D'acier, A. Is Ecological Speciation a Major Trend in Aphids? Insights from a Molecular Phylogeny of the Conifer-Feeding Genus *Cinara. Frontiers Zool.* **2013,** *10*(1), 56.

Kanavaki, O. M.; Margaritopoulos, J. T.; Katis, N. I.; Skouras, P.; Tsitsipis, J. A. Transmission of Potato Virus Y in Tobacco Plants by *Myzus persicae nicotianae* and *M. persicae* s. str. *Plant Dis.* **2006,** *90*(6), 777–782.

Kaper, J. M.; Waterworth, H. E. Cucummoviruses. In *Handbook of Plant Virus Infections: Comparative Diagnosis*; Kurstak, E., Ed.; Elsevier: North Holland, 1981; pp 257–332.

Karasev, A. V. Genetic Diversity and Evolution of Closteroviruses. *Ann. Rev. Phytopathol.* **2000,** *38*, 293–324.

Kato, K.; Hanada, K.; Kameya-Iwaki, M. Melon Yellow Spot Virus: A Distinct Species of the Genus Tospovirus Isolated from Melon. *Phytopathol.* **2000,** *90*, 422–426.

Kawada, K. Studies on Host Selection, Development and Reproduction of *Melanaphis sacchari* (Zehntner). *Bull. Res. Inst. Bioresources.* **1995,** *3*, 5–10.

King, A. M.; Adams, M. J.; Lefkowitz, E. J. Eds. Virus Taxonomy: Ninth Report of the International Committee on Taxonomy of Viruses. Elsevier: USA, 2012, 9.

Klingauf, F. A. Feeding, Adaptation and Excretion. In *Aphids: Their Biology, Natural Enemies, and Control*; Minks, A. K., Harrewijn, P., Ed.; Elsevier: Amsterdam, 1987; pp 225–253.

Klose, M. J.; Sdoodee, R.; Teakle, D. S.; Milne, J. R.; Greber, R. S.; Walter, G. H. Transmission of Three Strains of Tobacco Streak Ilarvirus by Different Thrips Species Using Virus infected Pollen. *J. Phytopathol.* **1996,** *144*, 281–284.

Kloth, K. J.; Ten Broeke, C. J.; Thoen, M. P.; Hanhart-Van Den Brink, M.; Wiegers, G. L.; Krips, O. E.; Noldus, L. P. J. J.; Dicke, M.; Jongsma, M. A. High-Throughput Phenotyping of Plant Resistance to Aphids by Automated Video Tracking. *Plant Meth.* **2015,** *11*(1), 4.

Koschier, E. H.; Sedy, K. A. Labiate Essential Oils Affecting Host Selection and Acceptance of *Thrips tabaci* Lindeman. *Crop Prot.* **2003,** *22*(7), 929–934.

Kuhn, C. W. Symptomatology, Host Range, and Effect on Yield of a Seed-Transmitted Peanut Virus. *Phytopathol.* **1965,** *55*(8), 880–884.

Labonne, G.; Yvon, M.; Quiot, J. B.; Avinent, L.; Llacer, G. Aphids as Potential Vectors of Plum Pox Virus: Comparison of Methods of Testing and Epidemiological Consequences. In XVI International Symposium on Fruit Tree Virus Diseases. *Acta. Hortc.* **1995,** *386,* 207–218.

Landis, D. A.; Van Der Werf, W. Early-Season Predation Impacts the Establishment of Aphids and Spread of Beet Yellows Virus in Sugar Beet. *Entomophaga.* **1997,** *42*(4), 499–516.

Lecoq, H.; Ravelonandro, M.; Wipf-Scheibel, C.; Monsion, M.; Raccah, B.; Dunez, J. Aphid Transmission of a Non-Aphid-Transmissible Strain of Zucchini Yellow Mosaic Potyvirus from Transgenic Plants Expressing the Capsid Protein of Plum Pox Potyvirus. *Mol. Plant-microbe Interact.* **1993,** *6*(3), 403–406.

Lei, H.; Tjallingii, W. F.; Lenteren, J. V. Probing and Feeding Characteristics of the Greenhouse Whitefly in Association with Host-Plant Acceptance and Whitefly Strains. *Entomol. Exp. Appl.* **1998,** *88*(1), 73–80.

Lian, S.; Lee, J. S.; Cho, W. K.; Yu, J.; Kim, M. K.; Choi, H. S.; Kim, K. H. Phylogenetic and Recombination Analysis of Tomato Spotted Wilt Virus. *PLoS ONE.* **2013,** 8(5), e63380.

Liu, Y.; Huang, C. J.; Tao, X. R.; Yu, H. Q. First Report of Tomato Zonate Spot Virus in *Iris tectorum* in China. *Plant Dis.* **2015,** *99*(1), 164–164.

Liu, Y.; Lu, X.; Zhi, L.; Zheng, Y.; Chen, X.; Xu, Y.; Li, Y. Calla Lily Chlorotic Spot Virus from Spider Lily (*Hymenocallis litteralis*) and Tobacco (*Nicotiana tabacum*) in the South-West of China. *J. Phytopathol.* **2012,** *160*(4), 201–205.

Livieratos, I. C.; Eliasco, E.; Müller, G.; Olsthoorn, R. C. L.; Salazar, L. F.; Pleij, C. W. A.; Coutts, R. H. Analysis of the RNA of Potato Yellow Vein Virus: Evidence for a Tripartite Genome and Conserved 3′-Terminal Structures Among Members of the Genus *Crinivirus.* *J. Gen. Virol.* **2004,** *85*(7), 2065–2075.

Madhubala, R.; Bhadramurthy, V.; Bhat, A. I.; Hareesh, P. S.; Retheesh, S. T.; Bhai, R. S. Occurrence of Cucumber Mosaic Viruson Vanilla (*Vanilla planifolia* Andrews) in India. *J. Biosci.* **2005,** *30*(3), 339–350.

Mandal, B.; Jain, R. K.; Chaudhary, V.; Varma, A. First Report of Natural Infection of *Luffa acutangula* by Watermelon Bud Necrosis Virus in India. *Plant Dis.* **2003,** *87,* 598.

Mandal, B.; Jain, R. K.; Krishnareddy, M.; Krishna Kumar, N. K.; Ravi, K. S.; Pappu, H. R. Emerging Problems of Tospoviruses (Bunyaviridae) and their Management in the Indian Subcontinent. *Plant Dis.* **2012,** *96*(4), 468–479.

Margaria, P.; Ciuffo, M.; Rosa, C.; Turina, M. Evidence of a Tomato Spotted Wilt Virus Resistance-Breaking Strain Originated through Natural Reassortment between two Evolutionary Distinct Isolates. *Virus Res.* **2015,** *196,* 157–161.

Margaria, P.; Miozzi, L.; Ciuffo, M.; Pappu, H.; Turina, M. The Complete Genome Sequence of Polygonum Ringspot Virus. *Arch. Virol.,* **2014,** *159*(11), 3149–3152.

Martelli, G. P.; Candresse, T. Closteroviridae. Els. **2014.** Online first: DOI: 10.1002/9780470015902.a0000747.pub3.

Martelli, G. P.; Agranovsky, A. A.; Bar-Joseph, M.; Boscia, D.; Candresse, T.; Coutts, R. H.; Dolja, V. V.; Falk, B. W.; Gonsalves, D.; Jelkmann, W. The Family closteroviridae Revised. *Arch. Virol.* **2002,** *147,* 2039–2044.

Mauck, K. E.; De Moraes, C. M.; Mescher, M. C. Deceptive Chemical Signals Induced by a Plant Virus Attract Insect Vectors to Inferior Hosts. *Proc. Natl. Acad. Sci. U. S. A.* **2010,** *107*(8), 3600–3605.

Mayo, M. A.; Ziegler-Graff, V. Molecular Biology of Luteoviruses. *Adv. Virus Res.* **1996,** *46,* 413–460.

Meena, R. L.; Venkatesan, T. R. S.; Mohankumar. S. Molecular Characterization of Tospovirus Transmitting Thrips Populations from India. *Am. J. Biochem. Biotechnol.* **2005,** *1,* 167–172.

Meier, R.; Zhang, G.; Ali, F. The Use of Mean Instead of Smallest Interspecific Distances Exaggerates the Size of the "Barcoding Gap" and Leads to Misidentification. *System. Biol.* **2008,** *57*(5), 809–813.

Miao, C. S.; Sunny, Y. Y. An Observation on the Predators of *Orius Minutes* Linn. on Some Insect Pests. *Insect Knowledge.* **1987,** *24,* 174–176.

Miyazaki, M.; Kudo, I. Bibliography and Host Plant Catalogue of Thysanoptera of Japan. Misc. Publ. Natl. Inst. Agro Environ. Sci. **1988,** *3,* 1–246.

Mona, M.; Oraby, A. M.; El-Borollosy, M. H.; Ghaffar, A. Molecular Characterization of a Sugar Beet Cucumber Mosaic Cucumovirus Isolate and its Control Via Some Plant Growth-Promoting Rhizobacteria *J. Biol. Chem. Environ. Sci.* **2008,** *3,* 147–167.

Morales, F. J. Distribution and Dissemination of Begomoviruses in Latin America and the Caribbean. In *Bemisia*: *Bionomics and Management of a Global Pest;* Springer: The Netherlands, **2010,** 283–318.

Morales, F. J.; Anderson, P. K. The Emergence and Dissemination of Whitefly-Transmitted Geminiviruses in Latin America. *Arch. Virol.* **2001,** *146*(3), 415–441.

Moreno-Delafuente, A.; Garzo, E.; Moreno, A.; Fereres, A. A Plant Virus Manipulates the Behavior of its Whitefly Vector to Enhance its Transmission Efficiency and Spread. *PLoS ONE.* **2013,** *8*(4), e61543. doi:10.1371/Journal. Pone.0061543.

Moriones, E.; Navas-Castillo, J. Tomato Yellow Leaf Curl Virus, an Emerging Virus Complex Causing Epidemics Worldwide. *Virus Res.* **2000,** *71*(1), 123–134.

Mound, L. A.; Collins, D. W. A South East Asian Pest Species Newly Recorded in Europe: *Thrips parvispinus* (Thysanoptera: Thripidae), Its Confused Identity and Potential Quarantine Significance. *Eur. J. Entomol.* **2000,** *97,* 197–200.

Nagata, T.; Almeida, A. C. L.; Resende, R. O.; De Aevila, A. C. The Competence of Four Thrips Species to Transmit and Replicate Four Tospoviruses. *Plant Pathol.* **2004,** *53,* 136–140.

Naidu, R. A.; Deom, C. M.; Sherwood, J. L. First Report of *Frankliniella fusca* as a Vector of Impatiens Necrotic Spot Tospovirus. *Plant Dis.* **2001,** *85*(11), 1211.

Nalam, V. J.; Keeretaweep, J.; Sarowar, S.; Shah, J. Root-Derived Oxylipins Promotegreen Peach Aphid Performance on Arabidopsis Foliage. *Plant Cell.* **2012,** *24,* 1643–53.

Nault, L. R. Arthropod Transmission of Plant Viruses: A New Synthesis. *Ann. Entomol. Soc. Am.* **1997,** *90*(5), 521–541.

Nault, L. R.; Madden, L. V. Phylogenetic Relatedness of Maize Chlorotic Dwarf Virus Leafhopper Vectors. *Phytopathol.* **1988,** *78*(12), 1683–1687.

Navas-Castillo, J.; Fiallo-Olive, E.; Sanchez-Campos, S. Emerging Virus Diseases Transmitted by Whiteflies. *Ann. Rev. Phytopathol.* **2011,** *49,* 219–248.

Nawaz-Ul-Rehman, M. S., Fauquet, C. M. Evolution of Geminiviruses and their Satellites. *FEBS Lett.* **2009,** *583*(12), 1825–1832.

Nawaz-ul-Rahman, M. S.; Fauquet, C. M. Evolution of Geminiviruses and Their Satellites. *FEBS Lett.* **2009,** *583*(12), 1825–1832.

Nielson, M. W. Leafhopper Vectors of Xylem-Borne Plant Pathogens. In *Leafhopper Vectors and Plant Disease Agents;* Maramorosch, K., Harris, K. F., Eds.; Academic Press: New York, 1979; pp 603–625.

Ohnishi, J.; Katsuzaki, H.; Tsuda, S.; Sakurai, T.; Akutsu, K.; Murai. T. *Frankliniella cephalica*, A New Vector for Tomato Spotted Wilt Virus. *Plant Dis.* **2006,** *90,* 685.

Ohshima, K.; Akaishi, S.; Kajiyama, H.; Koga, R.; Gibbs, A. J. Evolutionary Trajectory of Turnip Mosaic Virus Populations Adapting to a New Host. *J. Gen. Virol.* **2010,** *91*(3), 788–801.

Okuda, M.; Kato, K.; Hanada, K.; Iwanami, T. Nucleotide Sequence of Melon Yellow Spot Virus M RNA Segment and Characterization of Non-Viral Sequences in Subgenomic RNA. *Arch. Virol.* **2006,** *151*(1), 1–11.

Okuda, M.; Taba, S.; Tsuda, S.; Hidaka, S.; Kameya-Iwaki, M.; Hanada, K. Comparison of the S RNA Segments among Japanese Isolates and Taiwanese Isolates of Watermelon Silver Mottle Virus. *Arch. Virol.* **2001,** *146*(2), 389–394.

Oluwafemi, S.; Jackai, L. E. N. Stylet Penetration Behaviours of Four Cicadulina Leafhoppers on Healthy and Maize Streak Virus Infected Maize Seedlings. *Afr. Crop Sci. J.* **2013,** *21*(2), 161–172.

Oraby M, M.; El-Borollosy, M. H.; Ghaffar, A. Molecular Characterization of a Sugar Beet Cucumber Mosaic Cucumovirus Isolate and its Control via Some Plant Growth-Promoting Rhizobacteria. *J. Biol. Chem. Environ. Sci.* **2008,** *3,* 147–167.

Padidam, M.; Sawyer, S.; Fauquet, C. M. Possible Emergence of New Geminiviruses by Frequent Recombination. *Virol.* **1999,** *265*(2), 218–225.

Paguio, O. R.; Kuhn, C. W. Incidence and Source of Inoculum of Peanut Mottle Virus and its Effect on Peanut. *Phytopathol.* **1974,** *64*(1), 60–64.

Palukaitis, P.; Roosinck, M. J.; Dietzgen, R. G.; Francki, R. I. B. Cucumber Mosaic Virus. *Adv. Virus Res.* **1992,** *1,* 281–348.

Pappu, H. R.; Jones, R. A. C.; Jain, R. K. Global Status of Tospovirus Epidemics in Diverse Cropping Systems: Successes Achieved and Challenges Ahead. *Virus Res.* **2009,** *141*(2), 219–236.

Park, K. C.; Hardie, J. Electrophysiological Characterisation of Olfactory Sensilla in the Black Bean Aphid, *Aphis Fabae. J. Insect Physiol.* **2004,** *50,* 647–655.

Parrella, G.; Gognalons, P.; Gebre-Selassie, K.; Vovlas, C.; Marchoux, G. An Update of the Host Range of Tomato Spotted Wilt Virus. *J. Plant Pathol.* **2003,** *85,* 227–264.

Patil, B. S. Ecobiology and Management of Sorghum Aphid, *Melanaphis sacchari* (Zehntner) (Homoptera: Aphididae). M.Sc. (Ag) Thesis, University of Agriculture Science, Dharwad, Karnataka, 1992, 1–100.

Persley, D. M.; Thomas, J. E.; Sharman, M. Tospoviruses - An Australian Perspective. *Australasian Plant Pathol.* **2006,** *35*(2), 161–180.

Powell, G.; Tosh, C. R.; Hardie, J. Host Plant Selection by Aphids: Behavioral, Evolutionary, and Applied Perspectives. *Ann. Rev. Entomol.* **2006,** *51,* 309–330.

Power, A. G. Insect Transmission of Plant Viruses: A Constraint on Virus Variability. *Curr. Opin. Plant Biol.* **2000,** *3*(4), 336–340.

Power, A. G.; Flecker, A. S. Virus Specificity in Disease Systems: Are Species Redundant? In *The Importance of Species: Perspectives on Expendability and Triage;* Princeton University Press: Princeton, 2003, pp 330–346.

Premachandra, W. T. S. D.; Borgemeister, C.; Maiss, E.; Knierim, D.; Poehling, H. M. *Ceratothripoides claratris*, A New Vector of a Capsicum Chlorosis Virus Isolate Infecting Tomato in Thailand. *Phytopathol.* **2005**, *95*(6), 659–663.

Pirone, T. P.; Perry, K. L. Aphids: Non-Persistent Transmission. *Adv. Bot. Res.* **2002**, *36*, 1–19.

Radcliffe, E. B.; Ragsdale, D. W. Aphid-Transmitted Potato Viruses: The Importance of Understanding Vector Biology. *Am. J. Potato Res.* **2002**, *79*(5), 353–386.

Radcliffe, E. B.; Ragsdale, D. W.; Suranyi, R. A.; Difonzo, C. D.; Hladilek, E. E. Aphid Alert: How It Came To Be, What It Achieved, and Why It Proved Un-Sustainable. In *Area-wide Pest Management: Theory and Implementation*; CABI Press: Mussachusetts, 2008; pp 244–260.

Rajabaskar, D.; Bosque-Perez, N. A.; Eigenbrode, S. D. Preference by a Virus Vector for Infected Plants is Reversed after Virus Acquisition. *Virus Res.* **2014**, *186*, 32–37.

Rao, X.; Wu, Z.; Li, Y. Complete Genome Sequence of a Watermelon Silver Mottle Virus Isolate From China. *Virus Gen.* **2013**, *46*(3), 576–580.

Rasoulpour, R.; Izadpanah, K. Characterisation of Cineraria Strain of Tomato Yellow Ring Virus From Iran. *Aust. Plant Pathol.* **2007**, *36*(3), 286–294.

Ravi, K. S.; Kitkaru, A. S.; Winter, S. Iris Yellow Spot Virus in Onions: A New Tospovirus Record from India. *Plant Pathol.* **2006**, *55*(2), 288–288.

Rebijith, K. B.; Asokan, R.; Kumar, N. K.; Krishna, V.; Ramamurthy, V. V. Development of Species-Specific Markers and Molecular Differences in Mtdna of *Thrips palmi* Karny and *Scirtothrips dorsalis* Hood (Thripidae: Thysanoptera), Vectors of Tospoviruses (Bunyaviridae) in India. *Entomol. News.* **2012**, *122*(3), 201–213.

Reddy, D. V. R.; Ratna, A. S.; Sudarshana, M. R.; Poul, F.; Kumar, K. Serological Relationships and Purification of Bud Necrosis Virus, a Tospovirus, Occurring in Peanut (*Arachis hypogaea* L.) in India. *Ann. Appl. Biol.* **1992**, *120*, 279–286.

Reddy, M. S.; Reddy, D. V. R.; Appa Rao, A. A New Record of Virus Disease in Peanut. *Plant Dis. Rep.* **1968**, *52*, 494–495.

Redinbaugh, M. G.; Seifers, D. L.; Meulia, T.; Abt, J. J.; Anderson, R. J.; Styer, W. E.; Hogenhout, S. Amaize Fine Streak Virus, a New Leafhopper-Transmitted Rhabdovirus. *Phytopathol.* **2002**, *92*(11), 1167–1174.

Remaudiere, G.; Remaudiere, M. *Catalogue Des Aphididae Du Monde*; Homoptera- Aphidoidea. Institut National de la Recherche Agronomique: Versailles, INRA, Paris, 1997; pp 478.

Renukadevi, P.; Nakkeeran, S.; Gandhi, K.; Murugaiah, J.; Pappu, H. R.; Malathi, V. G. First Report of Tomato Spotted Wilt Virus Infection of Chrysanthemum in India. *Plant Dis.* **2015**, (Ja) http://Dx.Doi.Org/10.1094/PDIS–01–15–0126-PDN.

Revers, F.; Le Gall, O.; Candresse, T.; Le Romancer, M.; Dunez, J. Frequent Occurrence of Recombinant Potyvirus Isolates. *J. Gen. Virol.* **1996**, *77*(8), 1953–1965.

Richards, O. W.; Davies, R. G. *Imms's General Textbook of Entomology;* 10th Edn., Chapman & Hall: New York, 1977; Vol. 1–2.

Riley, D. G.; Joseph, S. V.; Srinivasan, R.; Diffie, S. Thrips Vectors of Tospoviruses. *J. Integ. Pest Manag.* **2011**, *2*(1), I1–I10.

Roossinck, M. J. Cucumoviruses (Bromoviridae) General Features. In: *Encyclopedia of Virology;* 2nd Edn., Granoof, R. G., Webster, Eds.; Academic Press: Sandiego, USA, 1999; pp 315–320.

Sakurai, T.; Inoue, T.; Tsuda, S. Distinct Efficiencies of Impatiens Necrotic Spot Virus Transmission by Five Thrips Vector Species (Thysanoptera: Thripidae) of Tospoviruses in Japan. *Appl. Entomol. Zool.* **2004**, *39,* 71–78.

Satyanarayana, T.; Gowda, S.; Reddy, K. L.; Mitchell, S. E.; Dawson, W. O.; Reddy, D. V. R. Peanut Yellow Spot Virus is a Member of a New Serogroup of Tospovirus Genus Based on Small (S) RNA Sequence and Organization. *Arch. Virol.* **1998**, *143,* 353–364.

Satyanarayana, T.; Reddy, K. L.; Ratna, A. S.; Deom, C. M.; Gowda, S.; Reddy, D. V. R. Peanut Yellow Spot Virus: A Distinct Tospovirus Species Based on Serology and Nucleic Acid Hybridisation. *Ann. Appl. Biol.* **1996,** *129,* 237–245.

Schenck, S. Factors Affecting the Transmission and Spread of Sugarcane Yellow Leaf Virus. *Plant Dis.* **2000,** *84,* 1085–1088.

Schneider, W. L.; Roossinck, M. J. Evolutionarily Related Sindbis-Like Plant Viruses Maintain Different Levels of Population Diversity in a Common Host. *J. Virol.* **2000,** *74*(7), 3130–3134.

Seeno, T. N.; Wilcox, J. A. Leaf Beetle Genera (Coleoptera: Chrysomelidae). *Entomography.* **1982,** *1,* 1–221.

Seepiban, C.; Gajanandana, O.; Attathom, T.; Attathom, S. Tomato Necrotic Ringspot Virus, a New Tospovirus Isolated in Thailand. *Arch. Virol.* **2011,** *156*(2), 263–274.

Setokuchi, O. Studies on the Ecology of Aphids on Sugarcane. I. Infestation of *Melanaphis sacchari* (Zehntner) (Homoptera: Aphididae). *Japanese J. Appl. Entomol. Zool.* **1988,** *32,* 215–218.

Sharma, P.; Sahu, A. K.; Verma, R. K.; Mishra, R.; Choudhary, D. K.; Gaur, R. K. Current Status of Potyvirus in India. *Arch. Phytopathol. Plant Prot.* **2014,** *47*(8), 906–918.

Simmons, H. E.; Holmes, E. C.; Stephenson, A. G. Rapid Turnover of Intra-Host Genetic Diversity
in Zucchini Yellow Mosaic Virus. *Virus Res.* **2011,** *155*(2), 389–396.

Singh, S. J.; Krishnareddy, M. Watermelon Bud Necrosis: A New Tospovirus Disease. *Tospoviruses and Thrips of Floral and Vegetable Crops.* **1995,** *431,* 68–77.

Sisterson, M. S. Effects of Insect-Vector Preference for Healthy or Infected Plants on Pathogen Spread: Insights from a Model. *J. Econ. Entomol.* **2008,** *101*(1), 1–8.

Sivaprasad, Y.; Reddy, B. B.; Kumar, C. V. M. N.; Reddy, K. R.; Gopal, D. V. R. S. Jute (*Corchorus capsularis*): A New Host of Peanut Bud Necrosis Virus. *New Dis. Rep.* **2011,** *23*(33), doi: 10.5197/j.2044-0588.2011.023.033.

Slipinski, S. A.; Leschen, R. A. B.; Lawrence, J. F. Order Coleoptera Linnaeus, 1758. Animal Biodiversity: An Outline of Higher-level Classification and Survey of Taxonomic Richness. *Zootaxa.* **2011,** *3148*(23), 203–208.

Stafford, C. A.; Walker, G. P.; Ullman, D. E. Infection with a Plant Virus Modifies Vector Feeding Behavior. *Proc. Natl. Acad. Sci. U. S. A.* **2011,** *108*(23), 9350–9355.

Stankovic, I.; Bulajic, A.; Vucurovic, A.; Ristic, D.; Milojevic, K.; Nikolic, D.; Krstic, B. First Report of Tomato Spotted Wilt Virus Infecting Onion and Garlic in Serbia. *Plant Dis.* **2012,** *96*(6), 918–918.

Sujitha, A.; Reddy, B. B.; Sivaprasad, Y.; Usha, R.; Gopal, D. S. First Report of Groundnut Bud Necrosis Virus Infecting Onion (*Allium cepa*). *Aust. Plant Dis. Notes.* **2012,** *7*(1), 183–187.

Sylvester, E. S. Beet Yellows Virus Transmission by the Green Peach Aphid. *J. Econ. Entomol.* **1956,** *49,* 789–800.

Takeuchi, S.; Shimomoto, Y.; Ishikawa, K. First Report of Melon Yellow Spot Virus Infecting Balsam Pear (*Momordica charantia* L.) in Japan. *J. Gen. Plant Pathol.* **2009,** *75*(2), 154–156.

Ten Broeke, C. J. M.; Dicke, M.; Van Loon, J. J. A. Performance and Feeding behaviour of Two Biotypes of the Black Currant-Lettuce Aphid, *Nasonovia ribisnigri,* On Resistant and Susceptible *Lactucasativa* Near-Isogenic Lines. *Bull. Entomol. Res.* **2013,** *103,* 511–21.

Timmerman-Vaughan, G. M.; Lister, R.; Cooper, R.; Tang, J. Phylogenetic Analysis of New Zealand Tomato Spotted Wilt Virus Isolates Suggests Likely Incursion History Scenarios and Mechanisms for Population Evolution. *Arch. Virol.* **2014,** *159*(5), 993–1003.

Tjallingii, W. F. Electrical Recording of Stylet Penetration activities. In *Aphids, Their Biology, Natural Enemies and Control;* Minks, A K., Harrewijn, P., Eds.; Elsevier: Amsterdam, 1988, 2B, pp 95–108.

Tjallingii, W. F. Membrane Potentials as an Indication for Plant cell Penetration by Aphid Stylets. Entomol. *Exp. Appl.* **1985,** *38,* 187–193.

Tjallingii, W. F. Salivary Secretions by Aphids Interacting with Proteins of Phloem Wound Responses. *J. Exp. Bot.* **2006,** *57*(4), 739–745.

Todd, J. C.; Ammar, E. D.; Redinbaugh, M. G.; Hoy, C.; Hogenhout, S. A. Plant Host Range and Leafhopper Transmission of Maize Fine Streak Virus. *Phytopathol.* **2010,** *100*(11), 1138–1145.

Todd, J. L.; Phelan, P. L.; Nault, L. R. Interaction between Visual and Olfactory Stimuli during Host-Finding by Leafhopper, *Dalbulus maidis* (Homoptera: Cicadellidae). *J. Chem. Ecol.* **1990,** *16*(7), 2121–2133.

Tomitaka, Y.; Abe, H.; Sakurai, T.; Tsuda, S. Preference of the Vector Thrips *Frankliniella occidentalis* for Plants Infected with Thrips-Non-Transmissible Tomato Spotted Wilt Virus. *J. Appl. Entomol.* **2015,** *139,* 250–259.

Tomlinson, J. A. Turnip Mosaic Virus. CMI/AAB Descriptions of Plant Viruses 8.2, 1970.

Trkulja, V.; Salapura, J. M.; Curkovic, B.; Stankovic, I.; Bulajic, A.; Vucurovic, A.; Krstic, B. First Report of Tomato Spotted Wilt Virus on Gloxinia in Bosnia and Herzegovina. *Plant Dis.* **2013,** *97*(3), 429–429.

Ullah, Z.; Chai, B.; Hammar, S.; Raccah, B.; Gal-On, A.; Grumet, R. Effect of Substitution of the Amino Termini of Coat Proteins of Distinct Potyvirus Species on Viral Infectivity and Host Specificity. *Physiol. Mol. Plant Pathol.* **2003,** *63*(3), 129–139.

Ullman, D. E.; Sherwood, J. L.; German, T. L. Thrips as vectors of Plant Pathogens. In *Thrips as Croppests;* Lewis, T., Ed.; CAB International: Wallingford, UK, 1997; pp 539–565.

Van Der Wilk, F.; Verbeek, M.; Dullemans, A. M.; Van Den Heuvel, J. F. J. M. The Genome-Linked Protein of Potato Leafroll Virus is Located Downstream of the Putative Protease Domain of The ORF1 Product. *Virol.* **1997,** *234,* 300–303.

Vandenberg, N. J. Family 93. Coccinellidae Latreille 1807. In *American Beetles. Polyphaga: Scarabaeoidea through Curculionoidea;* Arnett, R. H. Jr., Thomas, M. C., Skelley, P. E., Frank, J. H., Eds.; CRC Press LLC: Boca Raton, FL, 2002; Vol. 2. pp 371–389.

Varma, A.; Mandal B.; Singh, M. K. Global Emergence and Spread of Whitefly *Bemisia Tabaci* Transmitted Geminiviruses. In *The Whitefly, Bemisia tabaci (Homoptera: Aleyrodidae) Interaction with Geminivirus-Infected Host Plants;* Thompson, W. M. O., Ed.; Springer: The Netherlands, 2001; pp 205–292.

Vega, J.; Scagliusi, S. M.; Ulian, E. C. Sugarcane Yellow Leaf Disease in Brazil: Evidence for Association with a Luteovirus. *Plant Dis.* **1997,** *81,* 21–26.

Vijayalakshmi, K. Transmission and Ecology of *Thrips palmi* Karny, the Vector of Peanut Bud Necrosis Virus. Dissertation, Andhra Pradesh Agricultural University, Hyderabad, 1994.

Viswanathan, R. Sugarcane Yellow Leaf Syndrome in India: Incidence and Effect on Yield Parameters. *Sugarcane Intl.* **2002,** *5,* 17–23.

Viswanathan, R.; Chinnaraja C.; Malathi P.; Gomathi R.; Rakkiyappan P.; Neelamathi D.; Ravichandran V. Impact of Sugarcane Yellow Leaf Virus (Scylv) Infection on Physiological Efficiency and Growth Parameters of Sugarcane under Tropical Climatic Conditions in India. *Acta Physiol. Plant.* **2014,** *36,* 1805–1822.

Viswanathan, R.; Rao, G. P. Disease Scenario and Management of Major Sugarcane Diseases in India *Sugar Tech.* **2011,** *13*(4), 336–353.

Wallis, C. M.; Fleischer, S. J.; Luster, D.; Gildow, F. E. Aphid (Hemiptera: Aphididae) Species Composition and Potential Aphid Vectors of Plum Pox Virus in Pennsylvania Peach Orchards. *J. Econ. Entomol.* **2005,** *98*(5), 1441–1450.

Watson, M. A.; Robers, F. M. A Comparative Study of the Transmission of Hyoscyamus Virus 3, Potato Virus Y and Cucumber Virus 1 by the Vectors *Myzus persicae* (Sulz), *M. circumflexus* (Buckton), and *Macrosiphum gei* (K). *Proc. R. Soc. Lond.,* **1939,** B *127,* 543–577.

Webb, S.; Tsai, J.; Forrest, M. Bionomics of Frankliniella bispinosa and Its Transmission of Tomato Spotted Wilt Virus. In *Fourth International Symposium on Tospovirus and Thrips in Floral and Vegetable Crops*; Wageningen: The Netherlands, 1998; Abstract, 67.

Weimer, J. L. Alfalfa Mosaic. *Phytopathology.* **1931,** *21*(1), 122–123.

White, W. H.; Reagan, T. E.; Hall, D. G. *Melanaphis sacchari* (Homoptera: Aphididae), a Sugarcane Pest New to Louisiana. *Fla. Entomology.* **2001,** *84,* 435–436.

Whitfield, A. E.; Ullman, D. E.; German, T. L. Tospovirus-Thrips Interactions. *Ann. Rev. Phytopathol.* **2005,** *43,* 459–489.

Wijkamp, I.; Almarza, N.; Goldbach, R.; Peters, D.; Distinct Levels of Specificity in Thrips Transmission of Tospoviruses. *Phytopathology.* **1995,** *85,* 1069–74.

Wilbrink, G. An Investigation on Spread of the Mosaic Disease of Sugarcane by Aphids. *Medid Procfst. Java Suikerind.* **1922,** *10,* 413–456.

Wisler, G. C.; Duffus, J. E.; Liu, H. Y., Li, R. H. Ecology and Epidemiology of Whitefly-Transmitted Closteroviruses. *Plant Dis.* **1998,** *82,* 270280.

Xu, Y.; Lou, S. G.; Li, Q. F.; Hu, Q.; Sun, L. P.; Li, Z. Y.; Liu, Y. T. Molecular Characterization of the Hippeastrum Chlorotic Ringspot Virus L Segment and its Protein. *Arch. Virol.* **2014,** *159*(10), 2805–2807.

Yang, X.; Thannhauser, T. W.; Burrows, M.; Cox-Foster, D.; Gildow, F. E.; Gray, S. M. Coupling Genetics and Proteomics to Identify Aphid Proteins Associated with Vector-Specific Transmission of Polerovirus (Luteoviridae). *J. Virol.* **2008,** *82*(1), 291–299.

Yeh, S.; Sun, I.; Ho, H.; Chang, T. Molecular Cloning and Nucleotide Sequence Analysis of the S RNA of Watermelon Silver Mottle Virus. *Tospoviruses Thrips Floral Veg. Crops.* **1995,** *431,* 244–260.

Yokomi, R. K.; Garnsey, S. M. Transmission of *Citrus Tristeza Virus* by *Aphis gossypii* and *Aphis citricola* in Florida. *Phytophylactica.* **1987,** *19*(2), 169–172.

Zahniser, J. N.; Dietrich, C. H. Phylogeny of the Leafhopper Subfamily Deltocephalinae (Hemiptera: Cicadellidae) Based on Molecular and Morphological Data with a Revised Family-Group Classification. *System. Entomol.* **2010,** *35*(3), 489–511.

Zhou, J.; Tzanetakis, I. E. Epidemiology of Soybean Vein Necrosis-Associated Virus. *Phytopathol.* **2013,** *103*(9), 966–971.

Zitter, T. A.; Provvidenti. R. Vegetable Crops: Virus Diseases of Leafy Vegetables and Celery. *Vegetable MD Online, Fact Sheet,;* Coop. Ext. Dept. of Plant Pathology, Cornell University: Ithaca, NY, 1984; pp 737.

PART III
Adaptability and Dispersal of Phytopathogens

CHAPTER 11

CHANGES IN WHEAT PHYSIOLOGY AND BIOCHEMISTRY IN CONTEXT TO PATHOGEN INFECTION

PRAMOD PRASAD*, SUBHASH C. BHARDWAJ, HANIF KHAN, OM P. GANGWAR, and SUBODH KUMAR

ICAR-Indian Institute of Wheat and Barley Research, Regional Station Flowerdale, Shimla 171002, Himachal Pradesh, India

Corresponding author. E-mail: pramoddewli@gmail.com

CONTENTS

ABSTRACT

Biotic stresses including fungal, bacterial, and viral pathogens are a major constraint to wheat production world-wide. Significant changes in the pathotype situation within rusts along with emergence of new diseases such as wheat blast have occurred in recent years. Such example highlights the urgent need to better understand host–pathogen interactions and how this knowledge can be applied in better disease management strategies. The ability of the host plant to recognize the pathogen and activate these responses is regulated in gene-for-gene-specific manner. Pathogen infection triggers a wide variety of plant responses, ranging from cellular metabolism to cross-linking of cell-wall proteins, the activation of protein kinases and the increased expression of various defense-related genes. Understanding the physiological, biochemical, and molecular responses in host to pathogen infection is essential for a holistic perception of plant resistance mechanisms to biotic stresses. Plants have developed specific mechanisms to detect external signals with specific responses in order to survive under these challenges. Here, we discuss scientific advances which took place in the past for the understanding of the host–pathogen interactions in wheat diseases.

11.1   INTRODUCTION

Bread wheat (*Triticum aestivum* L.) is the third largest cereal produced in the world, after maize and rice with global production of 704 million tons (FAO, 2012). In terms of dietary intake, however, wheat comes second to rice as a main food crop, given the more extensive use of maize as animal feed. India is the second largest producer of the wheat after China in the world with 95.91 million tons production during 2013–2014 (Anonymous, 2014). Wheat provides 21% of the food calories and 20% of the protein to more than 4.5 billion people in 94 developing countries (Braun et al., 2010). The projected demand for wheat is estimated to increase by 60% by 2050 in developing countries; at the same time, climate change-induced temperature increases, diseases and other pests are expected to reduce wheat production by more than 29% in developing countries (Rosegrant et al., 1995). In addition to the abiotic factors including water imbalance, nutrients deficiencies, chemical toxicity and many other agronomic implications, there are huge number of biotic factors, which are contributing to lower wheat yield. Wheat is the host to many microbes causing a number of diseases such as rusts, smuts, bunts, leaf blight, powdery mildew, etc. (Table 11.1), which cause huge losses to the quality and quantity of the produce. Among the biotic stresses, rusts are the most widely prevalent and notorious pathogens on wheat, which pose serious threat to stability of its production.

TABLE 11.1 Important Wheat Diseases.

Disease	Causal agent
Bacterial	
Bacterial sheath rot	*Pseudomonas fuscovaginae*
Black chaff	*Xanthomonas campestris* pv. *translucens*
Bacterial leaf streak	*Xanthomonas translucens* pv. *undulosa*
Fungal	
Anthracnose	*Colletotrichum graminicola*
Alternaria leaf blight	*Alternaria triticina*
Ascochyta leaf spot	*Ascochyta tritici*
Black head (sooty molds)	*Alternaria* spp., *Cladosporium* spp., *Epicoccum* spp., *Sporobolomyces* spp., etc.
Black point (Kernel smudge)	*Alternaria* sp., *Fusarium* sp., *Helminthosporium* sp.
Common bunt (stinking smut)	*Tilletia tritici, Tilletia caries, Tilletialaevis, Tilletia foetida*
Ergot	*Claviceps purpurea, Sphacelia segetum* (anamorph)
Flag smut	*Urocystis agropyri*
Karnal (partial bunt)	*Tilletia indica, Neovossia indica*
Leaf rust (brown rust)	*Puccinia triticina*
Loose smut	*Ustilago tritici*
Powdery mildew	*Blumeria graminis, Oidiummonilioides* (anamorph)
Septoria blotch	*Septoria tritici, Mycosphaerella graminicola* (teleomorph)
Spot blotch	*Cochliobolus sativus* (teleomorph) *Bipolaris sorokiniana* (anamorph) or *Helminthosporiumsativum*
Stem (black) rust	*Puccinia graminis* f. sp. *tritici*
Stripe (yellow) rust	*Puccinia striiformis*
Take-all	*Gaeumannomyces graminis var. tritici*
Tan (yellow leaf) spot, red smudge	*Pyrenophora tritici-repentis, Drechslera tritici repentis* (anamorph)
Wheat blast	*Magnaporthe grisea*
Viral	
Barley stripe mosaic	genus *Hordeivirus*, Barley stripe mosaic virus (BSMV)

TABLE 11.1 *(Continued)*

Disease	Causal agent
Barley yellow dwarf	genus *Luteovirus*, Barley yellow dwarf virus (BYDV)
Maize streak	genus *Monogeminivirus*, Maize streak virus (MSV)
Rice black-streaked dwarf	genus *Fijivirus*, Rice black-streaked dwarf virus (RBSDV)
Wheat streak mosaic	genus *Tritimovirus*, Wheat streak mosaic virus (WSMV)
Wheat yellow mosaic	Wheat yellow mosaic virus
Aster yellows	Phytoplasma

Recent examples of wheat diseases/races include the worldwide emergence of new aggressive races (*Yr*9 virulence) of *Puccinia striiformis* f. sp. *tritici* (yellow rust), *Sr*31 virulences (Ug99), and other hypervirulent races of *Puccinia graminis* f. sp. *tritici* (stem rust) in Ug99 lineage and blast disease caused by *Magnaporthe oryzae* in Brazil. The emergence of Ug99 is considered a highly significant event having far reaching consequences not only for India but also for global wheat production due to susceptibility of nearly 20% of the wheat cultivars to Ug99. It has been estimated by Singh et al. (2008) that the area under the risk of Ug99 amounts to around 50 Mha of wheat grown globally, that is about 25% of the world's wheat area. The Ug99 race, as other races of the wheat-rust pathogens is evolving rapidly, till date eight variants have been documented as the member of the Ug99 race lineage (Singh et al., 2011). In Indian context, though the stem rust prone area is less than 25% of the total area, but the possible implications of entry of Ug99 race into the country or independent mutation for *Sr*31 cannot be ignored (Bhardwaj et al., 2014). In a study on diversity for stem rust resistance in Indian wheat, commendable diversity was observed (Bhardwaj et al., 2003).

11.2 ECONOMIC IMPORTANCE OF WHEAT DISEASES: A SNIPPET

Among the wheat diseases, rusts are the most important from historical perspective. Wheat rusts have appeared in epiphytotic form from time to time in many countries. These diseases have forced the farmers to change their cropping pattern in many parts of the world. In the southern parts of the USA, people eat corn rather than wheat which is consumed in northern USA in plenty, the reason behind this change in the food habit in southern part is that in these areas wheat does not grow well due to rust diseases. In ancient Roman religion, the Robigalia was a festival held to protect wheat fields from rust disease. In India rust had been considered to cause a loss of Rs. 4 crores every year before the introduction of Mexican dwarf wheat varieties, even in these varieties farmers may lose about 10% of yield due to rusts. Another disease of wheat, loose smut, is estimated to cause an average loss of 3% every year. Loose smut, along with other seed borne diseases like Karnal bunt, etc., has posed export restrictions of Indian wheat to other countries where these diseases are not reported. In 1722, 20,000 soldiers of the army of Peter the Great of Russia died from consuming bread made from ergot infected wheat. Outbreaks of "holy fire" occurred even during the twentieth century. In another incidence, in 1926–1927 in Russia, as many as 10,000 people were affected by the disease after eating bread made from ergot-contaminated wheat flour. Ergot is the fruiting structure produced by *Claviceps purpurea* and related fungi in place of the seed of the plant. Ergots contain a number of potent alkaloids and other biologically active compounds that affect primarily the brain and the circulatory system. The best known of the ergot alkaloids is lysergic acid diethylamide, the infamous LSD (Agrios, 2005).

11.3   PLANT INFECTION PROCESS

Plant infection by plant pathogens can occur in several ways. It could be passive, that is, accidental, by suction into the plant through natural plant openings such as stomata, hydathodes or lenticels, entrance through abrasions or wounds on leaves, stems or roots (Vidaver & Lambrecht, 2004). Infection of host plants by biotrophic plant pathogens generally involves a sequential development of specialized host–parasite interfaces, exemplified by those of haustoria, which are maintained over an extended period of time without causing significant cytological damage to host tissues in the infected region (Tariq & Jeffries, 1986). Using scanning electron microscopy (SEM), Hu and Rijkenberg (1998) identified key time points in the formation of infection structures by *Puccinia triticinia* (leaf rust of wheat) on susceptible and resistant lines of hexaploid wheat. Six hours after infection, the fungus formed appressoria over the stomatal openings. After 12 h, the fungus could successfully penetrate into the stroma, formed substomatal vesicles (SSVs), and primary infection hyphae are visible. After SSV formation, the primary infection hypha grows and attaches to a mesophyll or epidermal cell. At 24 h postinoculation, a septum appears separating the haustorial mother cell from the infection hypha after which the fungus forms haustorium and penetrates the cell. Garg et al. (2010) described the infection process in susceptible and resistant genotypes of *Brassica napus* against *Sclerotinia sclerotiorum*. They demonstrated that at cellular level resistance to *S. sclerotiorum*in in *B. napus* is a result of retardation of pathogen development, both on the plant surface and within host tissues. There are some indications that the infection process depends on the nutritional status of the inoculum (Garg et al., 2010). Indeed, previous studies have concluded that the presence of nutrients is essential for hyphal development, penetration and for subsequent establishment of a successful invasion of a susceptible host by the pathogen.

Histopathological studies have revealed that the development of leaf rust pathogen in susceptible wheat cultivars included several stages like germination of uredospores, formation of a series of structures such as appressoria, SSVs, primary and secondary infection hyphae, haustorial mother cells, haustoria, and finally the production of uredospores. Early at 8 hours post inoculation (hpi), a portion of appressoria produces a penetration peg to enter into the plant stoma, by 36 hpi; fungal colony formation completes at most infection sites. Thereafter, hypha grows quickly, extends, and forms more colonies (Huang et al., 2003). In case of spot blotch disease in wheat, *B. sorokiniana* starts infection in the plant with conidium germination on the leaf surface. A rudimentary appressorium forms from the apex of the germ tube and directly penetrates the cuticle and epidermal cell wall via infection hyphae that colonize the inter and intracellular leaf tissue. It produces non-selective toxins and several hydrolytic enzymes, which quickly destroy the leaf tissue to generate nutrients for continued fungal growth. Helminthosporol, the major non-specific toxin produced by *B. sorokiniana*, affects plasma membrane permeability so that the pathogen could feed upon the leaked electrolytes and then grow and colonize the remaining host tissue. *Septoria tritici* (teleomorph *Mycosphaerella graminicola*), the causal agent of septoria leaf blotch of wheat, penetrates host leaves through the stomata and grows slowly as filamentous hyphae in the intercellular spaces between the wheat mesophyll cells, typically up to 9–11 d. Subsequently, the fungus exhibits a rapid switch to necrotrophy immediately prior to symptom expression at 12–20 d after penetration.

11.4 POST INFECTION CHANGES IN PLANTS

Plants are source of food and shelter for a wide range of organisms, including fungi, bacteria, phytoplasma, insects, nematodes, viruses, and even other plants. The identification of potential pathogenic microbes by the plant leads to the activation of different defense responses which are designed to prevent further infection. In wheat leaf blotch, caused by *Septoria tritici*, the necrotrophic process post infection involves cell-wall degradation and accumulation of H_2O_2, resulting in massive collapse of mesophyll tissue, leakage of nutrients from dying plant cells into the apoplastic spaces, rapid increase of fungal biomass and sporulation in characteristic necrotic foliar blotches. Plant responses to infection are usually initiated by the specific recognition of the pathogen and the transmission of the signal via plasma membrane-bound receptors. The earliest reactions of plant cells include changes in plasma membrane permeability leading to calcium and proton influx (Fig. 11.1). This in turn leads to the production of reactive oxygen intermediates (ROI) such as superoxide (O_2^-) and hydrogen peroxide (H_2O_2) catalyzed by plasma membrane-located nicotinamide adenine dinucleotide phosphate (NADPH) oxidase or apoplastic peroxidases (Somssich & Hahl-brock, 1998). These initial ion fluxes and production of ROI also trigger the localized production of secondary messengers for the initiation of HR and defense gene expression. Other interacting components might include specifically induced phospholipases which act on lipid-bound unsaturated fatty acids within the membrane resulting in the release of linolenic acid, which in turn acts as substrate for the production of jasmonate (JA), methyl jasmonate (MeJA) and other related molecules (Creelman & Mullet, 1997). The expression of most of the inducible, defense-related genes are regulated by signal pathways involving one or more of the three key regulators namely jasmonate, ethylene, and salicylic acid.

In some plant–pathogen interactions, especially in diseases where biotrophs are the cause of the disease, the ability of the host plant to recognize the pathogen and activate these responses is regulated in a gene-for-gene-specific manner by the direct or indirect interaction between the products of a plants resistance (*R*) gene and a pathogen avirulence (*Avr*) gene. One of the earliest responses activated after the host plant recognition of an Avr protein or a non-host specific elicitor is the oxidative burst, in which levels of reactive oxygen species (ROS) rapidly increase. Other rapid responses include the cross-linking of cell-wall proteins, the activation of protein kinases, and the increased expression of various defense genes (Fig. 11.1). Some of these genes encode peroxidases, glutathione S-transferases, proteinase inhibitors, and various biosynthetic enzymes such as phenylalanine ammonia-lyase (PAL) and pathogenesis-related (PR) proteins. Activation of signal trans-duction networks after pathogen recognition results in the reprogramming of cellular metabolism, involving large changes in gene transcriptional activity while basic incompatibility frequently results in the expression of defense-related genes and localized host cell death. Plants contain many defense-related proteins. In addition to *R*-genes and genes encoding signal transduction proteins, they also possess downstream defense genes for the synthesis of enzymes involved in the generation of phytoalexins, oxidative stress protection, lignifications, and numerous other defense-related activities. Many of these genes are involved in the production of secondary metabolites, which are essential for the existence of shikimate and phenylpropanoid pathways.

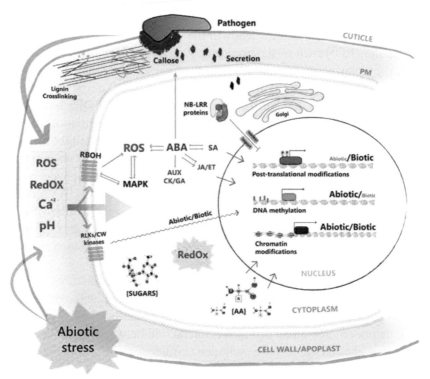

FIGURE 11.1 Schematic representation of signaling pathways of abiotic and biotic stress at the cellular (Source: Enhancing crop resilience to combined abiotic and biotic stress through the dissection of physiological and molecular crosstalk. (Copyright © 2014 Kissoudis, van de Wiel, Visser and van der Linden. In Front Plant Sci. 2014; 5: 207. Published under the Creative Commons Attribution License [CC BY]).

The resistance of wheat to *Puccinia graminis* f. sp. *tritici,* determined by the resistance gene *Sr5,* is associated with hypersensitive host cell death. In this system the hypersensitive response is observed visually as necrotic leaf area (Tiburzy & Reisener, 1990). Biochemical analysis reveals that complex metabolic changes occur in host plant post infection, leading to lignification and increased PAL activity in infected plant tissue (Tiburzy & Reisener, 1990). Intercellular fluids of a compatible interaction between *P. graminis* f. sp. *tritici* and wheat contain the elicitor with identical molecular mass. This indicates that the elicitor is released from fungal cell walls during infection. Infected wheat leaves frequently show a higher lignin content, independent of the compatibility grade. The lignification is induced by fungal penetration and is often limited to a few cells at the infection site. Since an enhanced lignification has often been observed in resistant wheat, its important role in the resistance has been assumed (Sander & Heitefuss, 1998). Induced lignification is accompanied by an increase in the activity of the key enzymes of the phenylpropanoid pathway such as phenylalanine ammonia-lyase (PAL), cinnamyl alcohol dehydrogenase (CAD), and peroxidases (POX) (Nicholson & Hammerschmidt, 1992).

11.5 EFFECT ON PHOTOSYNTHESIS

Photosynthesis is a process that converts solar energy to chemical energy in different organisms, ranging from plants to bacteria. It provides all the food we eat and all the fossil fuel we use. Photosynthesis of terrestrial higher plants is, however, constantly challenged by abiotic and biotic stresses. Pathogen infection often leads to the development of chlorotic and necrotic lesions and to a decrease in photosynthetic assimilate production in both compatible and incompatible interactions. Using chlorophyll fluorescence imaging, it has been reported that the changes in photosynthesis upon infection are local. At present, it is not clear if this stimulation of photosynthesis is due to the defense strategy of the plant. It is suggested that plants switch off photosynthesis and other assimilatory metabolism to initiate respiration and other processes required for defense. The primary reason for decline in the rate of photosynthesis is the reduction in total green area of the plant, because the photosynthetic surface of the plant is lessened, however, in some diseases toxins produced by the pathogen also contribute for the same. For example, *Pyrenophora tritici-repentis*, causing tan (yellow leaf) spot disease of wheat, produces the host-specific, chlorosis-inducing toxin Ptr ToxB, which is responsible for reducing rate of photosynthesis prior to the development of chlorosis. In some diseases plant pathogens reduce photosynthesis in the later stages of diseases, by affecting the chloroplasts and causing their degeneration. In some cases, the rate of photosynthesis may increase initially as Chang et al. (2013) observed increased photosynthetic rate in yellow rust infected wheat at 48 hpi in both the compatible and the incompatible interactions. The change in the photosynthetic rate was consistent with the change in the sucrose concentration in the infected wheat leaves during the early stages of the infection, which indicates that the sucrose accumulation is caused by enhanced photosynthesis.

Biotrophs affect photosynthesis in varying degrees, depending on the severity of the infection. A biotrophic infection site becomes a strong metabolic sink, changing the pattern of nutrient translocation within the plant and causing net influx of nutrients into infected leaves to satisfy the demands of the pathogen. The depletion, diversion, and retention of photosynthetic products by the pathogen stunt plant growth, and further reduced the plant's photosynthetic efficiency.

11.6   EFFECT ON RESPIRATION

Respiration is the process through which energy stored in organic molecules is released to do metabolic work. It is controlled by enzymes, and releases carbon dioxide and water. Respiration is a complex process involving a number of enzymes in every single living cell, with influential effects on the functions and existence of the cell. Respiration rate is usually increased in diseased plants. In the early stage of disease development, synthetic processes induce the rate of respiration, whereas in the late stages injury and decomposition of tissues lead to increase in respiration. In resistant wheat plant infected either by obligate or facultative parasites; initially O_2 consumption increases more rapidly, leading to the increase in the respiration rate. Later the rate of respiration gradually decreases.

In susceptible varieties, where no defense mechanism is mobilized quickly against a particular pathogen, there is slow and steady increase in respiration rate post infection and it continues to rise and remains at a high level for much longer periods. Several changes in the metabolism of the diseased plant accompany the increase in respiration after infection. Thus, the activity or concentration of several enzymes of the respiratory pathways seems to be increased. The accumulation and oxidation of phenolic compounds, many of which are associated with defense mechanisms in plants, are also greater during increased respiration. In diseased plants, there is an increase in the activity of pentose phosphate pathway which is the primary source of phenolic compounds synthesis, this increase also contributes in enhancing the respiration rate. Moreover, in many plant diseases, initially there is stimulation in metabolism, cell growth, protoplasmic streaming and material synthesis in the diseased area. The energy required for these activities derives from ATP produced through respiration, which further stimulates respiration. The increase in respiration rate is one of the early physiological responses to pathogen attack in mildew, rust, and blight infected wheat leaves (Shaw & Samborski, 1957). Respiratory rates in heavily rusted leaf tissues are usually two to three times the rate of uninfected tissue with wheat stem rust. This increase in wheat is first detected about 5d after inoculation. Increase in respiration in rust, mildew, and blight infected wheat is probably through tricarboxylic acid cycle (TCA cycle) coupled cytochrome electron transport.

11.7   EFFECT ON TRANSPIRATION

Transpiration is the evaporation of water from the surface of leaf cells in actively growing plants. It occurs chiefly on the leaves while their stomata are open for the passage of CO_2 and O_2 during photosynthesis. Transpiration is usually increased in plant diseases where the pathogen damages aerial parts of the plants. This is the result of increase in the permeability of leaf cells, the dysfunction of stomata, and destruction of the cuticle. It is observed that before producing visible disease symptoms, pathogens transiently or permanently enhance water loss by increasing the evaporating surface through the development of infection structures on the plant surface. In wheat diseases such as rusts, wherein numerous pustules form and rupture the epidermis of the host, leaf spots, where the cuticle, epidermis and all the other tissues, including xylem, may be damaged in the infected areas and in the powdery mildew, in which a large proportion of the epidermal cells are invaded by the pathogen and the injury of a considerable portion of the cuticle and epidermis of leaf and stem results in enormous loss of water from the affected areas.

11.8   EFFECT ON TRANSLOCATION OF ORGANIC NUTRIENTS

Organic nutrients, synthesized in leaf cells move through plasmodesmata into adjoining phloem elements. From phloem sieve tubes the nutrients move downwards and finally into the protoplasm of living non-photosynthetic cells, where they are directly used or stored. Plant pathogens may interfere with the movement of organic nutrients from the leaf cells to the phloem, with their translocation through the phloem elements, or, possibly, with their movement from the phloem into the cells that will utilize them. In wheat diseases like rusts and mildew, caused by obligate parasites, there is an accumulation of photosynthetic products, as well as inorganic nutrients, in the areas invaded by the pathogen. In rusts, powdery mildew and leaf blight infected areas there is significant reduction in photosynthesis and increase in respiration. However, translocation of organic nutrients from uninfected areas of the leaves or from healthy leaves toward the infected areas results in temporary increase in accumulation of starch, other compounds as well as dry weight.

11.9   HYPERSENSITIVE RESPONSE (HR)

The hypersensitive response (HR) is one of the most efficient forms of plant defense against biotrophic pathogens and is the result of signaling cascades that start by the interactions of a pathogen-provided ligand with intra or extra-cellular receptor encoded by plant disease resistance genes. The HR is a complex, early defense response that causes necrosis and cell death to restrict the pathogen growth. The HR cell death in plants shares similar characteristics of programmed cell death observed in animals. Direct physiological contact between the host and infecting parasite is necessary for the activation of HR. The HR was first described by Stakman (1915) to describe rapid host cell death in resistant wheat plants upon infection by rust fungi. The HR is accompanied by the intracellular accumulation of lignin or lignin-like material. Studies on the formation of fungal haustoria, the occurrence of hypersensitive host cell death, lignin biosynthetic pathway and inhibitor studies affecting different key enzymes in lignin biosynthesis indicate a direct involvement of cellular lignification in HR.

Plant immunity triggered by a gene-for-gene recognition is often correlated with HR. During the process, the host uses its resistance (R) genes to recognize the pathogen avirulent gene (Avr) and changes its own membrane potential and ion permeability of the plasma membrane. In the earlier stage of the response, the R genes trigger an increase in extra-cellular H^+ and K^+, while eliciting an influx of Ca^{2+} and H^+ ions into the cell. The outward K^+ and inward Ca^{2+} and H^+ ion flux are dependent and trigger the HR, resulting in cell death and formation of local lesions. At later stage, cells undergoing HR cause oxidative burst by producing ROS, including superoxide anions, hydrogen peroxide, and hydroxyl radicals (Fig. 11.1). Lipid peroxidation and lipid damage may be partially responsible for some of these cell changes and probably affect membrane function.

Studies on host responses in the wheat-*Septoria tritici* interaction revealed that the infection of wheat by *S. tritici* was associated with a large and early accumulation of H_2O_2 in incompatible interactions, coinciding with pathogen arrest and thus indicating a role for H_2O_2 in the active defense of wheat. However, some reports suggest that H_2O_2 levels increased dramatically during a compatible interaction too and peaked at the late necrotrophic stage, implying that this was likely a stress-response and not involved in defense.

Some investigators have described the HR as the process similar to apoptosis, the principal manifestation of programmed cell death in many animal cell types (Morel & Dangl, 1997). This definition of HR has now expanded to include defense gene expression in addition to cell death (Heath, 2000). In the interaction between wheat plant and biotrophic rust fungi and powdery mildews, the hypersensitivity is initiated in plant cells in which the pathogen is forming a haustorium to absorb nutrients. The hypersensitivity resistance in these interactions is, therefore, post haustorial. In wheat, almost all of the leaf rust resistance genes used in wheat breeding cause a hypersensitivity reaction (McIntosh et al., 1998), as are most of the powdery mildew resistance genes in wheat and in barley (Chen & Chelkowski, 1999).

11.10   STRUCTURAL CHANGES

11.10.1   CELLULAR CHANGES

Cellular modifications in the plant cell wall serve as an important defense mechanism operating in the defense response of plants against parasites. Reinforcement of the cell wall involves callose deposition at attempted penetration sites, making the cell wall less vulnerable to degradation by cell-wall degrading enzymes (CWDEs). Callose is a minor component of healthy plant tissues. Plants respond to infection by pathogens with the rapid deposition of callose (Fig. 11.1). The presence of callose is usually associated with poor development of the haustoria. Callose deposition starts at early stages of pathogen invasion to inhibit pathogen penetration. It is a high-molecular weight β-(1,3)-glucan polymer that is usually associated, together with phenolic compounds, polysaccharides and antimicrobial proteins but without cellulose and pectin, with cell-wall appositions, called papillae. Papillae are supposed to be effective barriers that are induced at the sites of pathogen attack. In the case of powdery mildew and rust fungi, the haustorium is surrounded by papillae to seize further invasion by the pathogens. The role of papillae in host defense mechanisms has been demonstrated in many resistant/incompatible host-pathogen interactions. The structural defense reactions such as cell-wall appositions, collar or papillae formations, and encasements of haustoria are markedly more expressed in the resistant wheat leaves infected with *Puccinia striiformis* as compared to the susceptible cultivar (Kang et al., 2003). In case of fusarium head blight infected lemma tissue of the resistant wheat cultivar, there is pronounced defense responses in the form of cell-wall appositions as compared to the susceptible one (Siranidou et al., 2002).

11.10.2 BIOCHEMICAL CHANGES

Besides morphological and physiological modifications in host plants there are numerous biochemical alterations taking place in response to the pathogen attack. Many of these biochemical alterations, involved in defense mechanisms include the synthesis and accumulation of pathogenesis-related (PR) proteins, phenols, defense-related enzymes like Lipoxygenase (LOX), Polyphenol oxidases (PPO), peroxidases (POX), phenylalanine ammonia-lyase (PAL), mitogen-activated protein kinase (MAPK) lignin, etc. Some of these components are described briefly here.

11.10.3 LIGNIFICATION

Lignin, a complex polymer deposited in plants secondary cell wall, is a complex polymer built from phenolic acids, which are reduced to the corresponding alcohols. Rapid lignification of plant cell walls appears to be an important host defense mechanism. Hydroxyproline-rich glycoprotein is known to be a matrix for deposition of lignin. Rapid lignification appears to be an important defense mechanism in many plant–pathogen interactions. Lignin content increased in wheat leaves inoculated with an avirulent strain of *Puccinia recondita* f. sp. *tritici*, but no such increase was observed in leaves inoculated with a virulent strain (Southerton & Deverall, 1990b).

The attempted cell penetration by the powdery mildew fungus elicits complex interrelated responses in the epidermis and underlying mesophyll cells of cereals including the production of autofluorogenic compounds. Autofluorescence of cell walls localized at *Blumeria graminis* germ tube contact sites and in papillae had been described in numerous investigations with cereal plants (Kunoh et al., 1982). Physical and biochemical studies indicate that these autofluorogens are phenolic compounds synthesized de novo in response to pathogen attack and they are probably products related to lignin biosynthesis (Carver et al., 1996). Infected wheat leaves frequently show a higher lignin content, independent of the compatibility grade, however, it is more pronounced in incompatible interactions. Lignin density remains higher in host cell walls of resistant cultivar of *Puccinia striiformis* infected wheat leaves than the susceptible cultivar (Kang et al., 2003). The lignification is induced by fungal penetration and is often limited to a few cells at the infection site. Since the enhanced lignification is more pronounced in resistant wheat, it is assumed that it contributes to resistance against pathogens. According to Moerschbacher et al. (1990) lignification is one of the defense responses in the hypersensitive reaction of wheat to *Puccinia graminis*. During such a response in wheat, lignin rich in syringyl units accumulates (Menden et al., 2007).

Induced lignification is accompanied by an increase in the activity of the key enzymes of the phenylpropanoid pathway such as PAL, CAD, and POX. Recent studies show that suppression of host phenyl-propanoid metabolism by inhibition of PAL and CAD reduces localized autofluorescence and increases penetration as well as haustorium formation of wheat powdery mildew fungus. These enzymes have been shown to play an integral part in disease resistance in wheat against leaf rust (Johnson & Lee, 1978) and spot blotch (Das et al., 2003). Expression of caffeic acid O-methyltransferase (COMT), an enzyme which is involved in lignin synthesis is elevated in wheat (*Triticum monococcum*) following inoculation with *Blumeria graminis* (Bhuiyan et al., 2007). Role of lignification is mentioned in many reports like Menden et al. (2007) (wheat rusts), Bishop et al. (2002) (fusarium head blight). In contrast, no changes in the lignin content occurred in wheat leaves infected with Wheat Streak Mosaic Virus (Kofalvi & Nassuth, 1995).

With most forms of resistance, use of enzyme inhibitors of phenylpropanoid synthesis AOPP (a-aminooxy-b-phenylpropionic acid) and OH-PAS {[[(2-hydroxyphenyl) amino] sulphinyl] acetic acid, 1,1-dimethylethyl ester} reduces local accumulation of lignin at sites of response to primary germ tubes and appressoria. This is often associated with increased success of attempted penetration (haustorium formation) by the powdery mildew fungus. Therefore, phenol synthesis appears to be a key factor in different forms of resistance to *Blumeria graminis tritici*.

The induction of lignification in wheat appears nearly specific for filamentous fungi (Pearce & Ride, 1978). Saprophytic and pathogenic fungi induce lignification, but yeasts and bacteria are poor elicitors of lignification. This suggests that a single elicitor might be held in common by taxonomically related microorganisms. Furthermore, a single receptor in wheat might serve for recognition of all filamentous fungi. Evidence from lignification experiments also suggests that more than one elicitor might interact with a single receptor and that two or more receptors might independently activate lignin mediated resistance mechanism.

11.11   PR PROTEIN

PR proteins are proteins produced in plants in the event of a pathogen attack. They are induced as part of systemic acquired resistance. Some of these proteins are antimicrobial, attacking molecules in the cell wall of bacterial or fungal pathogens. These proteins accumulate locally in the infected and surrounding tissues, and also in remote uninfected tissues. Production of PR proteins in the uninfected parts of plants can prevent the affected plants from further infection. Most PR proteins in the plant species are acid-soluble, low molecular weight, and protease-resistant proteins. PR proteins, depending on their isoelectric points may be acidic or basic proteins. Currently 17 families of inducible PR proteins have been recognized (Van Loon et al., 2006). The first PR protein to be purified and characterized extensively from any plant was the chitinase prepared from wheat germ by Molano et al. (1979). Since then, several other PR proteins belonging to almost all classes have been identified in wheat.

Among the PR proteins, two plant hydrolases namely, β-1,3-glucanase and chitinase are considered to play an important role in plant defense against invasion by fungal pathogens because these two plant hydrolases can degrade chitin and β-1,3-glucans in the fungal cell walls. The breakdown products of fungal wall components, caused by the hydrolases, also can act as elicitors of plant defense responses (Arlorio et al., 1992). Therefore, β-1,3-glucanase and chitinase have been intensively studied with regard to their accumulation and subcellular localization in the infected wheat plants and their synergistic action in plant defense reactions in different fungal pathogen–plant systems. Kang et al. (2003) reported increased activity of chitinase and β-1,3-glucanase in extracellularly localized mainly in the cell walls of the resistant wheat leaves infected with *P. striiformis* while the cytoplasm and organelles in host tissues were almost free of these enzymes. However, there was a slight increase in the activity of these two enzymes in the *P. striiformis* infected leaves of the susceptible plant as compared to the uninoculated leaves. β-1,3-glucanase gene, *PRP2* was found to over-express in barley, wheat, sorghum, and rice leaves upon infection with *Bipolaris sorokiniana* (Jutidamrongphan et al., 1991). This increase in the activity of β-1,3-glucanases and chitinases in the intercellular space of the resistant but not in the susceptible wheat plant may result in removal of fungal β-1,3-glucan cell-wall layers, which might then provide an easier access of the chitinases to their substrate in the inner wall layers of the intercellular hyphae. Thus, both enzymes may cause profound damage to the hyphal cell-wall ultrastructure.

11.12 LIPOXYGENASE (LOX)

The LOXs, identified in animals and plants are non-heme-iron-containing enzymes, which catalyze the oxidation of polyunsaturated fatty acids containing a cis,cis-1,4-pentadiene site. In higher plants, the natural substrates for these enzymes are linoleic acids and linolenic. Linoleic acid and linolenic are predominantly present as soluble proteins in the cytosol but are also found in membranes and in different organelles. An increase in LOX activity in response to infection has been reported for several plant–pathogen systems, and LOX activity has been correlated with plant resistance against pathogens (Slusarenko, 1996). It has been suggested that LOX is involved in the development of an active resistance mechanism known as the HR. LOX activity increases more rapidly and to greater levels in an incompatible response than in a compatible one. LOX isozyme profiles in the wheat-rust fungus pathosystem revealed that several LOX species were induced differentially during the HR evoked by the pathogen (Bohland et al., 1997). Jasmonates have been found to stimulate LOX gene expression, protein, and activity in plants. Ocampo et al. (1986) made an early attempt to investigate LOX in the incompatible wheat/*Puccinia graminis* f. sp. *tritici* interaction and observed a correlation between increased LOX activity and the HR. A crude elicitor preparation from germ tubes of the rust fungus was also effective in stimulating enzyme activity. Furthermore, the studies on systemic acquired resistance in wheat revealed that chemically induced LOX (WCI-2) gene expression correlated with the onset of resistance against *Erysiphe graminis* f. sp. *tritici* and in several other host–pathogen combinations.

11.13 POLYPHENOL OXIDASES (PPO) AND PEROXIDASES (POX)

Fungal, bacterial, and viral diseases lead to alterations in the biochemical constituents of hosts. One of the most marked changes in the physiology of the infected host is change in the activity of oxidizing enzymes which are involved in the defense mechanism against the infecting pathogen. PPOs and POXs are among the most important defense gene products. PPOs are nuclear-encoded, plastid-localized copper metalloenzymes that catalyze the oxygen dependent oxidation of *o*-dihydroxyphenols to more toxic *o*-quinones (Mayer & Harel, 1991). Polyphenol oxidase (tyrosinase) is a bifunctional oxidase having both catecholase and cresolase activity. Its activity is dormant until released from the thylakoid by any disruptive force including wounding, senescence and pests or pathogens. Tyagi et al. (2000) observed higher activities of PPO and POX in wheat infected with *Alternaria triticina*. They found similar isozyme pattern of PPO in control and infected plants but there was over expression of POX in plants exposed to the pathogen.

Peroxidase is a universal, haeme-containing glycoprotein catalyzing the oxidation of cellular components by either hydrogen peroxide (H_2O_2) or organic hydroperoxidases. Peroxidases have been implicated in a number of diverse phenomena observed in plants that is, lignification, suberization, cell elongation, growth and regulation of cell wall biosynthesis and plasticity, which diversified during disease development process (Gabaldon et al., 2005). In diseased plants peroxidase enzymes affect the pathogen growth by reinforcing the cell wall, by producing the toxic substances to the pathogen through the production of oxidative burst or by accelerating the necrotic response in the infected and nearby cells as in case of HR (Almagro et al., 2009). The H_2O_2 formed in the process is utilized for polymerization of monolignols into lignins. Besides lignin formation, the production of hydrogen peroxide leads to peroxidase-mediated cross-linking of structural proteins in the cell wall thereby reinforcing it against the pathogen invasion. The importance of the role of peroxidase in defense reactions is supported by the fact that the transgenic tobacco plants with suppressed expression of peroxidase are over responsive to pathogen attack (Mittler et al., 1999). Ryals et al. (1996) stated that the inhibition of oxidoreductases like peroxidase is important for lesion development by the infecting pathogen and hence is important in the resistance response by the host.

Induction of variety of peroxidase isozymes is shown following infection by a pathogen to a host plant or during ethylene induced leaf senescence in cucurbitaceae (Jakupovic et al., 2006). Southerton and Deverall (1990a) reported increase in the activity of peroxidase in *Puccinia recondite* f. sp. *tritici* infected wheat plants, which was further higher in resistant as compared to the susceptible lines. Yang et al. (1984) observed higher POX activity in a resistant wheat cultivar than in a susceptible cultivar following inoculation with *Erysiphe graminis* f. sp. *tritici*. In addition, they observed expression of an extra peroxidase isozyme in the highly resistant cultivar, which could be the expression of a resistant gene.

11.14   CHANGE IN PHENOLICS CONTENT

Phenolic compounds, the plant secondary metabolites, constitute one of the most common and widespread groups of substances in plants. Plants need phenolic compounds for a number of essential processes such as growth, reproduction, pigmentation, and more importantly for resistance to pathogens. They are frequently produced and accumulate in the subepidermal layers of stress and pathogen exposed plant tissues. Phenolics act as protective agents, inhibitors, natural animal toxicants, and pesticides against invading pests. A number of simple and complex phenolics accumulate in plant tissues and act as phytoalexins, phytoanticipins, and nematicides against soil-borne pathogens and phytophagous insects (Lattanzio et al., 2006). Therefore, phenolic compounds are proposed to work as useful alternatives to the chemical control of plant pathogens. The synthesis, release and accumulation of phenolics such as salicylic acid play the leading role in defense strategies employed by plants against microbial pathogens. Synthesis of defense-related phenolics starts after recognition of conserved pathogen-associated molecular patterns (PAMPs) of potential pathogens by plant receptor. As a result, the infection process is restricted, which eventually leads to PAMP-triggered immunity.

Total phenolic status of wheat has been correlated with host resistance to a variety of diseases, including rusts, karnal bunt, take-all disease and others. Gogoi et al. (2001) observed enhanced phenol content in karnal bunt (*Neovossia indica*) infected wheat, which was more prominent in resistant varieties. Anjum et al. (2012) conducted the biochemical analysis of wheat plants infected with smut (*Ustilago tritici*) and found that roots were most sensitive as considerable increase in both phenolic content and enzymes activity was recorded during disease development. Conversely in case of infected stem and leaves the enzyme activities were found lower at early disease stage in comparison to the control. However, the enzyme activities increased in both parts when checked again at crop maturity. Increase in total phenolics is observed in other wheat diseases like fusarium head blight (Siranidou et al., 2002) flag smut (Beniwal et al., 2008) and others.

KEYWORDS

- CWDE
- hypersensitive response
- lipoxygenases
- MAPK
- peroxidases
- PR protein

REFERENCES

Agrios, G. N. Plant Pathology, 5th Edn, Elsevier Academic Press, Burlington, Mass. 2005; p 37.

Almagro, L.; Ros, G.; Belchi-Navarro, S.; Bru, R.; Barcelo, A. R.; Pedren, M. A. Class III Peroxidases in Plant Defence Reactions. *J. Exp. Bot.* **2009**, *60*, 377–390.

Anjum, T.; Fatima, S.; Amjad, S. Physiological Changes in Wheat During Development of Loose Smut. *Tropical Plant Pathol.* **2012**, *37*(2), 102–107.

Anonymous. Progress Report of All India Coordinated Wheat & Barley Improvement Project 2013–14, Project Director's Report. Ed: Indu Sharma Directorate of Wheat Research, Karnal, India. p 120.

Arlorio, M.; Ludwig, A.; Boller, T.; Bonfante, P. Inhibition of Fungal Growth by Plant Chitinases and β-1,3-Glucanases: A Morphological Study. *Protoplasma.* **1992**, *171*, 34–43.

Beniwal, M. S.; Karwasra, S. S.; Chhabra, M. L. Biochemical Changes in Wheat Plants Infected with Flag Smut. *Indian Phytopath.* **2008**, *61*(2), 243–246.

Bhardwaj, S. C.; Nayar, S. K.; Prashar, M.; Jain, S. K.; Singh, S. B. Diversity of Resistance for *Puccinia graminis tritici* in Wheat (*Triticum aestivum*) and *Triticale* Mutant. *Indian J. Agric. Sci.* **2003**, *73*(12), 676–679.

Bhardwaj, S. C.; Prashar, M.; Prasad, P. Ug99-Future Challenges. In *Future Challenges in Crop Protection Against Fungal Pathogens, Fungal Biology;* Goyal, A., Manoharachary, C., Eds.; Springer Science and Business Media New York, 2014; pp 231–248.

Bhuiyan, N.; Liu, W.; Liu, G.; Selvaraj, G.; Wei, Y.; King, J. Transcriptional Regulation of Genes Involved in the Pathways of Biosynthesis and Supply of Methyl Units in Response to Powdery Mildew Attack and Abiotic Stresses in Wheat. *Plant Mol. Biol.* **2007**, *64*, 305–318.

Bishop, D. L.; Chyatterton, N. J.; Harrison, P. A.; Hatfield, R. D. Changes in Carbohydrate Partitioning and Cell Wall Remodelling with Stress-induced Pathogenesis in Wheat Sheaths. *Physiol. Mol. Plant Pathol.* **2002**, *61*, 53–63.

Bohland, C.; Balkenhohl, T.; Loers, G.; Feussner, I.; Grambow, H. J. Differential Induction of Lipoxygenase Isoforms in Wheat upon Treatment with Rust Fungus Elicitor, Chitin Oligosaccharades, Chitosan, and Methyl Jasmonate. *Plant Physiol.* **1997**, *114*, 679–685.

Braun, H. J.; Atlin, G.; Payne, T. Multi-location Testing as a Tool to Identify Plant Response to Global Climate Change. In *Climate change and crop production;* Reynolds, M. P., Ed.; CABI, London, 2010; pp 115–138.

Carver, T. L. W.; Zhang, L.; Zeyen, R. J.; Robbins, M. P. Phenolic Biosynthesis Inhibitors Suppress Adult Plant Resistance to *Erysiphe graminis* in Oat at 20^{0}C and 10^{0}C. *Physiol. Mol. Plant P.* **1996,** *49,* 121–141.

Chang, Q.; Liu, J.; Wang Q.; Han. L.; Liu, J.; Li, M.; Huang, L.; Yang, J.; Kang Z. The Effect of *Puccinia striiformis* f. sp. *tritici* on the Levels of Water Soluble Carbohydrates and the Photosynthetic Rate in Wheat Leaves. *Physiol. Mol. Plant P.* **2013,** *84,* 131–137.

Chen, Y.; Chelkowsky, J. Genes for Resistance to Wheat Powdery Mildew. *J. Appl. Genet.* **1999,** *40,* 317–334.

Creelman, R. A.; Mullet, J. E. Biosynthesis and Action of Jasmonates in Plants. *Annu. Rev. Plant Physiol. Plant Mol. Biol.* **1997,** *48,* 355–381.

Das, S.; Aggarwal, R.; Singh, D. V. Differential Induction of Defense Related Enzymes Involved in Lignin Biosynthesis in Wheat in Response to Spot Blotch Infection. *Indian Phytopath.* **2003,** *56*(2), 129–133.

FAO. Part 3–Feeding the World. Trends in the Crop Sector. In *FAO Statistical Yearbook 2012*; Prakash, A., Stigler, M., Eds.; World Food and Agriculture. Food and Agriculture Organization of the United Nations (FAO): Rome, Italy, 2012; pp 182–197.

Gabaldon, C.; Serrano, M. L.; Pedreno, M. A.; Barcelo A. R. Cloning and Molecular Characterization of the Basic Peroxidase Isoenzyme from *Zinnia elegans*, an Enzyme Involved in Lignin Biosynthesis. *Plant Physiol.* **2005,** *139,* 1138–1154.

Garg, H.; Li, H.; Sivasithamparam, K.; Kuo, J.; Barbetti, M. J. The Infection Processes of *Sclerotinia sclerotiorum* in Cotyledon Tissue of a Resistant and a Susceptible Genotype of *Brassica napus. Ann. Bot.* **2010,** *106,* 897–908.

Gogoi, R.; Singh, D. V.; Srivastava, K. D. Phenols as a Biochemical Basis of Resistance in Wheat Against Karnal Bunt. *Plant Pathol.* **2001,** *50,* 470–476

Heath, M. C. Hypersensitive Response-related Death. *Plant Mol. Biol.* **2000,** *44*(3), 321–334.

Hu, G.; Rijkenberg, F. H. J. Scanning Electron Microscopy of Early Infection Structure Formation by *Puccinia recondita* f. sp. *tritici* on and in Susceptible and Resistant Wheat Lines. *Mycol. Res.* **1998,** *102*(4), 391–399.

Huang, G.; Kang, Z.; Zhu, Z.; Li, Z. Histopathological and Ultrastructural Studies on Development of *Puccinia recondita* f. sp. *tritici* in a Susceptible Wheat Cultivar. *Acta. Phytopathologica. Sinica.* **2003,** *1,* 52–56.

Jakupovic, M.; Heintz, M.; Reichmann, P.; Mendgen, K.; Hahn, M. Microarray Analysis of Expressed Sequence Tags from Haustoria of the Rust Fungus *Uromyces fabae. Fungal Genet. Biol.* **2006,** *43,* 8–19.

Johnson, L. B.; Lee, R. F. Peroxidase Changes in Wheat Isolines with Compatible and Incompatible Leaf Rust Infections. *Physiol. Plant Pathol.* **1978,** *13,* 173–181.

Jutidamrongphan, W.; Andersson, J. B.; Mackinnon, G.; Manners, J. M.; Simpson, R. S.; Scott, K. J. Induction of β-1,3-Glucanase in Barley in Response to Infection by Fungal Pathogens. *Molec. Plant Microbe Interact.* **1991,** *4,* 234–238.

Kang, Z.; Huang, L.; Buchenauer, H. Subcellular Localization of Chitinase and β-1,3-Glucanase in Compatible and Incompatible Interactions Between Wheat and *Puccinia striiformis f. sp. tritici. J. Plant Dis. Prot.* **2003,** *110*(2), 170–183.

Kissoudis, C.; van de Wiel, C.; Visser, R. G. F.; van der Linden, G. Enhancing Crop Resilience to Combined Abiotic and Biotic Stress Through the Dissection of Physiological and Molecular Crosstalk. *Front. Plant Sci.* **2014,** *5,* 207.

Kofalvi, S. A.; Nassuth, A. Influence of Wheat Streak Mosaic Virus Infection on Phenyl Propanoid Metabolism and the Accumulation of Phenolics and Ligninin Wheat. *Physiol. Mol. Plant Pathol.* **1995,** *47,* 365–377. doi: 10.1006/pmpp.1995.1065.

Kunoh, H.; Yamamori, K.; Ishizaki, H. Cytological Studies of Early Stages of Powdery Mildew in Barley and Wheat. VIII. Autofluorescence at Penetration Sites of *Erysiphe graminis hordei* on Living Barley Coleoptiles. *Physiol. Plant Pathol.* **1982,** *21,* 373–379.

Lattanzio, V.; Lattanzio, V. M. T.; Cardinali, A. Role of Phenolics in the Resistance Mechanisms of Plants Against Fungal Pathogens and Insects. In *Phytochemistry Advances in Research;* Imperato, F., Ed.; India: Research Signpost, 2006; pp 23–67.

Mayer, A. M.; Harel, E. Phenol Oxidases and Their Significance in Fruit and Vegetables. In *Food enzymology;* Fox, P. F., Ed.; Elsevier, New York, USA, 1991; pp 373–398.

Mcintosh, R. A.; Hart, G. E.; Devos, K. M.; Gale, M. D.; Rogers, W. J. Proceedings of the 9th International Wheat Genetics Symposium, Volume 5: Catalogue of Gene Symbols for Wheat. Saskatoon, Saskatchewan, Canada, 1998; pp 235.

Menden, B.; Kohlhoff, M.; Moerschbacher, B. M. Wheat Cells Accumulate a Syringyl-Rich Lignin During the Hypersensitive Resistance Response. *Phytochemistry.* **2007,** *68,* 513–520.

Mittler, R.; Herr, E. H.; Orvar, B. L.; Camp, W.; Willekens, H.; Inze, D.; Ellis B. E. Transgenic Tobacco Plants with Reduced Capability to Detoxify Reactive Oxygen Intermediates are Hyper Responsive to Pathogen Infection. *Proc. Natl. Acad. Sci. U.S.A.* **1999,** *96,* 14165–14170.

Moerschbacher, B. M.; Noll, U.; Gorrichon, L.; Reisener, H. J. Specific Inhibition of Lignification Breaks Hypersensitive Resistance of Wheat to Stem Rust. *Plant Physiol.* **1990,** *93,* 465–470.

Molano, J.; Polacheck, I.; Duran, A.; Cabib, E. An Endochitinase from Wheat Germ. *J. Biol. Chem.* **1979,** *254,* 4901–4907.

Morel, J. B.; Dangl, J. L. The Hypersensitive Response and the Induction of Cell Death in Plants. *Cell Death Differ.* **1997,** *4*(8), 671–683.

Nicholson, R. L.; Hammerschmidt, R. Phenolic Compounds and Their Role in Disease Resistance. *Annu. Rev. Phytopathol.* **1992,** *30,* 369–389.

Ocampo, C.; Moerschbacher, B.; Grambow, H. J. Increased Lipoxygenase Activity is Involved in the Hypersensitive Response of Wheat Leaf Cells Infected with the Avirulent Rust Fungi or Treated with Fungal Elicitor. *Z. Naturforsch.* **1986,** *41C,* 559–563.

Pearce, R. B.; Ride, J. P. Elicitors of the Lignification Response of Wheat. *Ann. Appl. Biol.* **1978,** *89,* 306–307.

Rosegrant, M. W.; Agcaoili, S. M.; Perez. N. D. Global Food Projections to 2020: Implications for Investment. Food, Agriculture and the Environment Discussion, Paper 5. International Food Policy Research Institute, Washington, DC, 1995, pp 54.

Ryals, J. A.; Neuenschwander, U. H.; Willits, M. G.; Molina, A.; Steiner, H.; Hund, M. D. Systemic Acquired Resistance. *Plant Cell.* **1996,** *8,* 1809–1819

Sander, J. F.; Heitefuss, R. Susceptibility to *Erysiphe graminis* f. sp. *tritici* and Phenolic Acid Content of Wheat as Influenced by Different Levels of Nitrogen Fertilization. *J. Phytopathol.* **1998,** *146,* 495–507.

Shaw, M.; Samborski, D. J. The Physiology of Host-parasite Relations III. The Pattern of Respiration in Rusted and Mildewed Cereal Leaves. *Canadian J. Botany.* **1957,** *35,* 389–407.

Singh, R. P.; Hodson, D. P.; Huerta-Espino, J.; Jin, Y.; Bhavani, S.; Njau, P.; Herrera-Foessel, S.; Singh, P. K.; Singh, S.; Govindan, V. The Emergence of Ug99 Races of the Stem Rust Fungus is a Threat to World Wheat Production. *Ann. Rev. Phytopathol.* **2011,** *49,* 465–481.

Singh, R. P.; Hodson, D. P.; Huerta-Espino, J.; Jin, Y.; Njau, P. Will Stem Rust Destroy the World's Wheat Crop? *Adv. Agron.* **2008,** *98,* 271–309.

Siranidou, E.; Kang, Z.; Buchenaner, H. Studies on Symptom Development, Phenolic Compounds and Morphological Defence Responses in Wheat Cultivars Differing in Resistance to Fusarium Head Blight. *J. Phytopathol.* **2002,** *150,* 200–208.

Slusarenko, A. J. The Role of Lipoxygenase in Plant Resistance to Infection. In *Lipoxygenase and Lipoxygenase Pathway Enzymes;* Piazza, G., Ed., AOCS Press, Champaign, IL, 1996; pp 176–197.

Somssich, I.; Hahlbrock, K. Pathogen Defence in Plants – a Paradigm of Biological Complexity. *Trends Plant Sci.* **1998,** *3,* 86–90.

Southerton, S. G.; Deverall, B. J. Changes in Phenylalanine Ammonia Lyase and Peroxidase Activities in Wheat Cultivars Expressing Resistance to the Leaf-rust Fungus. *Plant Pathol.* **1990a,** *39,* 223–230.

Southerton, S. G.; Deverall, B. J. Histochemical and Chemical Evidence for Lignin Accumulation During the Expression of Resistance to Leaf Rust Fungi in Wheat. *Physiol. Plant Pathol.* **1990b,** *36,* 483–494.

Stakman, E. C. Relation Between *Puccinia graminis* and Plants Highly Resistant to its Attack. *J. Agric. Res.* **1915,** *4,* 193–299.

Tariq, V. N.; Jeffries, P. Ultrastructure of Penetration of *Phaseolus* spp. by *Sclerotinia sclerotiorum. Canadian J. Botany.* **1986,** *64,* 2909–2915.

Tiburzy, R.; Reisener, H. J. Resistance of Wheat to *Puccinia graminis tritici*; Association of the Hypersensitive Reaction with the Cellular Accumulation of Lignin-like Material and Callose. *Physiol. Mol. Plant Pathol.* **1990,** *36,* 109–120.

Tyagi, M.; Kayastha, M. A.; Sinha, B. The Role of Peroxidase and Polyphenol Oxidase Isozymes in Wheat Resistance to *Alternaria triticina. Biologia. Plantarum.* **2000,** *43,* 559–562.

Van Loon, L.; Rep, M.; Pieterse, C. Significance of Inducible Defense-related Proteins in Infected Plants. *Annu. Rev. Phytopathol.* **2006,** *44,* 135–162.

Vidaver, A. K.; Lambrecht, P. A. Bacteria as Plant Pathogens. *Plant Health Instructor.* **2004.** doi: 10.1094/PHI-I-2004-0809-01.

Yang, J. S.; Li, S. F.; Wu, W.; Cao, D. J.; Wu, Y. S.; Xue, Y. L. Varietal Resistance of Wheat to Powdery Mildew and its Relation to Peroxidases. *Acta. Phytopathologica. Sinica.* **1984,** *14,* 235–240.

CHAPTER 12

DISPERSAL AND ADAPTATION OF PHYTOPATHOGENS ON THE GLOBAL AND CONTINENTAL SCALES AND ITS IMPACT ON PLANT DISEASES

ASHOK K. MEENA*, SHANKAR L. GODARA, and ANAND K. MEENA

Department of Plant Pathology, College of Agriculture, SK Rajasthan Agricultural University, Bikaner 334006, India

Corresponding author. E-mail: ak_patho@rediffmail.com

CONTENTS

ABSTRACT

The second step in infection chain is the dissemination of plant pathogens. Some of the most striking and extreme consequences of rapid, long-distance aerial dispersal involve pathogens of crop plants. Long-distance dispersal of fungal spores by the wind can spread plant diseases across and even between continents and reestablish diseases in areas where host plants are seasonally absent. For such epidemics to occur, hosts that are susceptible to the same pathogen genotypes must be grown over wide areas, as is the case with many modern crops. The strongly stochastic nature of long-distance dispersal causes founder effects in pathogen populations, such that the genotypes that cause epidemics in new territories or on cultivars with previously effective resistance genes may be a typical. Similar but less extreme population dynamics may arise from long-distance aerial dispersal of other organisms, including plants, viruses, and fungal pathogens of humans.

12.1   INTRODUCTION

The unique combination of biological factors and human intervention in the life histories of wind-dispersed plant pathogens can result in such exceptional events as invasions of new territory on global or continental scales or the rapid spread of virulent genotypes on previously resistant cultivars. It is a considerable challenge to develop a predictive model that would describe such rare events. The combination of separate lines of research in epidemiology and population genetics has provided qualitative insights into the causes and consequences of these processes, and a goal for the next few years is to define with greater precision the limits of predictability of the occurrence of such unusual events.

"Transport of spores or infectious bodies, acting as inoculum, from one host to another host at various distances resulting in the spread of the disease, is called dispersal, dissemination or transmission of plant pathogens."

The dispersal of the pathogen or disease is important not only for spread of plant diseases but also for continuity of the life cycle and evolution of the pathogen. The knowledge of these methods of dispersal is essential for effective control of plant diseases because possibilities of preventing dispersal and thereby breaking the infection chain exist.

In fungi, productions of asexual and sexual spores follow the active vegetative growth of the fungus in or on the host tissues and are dispersed mechanically in time and space by various means. Natural populations generally show no evidence of dispersal on the scale of hundreds of kilometers,

except where human intervention has brought together aggressive pathogens and uniformly susceptible hosts, as in chestnut blight. This is because
of the patchiness and diversity of host populations, since spores of fungal
pathogens of wild plants presumably have the same dispersal characteristics as those of crops. The patchiness of populations following postglacial
expansion should be as fascinating a topic with fungi as it is with plants and
animals; only one case has been studied so far. *Coccidioides immitis*, an
aggressive pathogen of mammals, is endemic to North America but appears
to have codispersed from Texas to South America, which has compara-
tively little genetic diversity, with one of its hosts, probably humans, in the
late Pleistocene. Although *Coccidioides immitis* undergoes long-distance
dispersal by windblown arthroconidia, it is more likely that intercontinental
dispersal was by cysts in its human host.

In bacterial diseases, the bacterial cells come out on the host surface as
ooze or the tissues may be disintegrated so that the bacterial mass is exposed
and then dispersed by various physical and biological agencies.

In case of viral pathogens, which have no such organs and they are transmitted by insects, mites, phanerogamic parasites, nematodes, and human
beings. The two links in the infection chain of an animate pathogen, viz.,
survival through dormant structures and the dispersal of the pathogen are
very closely bound with each other. Actually the dormant structures provide
means of *dispersal in time*, that is, the pathogen is retained viable over a
period of time enabling it to be transported through physical agencies
without being harmed.

The dispersal of infectious plant pathogens in *space* occurs through two
ways:

1. Autonomous or direct or active dispersal.
2. Indirect or passive dispersal.

12.2 AUTONOMOUS OR DIRECT OR ACTIVE DISPERSAL

In this method, the dispersal of plant pathogens takes place through soil,
seed, and planting material during normal agronomic operations. There is
no major role of external agencies like insects, wind, water, etc. in this type
of dispersal.

12.2.1 SEED AS THE SOURCE OF AUTONOMOUS DISPERSAL

Since most of the cultivated crops are raised from seed the transmission
of diseases and transport of pathogens has much importance. The dormant

structures of the pathogen (e.g., seeds of *Cuscuta,* Sclerotia of ergot fungus, smut sori, etc.) are found mixed with seed lots and they are dispersed as seed contaminants. The bacterial cells or spores of fungi present on the seed coat (such as in smuts of barley, sorghum, etc.) are transported to long distances. Dormant mycelium of many fungi present in the seed is transmitted to long distances. There are three types of dispersal by seed, viz., contamination of the seed, externally seed borne, and internally seed borne.

12.2.1.1   CONTAMINATION OF THE SEED

Seed borne pathogens move in seed lot as separate contaminants without being in intimate contact with the viable crop seeds. The seeds of the pathogen or parasite and the host are mixed during harvest of the crop. In many cases, the identity of the seeds of the two entities (host and the pathogens) is difficult to separate. For example, smut of pearl millet and ergot of rye. Smut sori and ergots mix easily with the seed lots during harvest and threshing.

12.2.1.2   EXTERNALLY SEED BORNE

Close contact between structure of the pathogen and seeds is established, where the pathogen gets lodged in the form of dormant spores or bacteria on the seed coat during growth of the crop or at the time of harvest and threshing. For example, short smut of sorghum, bacterial blight of cotton, loose smut of barley, etc. In many pathogens the externally seed borne structures, such as smut spores can persist for many years due to their inherent capacity for long survival. For example, the spores of *Tilletia caries* (stinking smut of wheat) remain viable even after 18 years and those of *Ustilago avenae* (oat smut) for 13 years.

12.2.1.3   INTERNALLY SEED BORNE

The pathogen may penetrate into the ovary and cause infection of the embryo, while it is developing. They become internally seed borne. For example, loose smut of wheat.

12.2.1.4   DIFFERENTIATE SEED INFECTION AND INFESTATION

Seed infection: The seed in infected only when the pathogen has grown in or on it for some time and established its relationship with the seed tissues.

For example, loose smut of wheat, where the fungus grows in the embryonic tissues, and becomes dormant when the seed enters dormancy.

Seed infestation: When the fungus or the pathogen is present on the seed coat and in the seed lot, it is only transport of the pathogen, and the seed is infested.

12.2.2 SOIL AS A MEANS OF AUTONOMOUS DISPERSAL

Soil borne facultative saprophytes or facultative parasites may survive through soil. The dispersal may be by movement of pathogen in the soil or by its growth in soil or by movement of the soil containing the pathogen. The former is known as *dispersal in soil,* while the later is called *dispersal by soil.*

12.2.2.1 DISPERSAL IN SOIL

The following are the three stages of dispersal in soil.

i) *Contamination of soil*: Contamination of the soil takes place by gradual spread of the pathogen from an infested area to a new area.

ii) *Growth and spread of a pathogen in soil*: Once the pathogen has reached the soil it can grow and spread based on its ability to multiply and spread. Among characters of the pathogen its adaptability to soil environment including its saprophytic survival ability are most important. The survival ability of the pathogen is governed by high growth rate, rapid spore germination, better enzymatic activity, capability to produce antibiotics, and tolerance to antibiotics produced by other soil microorganisms. On the basis of this competitive saprophytic ability the pathogens in soil can be of three types. Specialized facultative parasites (Saprophytes) can pass their life in soil in the absence of host plants, but they depend more on the residues of the host plant (e.g., *Armillariella mellea, Ophiobolous graminis*, etc.). Unspecialized facultative parasites can pass their entire life in the soil (*Pythium* sp., and *Phytophthora* sp.). The soil borne obligate parasites, such as *Plasmodiophora brassicae, Synchytrium endobioticum* require the presence of active host.

iii) *Persistence of the pathogen in soil*: The pathogens persist in the soil as dormant structures like oospores (*Pythium, Phytophthora, Sclerospora*, etc.), Chlamydospores (*Fusarium*), smut spores (*Ustilago*), and sclerotia (*Rhizoctonia, Sclerotium*).

12.2.2.2   DISPERSAL BY THE SOIL

The pathogen is dispersed by the soil during cultural operations through the agricultural implements, irrigation water, workers feet, etc. Propagules of fungi and the plant debris containing the fungal and bacterial pathogens thus spread throughout the field. The transfer of soil from one place to another along with propagating materials is the most important method of dispersal of pathogen. For example, transfer of papaya seedlings from a nursery infested with *Pythium aphanidermatum* (causal agent of stem or foot rot of papaya) can introduce the pathogen in new pits for transplanting the seedlings. Similarly grafts of fruit trees transported with soil around their roots can transmit pathogens present in the nursery to the orchards.

12.2.3   THE PLANT AND THE PLANT ORGANS AS A MEANS OF AUTONOMOUS DISPERSAL

The plants, plant parts other than seed that are used for vegetative propagation, raw field produce, and plant debris that accumulates during the course of cropping constitute the third method of autonomous dispersal. For example, late blight of potato was introduced in North America and in Europe through seed tubers brought from the native source of the in South America. Citrus canker was introduced into California from Asia. The climatic conditions favored its epidemic in California.

12.3   PASSIVE OR INDIRECT DISPERSAL

Passive dispersal of plant pathogens happens through animate and inanimate agents.

12.3.1   ANIMATE AGENTS

12.3.1.1   INSECTS

Insects carry plant pathogens either externally (epizoic) or internally (endozoic). They can disseminate bacteria, fungi, viruses, mycoplasmas, spiroplasmas, rickettsia, etc.

Fungal diseases: The external transmission is of special interest in those fungi, which produce conidia, oidia, and spermatia in honey secretions having attractive odors, for example, sugary disease of sorghum. The spermatial oozings at the mouth of spermagonia in the ascomycetes attract various types of insects, flies, pollinating bees, and wasps, which play a dual role, viz., pollination and transmission of plant pathogens. Dutch elm disease (*Ceratostomella ulmi*) is transmitted internally by elm bark beetles.

12.3.1.2   GENERAL FEATURES OF SPORES

12.3.1.2.1 Spore Structure and Germination

Fungi are characterized by a wide variety of spores, possessing a complex morphology and specialized physiology, which make these microscopic bodies much more enduring then the vegetative thallus. Moreover, production at the climax of the fungal life cycle, spores have to perform embryonic functions for which they are equipped with the inherent potentiality to develop into a new generation. Spores being the principal agent of fungal dispersal have received consistent attention of researchers. On this aspect of spore germination in fungi has been discussed right from the time of De Bary (1987) and in many review articles as well as references including those by Gottlieb (1950); Lilly and Barnett (1951); Shaw (1964).

Germination of fungal spores is essentially process during which the normal metabolic and physiological activity is restored after a temporary halt or check in these activities in the resting spore, followed by a morphological transformation of the spore into a new thallus. In a general sense, however, germination usually implies the emergence of a definite germ tube. In some cases, germination does not produce any germ tube; rather it leads to the cleavage of the protoplasm in to a numbers of cells, which develop into zoospores as they emerge from the zoosporangium. However, the formation of a typical germ tube occurs in these cases also, when the zoospores undergo germination after they become non motile. Other morphological variations may also be noted in specific groups of fungi.

12.3.1.2.2   Dormancy

Dispersal of spores may lodge in an environment, which may or may not be conducive to their germination and the growth of the resultant thallus.

Some of the fungal spores initially remain indifferent to the environment around them and even if the conditions are favorable for their germination, they do not do some far a specified a period of time. Such spores are said to be dormant or it is also defined as a condition of suspended growth and reduced metabolism of an organism, generally induced by internal factors or environmental conditions as a mechanism of survival. During the period of dormancy, they are supposed to complete their maturation process, or if already mature this period is treated as an enforced period of rest. This kind of dormancy is obviously controlled by certain internal factors of the spore and is therefore, designated as constitutive dormancy. This type of dormancy is exhibited by the ascospores of certain ascomycetes, and has been studies in *Neurospora*. Doran (1922) considered that the internal factors controlling constitutive dormancy included the maturity of spores, its longevity animation as well as vitality. Shear and Dodge (1927) as well as Goddard (1935) found that the dormant ascospores of *Neurospora* could be induced to germinate by giving them a heat treatment for 20 min at 60 °C. Exogenous dormancy or the environmental dormancy is a more common type of dormancy, which is controlled or enforced upon the spore by some external environmental factors, unfavorable for germination, and the growth of the new thallus. Exogenous dormancy is obviously meant to tide over the hostile environment, in which the spore has been lodged, and it is broken as soon as the conditions for germination. Various environmental factors may enforce dormancy on a spore, which is otherwise ready to germinate. Physical factors, like temperature, humidity, pH, O_2/Co_2 balance, etc. may have definite influence over the germinability of fungal spores. Lockwood (1964) suggested that the living microorganisms compete for nutrients required for germination of fungal spores, which are thus unable to germinate unless the nutrients are made available from a newly incorporated substrate. According to this view, therefore no inhibitor as such is involved. However, specific inhibitors produced by soil microorganisms have been suggested to cause fungistasis by Waid (1960) and Park (1967).

Various fungal species, such as *Plasmodiophora brassicae*, cause club root disease of crucifer produces resting spores, chlamydospores that are able to survive in the soil for many years. Such spores must not germinate unless there is a suitable host nearby, and so spore germination is specifically linked with the chemicals produced by growing cabbage roots. Similarly, the resting spores of *Spongospora subterranean* are stimulated to germinate only by the roots of potato and a few other solanaceous host species. Recent work on mycorrhizal associations has given some insights into the specificity of fungi-host relationships, spore dormancy of this type is not thought to be common among the toadstools and mushrooms, but this may be partly a reflection of the present level of knowledge concerning this aspect of their biology.

Bacterial Diseases: The fire blight organism (*Erwinia amylovora*) and citrus canker bacterium (*Xanthomonas axonopodis* pv. *citri*) are transmitted by flies (bees) and ants and the later by leaf miner, respectively. The cucumber wilt bacterium, *Erwinia tracheiphila* is spread by the stripped cucumber beetles (*Acalymma vittata*), and the spotted cucumber beetle (*Diabrotica undecimipunctata*). When the beetles are feeding on the diseased plant, the bacterium contaminates the mouth parts and passes into the gut of the beetle, and over winters inside the beetle during the winter season. Thus the beetle helps the bacteria in two ways, that is, in their transmission and survival.

Viral Diseases: More than 80% of the viral and phytoplasmal diseases are spread by different types of insects. The insect which acts as specific carriers in disseminating the diseases are called *insect vectors*. Both Aphids (Aphididae) and leaf hoppers (Cicadellidae or Jassidae) in the order Homoptera contain largest number and the most important insect vectors of plant viruses viz., beet mosaic, lettuce mosaic, potato virus Y, bean common mosaic, citrus tristeza transmitted by aphids and rice tungro virus, beet curly top by leaf hoppers. Certain species of mealy bugs and scale insects (Coccoidae), whiteflies (Aleurodidae), and tree hoppers (Membracidae) in Homoptera also transmit virus diseases. The leaf curl virus diseases that is, okra yellow vein mosaic, okra leaf curl virus, chilli leaf curl, cotton leaf curl, and mung bean yellow mosaic are transmitted by white fly. Insect vectors of plant viruses are few in true bugs (Hemiptera), thrips (Thysanoptera), beetles (Coleoptera), and grasshoppers (Orthoptera).

Mycoplasma diseases: Plant MLO's are phloem inhabitants and those insects, which are feeding on phloem of plants transfer the MLO's. Mycoplasmal diseases are mostly transmitted by leaf hoppers, for example, sesamum phyllody (*Orosious albicinctus*) and little leaf of brinjal (*Hishimonas phycitis*).

12.3.1.3   MITES

Mites belonging to the families Eryophyiidae (eryophyiid mite) and Tetranychidae (spider mite) of class Arachnida transmit plant viruses. The genera *Abacarus, Aceria, Eriophyes,* and *Brevipalpus* are important. For example,

Aceria cajani transmits pigeonpea sterility mosaic virus *Aceria tulipae* transmits wheat streak mosaic.

12.3.1.4 FUNGI

Some soil borne fungal plant pathogens carry plant viruses in or on their resting spores and zoospores, and transmit them to susceptible hosts during the infection process (Table 12.1). *Tobacco necrosis virus* and *Cucumber mosaic virus* are carried outside the fungi, while lettuce big-vein virus is carried inside the zoospores. Many soil borne viruses are transmitted by the members of Chytridiales and Plasmodiophorales.

TABLE 12.1   Fungal Transmitted Viruses.

S. No.	Fungal vector	Disease
1.	*Olpidium brassicae*	Tobacco necrosis, tobacco stunt, Lettuce big-vein
2.	*Olpidium cucurbitacearum*	Cucumber necrosis
3.	*Polymyxa graminis*	Barley yellow dwarf mosaic, wheat soil borne mosaic, peanut clump
4.	*Polymyxa betae*	Beet necrotic yellow vein
5.	*Spongospora subterranean*	Potato mop top
6.	*Synchytrium endobioticum*	Potato virus X

12.3.1.5 NEMATODES

Several nematodes act as vectors for transmission of fungi, bacteria, and viruses (Table 12.2). The bacterium which causes yellow ear rot of wheat (*Corynebacterium tritici* or *Clavibacter tritici*) is disseminated by ear cockle nematode (*Anguina tritici*). If these two diseases appear together, a complex disease called tundu of wheat occurs. *Corynebacterium tritici* is not capable of dispersal and infection unless it is carried by *Anguina tritici*. Similarly, root rot and wilt pathogens, such as *Phytophthora, Fusarium, Rhizoctonia, Verticillium*, etc. are disseminated by nematodes. Plant nematodes play a vital role in transmitting certain virus diseases. Many soil borne viruses are known to be transmitted by the nematodes. *Xiphenema, Longidorous, Trichodorus,* and *Paratrichodorus* are the nematode genera belonging to Dorylaimoidea, which are known to transmit plant viruses. The nematode transmitted viruses are divided into two groups on the basis of shape of their particles: Nematode transmitted polyhedral viruses (NEPO) and nematode transmitted tubular viruses (NETU). NEPO viruses: These are nematode transmitted viruses with polyhedral particles. These are generally transmitted by species of *Xiphenema* and *Longidorus*: Tobacco ringspot virus, Tomato ringspot virus, Tomato black ring virus, and Arabis mosaic virus. NETU viruses: These are nematode transmitted viruses with tubular particles. NETU viruses are transmitted by *Trichodorus* and *Paratrichodorous*. Pea early browning virus (*Trichodorus* sp.), Tobacco rattle virus (*Trichodorus pachydermis*)

TABLE 12.2 Nematode Transmitted Viruses.

Genus	Species	Virus
Longidorus	*apulus*	*Artichoke Italian latent virus* (Italian isolate)
	arthensis	*Cherry rosette virus*
	attenuatus	*Tomato black ring virus*
	diadecturus	*Peach rosette mosaic virus*
	elongatus	*Raspberry ringspot virus*
		Tomato black ring virus
	fasciatus	*Artichoke Italian latent virus* (Greek isolate)
	macrosoma	*Raspberry ringspot virus*
	martini	*Mulberry ringspot virus*
Paralongidorus	*maximus*	*Raspberry ringspot virus*
Xiphinema	*americanum sensu lato*	*Cherry rasp leaf virus* (CRLV)
		Peach rosette mosaic virus (PRMV)
		Tobacco ringspot virus (TRSV)
		Tomato ringspot virus
	americanum sensu stricto	*Cherry rasp leaf virus* (CRLV)
		Tobacco ringspot virus (TRSV)
		Tomato ringspot virus
	bricolensis	*Tomato ringspot virus*

TABLE 12.2   *(Continued)*

Genus	Species	Virus
	californicum	*Cherry rasp leaf virus*
		Tobacco ringspot virus
		Tomato ringspot virus
	diversicaudatum	*Arabis mosaic virus*
		Strawberry latent ringspot virus
	index	*Grapevine fanleaf virus*
	intermedium	*Tobacco ringspot virus*
		Tomato ringspot virus
	rivesi	*Cherry rasp leaf virus*
		Tobacco ringspot virus
		Tomato ringspot virus
	tarjanense	*Tobacco ringspot virus*
		Tomato ringspot virus

12.3.1.6   HUMAN BEINGS

Human beings role in dissemination of plant pathogens is more direct than indirect. The ways and means in which human beings help in dispersal are as follows.

Transportation of seeds (seed trade): The import and export of contaminated seeds without proper precautions lead to movement of pathogens from one country to another or from one continent to another. The diseases which are amenable to such transmission are mainly those that are carried in or on the propagative parts and seed. For example, late blight of potato, Downy mildew of grapevine, Citrus canker, *Fusarium* wilt of banana, etc.

Planting diseased seed materials: Planting diseased bulbs, bulbils, corms, tubers, rhizomes, cuttings, etc., of vegetatively propagated plants, such as potato, sweet potato, cassava, sugarcane, banana, many ornamentals and fruit trees, etc., help in dispersal of pathogens from field to field, orchard to orchard, locality to locality, or from one country to another.

During adoption of normal farming practices: Human beings engaged in preparatory cultivation, planting, irrigation, weeding, pruning, etc., help in dispersal of plant pathogens. Spores and other external structures of fungi can be carried by workers clothing's, shoes, and hands, etc., from plant to plant and from field to field.

By use of contaminated implements: Pathogens are transferred from one area to another through implements used in various cultural operations (weeding, thinning, hoeing, etc.) in the field. Soil borne diseases such as root rot, wilt, etc. Cutting knives and pruning knives also help in dispersal from one plant to another (Bunchy top of banana).

By use of diseased grafting and budding material: Grafting and budding between healthy and diseased plants is the most effective method of distribution of pathogens of horticultural crops.

12.3.1.7 DISPERSAL BY PHANEROGAMIC PARASITES

Phanerogamic parasites transmit the viruses by acting as a bridge between the diseased and healthy plants.

Dodder (*Cuscuta california, C. campesris, C. subinclusa*, etc.) *Cuscuta subinclusa*—Cucumber mosaic virus, *Cuscuta California* —Tobacco mosaic virus, Tobacco rattle virus, Tomato spotted wilt virus, *Cuscuta campestri* – Tomato bushy stunt virus

12.3.1.8 DISPERSAL BY BIRDS

This mode of dispersal is important in dissemination of seeds of flowering parasites and certain fungi. In tropics, crows feeding on the fleshy, sticky, and gelatinous berries of gaint mistletoe (*Dendrophthoe* sp.) deposit the seeds on the other trees with excreta. Seeds of *Loranthus* are disseminated by birds by sticking on their beaks and also through excreta. Stem segments of dodder are carried by birds for preparing their nests and thus get transported to new areas. Moreover, spores of chestnut blight fungus, *Endothea parasitica* are disseminated by more than 18 species of birds. Cleistothecia of many powdery mildew fungi are carried by feathers of birds.

12.3.1.9 FARM AND WILD ANIMALS

Farm animals (cattle) while feeding on diseased fodder ingest the viable fungal propagules (spores or oospores or sclerotia) and pass out as such in the dung. This dung when used as manure spread in the field and act as source of inoculum. Further, soil inhabiting fungi especially sclerotia adhere to the hoofs and legs of animals and get transported to other places.

Although experimental evidence is lacking, the possibility does exist that birds feeding on insect vectors can transport pathogens to distances not ordinarily covered by the vector itself. Dispersal of spores by birds carrying them of their feather is also possible. Dispersal of phanerogamic parasite by birds is an established fact. Birds feeding on berries of dendropthoe deposit the seed with their excreta on the trees. Stem fragments of dodder are carried by birds preparing their nest and thus these get transported to new location. Rodents and fur bearing animals can also be source of transmission of pathogens. Among mammal cattle feeding on contaminate fodder often pass out viable fungal propagules in the dung which can act as a source of inoculam when used as manure. Thus, conidia of *Colletotricum falcatm* (red rot of sugarcane) and sclerotia of many fungi have been detected in cattle dung.

Various examples like (*Ceratocystus ulmi* Ascomycotina) fungus causes Dutch elm disease, is vectored by beetles of the genus *Scolytus*. The spores are dispersed when the sticky sporing structures of *C. ulmi* bearing very small spores in a droplet of mucilage project into the brood galleries and contaminate the next generation of emerging beetles. Many hypogeous fungi rely on mammals for spore dispersal, such as tuber spp., Ascomycetes, *Endogone* spp. and *Melanogaster, Leucogaster, Hymenogaster*, and *Rhizopogon* (Gasteromycetes).

12.3.2   INANIMATE AGENTS

12.3.2.1   WIND

The dispersal of pathogens by wind is known as anemochory. Wind transmission involves the upward air currents, velocity, and the downward movements of the wind. Wind acts as a potent carrier of propagules of fungi, bacteria, and viruses. Dispersal of fungal spores by wind is by far the most common method for terrestrial fungi. Wind–borne spores finally coming to rest by sedimentation, impaction, or rain-wash. Wind is most important source of dispersal of some important plant pathogenic fungi. For wind dispersal the fungus must have certain properties adapted to conditions in the air. To become air borne fungal spores must be able to cross the laminar air flow boundary near the surface of the host .The spores discharge occurs with sufficient force (viz. ascospores of sclerotinia) to enable them cross the boundary and reach the upper air current. The other requirements of spore dispersal by air are their lightness and number. The spores or other propagules that are disseminated by wind are usually small, often about 1–8 μm in diameter and produced in enormous numbers (viz., uredospores in rust fungi and conidia of powdery mildew fungi). Other infectious parts of fungi such as bits of mycelium are also some time disseminated by wind. In dispersal of fungus spores by air the altitude reached in the atmosphere, wind current speed, rate of fall, and influences of atmospheric conditions on factors determine the dispersal distance. The many rust spores (e.g., spores of *Puccinia gramminis*, it is the wheat stem rust fungus, strain Ug99) are remain able to germinate after travel many days long distance, while basidiospores soon lose their ability to germinate. Therefore, basidiospores infect the nearby host plant as compression to other spore.

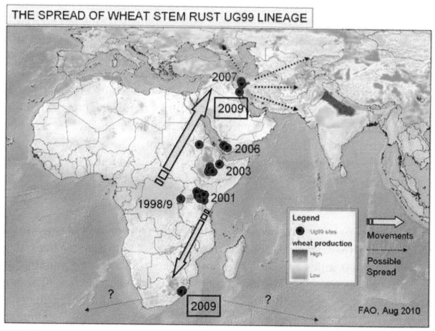

FIGURE 12.1 Fungi: Usually the fungal pathogens are light in weight and are well adapted to wind dispersal. The adaptations for wind dispersal in fungal pathogens include production of numerous spores and conidia, discharge of spores with sufficient force, production of very small and light spores so that they can move to long distances (powdery mildew, downy mildew, rusts, smuts, etc.).

Both short- and long-distance dissemination is possible by means of wind.

i) Spores adopted for short distance dissemination—sporangia of downy mildew fungi, conidia of powdery mildew fungi, and basidiospores of rust fungi. In the plains of northern India, the annual recurrence of cereal rusts is solely due to uredospores brought by wind from the source of survival in the hills in the far north (Himalayas) and south (Nilgiris).

ii) Spores adapted to long-distance dispersal—uredospores of rust fungi, Chlamydospores of smut fungi, and conidia of *Alternaria, Helmintho-sporium*, and *Pyricularia* uredial stages of the rust fungi travel long distances through air currents and thus are responsible for destructive epidemics over wide areas. For example, the uredospores of *Puccinia graminis* var. *tritici* have been detected as high as 14,000 feet above infected wheat fields (Christensen, 1942). Similarly, *Alternaria* spores at 8000 feet, *Puccinia recondita*, and *Cronartium ribicola* spores at 12,500 feet were reported. Long-distance dissemination of rust is a well established phenomenon. In the Indian subcontinent, rust spores are known to makes big jumps from the source areas to plain (Mehta, 1943; Nagarajan & Singh, 1974, 1975). Studies by Nagarajan and Singh (1973, 1974, and 1975) have shown that for south and central India the spores are wind borne, rain deposited and are transported from an altitude of around 700 mb (3030 m). The urediniospores get transported by upper winds from the Nilgiri and Palney hills and are then washed down over Central India by rain. Such a spread occurs to a distance of 200 m to more without infecting the fields in between.

Dispersal distance: In USA, uredospores of this fungus are blown from the far south (Mexico) into Dakota and Minnesota (far north) travelling more than 1000 miles in about two days without losing their viability. If the uredospores reach an altitude of 5000 feet, their distance dispersal in a 30 mile per hour wind could be about 1100 miles, without losing viability.

Nematodes: In addition to fungi, it also helps in the dissemination of the cysts of nematodes and also the seeds of phanerogamic parasites. For example, cysts of the nematode *Heterodera major*, which causes molya disease of wheat and barley, are carried by dust storms from Rajasthan to Haryana (Naik & Devika Rani, 2008).

Bacteria: Some pathogenic bacteria are carried along with the infected material to short distances by wind. For example, *Erwinia amylovora*, the causal agent of fire blight of apple and pear, produces fine strands of dried bacterial exudates, which may be broken off and are transmitted by wind (Naik & Devika Rani, 2008).

Viruses and phytoplasmas: are not directly transmitted by wind, but the insect and mite vectors that carry the viruses move to different directions and distances based on the direction and speed of the air (Naik & Devika Rani, 2008).

12.3.2.2 WATER

Transmission of plant pathogens by water is called as hydrochory. Water is less important than air in long-distance transport of pathogens, but it is more efficient as the pathogens land on the wet surface and can germinate immediately. Water dissemination occurs mainly through surface running water and rain splash. The surface flow of water after heavy rains or during irrigation from canals and wells carries the pathogens to short distances. For example, the mycelial fragments, spores or sclerotia of fungi, *Colletotrichum falcatum* (red rot of sugarcane), *Fusarium*, *Ganoderma*, *Macrophomina*, *Pythium*, *Phytophthora*, *Sclerotium*, etc., are transmitted through rain or irrigation water. Long-distance dispersal is also possible by water only when the floods cover larger areas or when the water flows from the sources of survival of pathogens to longer distances. Dissemination by rain splash is also called as splash dispersal. It is one of the efficient methods of dispersal of bacterial plant pathogens. Rain drops falling with force on sori, pustules, cankers or even soil surface may splash the propagules in small droplets, and enable them to land on neighboring healthy susceptible surfaces or the water droplets may be carried to long distances by air. For example, bacterial leaf spots of rice (*Xanthomonas campestris* pv. *oryzae*), bacterial leaf streak of rice (*Xanthomonas campestris* pv. *oryzicola*), and green ear of bajra (*Sclerospora graminicola*). Fungal spores and bacteria present in the air or plant surface are washed downward by rain splash or drops from overhead irrigation and are deposited on susceptible healthy plants.

12.3.2.3 SPORE DISPERSAL BY GRAVITY

Spore dispersal in space for many mushroom and toadstool species is influenced by the effects of gravity. When mature the four basidiospores of the field mushroom (*Agaricus campestris*) are violently discharged from the basidium to a distance of 0.1–0.3 μm. The four spores are liberated consecutively with about a minute or two between each discharge. In many species of fungal pathogen the discharge of spores is different accordingly to the species. In the basidium the spores are collapses soon after the discharge. While the mechanism of discharge is not fully understood, during the maturity of basidiospore exudes a minute drop of fluid, which grows to a definite size and then the spore and drop are discharged. Many factors are responsible for release the ascospores, such as air moisture content, temperature, light and air currents, etc. Various ascomycetes grow relatively close to the ground, the greater distance of discharge provided to ascospores enhances their chances of reaching the turbulent layer of air in the atmosphere where they stand a good chance of wide dispersal. Spore gravity is not a primary means of spore dispersal; evolutionary adaptations have been required of many fungi to overcome gravitational effects for effective spore dispersal.

12.3.3 ADAPTATION AND AGGRESSIVENESS OF THE PATHOGEN

Adaptation in fungi had been reviewed by Buxton (1960) and more recently by Person (1968). The possibility that the pathogens might adapt themselves to their host plants has been debated for many years. There seems to be no incontrovertible evidence that this occurs although the evidence for adaptation to chemicals is stronger. Apparent adaptation to host plants might be due to mutation or to the selection of pre-existing strains, which are better able to attack the host. Changes in the enzymatic capabilities of a pathogen may be mediated by the presence of appropriate substrates (adaptive enzyme) and this might modify its pathogenecity. Many years ago Ward (1903) suggested the concept of bridging host by means of which a pathogen is able to pass from a susceptible to a hitherto resistant host for example, in *Puccinia dispersa* the pathogen becomes adaptive to resistant species of Bromus. After propagation in some species of grass, somewhat similar changes were reported by Salmon (1904) in powdery mildews, and more recent work indicated that passage through some host plants can sometimes enhances the aggressiveness of a pathogen and vice versa, as has been demonstrated in several fungi and bacteria; that is, *Phytopathora infestans* (Ferris,1955) and *Xanthomonas stewarri* (Lincoln, 1940).

Changes in pathogenecity of *F. oxysporum* f. sp. *pisi* is brought about by exudates. Although there is increasing evidence in favor of parasitic adaptation. The mechanism involved remains obscure. Host selection of more aggressive biotypes from a mixture of biotypes, together with adaptation, mutation, and cytoplasmic variations may be involved in different cases.

Van der Plank (1963) defined aggressiveness as the non-specific component of pathogenicity. Later, Van der Plank (1971) illustrated this definition with data from an experiment (Paxman, 1963), in which isolates of *Phytophthora infestans* were grown for successive generations on a potato cultivar with no major resistance R-genes (i.e., all isolates were virulent, according to the gene-for-gene model). Environmental conditions, such as temperature, humidity, pH, ultra violet (UV), nutrient and salinity, have especially, critical effect establishment on fungal cellular composition and biochemical metabolism, which involve a role in virulence, pathogenicity, ecology, and colonization (Michea-Hamzehpour et al., 1980; Bennett et al., 1992; Fargues et al., 1997; Feder & Hofmann, 1999; Jessup et al., 2004; Fels & Kaltz, 2006; Toyoda et al., 2009). Several studies have reported that the temperature-responsive cellular component including tannin, phenol compounds and lipid content are involved in the synthesis of exported polysaccharides, and secondary metabolites (Leroi et al., 1994; Mejia et al., 1995; Li et al., 2003; Riehle et al., 2003; Garrett et al., 2006; Szeghalmi et al., 2006; Tharayil et al., 2011).

12.3.4 EVOLUTIONARY COMPONENTS OF THE HOST–PATHOGEN INTERACTION

To understand how adaptation of pathogens to their hosts and environments is translated in terms of aggressiveness, it is essential to be able to link the concept of aggressiveness to other concepts used in evolutionary epidemiology, that is, fitness and virulence (Galvani, 2003).

Fitness is generally defined as the *per capita* rate of increase of an individual or a gene copy (Futuyma, 1997). It is frequently measured as the average number of secondary infections produced from a single infected host in the absence of density-dependent constraints. Although focusing on among-host transmission seems natural for many animal pathogens, it may be more relevant in some plant pathogens to measure fitness as the average number of secondary lesions produced from a single initial lesion in the absence of density-dependent constraints, including alloinfection (among-host transmission) and autoinfection (within-host multiplication).

In evolutionary epidemiology, as well as in animal and human epidemiology, and unlike in plant pathology, virulence is defined as the quantity of damage induced by a pathogen on its host, and is measured in units of host fitness, and/or mortality (Read et al., 1999). Virulence is generally assumed to be a direct consequence of within-host–pathogen multiplication, although this direct causative relationship can be questioned (Day, 2002).

Aggressiveness, because it describes the ability of a pathogen to cause severe epidemics at the host population scale, combines both notions of pathogen fitness, and virulence. Both fitter and more virulent pathogens are generally considered as more aggressive since they will cause faster epidemics or more damage to the host population, respectively. However, aggressiveness cannot be considered strictly equivalent to pathogen fitness. Fitness related traits, such as spore viability or inter-seasonal survival are not usually considered aggressiveness traits. Similarly, aggressiveness in plant pathology is not a synonym for virulence in evolutionary epidemiology since many aggressiveness components (e.g., sporulation rate) do not quantify a decrease in host fitness or survival. In addition, other parameters that would be relevant to measuring pathogen virulence, such as decrease in host photosynthetic ability (usually larger than the effect accounted for by lesion size, see Bastiaans, 1991) and induced necrosis (pathogen-induced senescence, distinct from necrotic infected tissue, e.g., Magboul et al., 1992) are not usually measured. Aggressiveness can be broken down into several components and each of these components is likely to evolve (Lehman & Shaner, 2007). Recognizing that aggressiveness results from the expression of elementary quantitative components should make it possible to benefit from the predictions of evolutionary epidemiology models. For instance, most of these theoretical approaches assume that within-host multiplication harms hosts (i.e., causes virulence), so that fitness results from a trade-off between pathogen transmission and virulence (Day, 2003; Galvani, 2003). As a consequence, aggressiveness components that are more closely related to transmission are not expected to evolve under the same evolutionary forces as aggressiveness components linked to virulence. Establishing the correspondence between aggressiveness components, transmission, and virulence is not always easy, however. Lesion size can be related to virulence (Bastiaans, 1991). The latent period may have a twofold status, since a shorter latent period accelerates transmission but a longer latency could allow a greater development of the pathogen's organs within-host tissues and increase its ability to exploit the host. Infection efficiency, spore production rate, and infectious period are transmission traits, but also participate in within-host multiplication through autoinfection. Some difficulties met in this analysis obviously come from evolutionary epidemiology models making a distinction between within- and among-host scales, which is not always relevant in plant pathology. Although many animal diseases are caused by parasites with systemic effects that globally reduce the host viability, many plant parasites have localized effects and do not affect their host in a systemic manner. This is particularly true for foliar parasites, for which it has been demonstrated that the effect on the host is largely limited to a local reduction in photosynthetic capacity (Bastiaans, 1991; Robert et al., 2004). However, the pathogen can increase dramatically on an infected host leaf through autoinfection, which is analogous to within-host multiplication in animal diseases. The consequence is that the correspondence between aggressiveness components, transmission and virulence depends on the scale considered. At the scale of the lesion for instance, lesion growth may be considered as causing virulence (i.e., killing local host tissues), and other traits (infection efficiency, spore production, etc.) would correspond to transmission. Theoretical work on pathogen evolution progresses much faster than experimental work, and the lack of experimental evidence to evaluate theoretical predictions, and model hypotheses has been emphasized (Ebert & Bull, 2003). The growing body of experimental data on pathogen adaptation for quantitative traits, based on the study of crop pathogens, could offer opportunities to fill this gap, provided that the traits measured are clearly linked to parameters that underlie evolution. The extent of phenotypic variability and of heritability of these traits is likely to radically condition the durability of resistant cultivars. Collecting more information on the genetic architecture of aggressiveness components could bring valuable information for the development of quantitative resistance. Nonetheless, considering the huge evolutionary potential of plant pathogens, the design of quantitative resistance should take advantage of the potential trade-offs between aggressiveness components, in order to enhance the sustainability of crop resistance. Such trade-offs would reflect the constraint for the pathogen to simultaneously invest in different traits, such as sporulation or within-host growth. It is therefore most important that plant pathologists record, to the greatest extent possible, all aggressiveness components in pathogen adaptation studies, and check for any negative correlations between them. Further study on the survival of pathogens during intercropping, although difficult in most pathosystems, would bring additional valuable information for both understanding pathogen evolution and improving disease management.

12.3.5   EFFECTS OF ENVIRONMENT ON AGGRESSIVENESS

Most studies linking aggressiveness and climate are limited to the effect of temperature. It is well known that temperature influences pathogen development as well as the expression of host resistance. The effect of temperature on aggressiveness components has been established for many pathogen species and presents an optimum for spore germination, lesion development, and sporulation. However, the response to temperature may differ among individuals (Milus et al., 2006). For instance, Milus and Line (1980) showed that the spore production rate of two leaf rust isolates (*P. triticina*) was identical at 2–18 °C but different at 10–30 °C. Interestingly, differences in aggressiveness among pathogen isolates have sometimes been reported to be greater under non-optimal conditions: Differences in the latent period among isolates were more effectively observed at suboptimal temperatures for pathogen development in *P. triticina* and *P. striiformis* f. sp. *tritici* (Eversmeyer et al., 1980; Johnson, 1980; Milus et al., 2006).

12.3.6   EFFECT OF HOST PHYSIOLOGY

For several biotrophic parasites, high nitrogen content in host tissues results in increased infection efficiency, and spore production (Tiedemann, 1996; Jensen & Munk, 1997; Robert et al., 2004). Spore production of the biotropic parasite was also increased when photosynthesis activity stimulated. Milus and Line (1980) observed that the relative spore production of two *P. triticina* cultures changed with host growth stage (seedling or adult plant) on some cultivars. Turechek and Stevenson (1998) showed that the age of host tissues can have a strong effect on a tree disease such as pecan scab (caused by *Cladosporium caryigenum*) for aggressiveness components such as infection efficiency, incubation period, lesion size, and sporulation. Knott and Mundt (1991) found a significant difference between upper and lower leaves for latent period and infection efficiency when measuring the aggressiveness of field populations of *P. triticina* on wheat. On average, spore populations exhibited 25% higher infection efficiency and a 3–4% shorter latent period on the upper leaves than on the lower leaves. This was attributed either to greater susceptibility of the upper leaves or to physiological effects. Katsuya and Green (1967) even found significant differences in the latent period of wheat stem rust, depending on the position of the lesions along the leaves: Latent period was about 1 day shorter at the leaf base than toward the tip.

12.3.7 EFFECT OF LESION DENSITY

For many biotrophic pathogens, lesion size, and spore production are highly density-dependent (e.g., Robert et al., 2004), probably because of increased competition among lesions for host resources and available tissues. This density effect may have major consequences for experimental measurements of aggressiveness components, such as spore production and lesion size, particularly when differences in infection efficiency among isolates result in different lesion densities. In such cases, observed differences in spore production may result from a density effect rather than from genetic differences among isolates.

KEYWORDS

- **dispersal**
- **adaptation**
- **phytopathogens**
- **bacteria**
- **fungi**
- **virus and plant diseases**

REFERENCES

Bastiaans, L. Ratio between Virtual and Visual Lesion Size as a Measure to Describe Reduction in Leaf Photosynthesis of Rice due to Leaf Blast. *Phytopathology.* **1991,** *81,* 611–5.

Bennett, A. F.; Lenski, R. E.; Mittler, J. E. Evolutionary Adaptation to Temperature. I. Fitness Responses of *Escherichia coli* to Changes in its Thermal Environment. *Evolution.* **1992,** *46,* 16–30.

Buxton, E. W. Heterokaryiosis, Saltation, and Adaptation. In *Pl. Pathol. An Adv. Trea.*; Horsfall, J. G., Dimond, A. E. Eds.; Academic Press: New York, 1960; Vol. 2, pp 359–407.

Christensen, J. J. Long Distance Dissemination of Plant Pathogens. In *Aerobiology. Amer. Assoc. Adv. Sci.* Washington: D.C., 1942, Publ.17, 78–87.

Day, T. Virulence Evolution via Host Exploitation and Toxin Production in Spore-Producing Pathogens. *Ecol. Lett.* **2002,** *5,* 471–6.

Day, T. Virulence Evolution and the Timing of Disease Life-History Events. *Trends Ecol. Evolut.* **2003,** *18,* 113–8.

De Bary, A. Comparative Morphology and Biology of the Fungi Mycetozoa and Bacteria. Oxford University Press: Clarendon, UK, 1987.

Doran, W. L. *Bull. Torrey Bot. Club.* **1922** *,49,* 313–340.

Ebert, D.; Bull J. J. Challenging the Trade-Off Model for the Evolution of Virulence: Is Viru-lence Management Feasible? *Trends Microbiol.* **2003,** *11,* 15–20.

Eversmeyer, M. G.; Kramer, C. L.; Browder, L. E. Effect of Temperature and Host: Parasite Combination on the Latent Period of *Puccinia recondita* in Seedling Wheat Plants. *Phyto-pathology.* **1980,** *70,* 938–41.

Fargues, J.; Goettel, M. S.; Smits, N.; Ouedraogo, A.; Rougier, M. Effect of Temperature on Vegetative Growth of *Beauveria bassiana* Isolates from Different Origins. *Mycologia.* **1997,** *89* (3), 383–392.

Feder, M. E.; Hofmann, G. E. Heat-Shock Proteins, Molecular Chaperones, and the Stress Response: Evolutionary and Ecological Physiology. *Ann. Rev. Physiol.* **1999,** *6,* 243–282.

Fels, D.; Kaltz, O. Temperature-Dependent Transmission and Latency of *Holospora undu-lata*, a Micronucleus-Specific Parasite of the Ciliate *Paramecium caudatum. Proc. R. Soc. Biol. Sci.* **2006,** *273,* 1031–1038.

Ferris, V. R. Histological Study of Pathogen Suscept Relationship between *Phytopathora infestans* and Derivatives of *Solanum dermissum. Phytopathol.* **1955,** *45,* 546–552.

Futuyma, D. J. *Evolutionary Biology*; 3rd ed. Sinauer Associates: Sunderland, MA, USA, 1997.

Galvani, A. P. Epidemiology Meets Evolutionary Ecology. *Trends Ecol. Evol.* **2003,** *18,* 132–9.

Garrett, K. A.; Dendy, S. P.; Frank, E. E.; Rouse, M. N.; Travers, S. E. Climate Change Effects on Plant Disease: Genomes to Ecosystems. *Annu. Rev. Phytopathol.* **2006,** *44,* 489–509.

Goddard, D. R. *J. Gen. Physiol.* 1935, *19,* 45–60.

Gottlieb, D. *Botan. Rev.* 1950, *16,* 229–257.

Ingold, C. T. *The Biology of Fungi*; 5th ed. Hutchinson & Co. (Publishers) Ltd.: London, 1984.

Jensen, B.; Munk, L. Nitrogen-Induced Changes in Colony Density and Spore Production of *Erysiphe graminis* f. sp. *hordei* on Seedlings of Six Spring Barley Cultivars. *Plant Pathol.* **1997,** *46,* 191–202.

Jessup, C. M.; Kassen, R.; Forde, S. E.; Kerr, B.; Buckling, A.; Rainey, P. B.; Bohannan, B. J. Big Questions, Small Worlds: Microbial Model Systems in Ecology. *Trends Ecol. Evol.* **2004,** *19,* 189–197.

Johnson, D. A. Effect of Low Temperature on the Latent Period of Slow and Fast Rusting Winter Wheat Genotypes. *Plant Dis.* **1980,** *64,* 1006–8.

Katsuya, K.; Green, G. J. Reproductive Potentials of Races 15B and 56 of Wheat Stem Rust. *Can. J. Bot.* **1967,** *45,* 1077–91.

Knott, E. A.; Mundt, C. C. Latent Period and Infection Efficiency of *Puccinia recondita* f. sp. *tritici* Populations Isolated from Different Wheat Cultivars. *Phytopathology.* **1991,** *81,* 435–439.

Lehman, J. S.; Shaner, G. Heritability of Latent Period Estimated from Wild-Type and Selected Populations of *Puccinia triticina. Phytopathology.* **2007,** *97,* 1022–9.

Leroi, A. M.; Bennett, A. F.; Lenski, R. E. Temperature Acclimation and Competitive Fitness: An Experimental Test of the Beneficial Acclimation Assumption. *Proc. Natl. Acad. Sci. U.S.A.* **1994,** *91,* 1917–1921.

Li, Y.; Cole, K.; Altman, S. The Effect of a Single, Temperature-Sensitive Mutation on Global Gene Expression in *Escherichia coli. RNA.* **2003,** *9,* 518–532.

Lilly, V. G.; Barnett, H. L. *Physiology of the Fungi;* Mcgraw Hill: New York, 1951; pp 464.

Lincoln, R. E. Bacterial Wilt Resistance and Genetic Host Parasite Interaction in Maize. *J. Agr. Res.* **1940,** *60,* 217–239.

Lockwood, J. L. Soil Fungistasis. *Annu. Rev. Phytopathol.* **1964,** *2,* 341–362.

Magboul, A. M.; Geng, S.; Gilchrist, D. G.; Jackson, L. F. Environmental Influence on the Infection of Wheat by *Mycosphaerella graminicola. Phytopathology.* **1992,** *82,* 1407–13.

Mehta, K. C. Further Studies on Wheat Rust in India. *Sci. Monog. Coun. Agr. Res. India.* **1943,** *14,* 224.

Mejia, R.; Gomez-Eichelmann, M. C.; Fernandez, M. S. Membrane Fluidity of *Escherichia coli* during Heat-Shock. *Biochim. Biophys. Acta.* **1995,** *1239,* 195–200.

Michea-Hamzehpour, M.; Grance, F.; Ton, T. C.; Turian, G. Heat-Induced Changes in Respiratory Pathway and Mitochondrial Structure during Conidiation of *Neurospora crassa. Arch. Microbiol.* **1980,** *125,* 53–58.

Milus, E. A.; Seyran, E.; McNew, R. Aggressiveness of *Puccinia striiformis* f. sp. *tritici* Isolates in the South-Central States. *Plant Dis.* **2006,** *90,* 847–52.

Milus, E. A.; Line, R. F. Characterization of Resistance to Leaf Rust in Pacific Northwest Wheat Lines. *Phytopathology.* **1980,** *70,* 167–72.

Nagarajan, S.; Singh, H. Satellite Television Cloud Photography as a Possible Tool of Plant Disease Spread. *Curr. Sci.* **1973,** *42,* 273–274.

Nagarajan, S.; Singh, H. Satellite Television Cloud Photography—a New Method to Study Wheat Rust Dissemination. *Indian J. Gen.* **1974,** *34* (A), 486–489.

Nagarajan, S.; Singh, H. Indian Stem Rust Rules – an Epidemiological Concept on the Spread of Wheat Stem Rust. *Pl. Dis. Reptr.* **1975,** *59,* 133–136.

Naik, M. K.; Devika Rani, G. S. *Advances in Soil Borne Plant Diseases.* New Delhi Publishing Agency: New Delhi, 2008; pp 261–283.

Park . In Burges, A. and R. Row., Eds.; Soil Biology Academic Press: New York. 1967; pp 435–447.

Paxman, G. J. Variation in *Phytophthora infestans. Europ. Potato J.* **1963,** *6,* 14–23.

Person, C. Genetical Adjustment of Fungi to their Environment; Anisworthand, G. C., Sussman, A. S., Eds.; Academic Press: London, 1968; Vol. III, pp 395–415.

Read, A. F.; Aaby, P.; Antia, R. Group Report: What can Evolutionary Biology Contribute to Understanding Virulence? In *Evolution in Health and Disease;* Stearns, S. C., Ed.; Oxford University Press: Oxford, UK; 1999; pp 205–15.

Riehle, M. M.; Bennett, A. F.; Lenski, R. E.; Long, A. D. Evolutionary Changes in Heat-Inducible Gene Expression in Lines of *Escherichia coli* Adapted to High Temperature. *Physiol. Genomics.* **2003,** *14* (1), 47–58.

Robert, C.; Bancal, M. O.; Lannou, C. Wheat Leaf Rust Uredospore Production on Adult Plants: Influence of Leaf Nitrogen Content and *Septoria tritici* Blotch. *Phytopathology.* **2004,** *94,* 712–21.

Salmon, E. S. Recent Research on the Specialization of Parasitism in Erysiphaceae. *New Phytopathol.* **1904,** *3,* 55–60.

Shaw, M. *Phytopathology.* 1964, *50,* 159–80.

Shear, C. L.; Dodge, B. O. *J. Agric. Res.* 1927, *34,* 1019–1942.

Sussman, Encyclopaedia of Plant Physiology.,15 (2): 934–1025, W-Ruhland, ed.

Symposium, Liver pool Univ Press: Liverpool. 1965; pp 55–75.

Szeghalmi, A.; Kaminskyj, S.; Gough, K. M. A Synchrotron FTIR Microspectroscopy Investigation of Fungal Hyphae Grown under Optimal and Stressed Condition. *Anal. Bioanal. Chem.* DOI 10.1007/s00216–006–0850–2, **2006.**

Tharayil, N.; Suseela, V.; Triebwasser, D. J.; Preston, C. M.; Gerard, P. D.; Dukes, J. S. Changes in Structural Composition and Reactivity of *Acer rubrum* Leaf Litter Tannins Exposed to Warming and Altered Precipitation: Climatic Stress-Induced Tannins are More Reactive. *New Phytologist*. DOI 10.1111/j.1469–8137.2011.03667. **2011.**

Tiedemann, A. V. Single and Combined Effects of Nitrogen Fertilization and Ozone on Fungal Leaf Diseases on Wheat. *J. Plant Dis. Prot.* **1996,** *103,* 409–419.

Toyoda, T.; Hiramatsu, Y.; Sasaki, T.; Nakaoka, Y. Thermo-Sensitive Response Based on the Membrane Fluidity Adaptation in *Paramecium multimicronucleatum. J. Exp. Biol.* **2009,** *212,* 2767–2772.

Turechek, W. W.; Stevenson, K. L. Effects of Host Resistance, Temperature, Leaf Wetness, and Leaf Age on Infection and Lesion Development of Pecan Scab. *Phytopathology.* **1998,** *88,* 1294–301.

Van der Plank, J. E. *Plant Diseases: Epidemics and Control*; Academic Press: New York, USA, 1963.

Van der Plank, J. E. Stability of Resistance to *Phytophthora infestans* in Cultivars without *R* Genes. *Potato Res.* **1971,** *14,* 263–270.

Waid, J. S. *The Ecolgy of Soil Fungi*; In: Park Son, D., Waid, J. S., Eds., Liverpool University Press: Liverpool, 1960.

Ward, H. M. Further Observations on the Brown Rusts of the Bromes *Puccinia dispersa* (Erikss.) and its Adaptive Parasitism. *Ann. Mycol.* **1903,** *1,* 132–151.

CHAPTER 13

NATURE, DISSEMINATION AND EPIDEMIOLOGICAL CONSEQUENCES IN CHARCOAL ROT PATHOGEN *MACROPHOMINA PHASEOLINA*

CHANDA KUSHWAHA, NEHA RANI, and ARUN P. BHAGAT

Department of Plant Pathology, Bihar Agricultural University, Sabour 813210, Bhagalpur, India

CONTENTS

ABSTRACT

This chapter focuses on evolutionary biology of the globally devastating pathogen *Macrophomina phaseolina* causing huge losses in many crops worldwide. It has been indicated as an emerging phytopathogen under high temperatures and drought stress. The chapter discusses about pathometry, basic morphological, cultural and physiological aspects of the pathogen. It summarizes various kinds of research activity in the past that were directed towards host range, variation in symptom expression and occurrence and distribution of this pathogen worldwide. Various abiotic factors responsible for the successful pathogenecity of the pathogen. The present discussion revolves around variability in the pathogen with reference to survival, dissemination and evolutionary potential of *M. phaseolina* that are critical for prevalence of the pathogen in future and predicting its potential threats to crop species. Various aspects of genetic resistance existing in the crop species and its potential for exploitation in managing this dreaded pathogen is also discussed.

13.1   INTRODUCTION

Macrophomina phaseolina is an emerging and devastating fungal pathogen that causes significant losses in crop production under high temperatures and drought stress. An increasing number of disease incidence reports highlight the wide prevalence of the pathogen around the world and its contribution toward crop yield suppression. *Macrophomina phaseolina* a fungal pathogen belonging to family Botryosphaeriaceae is of ubiquitous nature and causes charcoal rot on many plant species including agriculturally important crops like corn, soybean, groundnut, pigeon, and pea. *M. phaseolina* is considered to be a globally devastating necrotrophic fungal pathogen causing charcoal rot disease in more than 500 host plants.

The success of this pathogen as a necrotroph could be attributed to factors like wide adaptability to environments, and its ability to infect wide range of hosts belonging to diverse crop families. With the aim of understanding nature, dissemination, and epidemiological consequences of this devastating pathogen following informations has been compiled for devising better management strategies against this pathogen. The following text would comprise of taxonomical position of *Macrophomina phaseolina*, its description, symptomatology, survival, and dissemination, sources of variation and its epidemiological consequences and methods and means of management of this dreaded pathogen.

13.2 *MACROPHOMINA PHASEOLINA*—THE PATHOGEN AND PATHOMETRY

Macrophomina phaseolina belong to the class deuteromycetes, or the "fungi imperfecti" group. Since the sexual stage is seldom observed. In the case of *M. phaseolina* a teleomorph has initially being described as *Orbilia obscura* by Ghosh et al. (1964) (Holliday, 1980), although it could not be confirmed later (Mihail, 1992). This fungus is by various names and described from time to time as *Macrophomina phaseolina* (Tassi) Goid. Synonyms are: *Macrophoma phaseolina* Tassi (1901), *Macrophoma phaseoli* Maubl. (1905), *Sclerotium bataticola* Taub (1913), *Rhizoctonia bataticola* (Taub) Briton-Jones (1925), *Macrophomina phaseoli* (Maubl.) Ashby (1927), and *Macrophomina phaseolina* (Tassi) Goid. (1947). Some of these are still in use along with the present convention of *Macrophomina phaseolina* (Tassi) Goid. (1947) but are more likely to be seen in many of the older published reports on the subject.

Due to presence of sclerotial stage, this pathogen was first named as *Sclerotium bataticola* by Taubenhaus (1913). Later on it was transferred under genus *Rhizoctonia bataticola* (Taub.) by Butler (1918). The pycnidial stage was described by Maublanc as *Macrophomina phaseolina* (Maubl.). In 1905 Ashby (1927) showed that the fungus produces a pycnidial stage corresponding to *Macrophomina philippinensis* Petrak; the type species of the genus *Macrophomina*. Goidanich suggested on earlier name; *Macrophomina phaseolina*. Tassi and proposed the combination; *Macrophomina phaseolina* (Tassi) Goid. for the fungus. Thus this came in the existence of the present name of the pathogen as *Macrophomina phaseolina* (Tassi) Goid. with perfect stage *Rhizoctonia bataticola*.

13.3 MORPHOLOGICAL AND PHYSIOLOGICAL CHARACTERIZATION OF THE PATHOGEN

Various studies were undertaken in recent past to characterize morphological and physiological parameters of this pathogen in order to assess the diversity in this pathogen. The colony color ranged from off white – grey to black in color. Reports indicated that the fungus at 30°C produced grayish white fluffy mycelia which were pro-geotropic in growth response Thirumal-achar (1955). Deshpande et al. (1969) observed white to brown fluffy mycelium in culture. Mycelium was branched and sparsely septate. Sclerotia connected by short strands or fibrilla with the rest of mycelium and were brown to black in color and globose to irregular in shape. It measures 66.4–315 μ in diameter (average 136.12 μ); the irregular ones were 166.4–498 μ in size.

Some of the isolates of *M. phaseolina* produce numerous microsclerotia. Brooks (1928) observed smooth, hard, and black sclerotia measuring 50–1000μ in diameter which occurred mainly on the infected roots. Smits and Nogeura (1999) reported that the formation of sclerotia in *M. phaseolina* began with branching and inter-winning of adjacent hyphal filaments. Subramanayam (1971) reported that sclerotia were jet-black, minute smooth, externally composed of anastomosed black hyphae, interior light to dark brown matrix, composed of free thick-walled cells. Sclerotia were variable in shape, globose, oval, oblong, elliptical, curved or even forked, varying in size 25 × 22–152 × 32 μ, produced abundantly in the infected host tissues.

The pycnidial stage began with the merging of hyphal filaments toward a common point, followed by the development of ringed primordia and finally pycnidia. Matured pycnidia are subglobose with a reticulate appearance, a short neck, and a circular ostiole. The pycnidia formed on the stems were variable in size ranging from 16–30 × 5–10 μ spores are single celled, hyaline somewhat elongated or cylindrical, 16–30 × 5–10 μ in size.

13.4 CULTURAL AND PHYSIOLOGICAL STUDIES

Numerous studies were undertaken to characterize morphological, cultural, and physiological parameters of *Macrophomina phaseolina* on different artificial media to identify optimized growth conditions with respect to nutrient requirements as well as diversity among the isolates. Such studies provide basic information for genotype screening against this pathogen.

Linear growth of this pathogen as an indicator of nutritive status of culture medium was studied by Paracer and Bedi (1962) were rich medium promoted rapid growth of mycelium on artificial medium. Sclerotial formation started as early as 48 hours after inoculation on Brown's agar re-informed with 2% potato (starch), whereas it took 168 hours or one week to start on the poorest medium like water agar. Rich media like Richard's agar and nutrient glucose agar, the formation of sclerotia were delayed considerably. *M. phaseolina* preferred Richard's medium at pH 5, with dextrose and asparagines as carbon and nitrogen sources for excellent mycelia growth and sclerotial induction (Gupta & Kolte, 1981; Jha, 1996; Shanmugam & Govindaswamy, 1973).

Okwulehie (2004) reported that the best synthetic medium for the growth of *M. phaseoli* [*M. phaseolina*] and standardized inoculation technique. The physiological changes in groundnuts under infection with the pathogen were analyzed. Four different media screened were Potato Dextrose Agar (PDA), Peanut Leaflet Oatmeal Agar (POMA), Czapek-Dox Agar (CDA), and Corn Meal Agar (CMA). The best growth was observed on POMA and PDA media.

The use of semi synthetic media for the growth of the pathogen was studied by Okwulehie (2001). The fungus grew in both solid and liquid synthetic media; however, the best growth was recorded in POMA, and PDA. Such informations were necessary for rapid mass multiplication of the pathogen for in vitro assessments. It also improves our understanding of the specific nutrition requirements of the pathogen.

13.5　SYMPTOMS AND HOST RANGE

Symptoms indicative of this pathogen's infection are described by many workers on various host crops. These symptoms were similar in terms of its black dark colored appearance which is termed as charcoal rot, dry root rot when infection takes place on roots under drier conditions. Thirumal-achar (1953) observed grey discoloration in the beginning at the infection site followed by blackening of the stems, which bear numerous sclerotia of the fungus, as characteristic symptoms of the disease. However, Jain and Kulkarni (1965) observed different types of symptoms at different stages of growth of sesame plant. In seedling stage, the roots may become brown and rot resulting in death of the whole plant. In other plants, the fungus attacks on the collar region of the stem showing brown coloration which later extends upward from few mm to few inches. Slowly and slowly, the whole plants becomes brown colored and small dot-like black pycnidial structures containing fungal spores are seen on stem, branches, capsules, and seeds. In cluster bean, *M. phaseolina* infected seedling showed seed rot, dark brown color patches on root shoot transitional zone and brown or black circular spot on cotyledonary leaves leading to root rot symptoms (Singh & Chouhan, 1973). Groundnut crop affected with *M. phaseolina*, in Nigeria was studied to find out various anatomical changes associated with this phytopathogen. Anatomically significant changes were observed in the hypocotyl, root, and stem. Hypocotyl tissues were completely destroyed and internal necrosis extended into the stem tissue. Plugging of vascular tissues, especially xylem vessels, was observed which probably induced wilting. Cells were hypertrophied while those of the root cortex were destroyed (Okwulehie, 2002). Rasheed et al. (2004) also reported pre- and post-emergence seedling damping off in. Smith and Carvel (1997) for soybean: In soybean, charcoal rot disease symptoms appear in midsummer during high temperature and low soil moisture when the plant is stressed. Initial infections occur at seedling stage but remain latent until the soybean plant approaches maturity. Plants may wilt and die. Sloughing of cortical and taproot tissues occur with characteristic black speckled appearance due to the presence of sclerotia (Fig. 13.1).

Browning at collar region due to *Macrophomina phaseolina* in groundnut   Blackening and rotting of roots in strawberry due to *Macrophomina phaseolina*

FIGURE 13.1   Typical symptoms of infection by *Macrophomina phaseolina* in groundnut and strawberry.

13.6   HOST RANGE

13.6.1   IN INDIA

The pathogen has been described as a cosmopolitan pathogen there are many hosts to this particular pathogen and many new hosts are also reported from time to time. Butler and Bishy (1931) cited many records on *Solanum tuberosum* L. *Gossypium* spp., *Corchorus capsularis* L., *C. olitorius* L.

Cajanus cajan L. Millsp., *Arachis hypogaea* L., *Alysicarpus* spp., *Carica papaya* L., *Citrullus vulgaris schrad.* L. Millsp., *Crotalaria juncea* L. *Cucurbita maxima* Duchesne, *Dolichos biflorus* L., *D. lablab* L., *Hibiscus cannabinus* L., *Lycopersicon esculentum* L., *Medicago sativa* L., *Morus alba* L., *Nicotiana tabacum* L., *Phaseolus lunatus* L., *P. aureus* Roxb., *Solanum melongena* L., and *Vigna sinensis* L. as a root, stem, and tuber parasite throughout India.

This pathogen was reported from almost all parts of India. From Tamil Nadu it was isolated from black cotton soil from Udamalpet, and Vandalur state (Subramanian, 1952; Ramakrishnan, 1955). Bhattacharya and Baruah (1953) reported this fungus on *Camellia sinensis* L. form Assam. Nema and Agrawal (1960) reported it on roots of *Cicer arietinum* L. and *Pisum sativum* L. from Jabalpur (M.P.). Singh and Nene (1990) reported that *Rhizoctonia* incited diseases in wide range of hosts, especially under high temperature and drought stress conditions. Jayati-Bhowal et al. (2006) observed the phytopathogenic fungus *M. phaseolina* infects many plants, for example jute (*Corchorus capsularis*), soybean (*Glycine max*) and groundnut (*Arachis hypogaea*). New reports of occurrence of this phytopathogen was observed on marigold (*Tagetes erecta*), cantaloupe (*Cucumis melo* var. *cantalupensis*), cumin (*Cuminum cyminum*), hemp (*Cannabis sativa*), mung bean (*Vigna radiata*), okra (*Abelmoschus esculentus*), tomato (*Lycopersicon esculentum*), turnip (*Brassica rapa*), and watermelon (*Citrullus lanatus*) (Mahdizadeh et al., 2011a).

The pathogen occurs on many diverse hosts both across monocots as well as dicot crops. Winter rape, sesame, saffron, squash, lucerne, cotton, potato, sorghum, cucumber, okra, and capsicum have been reported as new hosts for *M. phaseolina* in Romania (Ionita et al., 1995). Eleven new hosts of *M. phaseolina* were reported from the ICRISAT, Hyderabad that included *Celosia argentea, Corchorus trilocularis, Cyanotis axillaris, Dactyloctenium aegyptium, Digera muricata. Digitaria ciliaris, Echinochloa colonum, Eclipta prostrata, Indigofera glandulosa, Trianthema portulacastrum, and Tridax procumbens* (Singh et al., 1990).

13.7 OCCURRENCE AND DISTRIBUTION

Macrophomina phaseolina (Tassi) Goidanich causes seedling blight, root rot, and charcoal rot of more than 500 crop and non-crop species (Smith & Carvel, 1997). It has a very wide distribution covering most of the tropics and subtropics, as well as extending into some parts of temperate zones with occurrence reported from far north of UK and far south of New Zealand (Songa, 1995). Patel and Patel (1990a, 1990b) observed that the optimum temperature for growth and sclerotial formation by *Macrophomina phaseolina* was at 35°C, which declined at temperatures below 15°C and above 40°C. Under field conditions disease intensity increased with a progressive rise in temperature and decrease in relative humidity. They observed maximum disease occurrence at 35°C and 76% relative humidity. It is an important pathogen of crops particularly where high temperatures and water stress occurs during the growing season. The typical symptoms of the disease are like those described by necrotic spots may also appear on the leaves of some plants due to a translocatable toxin (Chan & Sackston, 1973; Day, 1993) that may be responsible for the rotting symptoms sometimes without being able to see other signs. Such toxins from cultures and diseased plants have been shown to produce typical rot symptoms (Dhingra & Sinclair, 1974; Day, 1993). The disease is very destructive in nature and under epidemic conditions may cause huge losses and has been reported from many parts of the world, for example India (Pearl, 1923), Burma and Ceylon (Small, 1927a & 1927b), Palestine (Reichert, 1930), Cyprus (Nattrass, 1934), Greece (Sarejanni & Cortzas, 1935), Uganda (Hansford, 1940), Turkey (Bremer, 1944), Pakistan (Prasad, 1944), Nigeria (Anon, 1955) Syria (Al-Ahmad & Saidawi, 1988), Iran (Mahdizadeh, et al., 2011b).

13.8 INFLUENCE OF ABIOTIC FACTORS ON INCIDENCE OF DRY ROOT ROT

Disease is an outcome of interaction between host, pathogen, and its environment, where by abiotic factors play an equally important role in occurrence of any disease. Even in presence of host and pathogen if the environmental conditions prevailing are not conducive; will result in non-occurrence of the disease incidence. Singh et al. (1993) reported that the severity of stem rot caused by *Rhizoctonia bataticola* (*Macrophomina phaseolina*) was reduced by sowing sesame between 10–20 July, resulting in increased yield as compared with crop sown on 1st July. The earlier planting recorded maximum charcoal rot (*M. phaseolina*) incidence in sorghum and it reduced with delayed sowing (Lukade, 1995). Singh and Sandhu (1995) found that low and higher moisture levels of the soil induced pre-emergence charcoal rot of cowpea but seedling mortality was severe only under stress soil moisture (below 40%) and decreased with increasing moisture level. Disease was reported during 1986–1989 when collar rot (caused by *Aspergillus niger*) was more prominent than, *Rhizoctonia bataticola* [*Macrophomina phaseolina*] infection in Andhra Pradesh Rao et al. (1997). Mixed cropping system modulated the incidence of disease as reported by Mittal (1997) who observed in general, that disease incidence was less in the mixed than in the sole crop and gradually declined from the first to last sowing date. Sowing on 19th October was most effective in reducing the disease and increasing yield of lentil.

Pathak and Barman (1998) reported that the highest root rot incidence was found in the May sown crop of Jute. Among the varieties screened, JRO-878 was found to be susceptible to the disease and recorded the highest disease incidence. High temperature, high relative humidity, and rainfall favored disease development in late sown crop.

Moisture holding capacity (MHC) of the soil influences the saprophytic survival of *M. phaseolina*. Maximum survival was noted at low moisture levels of 40% recording a mean of 65.9% survival against 15.9 and 7.9% survival at 60 and 80%, respectively. At 65 days of incubation, the saprophytic ability was 3.4% at 80% MHC compared with 73.4% at 40% MHC while, it decreased progressively with increase in incubation period at 60 and 80% MHC at both low and high inoculum levels (Maheswari & Ramakrishnan, 2000). Khan et al. (2003) showed occurrence of the disease but Sahiwal showed high incidence and severity in 1999. Distribution of the disease in Sindh, Punjab, and NWFP was 85, 83, and 48%, respectively. Among provinces NWFP showed highest incidence 57% and Punjab exhibited highest severity 2.62 according to 0–5 severity rating scale. Continuous increasing trend of charcoal rot is alarming for farmers and authorities engaged in sunflower business. Cardona (2006) observed the influence of temperature and humidity on the vertical distribution of *Macrophomina phaseolina* sclerotia in a naturally infested soil in Turen, Portuguesa state, and Venezuela. Samples were collected from the soil at depths of 0–5, 5–10, and 10–20 cm in plots sown after maize-sesame. Linear regression analysis showed that the quantity of sclerotia in the soil was negatively correlated with humidity but positively correlated with temperature. The quantity of sclerotia decreased, the humidity increased and the difference between maximum and minimum temperatures decreased at higher soil depths.

Altering sowing time and irrigation conditions may influence incidence of charcoal rot as reported by Sağır et al. (2010) in diverse lines of sesame. The lowest disease incidence was recorded from B-60 line (40.60%), and the highest from C-36 line (48.98%). Irrigated and late sowing in sesame promoted lesser disease development as evident from the finds where lowest disease percentage were recorded from irrigated (27.56%) and late sowing time (34.73%), the highest from in dry condition (57.98%) and early sowing time (50.82%). Early sowing time insured concurrence of flowering stage with drier conditions.

The disease attributes toward high losses in yield in groundnut according to a previous survey at Main oilseeds Research station, Junagadh Agricultural University, Junagadh during 2002–2003. The maximum plant mortality due to dry root rot was of 29.3% caused by *Macrophomina phaseolina* with highest yield loss of 435 kg/ha was found in Keshod tehsil of Junagadh district of Saurashtra region. The survey in different locations of Cuddalore district revealed the endemic nature of the root rot disease incidence with the maximum incidence of the disease (31.68%) registered in Vengatakuppam location. The disease incidence was more in improved cultivars like, VRI2, JL24; and more severe in sandy loam soils and rainfed conditions (Raja Mohan & Balabaskar, 2012). During 2009–2010 and 2010–2011, disease incidence ranged from 0 to 97% and 3 to 91% with the highest in Gorgan and Aliabad (valley), respectively (Taliei et al., 2013).

13.9   PATHOGENESIS AND VARIABILITY IN THE POPULATIONS OF *MACROPHOMINA PHASEOLINA*

Infection process depends upon production of many substances like endogenous lectin in jute seedlings (Jayati-Bhowal, 2006). More recent report on *Macrophomina phaseolina* infection in *Corchorus capsularis* (jute) plants suggested an elevated levels of nitric oxide, reactive nitrogen species and S-nitrosothiols production in infected tissues (Sarkar et al., 2012).

Malachi and Sabitha Doraisamy (2003a, 2003b) observed increased growth of both pathogen (*M. phaseolina*), and growth, sporulation and biomass production of the fungal antagonist (*Trichoderma* spp.) were observed between 25 and 35°C.

Morphological and pathogenic variability in the pigeon pea isolate of *M. phaseolina* had been established (Kaur et al., 2013). Genetic variability among the isolates of *M. phaseolina* was assessed through PCA analysis that extracted three main components of variation that is microsclerotia, texture and color. Color of the isolates and the presence of the microsclerotia have a significant effect on aggressiveness of the pathogen as indicated by area under disease progress curve. Isolates with the production of microsclerotia

(M^+) were more aggressive as compared to isolates with no production of microsclerotia (M^-).

Macrophomina phaseolina (Tassi) Goid. causes charcoal disease of oilseed plants. In the study 24 isolates, which were obtained from sunflower, soybean and sesame were compared based on chlorate phenotypes and pathogenicity tests. For chlorate phenotypes, the isolates were grown on potassium chlorate and stored at 30°C in darkness. Results of pathogenicity test showed that there was significant difference ($p < 0.01$) between the isolates. The results confirmed that the feathery like pattern of the isolates was more virulent on soybean and sunflower (Slavish Rayatpanah & Seyed Alireza Dalili, 2012). Significant pathogenic and genetic variability within the Iranian isolates of *M. phaseolina* based on molecular markers like RAPD was reported (Siavosh Rayatpanah et al., 2012).

Genetic variability among soybean isolates has been observed but the effect of host specialization on genetic variability has not been reported. In this work, isolates from soybean, corn, and sunflower were evaluated based on cross inoculations and number of microsclerotia/g of roots. Results suggest that genetic differentiation of *M. phaseolina* can be altered by crop rotation (Almeida et al., 2008).

In an another study genetic diversity analysis of *Macrophomina phaseolina* isolates obtained from different host range and diverse geographical locations in India was carried out using RAPD finger printing. Of the thirteen 10-mer random primers used, primer OPB-08 gave the maximum polymorphism and the UPGMA clustering could separate 50 isolates into ten groups at more than 65% similarity level. The ten clusters correlated well with the geographical locations with exceptions for isolates obtained from Eastern and Western Ghats. There was a segregation of isolates from these two geographical locations into two clusters thus, distributing 10 genotypes into eight geographical locations. All the isolates of *M. phaseolina* irrespective of their host and geographical origin, exhibited two representative monomorphic bands at 250 bp and 1 kb, presence of these bands suggests that isolates might have evolved from a common ancestor but due to geographical isolation followed by natural selection and genetic drift might have segregated into subpopulations. Genetic similarity in the pathogenic population reflects the dispersal of single lineage in all locations in India (Babu et al., 2010).

In addition, 65 isolates of *Macrophomina phaseolina* from different agroecological regions of Punjab and Khyber Pakhtunkhwa provinces of Pakistan were analyzed for morphological and pathogenic variability. Regardless of their geographic origins, significant differences were detected among 65 isolates in their radial growth, sclerotial size, and weight as well as in pathogenicity. Sixteen isolates were rated as fast growing, 11 as slow growing, and the rest of the isolates as medium growing. Nine isolates were classified as large sized, 26 as small sized, and the remaining 30 isolates as medium sized. Thirty-five isolates were ranked as heavy weight, 12 as low weight, and the rest of isolates were grouped as medium weight. Ten fungal isolates appeared to be least virulent, whereas eight isolates of diverse origin proved to be highly virulent against mungbean cultivars. The remaining isolates were regarded as moderately virulent. No relationship was found among the morphological characters and pathogenicity of the isolates (Iqbal & Mukhtar, 2014).

13.10 SURVIVAL, DISSEMINATION AND EVOLUTIONARY POTENTIAL OF *MACROPHOMINA PHASEOLINA*

Survival of this pathogen in soil has been reported to be over 17 years. Such long-term persistence in soil indicates its ability to deter various soil degradation mechanisms. One of the substances that can resist environmental stresses in form of UV light, soil degradation is melanin. Presence of highly melanized mycelium and sclerotia are two key aspects that resist degradation and promotes survival. Even highly virulent strains with low survival will have low probabilities of dominance. Therefore, under long-term situations ability of a pathogen to survive adverse climatic condition are more deciding for the existence and dominance a pathogen in nature. The nature of soil and amendments added to it directly affects the inoculums density in the soil. Reports suggest positive interactions between soil texture and weedicides being used. Five weedicides namely EPTC, dinoseb, alachlor, fluorodifen, and fluometuron, directly lowers the inoculums levels in the soil (Filho & Dhingra, 1980).

Another, aspects that decides the successful parasitism is the pathogen's ability to ward of self-competition and increased probability of pathogen coming in contact with its host. This is possible only with efficient dissemination of the pathogen. In *M. phaseolina* dissemination is low but man assisted movement of inoculums with seeds, soil, and other planting material is high (Fig. 13.2). A report suggests that high levels of seed infection are common when rainfall occurs during pod fill in Mungbean. It also suggests systemic infection by *M. phaseolina* (Fuhlbohm et al., 2013). Due to lack of numerous asexual spores such rapid dissemination through air and water is low. Seed infection in pigeon pea has also been documented where the pathogens upon artificial inoculation can reside in seed coat and cotyledons as evident from presence of mycelium and formation of microsclerotia within the seed coat (Maurya et al., 2013). Similar reports were also presented in soybean where sclerotia were formed under latent infection of cotyledons after 4–5 days of infection of hypocotyls under field conditions (Kunwar et al., 1986).

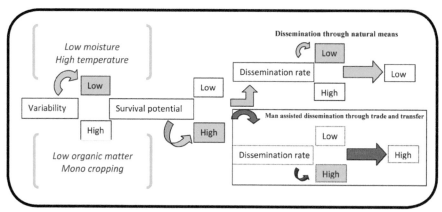

FIGURE 13.2 Schematic diagrams for prediction of risk and evolutionary potential of *Macrophomina phaseolina* under drier ecologies.

13.11 EVOLUTIONARY POTENTIAL AND RISK ASSESSMENT IN *MACROPHOMINA PHASEOLINA*

Recent reports on *M. phaseolina* suggest that *Macrophomina* is gradually evolving into a high-risk pathogen of diverse crops under high temperature drought prone ecologies (Fig. 13.2). Reported management strategies are not producing desired levels of crop protection yielding huge losses in crop produce. Of the reported strategies, uses of bioagents have been advocated by various workers. Ahmed and Gangopadhayay (2008) emphasized efficient use of bioagents like *Pseudomonas fluorescens* in controlling root rot of chickpea caused by *M. phaseolina*. Similarly, effective use of *Trichoderma* sp. for seed treatment as a measure for managing root rot pathogen in chickpea is also advocated (Malachi & Sabetha Doraisamy, 2003a, 2003b; Sreedevi et al., 2011; Nageswararao et al., 2012).

Management of this disease through soil microflora was evident when higher disease incidence was noted in sterilized soil when compared with unsterilized soil at 15–45 days after sowing. A significant increase in disease incidence (48.7 and 98.5%) was observed with increasing inoculum level from 500 to 1000 mg/kg of unsterilized and sterilized soil, respectively. In unsterilized soil, 50% reduction in root incidence due to the antagonistic activity of bacteria and actinomycetes was observed compared to sterilized soil (Umamaheswari et al., 2001). Influence of physical methods of disease management like use of mulches on disease incidence caused by *M. phaseolina* was also studied in groundnut root rot; where effect of seed treatment was higher in polythene mulch as evident by low disease incidence (Thakare et al., 2002) reported that Seed treatment Root rot incidence in groundnut was more in non-mulch with no seed treatment as compared to polythene mulch. Percent root rot caused by *Sclerotium rolfsii* [*Corticium rolfsii*] and *Rhizoctonia batati-cola* [*Macrophomina phaseolina*] reduced by 41% in mulches. Storage condition had a direct influence on the existence of seed mycoflora as reported by Jain and Sharma (2009) were 8% seed moisture for six months displayed a minimum incidence of *M. phaseolina* and maximum seed germination. Storage temperature of 40°C and RH of 90% for six months of storage restricted the incidence of *M. phaseolina* in cluster bean. Kale et al. (2009) reported that TG 51 also had a lower dry root rot (*Macrophomina phaseolina*) incidence (13.0%) than TAG 24 (43.0%) at Kadiri during the Rabi (winter) season of 2003–2004.

There are not many fungicides for management of this pathogen due to its soil borne nature where some more work is required. It is also observed that those conditions conducive for bioagent growth depreciate the growth of *M. phaseolina*. But under high temperature stress and drought or drought like conditions; where growth of beneficial soil microflora is limited threats from successful pathogenesis by *M. phaseolina* are high owing to its high levels of survival potential due to presence of substances like melanin. Being a soil borne pathogen disseminations through soil is limited and thereby posing limitations on its evolutionary potential. With increase in trade and transport in recent times, there is rapid and across the geographical boundaries exchange of seed and planting material is exchanged and along with it there is potential risk of movement of pathogen as seed contaminants thereby enhancing the levels of evolution in the pathogen as well as risk posed by the pathogen under drier ecologies. In light of climatic abreactions and general rise in temperature across the world with increasing mono cropping practices; this pathogen may become one of the important pathogen across many cereals, legumes, and fruits crops under drier ecologies.

13.12 GENETIC BASIS OF RESISTANCE AGAINST MACROPHOMINA PHASEOLINA

Literature suggests resistance to *M. phaseolina* in beans is governed by two dominant complementary genes *Mp-1* and *Mp-2*. Two RAPD markers were indicated to be linked with resistance to *M. phaseolina* on bean variety BAT-447 (Olaya et al., 1996).

Charcoal rot [*Macrophomina phaseolina* (Tassi) Goid.] causes significant yield losses to common bean (*Phaseolus vulgaris* L.) in Mexico when drought and high temperature stress prevails particularly at flowering stage. In this work, in vitro reactions to *M. phaseolina* in the cotyledons of 100 F2 plants derived from the cross between BAT 477 (resistant) × Pinto UI- 114 (susceptible) and the parents were measured, and AFLP molecular markers associated with BAT 477 *M. phaseolina* resistance were found. The AFLP analysis consisted of selective amplification using 20 EcoRI + 3/MseI + 3 selective oligo-nucleotides. Charcoal rot resistance is controlled by two dominant genes with double recessive epistatic effects. In conclusion, BAT 477 resistance to charcoal rot was controlled by two dominant genes with double-recessive epistasis and one possible QTL at LG1 was detected in BAT 477 (Delgado et al., 2009).

The experimental material consisted of the 45 F_1s from a 10-parent diallele with no reciprocals, the 10 parents, and 5 hybrid checks grown in two years under two water regimes. The inoculated and non-inoculated plants in a plot were rated for stalk disintegration on a scale of 1 (resistant) to 6 (susceptible). Hybrids differed significantly and consistently across environments for the two inoculation treatments but not for the non-inoculated. General combining ability was significant in both dryland environments for both inoculation treatments; specific combining ability was significant for *Fusarium moniliforme* in both years but only in 1984 for *M. phaseolina*. The expression of resistance seems to depend upon the environment, especially for *F. moniliforme*.

Nine quantitative trait loci (QTLs), accounting for between 6.1 and 40.0% of the phenotypic variance (R^2), were identified using plant mortality data taken over three years in field experiments and disease severity scores taken from two greenhouse experiments. Based on annotated genic SNPs as well as synteny with soybean (*Glycine max*) and *Medicago truncatula*, candidate resistance genes were found within mapped QTL intervals. QTL *Mac-2* explained the largest percent R^2 and was identified in three field and one greenhouse experiments where the QTL peak co-located with a SNP marker derived from a pectin esterase inhibitor encoding gene. Maturity effects on the expression of resistance were indicated by the co-location of *Mac-6* and *Mac-7* QTLs with maturity-related senescence QTLs *Mat-2* and *Mat-1*, respectively. Homologs of the *ELF4* and *FLK* flowering genes were found in corresponding syntenic soybean regions. Only three *Macrophomina* resistance QTLs co-located with delayed drought-induced premature senescence QTLs previously mapped in the same population, suggesting that largely different genetic mechanisms mediate cowpea response to drought stress and *Macrophomina* infection (Muchero et al., 2011).

The inheritance of resistance to the causal agent of charcoal stem rot *M. phaseolina* and water content in the internodes below the ear was analyzed by Generation Mean Analysis over two years at two locations in Vojvodina. Chosen for the study were six crosses that had previously been determined to be different with regard to their resistance to the above pathogen. Different generations (P1, P2, F1, F2, BC1, and BC2) of one cross were included in the study in order to determine the values of parameters for additive and dominant gene effects and effects of two-gene epistasis. A randomized complete block design was used. Gene effects and the mode of inheritance were assessed from the analysis of generation means from a cross between two parents. Plants were artificially inoculated in the field by the toothpick method. It was found that additive genes played the most important role in the inheritance of both stem resistance to the causal agent of charcoal stem rot and water content in the internodes below the ear (Bizana et al., 2000).

13.13   CONCLUSION

Reports suggest that *Macrophomina phaseolina* owing to is extensive ability to survive and cause infection, is emerging as a devastating fungal pathogen that may lead to huge losses in crop production under high temperatures and drought stress conditions. With changes in climate over a period of time, global warming, intensification of cropping systems accompanied by deterioration in soil health this pathogen may pose a serious threat to production in numerous crops. Therefore, more focused studies are required to breed genotypes resistant to this pathogen. Pre-breeding efforts may be required to identify genotypes with resistance to this pathogen under drought and drought like situations. Such studies are limited at present and our understandings in host parasite interactions are few. However, in recent times emphasis has been given on the issue and more strengthening information would be generated in years to come. Simultaneously, an efficient management strategy is also required in order to contain the disease and combat the threat to yield losses due to this disease. Emphasis should also be given to eco-friendly means of disease control through bio agents, crop rotation and incorporation of crop residue to improve soil heath which would ultimately have impact on limiting the inoculums load in soil and there by managing the disease in field conditions.

KEYWORDS

- **biology**
- **dispersal**
- ***Macrophomina***
- **management**
- **pathogenesis**
- **variability**

REFERENCES

Álvaro, M. R. Almeida; Daniel, R. Sosa-Gomez; Eliseo, Binneck; Sylvan R. R. Marin; Maria, I. Zucchi; Ricardo, V. Abdelnoor; Eliezer, R. Souto. Effect of Crop Rotation on Specialization and Genetic Diversity of *Macrophomina phaseolina*. *Trop. Plant Pathol.* **2008**, *33*, 257–264.

Ashby, S. F. *Macrophomina phaseoli* (Maul) Comb. Nov. Pycnidial Stage of *Rhizoctonia bataticola* (Taube) *Butler. Trans Brit. Myc. Soc.* **1927**, *12*, 141–147.

Badinage, P. S.; Tripathi, B. P.; Khari, N. Effect of Chemical and Biological Control of *Macrophomina phaseolina in Vitro* and *in Vivo*. *Environ. Ecol.* **2007**, *25*(1), 29–31.

Babu, B. k.; Reddy, S. S.; Yadav, K. Mukesh; Sukumar, M.; Vijendra Mishap; Saxena, A. K.; Dilip, K. Arora. Genetic Diversity of *Macrophomina phaseolina* Isolates from Certain Agro-Climatic Regions of India by Using RAPD Markers. *Indian J. Microbiol.* **2010**, *50*(2), 199–204.

Bizana, P.; Goran, B.; Melissa, S.; Đorđe, J. Genetic Analysis of Resistance to *Macrophomina phaseolina* (Tasso) Good in Maize (Zeal Mays L.). *Genetic.* **2000**, *32*, 507–515.

Bremer, H. Umber Welkrantheiten in Sudwest-Anatolien (On Wilt Diseases in South-West Anatolia). *Rev. Appl. Mycol.* **1944**, *25*, 255.

Brooks, F. T. *Plant Disease*; Oxford Univ. Press: New York, 1928; pp304.

Brown, W. Experiments on the Growth of Fungi in Culture Media. *Ann. Bot.* **1928**, *37*, 105–109.

Brummel-Cox, P. J.; Stein, I. S.; Rodgers, D. M.; Catlin, L. E. Inheritance of Resistance to *Macrophomina phaseolina* (Tassi) Goid. and *Fusarium moniliforme* Sheldon in Sorghum. *Crop Sci.* **1987,** *28,* 37–40.

Butler, E. J. *Fungi and Diseases in Plants.* Thacker, Spink and Co.: Calcutta; 1918.

Butler, E.J.; Bisbee, G. R. *The Fungi of India,* ICAR Sci. Monograph, No. II, New Delhi, 1931, pp 158.

Cardona, R. Vertical Distribution of Sclerotia of *Macrophomina phaseolina* in a Natural Infested Soil in Portuguesa State. *Revisit. De. La. Faulted de. Agronomic, Universidad del. Zulia.* **2006,** *23*(3), 285–293.

Chaudhary; Sumuti, P. A.; reek Salvia; Sabena Jota. Efficacy of Bio Control Agent Singly and in Combination against Dry Root Rot of *Macrophomina phaseolina* of Moonbeam. *J. Mica. Pl. Patrol.* **2010,** *40*(1), 141–144.

Choudhary, C. S.; Singh, S. N.; Prasad, S. M. *In vitro* Effectiveness of Chemicals to Control *Macrophomina phaseolina* (Tassi.) Goid, Causing Stem and Root Rot of Sesame. *J. Appl. Biol.* **2004,** *14*(1), 46–47.

Choudhary, C. S.; Prasad, S. M.; Singh. S. N. Effect of Sowing Date and Fungicidal Spray on *Macrophomina* Stem and Root Rot and Yield of Sesame. *J. Appl. Biol.* **2004,** *14*(2), 51–53.

Choudhary, C. Studies on Stem and Root Rot of Til (*Sesamum indium L.*) Caused by *Macrophomina phaseolina* (Tassi) Goid. M.Sc. (Ag) Thesis. Dept. Of Mycol. & PI, Pathos RAU, 2000.

Christopher, B. J.; Usharani, S.; Kumar, R.; Udhaya. Management of Dry Root Rot (*Macrophomina phaseolina* (Tassi.) (Goid)) of Urn Bean (*Vegan Mango* (L.) Hipper) by the Integration of (*Trichoderma vixens*) and Organic Amendments. *Mysore J. Argil Sci.* **2008,** *42*(2), 241–246.

Christopher, D. J.; Usharani, S. Kumar; Udhaya, R. Management of Dry Root Rot (*Macrophomina phaseolina* (Tassi.) Goid) of Peanut (*Rachis Hypogeal L.*) by the Integration of Antagonistic (*Trichoderma vixens*) and Organic Amendments. *Adv. Pl. Sci.* **2008,** *21*(2), 389–392.

Day, P. *In Vitro* Studies into Host-Pathogen Interactions of Sunflower and *Macrophomina phaseolina*. P. Day. Ph.D. Thesis. University of Cambridge, 1993.

Deshpande, A. L.; Arrayal, J. P.; Mathura, B. N. *Rhizoctonia bataticola* Causing a Root Rot of Opium in Rajasthan. *Indian Psychopath.* **1969,** *22*(4), 510.

Filho, E. S.; Dhingra, O. D. Effect of Herbicides on Survival of *Macrophomina phaseolina* in Soil. *Trans. Br. Mycol. Soc.* **1980,** *74*(1), 61–64.

Rangswami, G.; Mahadevan, A. *Diseases of Crop Plants in India*; PHI Learning Private Limited: New Delhi, **2008**; pp 275–278.

Gamesman, S.; Sear, R. Bio Control Mechanism of Groundnut (*Arachis hypogea L.*) Diseases- Trichoderma System. In: *Biotechnological Applications in Environmentand Agriculture*, G. R., Pithead, P. K., Eds.; ABD Pub.: Jaipur, 2004; pp 312–327.

Genic SNP markers and legume syntonic reveal candidate genes underlying QTL for *Macrophomina phaseolina* resistance and maturity in cowpea [*Vigana unguiculata* (L) Wall.]

Gibson, I. A. S. Crown Rot, a Seedling Disease of Groundnut Caused by *Aspergillums niger.* II. An Anomalous Effect of Organo-Mercurial Seed Dressings. *Trans. Brit. Mycol. Soc.* **1953,** *36,* 19–122.

Gopal, K.; Ahamed, S. K.; Babu, G. P. Relative Resistance in Groundnut Genotypes to Pod Rot Disease. *Legume Res.* **2006,** *29*(3), 205–208.

Gupta, S. C.; Kolte, S. J. Cultural Characteristics of Leaf and Root Isolates of *M. phaseolina* (Tassi) Goid from Groundnut. *Indian J. Microbial.* **1981,** *21*(4), 345–346.

Hangar, O. P; Sinclair, J. B. Biology and Pathology of *Macrophomina phaseolina*. Universidad Federal de Viscose. 1978; pp 166.

Hedge, Y. R.; Chauhan, T. L. Management of Root Rot of *Jatropha curcas* in Karnataka. *Int. J. Pl. Protec.* **2010,** *2*(2), 243–244.

Delgado, S. H.; Reyes-Valdés, M. H.; Rosales-Serna, R.; Mayer-Pérez; N. Molecular Markers Associated with Resistance to *Macrophomina* phaseolina (Tassi) Goid. In common bean. *J. Plant Pathol.* **2009,** *91*(1), 163–170.

Holliday, P. *Fungal Diseases of Tropical Crops*; Dover Publications: Minneola. 1980; pp 254–255.

Jaiman, R. K.; Jain, S. C. Effect of fungicides on Root Rot of Cluster Bean Caused by *Macrophomina phaseolina*. *Environ. Ecol.* **2010,** *28*(2A), 1138–1140.

Jaiman, R. K.; Jain S. C.; Sharma Pankaj. Effect of Storage Condition on *Macrophomina phaseolina* Root Rot in Cluster Bean. *J. Myco. Pl. Pathol.* **2009,** *39*(1), 82–85.

Jain, A. C.; Kulkarni, S. N.; Root and Stem Rot of Sesame. *Indian oilseed J.* **1965,** 9(3), 201–203.

Jayati-Bhowal, Suita-Ghosh, Gotha, A. K.; Chatterjee, B. P. Infection of Jute Seedlings by the Phytopathogenic Fungus *Macrophomina phaseolina* Mediated by Endogenous Lection. *Res. J. Micro bio.* **2006,** *1*(1), 51–60.

Kale, D. M.; Marty, G. S. S.; Badigannavar, A. M.; New Trombay Groundnut Variety TG 38 Suitable for the Residual Moisture Situation in India. *J. SAT-Argil. Res.* **2007,** *3*(1), 1–2.

Kale, D. M.; Marty, G. S. S.; Badigannavar, A. M.; Dhal, J. K. New Trombay Groundnut variety, TG 51, for commercial cultivation in India. *J. SAT-Argil. Res.* **2009,** 7, 1–2.

Kaur, S.; Dillon, G. S; Chauhan, V. B. Morphological and Pathogenic Variability in *Macrophomina phaseolina* Isolates of Pigeon Pea (*Cajuns cajon* L.) and Their Relatedness using Principle Component Analysis. Arch. Physiopathol. Plant Protect. 2013, *46*, 2281–2293.

Ahmed, M. A.; Gangopadhaya. Efficiency of *Pseudomonas fluorescens* in Controlling Root Rot of Chickpea Caused by *Macrophomina phaseolina*. *J. Mica. Pl. Patrol.* **2008,** *38*(3), 580–587.

Khan, S. N.; Aye, N.; Ahmed, I. A Re-Evaluation of Geographical Distribution of Charcoal Rot on Sunflower Crop in Various Agro Ecological Zones of Pakistan. *Sympathy.* **2003,** *1*(1), 63–66.

Kunwar, I. K.; Singh, T; Machado, C. C; Sinclair, J. B. Histopathology of Soybean Seeds and Seedling Infection by *Macrophomina phaseolina*. Phytopathology. **1986,** *76*, 532–535.

Lukade, G. M. Effect of Sowing Time on Incidence of Charcoal Rot and Yield of Released Sorghum Cultivars. *Indian J. Mycol. Pl. Patho.* **1995,** *25*(1&2), 53.

Lewis, P.R.; Knight, D.P.; *Staining Methods for Sectioned Material*; North Holland Publishing Company: Amsterdam, 1977.

Li ZhengChao; Qiu QingShu. Huayu 16: A New High-Yielding, Improved Quality Groundnut Cultivar with Wide Adaptability for Northern China. *Int. Arachis Newsletter.* **2000,** *20,* 31–32.

Fuhlbohm, M. J.; Ryleyand, E. A.; Aitken, B. Infection of Mungbean Seed by *Macrophomina phaseolina* is More Likely to Result from Localized Pod Infection than from Systemic Plant Infection. *Plant Pathol.* **2013,** *62*(6), 1271–1284.

Mahdizadeh V.; Safari N.; Agbayani M. A. New Hosts of *Macrophomina phaseolina* in Iran. *J. Pl. Pathology.* **2011b,** *93*, S463–S489.

Mahdizadeh, V.; Safari, N.; Goltapeh, E. M. Diversity of *Macrophomina phaseolina* Based on Morphological and Genotypic Characteristics in Iran. *Plant Pathos. J.* **2011a**, *27*(2), 128–137.

Maheswari, C. U.; Ramakrishna, G. Factors Influencing the Competitive Saprophytic Ability of *Macrophomina phaseolina* in Groundnut. *Madras Argil. J.* **2000**; *86*(10/12), 552–553.

Malachi, P.; Sabetha Doraisamy. Effect of Temperature on Growth and Antagonistic Activity of *Trichoderma spp.* against *Macrophomina phaseolina. J. Boil Control.* **2003a**, *17*(2):153–159.

Malachi, P.; Sabetha Doraisamy. Compatibility of *Trichoderma harzianum* with Fungicides against *Macrophomina phaseolina. Pl. Dis. Res. (Ludhiana).* **2003b**, *18*(2), 139–143.

Maryam Aghakhani; Dubai, S. C. Morphological and Pathogenic Variation Among Isolates of *Rhizoctonia bataticola* Causing Dry Root Rot of Chickpea. *Indian phytopath.* **2009**, *62* (2), 183–189.

Mathura, S. B.; Confer.B. M. *Seed Borne Diseases and Seed Health Testing of Wheat.* Danish Government Institute and Pathology for Developing Countries: Copenhagen, 1993; pp 168.

*Maurya, S.K.; Kaur, S.; Chauhan V. B. H*istopathology of Macrophomina Stem Canker Disease in Pigeonpea (*Cajanus cajan* l.). *Int J Phytopathol.* **2013**, *2*(3), 187–192.

Mayee, C. D. Current Status and Future Approaches for Management of Goundnut Disease in India. *Indian Phytopath.* **1995**, *48*, 389–401.

Mayee, C. D.; Datar.V. V. Diseases of Groundnut in the Tropics Review. *Trop. Pl. Path.* **1988**, *5*, 169–198.

Mayee, C.D. Diseases of Groundnut and Their Management. In: *Plant Protection in Field Crops*; Rao, M. V. N., Sitanantham, S., Eds.; PPSI: Hyderabad, 1987; pp 235–243.

Microscopy resin. TAAB Laboratories Equipment Ltd.

Mihail, J. D. Macrophomina. In *Methods for Research on Soil Borne Phytopathogenic Fungi*; Singleton, L. L., Rush, C. M., Eds.; APS press: St. Paul, Minnesota, 1992; pp134–136.

Mittal, R. K.; Effect of Sowing Dates and Disease Development in Lentil as Sole and Mixed Crop with Wheat. *J. Mica. Pl. Pathos.* **1997**, *27* (2), 203–209.

Moore, D. Tissue Formation. In: *The Growing Fungus;* Grew, N. A. R., Gad, G. M., Eds.; Chapman & Hall: London, 1995; pp 432–433.

Moravia, A. M. Management of *Macrophomina phaseolina* in Groundnut through Systemic Fungicides. *Internat. J. agric. Sci.* **2011**, *4*(1), 212–213.

Moravia, A. M. Effect of Source of Varietal Resistance against of *Macrophomina phaseolina* on Groundnut. *Int. J. Pl. Protec.* **2012**, *5*(2), 438–439.

Moravia, A. M.; Khaddar, R. R. Loss of Yield of Groundnut (*Arachis hypogea* L.) Due to Dry Root Rot (*Macrophomina phaseolina*) and Their Management Under in Vivo Condition. *Internat. J. agric. Sci.* **2011**, *7*(2), 282–285.

Nageswararao, G.; Patibanda, A. K.; Ranganathswam, Y. M. Studies on the Efficacy of Fungicides, Organic Amendments and Bio Control Agent on Dry Root Rot (*Rhizoctonia bataticola*) of Groundnut in Vivo. *J. Mycopathy. Res.* **2012**, *50*(2), 285–289.

Nema, K. G.; Agarwal, G. P. *M. phaseolina* on Roots of *Cicer arietinum* L. and Possum Stadium L. *Proc. Nat. Acad. Sci. India,* **1960**, *30,* 57.

Okwulehie, I. C. Anatomical Changes in Groundnut due to Infection with *Macrophomina phaseolina* (Maub) Ashby. *J. Sustain. Argil. Environ.* **2002**, *4*(2), 249–257.

Okwulehie, I. C. Physiological Studies in Groundnuts *Arachis hypogea* L. Infected with *Macrophomina phaseoli* (Maub.) Ashby. *Int. J. Tropical Plant Dis.* **2001**, *19*(1/2), 25–37.

Okwulehie, I. C. Studies on *Macrophomina phaseoli* (Maub.) Ashby Growth and Some Physiological Aspect of Groundnut *(Arachis hypogea L)* Plant Infected with the Fungus. *Global J. Pure Appl. Sci.* **2004,** *10*(1), 23–29.

Olaya, G.; Abaci, G. S.; Weeded, N. F. Inheritance of the Resistance to *Macrophomina phaseolina* and Identification of RAPD Markers Linked to Resistance Genes in Beans. Phytopathology. **1996,** *86,* 674–679.

Patel, K. K.; Patel, A. J. Control of Charcoal Rot of Sesame. *Indian J. Mycol. Pl. Pathos.* **1990a,** *20*(1), 62–63.

Patel, K. K.; Patel, A. J. Meteorological Correlation of Charcoal Rot of Sesame. *Indian J. Mycol. Pl. Pathos.* **1990b,** *20*(1), 64–65.

Pathak, D.; Barman, B. Effect of Dates of Sowing on the Incidence of Root Rot Disease of Jute Caused by *Macrophomina phaseolina* (Tassi.) Gold. *Ann. Biol. (Ludhiana).* **1998,** *14*(2), 185–187.

Prasad, Mohan Vijay.; Kudada, N.; Baranwal, S. M. Varietal Screening of Chickpea Against Dry Root Rot Disease. J. *Res. Birsa. Agril. University.* **2006,** *18*(1), 145–147.

Prasad, V. M.; Barnwal, S. M.; Kudada, N. Fungicidal Management of Dry Root Rot Disease and Yield of Chickpea. *J. Appl. Bio.* **2006,** *16,* 42–44.

Prasad, N. Studies on Root Rot of Cotton in Send. II Relation of Root-Rot of Cotton with Root-Rot of Other Crops. *Indian J. Agric. Sci.* **1944,** *14,* 388–399.

Prasanthi, L. Evaluation of Indian Bean (Field Bean) Lines for Resistance to Dry Root Rot Caused by *Rhizoctonia bataticola. J. Arid. Legumes.* **2007,** *4* (2), 154–155.

Rai, M.; Veda Ratan Srivastava, S. S. L. Management of Root Rot of Urn Bean [*Vegan mango* (L.) Hipper] Caused by *Macrophomina phaseolina. Farm Sci. J.* **2005,** *14*(1), 89.

Raja Mohan, K.; Balabaskar, P. Survey on the Incidence of Groundnut Root Rot Disease in Cuddlier District of Tamilnadu and Assessing the Cultural Characters and Pathogen City of *Macrophomina phaseolina* (Tassi.) Giad. *Asian J. Sic. Tech.* **2012,** *3*(4), 90–94.

Rajani, V. V. Parochial Management of Root Rot Disease (*Macrophomina phaseolina*) of Castor (Ri) with Soil Amendments and Bio Control Agent. *J. Myco. Pl. Pathos.* **2009,** *39*(2), 290–293.

Rao, V. V. R.; Rao, K. C.; Shantaram, M. V.; Reddy, M. S. Management of Seed and Seedling Diseases of Groundnut (*Arachis hypogeal L.*) Through Seed Treatment. *Indian J. Pl. Protec.* **1998,** *26*(1), 1–8.

Rao, V. V. R.; Reddy, M. S.; Santarem, M. V.; Rao, K. C. Survey on the Occurrence of Seedling Diseases of Groundnut (*Arachis hypogaea L.*). *Indian J. Pl. Protec.* **1997,** *25*(1), 75–77.

Reichert, I. Palestine, Root Disease Caused by *Rhizoctonia bataticola. Int. Bull. Plant. Prot.* **1930,** *4,* 17.

Reis, Erlei Melo; Boaretto, Cristiane; Danelli, Anderson Luiz Durante. *Macrophomina phaseolina*: Density and Longevity of Microsclerotia in Soybean Root Tissues and Free on the Soil, and Competitive Saprophytic Ability. *Summa phytopathol.* **2014,** *40,* 128–133.

Sadashivaiah, A. S.; Ranganathaiah, K. G.; Gouda, D. N. Seed Health Testing of *Helianthus Annuls* with Special Reference to *Macrophomina phaseolina. Indian Phytopathol.* **1986,** *39,* 445–447.

Sağır, P.; Sağır, A.; South, T. The Effect of Sowing Time and Irrigation on Charcoal Rot Disease *(Macrophomina phaseolina)*, Yield and Yield Components of Sesame. *Batik Korma Bulletin.* **2010,** *50* (4), 157–170.

Shazia Rushed, Shahnaz Dewar; Abdul Gaffer. Location of Fungi in Groundnut Seed. *Pak. J. Bot.* **2004,** *36*(3), 663–668.

Shelby, S. I. M. Effect of Fungicidal Treatment of Sesame Seeds on Root Rot Infection, Plant Growth and Chemical Components. *Bull. Faculty Argil. Univ. of Cairo.* **1997,** *48*(2), 397–411.

Siavosh Rayatpanah; Siranoush G. Nanagulyan; Seyed V. Alav; Mohammad Razavi; Abbas Ghanbari-Malidarreh. Pathogenic and Genetic Diversity Among Iranian Isolates of *Macrophomina phaseolina. Chil. J. Agric. Res.* **2012,** *72*(1), 40–44 .

Singh Permit; Gupta, T. R.; Singh, P. Effect of Sowing Dates on the Development of Disease and Seed Yield in Sesame (*Sesame indium* L.) *Plant Dis. Res.* **1993,** *8*(1), 61–63.

Singh, R. D.; Sandhu, A. C. Factors Affecting Development of Charcoal Rot of Cowpea. *Indian J. Mycale Pl. Pathol.***1995,** *25*(1&2), 74.

Singh, S. K.; Nene, Y. L.; Reddy, M. V. *Additions to the Host Range of Macrophomina phaseolina. Plant Dis.* **1990,** *74*(10), 828.

Singh, S. K.; Nene, Y. L. Cross Inoculation Studies on *R. bataticola* Isolate from Different Crops. *Indian Phytopath.* **1990,** *43*, 446–448.

Slavish Rayatpanah; Sayed Alireza Dalili. Diversity of *Macrophomina phaseolina* (Tassi) Goid Based on Chlorate Phenotypes and Pathogenicity. *Int. J. Biol.* **2012,** *4*(2), 54–63.

Small, W. Further Notes on *Rhizoctonia* (Taub) Butler. *Trop. Agriculturist.* **1927**a, *69*(1), 9–12.

Small, W. Further Occurrence of *Rhizoctonia* (Taube) Butler. *Trop. Agriculturist.* **1927**b, *69*(4), 202–203.

Smith B. W. Foliar Diseases. In *Compendium of Peanut Disease.* American Photo-pathological Society: St. Paul, 1994; pp 55–57.

Smith, G. S.; Carvel, O. N. Field Screening of Commercial and Experimental Soybean Cultivars for Their Reaction *to Macrophomina phaseolina. Plant Disease.* **1997,** *81*(4), 363–368.

Smits, B. G.; Nogeura, R. The Ontogeny and Morphogenesis of Sclerotic & Pycnidia of *M. phaseolina. Agron. Trop. (Maracay).* **1999.**

Sreedevi, B.; Charisma Devi, M.; Saigopal, D. V. R. Isolation and Screening of Effective *Trichoderma* spp. Against the Root Rot Pathogen *Macrophomina phaseolina. J. Agric. Techno.* **2011,** *3*, 623–635.

Subhra Sarkar; Prandial Biswas; Subrata Kumar; Tuhin Ghosh; Ghosh Sanjay. Nitric Oxide Production by Necrotrophic Pathogen *Macrophomina phaseolina* and the Host Plant in Charcoal Rot Disease of Jute: Complexity of the Interplay between Necrotroph–Host Plant Interactions DOI: 10.1371/journal.pone.0107348.

Subrahmanyam, P.; Mehan, V. K.; Nevill D. J.; Donald. D. Mac Research on Fungal Diseases of Groundnut. ICRISAT. In *A Proceeding of an International Workshop on Groundnut,* Patancheru, A. P., Ed., ICRISAT: India, 1980; pp 189–197.

Subramanayam, C. V. *Hyphomycetes.* ICAR: New Delhi, 1971; pp 881–82.

Subramanian, C. V. *M. phaseolina* from Black Cotton Soil, Udamalpet, Madras state. *J. Madras Unit.* **1952**. 22, 220.

Sundararaman, S. Administration Report of the Mycologist for the year 1929. Deptt Agric: Madras, 1931; pp 30.

Sundararaman, S. Administration Report of the Mycologist for the year 1930. Dept Agric: Madras, 1932; pp 20.

Songa, A. Variation and Survival of *Macrophomina phaseolina* in Relation to Screening Common Bean (*Pharsalus vulgaris* L.) for Resistance. Sunga. A. Ph.D. Thesis, University of Reading 1995, 7–32, 89–90..

Tandel, D. H.; Sabalpara, A. N.; Pandya, H. V.; Naik, R. M. Effect of Leaf Blight [*Macrophomina phaseolina* (Tassi.) Goid.] On Growth Parameters and Yield of Green Gram and Its Chemical Control. *Int. J. Pl. Protect.* **2010**, *3*(2), 329–331.

Taubenhaus, J. J. *Sclerotium bataticola Phytopathology*, 1913, *3*, 164f.

Thakare, A. R.; Chavan, P. N.; Rout, B. T.; Tine-Pillai; Paulkar, P. K. Effect of Polythene Mulch and Seed Treatment on Diseases and Yield of Groundnut. *Res. Crops.* **2002**, *3*(1), 159–163.

Thirumalachar, M. J. Incidence of Charcoal Rot of Potato in Bihar (India) in Relation to Cultural Conditions. *Psychopath.* **1955**, *45*(2), 91–93.

Thirumalachar, M. J. Pycnidial Stage of Charcoal Rot Inciting Fungus with a Discussion of its Nomenclature. *Phytopath.* **1953**, *43*, 608–610.

Umamaheswari, C.; Ramakrishnan, G.; Nallathambi, P. Role of Inoculum Level on Diseases Incidence of Dry Root Rot Caused by *Macrophomina phaseolina* in Groundnut. *Madras Agril. J.* **2001**, *87*(1/3), 71–73.

Umer Iqbal; Tariq Mukhtar. Morphological and Pathogenic Variability among *Macrophomina phaseolina* Isolates Associated with Mungbean (*Vigna radiata* L.) Wilczek from Pakistan. *Sci. World J.* **2014**.

Muchero, W.; Jeffrey D. Ehlers; Timothy J. Close; Philip A. Roberts. Genic SNP Markers and Legume Synteny Reveal Candidate Genes Underlying QTL for *Macrophomina phaseolina* resistance and Maturity in Cowpea [*Vigna unguiculata* (L) Walp.] *BMC Genomics.* **2011**, *12*, 8.

Willetts, H. J. *The Filamentous Fungi.* Smith, J. E., Berry, D.R., Eds.; Edward Arnold: London, 1978; pp 197–213.

Willison, J. H. M.; Rowe, A. J. *Replica, Shadowing and Freeze-Etching Techniques.* Audrey M. Glauert. P, Eds.; North Holland Publishing Company: Amsterdam, 1980, pp 59–93.

Wisniewska, H.; Chelkowski J. Influence of Exogenic Salicylic Acid on *Fusarium* Seedling Blight Reduction in Barley. *Acta. Physiologiae Plantarum.* **1999**, *21*, 63–66.

Wyllie, T. D. Charcoal Rot of Soybean-Current Status. *In Soybean Diseases of the North Central Region.* Wyllie, I. D., Scott, K. H., Eds.; The American Phytopathology Society: St. Paul, 1988; pp 106–113.

Young D. J.; Gilbertson, R. L.; Alcorn S. M. A New Record for the Longevity of *Macrophomina phaseolina* Sclerotic. *Mycological.* **1982**, *74*(3), 504–505.

PART IV
Diversity of Phytopathogens in Natural Ecosystems

CHAPTER 14

SOIL MICROBIAL COMMUNITY AND THEIR POPULATION DYNAMICS: ALTERED AGRICULTURAL PRACTICES

SANTOSH KUMAR[1*], RAVI R. KUMAR[2], MAHENDRA SINGH[3], ERAYYA[1], MAHESH KUMAR[2], MD. SHAMIM[2], NIMMY M. SUBRAMANIAN[4], VINOD KUMAR[2], and AMARENDRA KUMAR[1]

[1]*Department of Plant Pathology, Bihar Agricultural University, Sabour 813210, Bihar, India*

[2]*Department of Molecular Biology and Genetic Engineering, Bihar Agricultural University, Sabour 813210, Bihar, India*

[3]*Department of Soil Science and Agricultural Chemistry, Bihar Agricultural University, Sabour 813210, Bihar, India*

[4]*National Research Centre on Biotechnology, Pusa Campus, New Delhi 110012, India*

Corresponding author. E-mail: santosh35433@gmail.com

CONTENTS

ABSTRACT

Microbial communities play an important role in nutrient cycling by the process of mineralization and decomposition of organic waste material, which are released into the soil as nutrients that are essential for plant growth. These communities can influence nutrient availability under the processes of solubilisation, chelation, and oxidation/reduction. In addition, soil microorganisms may affect nutrient uptake and plant growth by the release of growth stimulating or inhibiting substances that influence root physiology and root architecture in the rhizosphere zone. The rhizosphere is one of the most complex environments with thousands of interactions that play critical roles in changing the microbial population (microbial population dynamics) which lead to the plant health and its development. A number of factors such as such as quantity and quality of root exudates including plant species developmental stage and nutritional status of the soil have been shown to affect the population dynamics of microbial community. Knowledge and understanding of mixed microbial communities in soil can be analyzed by application of several conventional and molecular tools. Molecular techniques have contributed significantly to the detection and identification of microorganisms especially in non-culturable organisms as well as the quantification of the relative abundance of organisms from soil. Currently, there is no single technique that can adequately describe the entire microbial diversity and the associated catabolic genes from agricultural soil.

14.1 INTRODUCTION

One of the biggest challenges in agriculture nowadays is to increase yield and sustainability of crop production as the global population is approaching nine billion people by 2050 (Evans, 2013). Another major concern is climate change, global warming, and human health issues, for which major funds of the world are diverted toward this direction, for its control measure. According to projections of the Food and Agriculture Organization of the United Nations (FAO), the demand for soil quality is believed to be an integrative indicator of environmental quality, food security, and economic viability. Soil quality may be defined as "the capacity of a specific kind of soil to function, within natural managed ecosystem boundaries, to sustain plant and animal productivity, maintain or enhance water and air quality, and support human health and habitation" (Karlen et al., 1997). The qualities of the soil are improved by microorganism, as soils are the reservoir of tremendous microbial diversity. About 1 g of soil or sediment may contain more than 10^{10} bacteria (Torsvik et al., 1990). Among soil microorganisms with significant positive effects on plant, the legume-nodulating, phosphorus solubilizing, and iron utilization bacteria are the most studied (McInnes et al., 2004). Legume-nodulating bacteria are capable of fixing atmospheric nitrogen in specific root located structures, named nodules. Most of these bacteria described so far, and known as "rhizobia," have been classified as belonging to the alpha proteobacteria and to one of the six genera: *Rhizobium, Sinorhizobium, Bradyrhizobium, Azorhizobium, Mesorhizobium*, and *Allorhizobium* (Garrity & Lilburn, 2003; Rivas et al., 2009).

Plants require at least 17 different minerals for adequate nutrition. Several factors including soil, plant species, microbial interactions, and environment can affect the acquisition of these nutrients. Microorganisms are very diverse and include all the bacteria including archaea and eubacteria as well as some fungi, algae and certain animals such as rotifers. Human activities, especially, the application of pesticides and fertilizers, and pollution could affect the soil microbial communities, with significant consequences on plants and animals health.

Microbial communities play an important role in nutrient cycling by mineralizing and decomposing organic material, which are released into the soil as nutrients that are essential for plant growth. These communities can influence nutrient availability by solubilization, chelation, and oxidation/ reduction processes. In addition, soil microorganisms may affect nutrient uptake and plant growth by the release of growth stimulating or inhibiting substances that influence root physiology and root architecture in the rhizosphere zone.

The rhizosphere is one of the most complex environments with thousands of interactions that play critical roles in changing the microbial population (microbial population dynamics) which lead to the plant health and its development. A number of factors have been shown to effect the population dynamics of microbial community, such as quantity and quality of root exudates including plant species (Dennis et al., 2010), soil type (Berg, 2009; Bulgarelli et al., 2012), developmental stage and nutritional status (Houlden et al., 2008) of the soil.

Diversity of microbial population present in the soil is determined by various factor such as soil depth of soil, organic matter present in soil, soil porosity, oxygen and carbon dioxide concentration, soil temperature, soil pH, etc. Factors that influence microorganism role in nutrient building and cycling in soil and organic matter decomposition are of unique interest.

Microorganisms decompose organic matter, detoxify the toxic substance, fixing the nitrogen, transformation of nitrogen, phosphorous, potassium and other secondary and micro nutrients are the major biochemical activities performed by microbes in soil. Here, we explain most important factors that can influence the population dynamics of microbial communities in briefly, in flowing sub headings.

14.2 QUANTITY AND QUALITY OF ROOT EXUDATES AND MICROORGANISM POPULATION

Plant heath and its growth promotion can be achieved directly by the interaction between the microbe and roots exudates, as well as indirectly, due to antagonistic activities against plant pathogens. Innumerable interacting microbes produce phytohormones, which have been shown to inhibit or promote root growth, protect plants against biotic or abiotic stress, and improve nutrient acquisition by roots (Berg & Smalla, 2009; Davies, 2010). Soil microbial community promotes an environmentally sustainable alternative to increase crop production.

An interesting example, interaction between rhizosphere and fluorescent pseudomonas; plants reduce soil iron (Fe) availability by acquiring iron and releasing exudates which attract to the rhizosphere microbes that also utilizes Fe. In Fe-stressed environment, siderophore-producing bacterial populations are enriched, which then suppress pathogens such as fungi and oomycetes through competition for Fe. The plants, however, are able to utilize siderophore-bound iron, which enhances their growth (Lemanceau et al., 2013). In a soil suppressive to the fungal pathogen *Rhizoctonia solani*, Proteobacteria, Firmicutes, and Actinobacteria were prominent taxa found to be involved in disease suppression (Mendes et al., 2011). There is also evidence to suggest that plants may use microbial communities to their own benefit to avoid infections (Mendes et al., 2011). The presence of potentially toxic compounds, low availability of essential minerals and pathogens in the soil often restrict crop production. Some reports available on the specific genes such as *NRT2* gene from *Arabidopsis thaliana* link to the biotic and abiotic stress (Dechorgnat et al., 2012). Damiani et al. (2012) shown in their finding involvement of the microbes in nutrient uptake. Microbial effector molecules from the root exudates act as defense suppression against pathogens.

14.3   ORGANIC MATTER AND MICROORGANISM POPULATION

Soil contains diverse micro and macro flora and fauna as long as there is a carbon source for energy. A large number of bacteria exist in the soil but having lesser biomass due to their small size (Table 14.1). For example, actinomycetes are ten times less in number but are of large in size so their biomass is similar to bacteria present in soil (Schmidt & Gier, 1989). Although, actinomycetes are heterogeneous group of gram-positive, mainly anaerobic bacteria famed for a filamentous and branching ontogeny pattern that results, in most forms, in an extended colony, or mycelium (Rao, 2005), due to of this character they often consider separately in the group. Fungal populations are smaller but they dominate the soil biomass (Cocks & Torsvik, 2002). Bacteria, actinomycetes, and protozoa can tolerate more soil disturbance than fungal populations so they dominate in tilled soils while fungal and nematode populations tend to dominate in untilled (Janusauskaite et al., 2013 & Silva et al., 2013). Soils contain about 8 to 15 tons of bacteria, fungi, protozoa, nematodes, earthworms, and arthropods (Brady & Weil, 2012). According to the study (Watson & Kelsey, 2006), agricultural fuel containing 4% organic matter has been following count of soil organism.

TABLE 14.1   Estimated Count of Organism in Soil (Source: Bhattarai et al., 2015).

Animal (2%)	Biomass No. (Kg/ha)	Plant (98%)	Biomass No. (Kg/ha)
Micro protozoa	$100(10^4–10^5/\text{gm.})$	Microbacterial	$5000(10^8–10^9/\text{gm.})$
Macro earthworm	$50(3.3×10^5/\text{HFS})$	Fungi	$5000(10^5–10^6/\text{gm.})$
Nematode	$2–100(10–10^2/\text{gm.})$	Actinomycete	$1500(10^7–10^8/\text{gm.})$
Myriapoda	$40(5.5×10^6/\text{HFS})$	Algae	$10(10^4–10^5/\text{gm.})$
Insects	10–100 (50/HFS)	Macro: Plant roots	4000
Rodents	5	–	–

14.4   SOIL AERATION AND MICROORGANISM POPULATION

Microorganism population in soil is limited by soil porosity; more the pore space higher is the count of microbes. Well-tilled soil is well aerated and favors microorganism growth. The microbial population is found to be more in O_2 rich soil compared to CO_2 (McNabb & Startsev, 2009). A major contributor to poor aeration is soil compaction (Day & Bassuk, 1994). Soil compaction can create problems by alterations in the physical properties of the soil including: decreases in total pore space, decreasing soil oxygen content, reductions in water infiltration and percolation rates, increases in soil strength and density, and increase water retention (Brady & Weil, 2012).

14.5   SOIL DEPTH AND MICROORGANISM POPULATION

Huge diversity of micro flora and fauna are found in soil horizons. Soil profiles are many meter deep and soil varies from place to place. The microorganism population also varies with the depth. The study conducted by Hoorman and Islam at Ohio State University shown following result on relative number and biomass microorganism species at 0–15 cm depth of soil (Table 14.2).

TABLE 14.2   Microorganism Species at 0–15 Depth of Soil (Source: Bhattarai et al., 2015).

Microorganism present in the soil	Number/g of soil	Biomass (g/m²)
Bacteria	10^8–10^9	40–500
Actinomycetes	10^8–10^9	40–500
Fungi	10^8–10^9	100–1500
Algae	10^8–10^9	1–50
Protozoa	10^8–10^9	Varies
Nematodes	10^8–10^9	Varies

Agricultural soils would more closely resemble soil of natural ecosystems if management practices like heavy machinery and general biocides would be reduced or eliminated in cultivation whereas incorporation of perennial crops and organic material synchronizes nutrient release and water availability with plant demand will be carried out. In order to achieve these goals, research must be undertaken to develop methods for successful application of organic materials and their associated microorganisms, synchronization of management practices with crop and soil biota phenology, and improvement of our present knowledge of the mechanisms linking species to ecosystem processes.

14.6   MICROBES IN SOIL FERTILITY

Soil fertility has long been appreciated as a major driver of vegetation diversity, and the unimodal or humpback relationship is the dominant conceptual model for this driver (Grime, 1979). It is well accepted that system productivity will increases if we increasing soil fertility, and the humpback model describes a relationship between productivity and plant community diversity for which diversity first increases and then decreases with increasing productivity (Grace, 1999).

Agriculturally important microorganisms consist of phenomenal diversity which includes plant growth promoting N-fixing *Cyanobacteria*, *Rhizobacteria*, Mycorrhiza, plant disease suppressive beneficial bacteria, stress tolerance entophytes, and bio-degrading microbes. In no tillage or minimum tillage soil, the count of *Azotobacter, Azospirillum, Rhizobium, Cyanobacteria*, phosphorus and potassium solubilizing microorganisms, and mycorrhizae can be found significantly higher (Bhardwaj et al., 2014). Bacteria are the important soil microorganism responsible for many enzymatic transformations like nitrification, ammonification, etc. *Azospirillum* is a micro aerobic organism that fixes the nitrogen in association with roots of grasses. Inoculation of *Azospirillum* to the grass crops has positive hormonal effect on roots and plant growth (Dastager et al., 2010). *Rhizobium* alone in symbiotic association with legume fixes about 50–200 kg of N_2 per hectare. Table 14.3 shows the amount of nitrogen fixed by the different microorganism with legume fixes about 50–200 kg of N/ha. The table also shows the amount of nitrogen fixed by the different microorganism with different spp. of plants.

TABLE 14.3 Amount of Nitrogen Fixed by the Different Microorganism (Source: Bhattarai et al., 2015).

Crop/plant	Organism association	Level of N^2 fixation (Kg/ha/years)
Leucaena leucocephala	*Rhizobium*	100–500
Medicago sativa	*Rhizobium*	100–500
Trifolium pratense	*Rhizobium*	100–150
Vigna unguiculata	*Bradyrhizobium*	50–100
Cajanus cajan	*Bradyrhizobium*	150–280
Alnus	*Frankia*	50–150
Species of Gunnera	*Nostac*	10–20
Azolla	*Anabaena*	150–300

14.7 MAJOR ROLE OF SOIL MICROBES

Soil microbes are increasingly appreciated as important drivers of vegetation structure and dynamics. Two distinct conceptual models of plant–microbe interactions are microbially mediated niche partitioning (Reynolds et al., 2003) and feedback dynamics. Microbially mediated niche partitioning is an explicitly niche-based model, whereby coexistence of plant species is promoted by associations with microbes that allow different plant species to access different sources of soil N or other soil nutrients (Reynolds et al., 2003). This model identifies the vital role played by microbes, through their enzymatic capacities, which helps in making soil nutrients available to plants (Fig. 14.1).

FIGURE 14.1 Conceptual representation of three important soil based drivers of plant community structure and dynamics, showing areas of two- and three-way overlap.

In the case of N partitioning, individual plant species might obtain ammonium from different nitrogenous organic sources, for example urea, proteins, or chitins, via differential associations with such enzymes as ureases, proteases, or chitinases. A similar concept can be envisioned for phosphorus (P) or other limiting soil nutrients. In contrast to microbially mediated niche partitioning, feedback dynamics require no assumptions of abiotic resource partitioning among plant species. Feedback dynamics can in fact represent alternatives to abiotic niche partitioning and abiotic resource competition as mechanisms of, respectively, coexistence (in the case of negative feedbacks) and exclusion (in the case of positive feedbacks) (Bever et al., 1997). Plant–microbe feedbacks arise from the species-specificity of plant–microbe associations, whereby soil communities respond differently to different plant species and plant species, in turn, can be affected in positive or negative ways by their associated microbial communities (Bever et al., 1997). Positive feedback dynamics (e.g., from a plant host's accumulation of beneficial soil microbes) can lead to loss of local diversity, as the abundance of the plant species experiencing the greatest positive feedback is continuously reinforced. In contrast, negative feedback (e.g., from a plant host's accumulation of harmful pathogens) should prevent competitive dominance and increase plant community diversity, as well as cause species replacement over time (Bever, 2002, 2003). It has been hypothesized that the relative importance of positive and negative feedbacks changes over succession, with positive feedbacks dominating early in succession, when conditions can be harsher (favoring microbial mutualists) and plant densities lower (inhibiting the spread of negative feedback from pathogens), and negative feedbacks increasing as conditions moderate and plant densities build (Reynolds et al., 2003).

Bacteria in soil for crop production have been used more than ten years back. The functions of these bacteria (Davison, 1988) are (a) nutritional supply to crops, (b) plant growth stimulation, for example, through the production of plant hormones, (c) by controlling and inhibiting the activity of plant pathogens, (d) to improve soil structure, and (e) bioaccumulation or microbial leaching of organics. Recently, these microorganisms have also been used for soil mineralization of organic pollutants, a natural process, called bioremediation (Zhuang et al., 2007; Zaidi et al., 2008). For sustainable crop production, the interactions of microbes with plants in the rhizosphere play pivotal role in transformation, mobilization, solubilization, and much more of nutrients from a limited nutrient pool, and subsequently uptake of essential nutrients by plants to realize their full genetic potential. At present, use of biological mediated application is getting more popular in combination with chemical fertilizers for improving crop yield in an integrated plant nutrient management system.

In this regard, plant growth promoting rhizobacteria (PGPR) also have potential role in developing sustainable systems in crop production (Shoebitz et al., 2009). With the aim of enhancing plant productivity, various symbiotic (*Rhizobium* sp.) as well as non-symbiotic bacteria (*Azotobacter, Azospirillum,* and *Klebsiella,* etc.) are now being used worldwide (Burd et al., 2000; Cocking, 2003). Beneficial free-living soil bacteria to plant growth are usually referred to as PGPR, which are capable of promoting plant growth by colonizing the plant root. It is also known as PHPR or nodule promoting rhizobacteria (NPR) and are associated with the rhizosphere which is an important soil ecological environment for plant–microbe interactions. PGPR term was first introduced by Kloepper in late 70s. According to the definition, rhizobacteria include free-living bacteria except nitrogen-fixing *Rhizobia* and *Frankia.* Hence, growth stimulation resulting from the biological dinitrogen fixation by *rhizobia* in legume nodules or by *Frankia* in nodules of *Alnus* spp. is not considered as a PGPR mechanism of action. PGPR are grouped into two according to their relationship with the plants, symbiotic bacteria and free-living rhizobacteria (Khan, 2005) and they are again divided into two groups according to their residing sites, iPGPR (i.e., symbiotic bacteria), the one which lives inside the plant cells, produces nodules, and are localized inside the specialized structures), and the one which live outside the plant cells and do not produce nodules, but still prompt plant growth called ePGPR (i.e., free-living rhizobacteria (Gray & Smith, 2005). Rhizobia are known as best iPGPR, which produce nodules in leguminous plants. To improve the supply and enrichment of nutrients to crop plants, a variety of bacteria have been used as soil inoculants. Species of *Rhizobium* (*Rhizobium, Bradyrhizobium, Mesorhizobium, Allorhizobium, Azorhizobium*, and *Sinorhizobium*) have been successfully used worldwide to permit an effective establishment of the nitrogen-fixing symbiosis with leguminous crop plants (Bottomley & Maggard, 1990). On the other hand, with the goal of enhancing crop productivity non-symbiotic nitrogen fixing bacteria such as *Azotobacter, Azospirillum, Bacillus*, and *Klebsiella* sp. are also used to inoculate a large area of arable land worldwide (Lynch, 1983). In addition, phosphate-solubilizing bacteria such as species of *Bacillus* and *Paenibacillus* (formerly *Bacillus*) have been applied to soils to specifically enhance the phosphorus status of plants (Brown, 1974). PGPR have the potential to contribute in the development of sustainable agricultural systems. Generally, PGPR function in three different ways (Glick, 2001): Synthesizing particular compounds for the plants, facilitating the uptake of certain nutrients from the soil and lessening or preventing the plants from diseases. The mechanisms of PGPR-mediated enhancement of plant growth and yield of many crops are yet to be understood completely. However, the possible explanation includes (a) the ability to produce a vital enzyme, 1-aminocyclopropane-1-carboxylate (ACC) deaminase which reduces ethylene production in the root of growing plants results in increasing the root length and (b) the ability to produce hormones like auxin, that is, indole acetic acid (IAA) , abscisic acid (ABA), gibberellic acid (GA), and cytokinins, (c) a symbiotic nitrogen fixation (d) antagonism against phytophatogenic bacteria by producing siderophores, ß-1,3-glucanase, chitinases, antibiotic, fluorescent pigment, and cyanide, (e) solubilization and mineralization of nutrients, mainly phosphates mineral, (f) enhanced resistance to drought, salinity, waterlogging and oxidative stress and (g) production of water-soluble group B vitamins *viz.* niacin, pantothenic acid, thiamine, riboflavin, and biotin (Revillas et al., 2000). The application of PGPR has also been extended to remediate contaminated soils in association with plants (Zhuang et al., 2007). Thus, for sustainable agricultural production system, it is the need of the time to enhance the efficiency of meager amounts of external inputs by employing the best suitable combinations of beneficial bacteria. Effects of increasing soil organic matter and overall soil fertility can be achieved by soil organic carbon improvement.

14.8   PLANT GROWTH PROMOTING RHIZOBACTERIA

About 2–5% of rhizobacteria, when reintroduced by plant inoculation in a soil containing competitive microflora, exert a beneficial effect on plant growth. Free-living soil and rhizosphere bacteria that are beneficial to plants are often referred to as PGPR. This term was first introduced by Kloepper in late 1970s. According to the original definition, rhizobacteria include free-living bacteria except nitrogen-fixing rhizobia and *Frankia*. Hence, growth stimulation resulting from the biological dinitrogen fixation by rhizobia in legume nodules or by *Frankia* in nodules of *Alnus* spp. is not considered as a PGPR mechanism of action. These microbes have the ability to colonize rhizosphere aggressively, benefit plant, and inhibit minor pathogens. These, in broad sense, include symbionts (*Rhizobium, Bradyrhizobium,* certain actinomycetes) as well as free-living bacteria and some of them invade the tissues of living plants and cause unapparent and asymptomatic infections. A number of different bacteria may be considered to be PGPR including *Azotobacter, Azospirillum, Pseudomonas, Acetobacter, Burkholderia, Enterobacter*, and *Bacilli*.

Based on their activities, PGPR have been classified as biofertilizers (increasing the availability of nutrients to plant), phytostimulators (plant growth promoting, usually by the production of phytohormones), rhizoremediators (degrading organic pollutants), and biopesticides (controlling diseases, mainly by the production of antibiotics and antifungal metabolites).

14.9   MICROBIAL AND FUNCTIONAL DIVERSITY IN SOIL AND THE ROLE OF ECOLOGICAL THEORIES

Microbial diversity is usually taken as the number of individuals assigned to different taxa and their distribution among taxa (Atlas & Bartha, 1998). To date ecological theories have been based on the study of aboveground ecosystems. Despite the fact that the soil biota plays a fundamental role in ecosystem functioning, through nutrient cycling, decomposition, and energy flow, soil organisms have had a negligible influence on the development of contemporary ecological theories (Wardle & Giller, 1996). Microbial diversity is usually taken as the number of individuals assigned to different taxa and their distribution among taxa (Atlas & Bartha, 1998).

On a landscape level, diversity may be viewed at different levels of resolution. Whittaker (1972) proposed into distinguish between diversity of species within a community of a habitat (α-diversity), rate and extent in change of species along a gradient of habitats (β-diversity) and richness of species over a range of habitats (γ-diversity). This concept, plausible for traditional habitat diversity, may also be used to describe soil microbial diversity concepts. However, soil biota is characterized by a spatial diversity with possible differences between rhizosphere and bulk soil, macroaggregates and microaggregates, macropores and micropores, different horizons, etc. (Fig. 14.2). Indeed, within soil there are several microhabitats, for example, the rhizoplane, the rhizosphere, aggregates, decaying organic matter, or the bulk soil. Typically, soils are also largely stratified habitats, with distinct horizons; each of them may be regarded as a separate entity. How the diversity of these microhabitats can be incorporated in a general soil microbial diversity concept is not known.

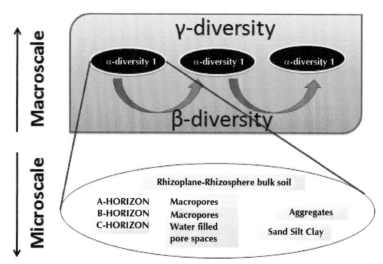

FIGURE 14.2 Different levels (α, β, and γ diversity) of ecosystem diversity (Source: J.M. Lynch, "Microbial diversity in soil: ecological theories, the contribution of molecular techniques and the impact of transgenic plants and transgenic microorganisms," in Biology and Fertility of Soils December 2004, Volume 40, Issue 6, pp 363–385. With permission from Springer.).

Ecosystem diversity viewed at different levels of resolution by Whittaker (1972). According to him, diversity of species within a community of a habitat (α-diversity), rate, and extent in change of species along a gradient of habitats (β-diversity) and richness of species over a range of habitats (γ-diversity) was categorized. In addition, in soil, spatial variability in biodiversity is also important with differences between rhizosphere and bulk soil, macroaggregates and microaggregates, macropores and micropores, different horizons, etc.

14.10   METHODS FOR MEASURING MICROBIAL DIVERSITY OF SOIL

Soil is having tremendous microbial diversity which has great potential to enrich the soil by improving soil fertility and providing eco friendly environment for growing crop species. About 1 g of soil or sediment may contain more than 10^{10} bacteria as counted in fluorescence microscope after staining with a fluorescent dye (Torsvik et al., 1990). Torsvik et al. (1990) also estimated that in 1 g of soil there are 4000 different bacterial "genomic units" based on DNA–DNA reassociation. Still, our present knowledge of soil microbial diversity is rather limited in part by our inability to study soil microorganisms. Approximately 1% of the soil bacterial population can only be cultured by standard laboratory practices. It is not known if this 1% is representative of the bacterial population (Torsvik et al., 1998). An estimated 1,500,000 species of fungi exist in the world (Giller et al., 1997). But unlike bacteria, many fungi cannot be cultured by current standard laboratory methods (Van Elsas et al., 2000). Although several methods have been used to study soil bacterial communities, very little research has been undertaken for soil fungi (Kirk et al., 2008).

An extensive study on microbial diversity is important in order to increase the knowledge of the diversity of genetic resources, understanding their functional role of diversity, distribution of organisms and consequences of diversity.

Methods to measure microbial diversity in soil can be categorized into two groups, that is, (a) conventional and biochemical techniques and (b) molecular techniques.

14.10.1   CONVENTIONAL AND BIOCHEMICAL TECHNIQUES

Both conventional and biochemical methods are of high significance in the study of microbial diversity. The methods described below can be used for either bacteria or fungi, although some are specific to one or the other.

14.10.1.1   PLATE COUNTS

The most conventional method for assessment of microbial diversity is selective and differential plating and subsequent viable counts. These methods are fast, inexpensive, and can provide information on the active, heterotrophic component of the population. Plate count methods select microorganisms with faster growth rate and fungi producing large number of spores. The major limitations of these methods are inability to culture a large number of bacterial and fungal species, difficulty in dislodging bacteria or spores from soil particles or biofilms, growth medium selections, growth conditions (pH, temperature, and light).

14.10.1.2   SOLE-CARBON-SOURCE UTILIZATION (SCSU)/ COMMUNITY LEVEL PHYSIOLOGICAL PROFILING (CLPP)

This technique was introduced by Garland & Mills (1991). This was initially developed as a tool for identifying pure cultures of bacteria to the species level, based upon a broad survey of their metabolic properties. SCSU examines the functional capabilities of the microbial population, and the resulting data can be analyzed using multivariate techniques to compare metabolic capabilities of communities. Slow growers microbial communities cannot be included in this analysis. Growth on secondary metabolites may also occur during incubation.

Advantages of SCSU include its ability to differentiate between microbial communities, relative ease of use, reproducibility, and production of large amount of data describing metabolic characteristics of the communities. SCSU selects only culturable portion of the microbial community which limits its application (Garland & Mills, 1991), favors fast growing microorganisms (Yao et al., 2000), is sensitive to inoculum density, reflects the potential, and not the *in situ*, metabolic diversity.

14.10.1.3   PHOSPHOLIPID FATTY ACID (PLFA) ANALYSIS

Another widely used biochemical method of characterizing microbial communities in the soil and in compost is ester-linked phospholipid fatty acid analysis (PLFA). PLFAs are constituents of all cell membranes, and are compounds with no storage function, which, therefore, represent a relatively constant fraction of the cell mass (Malik et al., 2008). Taxonomically, fatty acids in the range C2 to C24 have provided the greatest information and are present across a diverse range of microorganisms (Banowetz et al., 2006). The fatty acid composition is stable, and is independent of plasmids, mutations, or damaged cells. The method is quantitative, cheap, robust, and with high reproducibility. However, it is important to notice that the bacterial growth conditions are reflected in the fatty acid pattern. This method is also known as the fatty acid methyl ester (FAME) analysis.

Microorganisms have the ability to change the lipid composition of their membranes in response to environmental conditions such as chemical stress (Frostegard et al., 1993) and temperatures fluctuations (Bartlett, 1999). PLFA rapidly degrade upon cell death thus making it good indicator of living organisms (Drenovsky et al., 2004) and changes in PLFA patterns under environmental stress conditions are a useful biomarker tool to describe the community structure and physiological state of certain microbial taxa.

Changes in phospholipid profiles are generally related to the variation in the abundance of microbial groups and this can be interpreted by reference to a database of pure cultures and known biosynthetic pathways. The extracted fatty acids are quantitatively analyzed by gas chromatography equipped with mass spectrometry (Zelles & Bai, 1993), while comparison of data with information on fatty acids database allows for the identification of extracted PLFAs.

Frostegard et al. (1993) examined changes in microbial population profiles in soils artificially polluted with cadmium, copper, nickel, lead, or zinc using PLFA. They observed that certain fatty acid patterns characteristic of Gram-positive bacteria were reduced in both forest and arable soils spiked with metals and replaced by PLFA patterns indicative of a Gram-negative bacteria populations.

14.10.1.4   MOLECULAR TECHNIQUES

Traditional methods for characterizing microbial communities have been based on analysis of the culturable portion of the bacteria. Due to the non-culturability of the major fraction of bacteria from natural microbial communities, the overall structure of the community has been difficult to interpret (Dokić et al., 2010). Recent studies to characterize microbial diversity have focused on the use of methods that do not require cultivation, yet provide measures based on genetic diversity. Few numbers of approaches have been discussed below to study molecular microbial diversity.

14.10.1.5   MOLE PERCENTAGE GUANINE + CYTOSINE (MOL% G+C)

The first property of DNA used for taxonomical purpose was the base composition expressed as mole percentage guanine + cytosine (mol% G+C). Within bacteria, this value ranges from 25 to 75%, though a value is constant for a certain organism. Closely related organisms have fairly similar GC profiles and taxonomically related groups only differ between 3 and 5%.

Molecular percent of G+C can be determined by thermal denaturation of DNA. Advantages of G+C analysis are that it is not influenced by polymerase chain reaction (PCR) biases, it includes all DNA extracted, it is quantitative and can uncover rare members in the microbial populations. It does, however, require large quantities of DNA (up to 50 μg).

14.10.1.6   NUCLEIC ACID HYBRIDIZATION

Nucleic acid hybridization using specific probes is an important qualitative and quantitative tool in molecular bacterial ecology (Clegg et al., 2000). These hybridization techniques can be done on extracted DNA or RNA, or *in situ*. Oligonucleotide or polynucleotide probes designed from known sequences ranging in specificity from domain to species can be tagged with markers at the 5'-end. Dot-blot hybridization with specific and universal oligonucleotide primers is used to quantify rRNA sequences of interest relative to total rRNA. The relative abundance may represent changes in the abundance in the population or changes in the activity and hence the amount of rRNA content (Fakruddin & Mannan, 2013).

One of the most popular DNA hybridization methods is FISH (Fluorescent *in situ* hybridization). FISH is a method used to quantify the presence and relative abundance of microbial populations in a community sample. Microbial cells are treated with fixative, hybridized with specific probes (usually 15–25 bp oligonucleotide-fluorescently labeled probes) on a glass slide, then visualized with either epiflourescence or confocal laser microscopy. Spatial distribution of bacterial communities in different environments such as biofilms can be determined using FISH (Schramm et al., 1996). Lack of sensitivity of hybridization of nucleic acids extracted directly from environmental samples is the most notable limitation of nucleic acid hybridization methods. If sequences are not present in high copy number, such as those from dominant species, probability of detection is low.

14.10.1.7   DNA MICROARRAY

DNA–DNA hybridization has been used together with DNA microarrays to detect and identify bacterial species (Cho & Tiedje, 2001) or to assess microbial diversity. This tool could be valuable in bacterial diversity studies since a single array can contain thousands of DNA sequences (Cho & Tiedje, 2001) with high specificity. The microarray can either contain specific target genes such as nitrate reductase, nitrogenase, or naphthalene dioxygenase to provide functional diversity information or can contain a sample of environmental "standards" (DNA fragments with less than 70% hybridization) representing different species found in the environmental sample. Reverse sample genome probing (RSGP) is a method used to analyze microbial community composition of the most dominant culturable species and uses genome microarrays. RSGP has four steps: (a) Isolation of genomic DNA from pure cultures; (b) cross-hybridization testing to obtain DNA fragments with less than 70% cross-hybridization. DNA fragments with greater than 70% cross-hybridization are considered the same species; (c) preparation of genome arrays onto a solid support; and (d) random labeling of a defined mixture of total community DNA and internal standard (Greene & Voordouw, 2003). RSGP is a useful technique when diversity is low, but several authors have had difficulty when assessing community composition of diverse habitats. If diversity is high, then cross-hybridization can be a problem or interpretation of the results is difficult. Like DNA–DNA hybridization, the use of RSGP and microarrays has the advantage that it is not confounded by PCR biases and microarrays can contain thousands of target gene sequences. Using genes or DNA fragments instead of genomes on the microarray offer the advantages of eliminating the need to keep cultures of organisms growing as genes can be cloned into plasmids or PCR used to continually amplify the DNA fragments.

14.10.1.8   CONVENTIONAL PCR

The polymerase chain reaction (PCR) is one of the most important molecular biology tools for the detection of plant pathogens. It allows the amplification of millions of copies of specific DNA sequences by repeated cycles of denaturation, polymerization and elongation at different temperatures using specific oligonucleotides (primers), deoxyribonucleotide triphosphates (dNTPs) and a thermostable DNA polymerase from Thermus aquaticus (Taq polymerase) in the adequate buffer. Conventional PCR may be used for the identification of fungal pathogens at different taxonomic levels (genus, species, or strain) depending on the specificity of the primers. However, most applications have analyzed genes encoding the small subunit (SSU) of ribosomal RNA. Analysis of 16S rRNA genes is now widely used for analysis of bacterial populations, and analysis of 18S rRNA genes and internal transcribed spacer (ITS) region is increasingly being used to analyze fungal populations. Ribosomal rRNA genes are ideal for this purpose in that they possess regions with sequences conserved between all bacteria or fungi, facilitating alignment of sequences when making comparisons, while other regions exhibit different degrees of variation, enabling distinction between different groups. These differences provide the basis for a phylogenetic taxonomy and enable quantification of evolutionary differences between different groups. Discrimination of bacteria, using 16S rRNA gene sequences, is greater than that that for fungi, using 18S rRNA sequences, but finer scale information may be obtained by analysis of ITS regions.

Most information about fine scale discrimination between groups can be obtained by cloning the amplified rRNA genes and sequencing members of the clone library. Comparison of sequences with those in databases determines which phylogenetic groups are present and, in many cases, enables more detailed identification. This approach is particularly useful for studies of bacteria, as 16S rRNA databases are now extensive and comprehensive. They contain sequences of large numbers of laboratory cultures and also of clones obtained from a range of environments, which are not represented in laboratory cultures. Fungal databases are increasing in size, but are currently less extensive and informative, except for phylogenetic groups that have been the subject of detailed taxonomic study (Prosser, 2002).

14.10.1.9   RIBOSOMAL INTERGENIC SPACER ANALYSIS (RISA)

RISA is a PCR-based technique that amplifies the region between the 16S and 23S rRNA operons. The intergenic spacer region, depending on the species, has both sequence and length (50–1500 bp) variability (Ranjard et al., 2001) and this unique feature facilitates taxonomic identification of organisms. In RISA, the intergenic spacer (IGS) region between the 16S and 23S ribosomal subunits is amplified by PCR, denatured, and separated on a polyacrylamide gel under denaturing conditions. This region may encode tRNAs and is useful for differentiating between bacterial strains and closely related species because of heterogeneity of the IGS length and sequence (Fisher & Triplett, 1999). RISA has been used to distinguish between different strains and closely related species. This technique is widely used for the detection of soil microbial diversity. RISA is a very rapid and simple rRNA fingerprinting method but its application in microbial community analysis from contaminated sources is limited partly due to the limited database for ribosomal intergenic spacer sequences is not as large or as comprehensive as the 16S sequence database.

14.10.1.10   RESTRICTION FRAGMENT LENGTH POLYMORPHISM (RFLP)

Restriction fragment length polymorphism (RFLP)/amplified ribosomal DNA restriction analysis (ARDRA) is another tool used to study microbial diversity. This method relies on DNA polymorphisms. In the last couple of years, RFLP applications have also been applied to estimate diversity and community structure in different microbial communities (Moyer et al., 1996). In this method, electrophoresis of DNA restriction digests in agarose gel is blotted onto nitrocellulose or nylon membranes and hybridized with appropriate probes prepared from cloned DNA segments of related organisms. RFLP has been found to be very useful particularly in combination with DNA–DNA hybridization and enzyme electrophoresis for the differentiation of closely related strains (Palleroni, 1993), and the approach seems to be useful for determination of intra species variation (Kauppinen et al., 1994). RFLPs may provide a simple and powerful tool for the identification of bacterial strains at and below species level. This method is useful for detecting structural changes in microbial communities but not as a measure of diversity or for detection of specific phylogenetic groups (Liu et al., 1997). Banding patterns in diverse communities become too complex to analyze using RFLP since a single species could have four to six restriction fragments (Tiedje et al., 1999).

    This method is useful for detecting structural changes in microbial communities but not as a measure of diversity or detection of specific phylogenetic groups. Banding patterns in diverse communities become too complex to analyze using RFLP since a single species could have four to six restriction fragments.

14.10.1.11   TERMINAL-RESTRICTION FRAGMENT LENGTH POLYMORPHISM (T-RFLP)

Terminal-restriction fragment length polymorphism (T-RFLP) is a technique that addresses some of the limitations of RFLP (Tiedje et al., 1999). It follows the same principle as RFLP except that one PCR primer is labeled with a fluorescent dye, such as TET (4,7,2V,7V-tetrachloro-6-carboxyfluorescein) or 6-FAM (phosphoramidite fluorochrome-5-carboxyfluorescein). This allows detection of only the labeled terminal restriction fragment (Liu et al., 1997). This simplifies the banding pattern, thus allowing the analysis of complex communities as well as providing information on diversity as each visible band represents a single operational taxonomic unit or ribotype (Tiedje et al., 1999). The banding pattern can be used to measure species richness and evenness as well as similarities between samples (Liu et al., 1997). T-RFLP has also been thought to be an excellent tool to compare the relationship between different samples (Dunbar et al., 2000). T-RFLP has been used to measure spatial and temporal changes in bacterial communities, to study complex bacterial communities, to detect and monitor populations and to assess the diversity of arbuscular mycorrhizal fungi (AMF) in the rhizosphere of *Viola calaminaria* in a metal-contaminated soil (Tonin et al., 2001; Lukow et al., 2000).

14.10.1.12   DENATURING GRADIENT GEL ELECTROPHORESIS (DGGE)/TEMPERATURE GRADIENT GEL ELECTROPHORESIS (TGGE)

Denaturing gradient gel electrophoresis (DGGE) and temperature gradient gel electrophoresis (TGGE) were developed to separate PCR-amplified ribosomal DNA fragments of DNA with the same length but with varia-tion in nucleotide composition. Over the years, these methods were adapted to analyze bacterial community structure. The separation principle for both methods is applying a linear gradient of DNA denaturing agents (such as a mixture of formamide and urea in DGGE), or TGGE on polyacrylamide gels to influence the electrophoretic mobility of partially melted double-stranded DNA. Melting temperatures are associated to the sequence, and DNA fragments stop migrating when regions of base pairs with the lowest melting temperature reach this temperature. This occurs due to a transition of conformation from helical to partially melted, and consequently the movement along the electric field will stop. A GC clamp (GC rich sequence) attached to the 5′-end is used as a special primer to anchor the PCR fragments and prevent them from completely dissociating. Soil bacterial dynamics, structure, and diversity are still being assessed through these methods but have also been increasingly replaced by the advent of high-throughput sequencing platforms. This is because DGGE and TGGE can only detect the most abundant organisms present in the bacterial community. Advantages of DGGE/TGGE include reliability, reproducibility, rapidness, and low expense. As multiple samples can be analyzed simultaneously, tracking changes in microbial population in response to any stimuli or adversity is possible by DGGE/TGGE (Muyzer, 1999).

14.10.1.13   SINGLE-STRAND CONFORMATION POLYMORPHISM (SSCP)

Single strand conformation polymorphism (SSCP) also relies on electrophoretic separation based on differences in DNA sequences and allows differentiation of DNA molecules having the same length but different nucleotide sequences. This technique was originally developed to detect known or novel polymorphisms or point mutations in DNA (Peters et al., 2000). In this method, single-stranded DNA separation on polyacrylamide gel was based on differences in mobility resulted from their folded secondary structure (Heteroduplex). As formation of folded secondary structure or heteroduplex and hence mobility is dependent on the DNA sequences. This technique has been used for the rapid profiling of soil microbial communities and phylogenetic studies. An interesting study on diversity and distribution of polyhydroxyalkanoate-producing bacteria has used SSCP as a culture-independent approach. This method helped to confirm that rhizosphere is an attractive reservoir for bacteria, which are producers of polyhydroxyalkanoate (Gasser et al., 2009).

14.10.1.14   QUANTITATIVE/SEMI QUANTITATIVE PCR

Real-time quantitative PCR (Q-PCR) or reverse transcriptase Q-PCR (RT-PCR) is a technique that collects amplification data while the PCR occurs. Reverse transcriptase-PCR is generally used for the microorganisms having RNA genome. Different fluorescence chemistries are available, including SYBR green and TaqMan for real-time PCR. The first dye binds to any double-stranded DNA. Latter requires pre-designed probes that will be hydrolyzed given the 5′ nuclease ability of the DNA polymerase during the extension step and fluorescence emission will be consequently higher. The PCR cycle where amplification is first detected is known as cycle threshold (CT) and can be identified when the background fluorescence is lower than the fluorescence intensity. Real-time has been used in several rhizosphere studies such as in the evaluation of soil acidobacterial communities' responses from soybean croplands and adjacent Amazon forest (Rincon-Florez et al., 2013). In addition, the effect of long-term fertilization on the activity of ammonia oxidizers communities in the rhizosphere of a fluvo-aquic soil was also assessed through this method. Many are the advantages of real-time PCR over conventional PCR that are mentioned in following:

i.   It does not require the use of post PCR processing (electrophoresis, colorimetric reaction or hybridization),
ii. Avoiding the risk of cross contamination,
iii. Reduction of the assay labor and material costs, and
iv. Increase the sensitivity and specificity and allows the accurate quantification of the target microorganisms (Smith & Osborn, 2009).

14.10.1.15   HIGH THROUGHPUT SEQUENCING TECHNIQUES

DNA sequencing is a process of determining the order of nucleotides in a stretch of DNA. The first sequencing methods were published in the 1970s when first Sanger and colleagues and then Maxam and Gilbert published their sequencing methods (Sanger & Coulson, 1975; Maxam & Gilbert, 1977). Sanger sequencing method was the first sequencing technique applied in metagenomic studies of natural environments (Hugenholtz & Tyson, 2008). Despite the automation in methodology, the list of required steps in protocol is long and the stages laborious: The gene of interest needs to be amplified or the environmental DNA fragmented. Then the clone libraries are constructed of the resulting amplicons or DNA fragments and these fragments are sequenced individually (Metzker, 2005). However, Sanger sequencing method has been widely applied with functional genes and community structure and diversity studies using 16S rRNA gene (Rastogi & Sani, 2011), and it is still in use in many laboratories. Yet, it is not feasible to sequence tens of thousands or millions sequences per project using Sanger sequencing and, therefore, the rare microbial groups in studied environment are easily overlooked (Kosinen, 2013).

Over the last few years, the next generation sequencing (NGS) technologies have revolutionized the field of genome and community sequencing, and consequently microbial ecology (Metzker, 2010). NGS technologies typically pursue high throughput by parallelizing the sequencing process, producing thousands to millions of sequences concurrently. The major advance in NGS is indeed the huge amount of data produced at considerably low cost compared to Sanger method (Rastogi & Sani, 2011; Metzker, 2010). These technologies play a major role in metagenomic (DNA-based), and metatranscriptomic (RNA-based) approaches, which provide a comprehensive picture of potential and active functions of microbial communities, respectively. The most widely-used platforms for massive parallel sequencing for assessing soil microbial diversity are Roche 454 Genome Sequencer (Roche Diagnostics Corp., Branford, CT, U.S.A.), HiSeq 2000 (Illumina Inc., San Diego, CA, U.S.A.), and AB SOLiD™ System (Life Technologies Corp., Carlsbad, CA, U.S.A.). Other commonly used high throughput sequencing systems that have been applied to other approaches including metatranscriptomics and whole genome re-sequencing are also described below for comparison, which includes Ion Personal Genome Machine (Life Technologies, South San Francisco, CA, U.S.A.), Heliscope (Helicos Bioscience Corp., Cambridge, MA, U.S.A.), and PacBio RS SMRT system (Pacific Bioscience, Menlo Park, CA, U.S.A.). These methods have been extensively used to characterize composition and diversity of soil microbial communities and have also been applied to understand the effect of heavy metals and disturbances on soil microbial communities (Carvalhais et al., 2013; Dohrmann et al., 2013; Eilers et al., 2012; Suleiman et al., 2013). Pyrosequencing has been used to evaluate the effect of conventional and organic systems on bulk soil bacterial communities (Li et al., 2012). Another study found differences in soil bacterial community composition in the rhizosphere of plants with activated jasmonate signaling pathway. A combined approach using metagenomic analysis and functional assays by Illumina sequencing platform provided evidence to suggest that soil resource availability and soil stratification had an effect on functional diversity and to a lower degree on taxonomic diversity (Uroz et al., 2013).

14.11   METAGENOMICS

Metagenomics refers to the study of genomic DNA obtained from microorganisms that cannot be cultured in the laboratory. The term metagenomics, the genomic analysis of a population of microorganisms, was coined by

TABLE 14.4 Advantages and Disadvantages of Different Methods of Studying Soil Microbial Diversity.

Sl. No.	Method	Advantages	Disadvantages
1.	Plate count	Fast, simple and inexpensive	Unculturable microorganisms not detected, bias toward fast growing individuals, bias toward fungal species that produce large quantities of spores
2.	SSCU	Fast, highly reproducible, relatively inexpensive, able to differentiate microbial communities, generates large amount of data, option of using bacterial, fungal plates or site specific carbon sources (Biolog)	Only represents culturable fraction of community, Favors fast growing organisms, only represents those organisms capable of utilizing available carbon sources, potential metabolic diversity, not *in situ* diversity, sensitive to inoculum density
3.	PLFA/FAME	Culturing of microorganisms is not required, direct extraction from soil, follow specific organisms or communities	If fungal spores are used, more material is needed, can be influenced by external factors, results can be confounded by other microorganisms is possible
4.	%G+C	Not influenced by PCR biases, includes all DNA extracted, quantitative, includes rare members of community	Requires large quantities of DNA, dependent on lysing and extraction efficiency, coarse level of resolution
	Nucleic acid hybridization	Total DNA extracted, not influenced by PCR biases, can study DNA or RNA, can be studied *in situ*	Lack of sensitivity, sequences need to be in high copy number for detection, dependent on lysing and extraction efficiency
	DNA microarray	Same as nucleic acid hybridization, thousands of genes can be analyzed if using genes or DNA fragments, increased specificity	Only detect the most abundant species, need to culture organisms, only accurate in low diversity systems
	Conventional PCR	Simple, rapid, and inexpensive; useful to study phylogenetic relationship	Less extensive and informative in case of Fungi,
	RISA	Highly reproducible community profiles	Requires large quantities of DNA, PCR biases
	RFLP	Detect structural changes in microbial community	Banding patterns often too complex

Conventional and biochemical methods

TABLE 14.4 (Continued)

Sl. No.	Method	Advantages	Disadvantages
	T-RFLP	Simpler banding patterns than RFLP, can be automated large number of samples, highly reproducible, ability to compare differences between microbial communities	Dependent on extraction and lysing efficiency, type of *Taq* can increase variability, choice of restriction enzymes will influence community fingerprint
	DGGE/ TGGE	Large number of samples can be analyzed simultaneously Reliable, reproducible and rapid	PCR biases, dependent on lysing and extraction efficiency, way of sample handling can influence community, that is, the community can change if stored too long before extraction, one band can represent more than one species (co-migration), only detects dominant species
	SSCP	Same as DGGE/TGGE, no GC clamp, No gradient	PCR biases, some ssDNA can form more than one stable conformation
	Quantitative/ semi quantitative PCR	Quick, accurate and highly sensitive method for sequence quantification that can also be used to quantify microbial groups, Relatively cheap and easy to implement, specific amplification can be confirmed by melting curve analysis	Can only be used for targeting of known sequences, DNA impurities and artifacts may create false-positives or inhibit amplification
	High throughput sequencing	Rapid method to assess biodiversity and abundance of many species/organizational taxonomic units simultaneously and at a considerable depth compared to the methods that have been available so far	Relatively expensive, replication and statistical analysis are essential, computational intensive, challenging in terms of data analysis
	Metagenomics	Biodiversity can be studied in more detail, captures polymorphism in microbial communities, reveals the presence of thousands of microbial genomes simultaneously, provides information about the functions of microbial communities in a given environment	High cost, data analysis is challenging and time consuming, difficult to use for low-abundance communities, the high biodiversity in soil leads to many incomplete genomes, current sequencing methods and computing power still in its infancy to the high biodiversity in soil

Molecular methods

Handelsman et al. (1998). With a notion to analyze a collection of similar, but not identical items, as in the statistical concept of meta-analysis is observed. It has opened new horizons in the development of biotechnology based on the exploitation of uncultivated microbial species. The vast majority of microorganisms being unculturable, metagenomics has resulted in discoveries that remained hidden from the traditional culturing techniques.

Metagenome sequencing, also called shotgun sequencing, based on sequencing of DNA fragments extracted from microbial populations. Because this approach captures the complete genomes of all the organisms in the population, mosaicism and biases have little impact. The comprehensive information obtained by this approach enables accurate phylogenetic inferences of close and distant relatives. However, the most substantial advantage is the information it provides about the genes present in the bacterial population, without assembling the individual bacterial genomes. The metatranscriptome is the identity and quantity of a complete set of transcripts in a population of cells. While metagenomics tells us who is there and what they are capable of, based on their gene complement, metatranscriptomics tells us what they are doing at that moment. Unlike hybridization-based techniques, such as PCR, northern blotting, or microarrays, RNA-Seq information is matched to genes by sequence alignment. This approach offers several advantages like; no prior knowledge of the genome sequence is required, accurate mapping etc.

Soil metagenomics has the potential to substantially impact on antibiotic production (Ghazanfar et al., 2010). Several studies reported the successful screening of soil metagenomic libraries for indirubin (MacNeil et al., 2001; Lim et al., 2005) while a range of novel antibiotics have been detected in metagenomic libraries (Brady et al., 2004; Bashir et al., 2014). A clone found in a soil metagenomic library produces deoxyviolacein and the broad-spectrum antibiotic violacein (Brady et al., 2001).

Conventional and molecular tools applied for the analyses of mixed microbial communities from soil have undoubtedly advanced our knowledge and understanding about microbial diversity. Molecular techniques have contributed significantly to the detection and identification of microorganisms especially in non-culturable organisms as well as the quantification of the relative abundance of organisms from soil. Currently, there is no single technique that can adequately describe the entire microbial diversity and the associated catabolic genes from agricultural soil (Table 14.4). Each technique has its own limitations with respect to the introduction of biases for the investigation of microbial diversity.

KEYWORDS

- rhizosphere
- phytohormone
- siderophore
- population dynamics
- agricultural practices

REFERENCES

Atlas, R.; Bartha, R. *Microbial Ecology. Fundamentals and Applications;* 4th edn. Addison-Wesley: Reading, 1998.

Bartlett, D. H. Microbial Adaptations to the Psychrosphere/Piezosphere. *J. Mol. Microbiol. Biotechnol.* **1999,** *1,* 93–100.

Bashir, Y.; Singh, S. P.; Konwar, B. K. Metagenomics: An Application Based Perspective, *Chin. J. Biol.* **2014,** *8,* 1–7, http://dx.doi.org/10.1155/2014/146030

Berg, G. Plant–Microbe Interactions Promoting Plant Growth and Health: Perspectives for Controlled Use of Microorganisms in Agriculture. *Appl. Microbiol. Biotechnol.* **2009,** *84,* 11–18.

Berg, G.; Smalla, K. Plant Species and Soil Type Cooperatively Shape the Structure and Function of Microbial Communities in the Rhizosphere. *FEMS Microbiol. Ecol.* **2009,** *68,* 1–13.

Bever, J. D. Negative Feedback within a Mutualism: Host-Specific Growth of Mycorrhizal Fungi Reduces Plant Benefit. *Proc. Roy. Soc. Lond. B.* **2002,** *269,* 2595–2601.

Bever, J. D. Soil Community Feedback and the Coexistence of Competitors: Conceptual Frameworks and Empirical Tests. *New Phytologist.* **2003,** *157,* 465–473.

Bever, J. D.; Westover, K. M.; Antonovics, A. Incorporating the Soil Community into Plant Population Dynamics: The Utility of the Feedback Approach. *J. Ecol.* **1997,** *85,* 561–573.

Bhardwaj, D.; Ansari, M. W.; Sahoo, R. K.; Tuteja, N. Biofertilizers Function as Key Player in Sustainable Agriculture by Improving Soil Fertility Plant Tolerance and Crop Productivity. *Microb. Cell. Fact.* **2014,** *13,* 66.

Bhattarai, A.; Bhattarai, B.; Pandey, S. Variation of Soil Microbial Population in Different Soil Horizons. *J. Microbiol. Exp.* **2015,** *2*(2), 00044.

Bottomley, P. J.; Maggard, S. P. Determination of Viability within Serotypes of a Soil Population of *Rhizobium leguminosarum* biovar *trifolii. Appl. Environ. Microbiol.* **1990,** *56,* 533–540.

Brady, S. F.; Chao, C. J.; Clardy, J. Longchain N-Acyltyrosine Synthases from Environmental DNA. *Appl. Environ. Microbiol.* **2004,** *70,* 6865–6870.

Brady, S. F.; Choa, C. J.; Handelman, J.; Clardy J. Cloning and Heterologous Expression of a Natural Product Biosynthetic Gene Cluster from eDNA. *Org. Lett.,* **2001,** *3,* 1981–1984.

Brady, N. C.; Weil, R. C. *The Nature and Properties of Soils;* 14th edn. Dorling Kindersley India: Noida, India, 2012.

Brown, M. E. Seed and Root Bacterization. *Annu. Rev. Phytopathol.* **1974,** *12,* 181–197.

Bulgarelli, D.; Rott, M.; Schlaeppi, K.; Van Themaat, E. V. L.; Ahmadinejad, N.; Assenza, F.; Rauf, P.; Huettel, B.; Reinhardt, R.; Schmelzer, E. Revealing Structure and Assembly Cues for *Arabidopsis* Root-Inhabiting Bacterial Microbiota. *Nature.* **2012,** *488,* 91–95.

Burd, G.; Dixon, D. G.; Glick, B. R. Plant Growth Promoting Bacteria that Decrease Heavy Metal Toxicity in Plants. *Can. J. Microbiol.* **2000,** *46,* 237–245.

Carvalhais, L. C.; Dennis, P. G.; Badri, D. V.; Tyson, G. W.; Vivanco, J. M.; Schenk, P. M. Activation of the Jasmonic Acid Plant Defence Pathway Alters the Composition of Rhizosphere Bacterial Communities. *PLoS One.* **2013,** *8,* e56457.

Cho, J. C.; Tiedje, J. M. Bacterial Species Determination from DNA–DNA Hybridization by Using Genome Fragments and DNA Microarrays. *Appl. Environ. Microbiol.* **2001,** *67,* 3677–3682.

Clegg, C. D.; Ritz, K.; Griffiths, B. S. %G+C Profiling and Cross Hybridisation of Microbial DNA Reveals Great Variation in Below-Ground Community Structure in UK Upland Grasslands. *Appl. Soil Ecol.* **2000,** *14,* 125–134.

Cocking, E. C. Endophytic Colonization of Plant Roots by Nitrogen-Fixing Bacteria. *Plant Soil.* **2003,** *252*(1), 169–175.

Cocks, L. R. M.; Torsvik, T. H. Earth Geography from 500 to 400 Million Years Ago: Faunal and Palaeomagnetic. *Rev. J. Geol. Soc. Lond.* **2002,** *159,* 631–644.

Damiani, I.; Baldacci-Cresp, F.; Hopkins, J.; Andrio, E.; Balzergue, S.; Lecomte, P.; et al. Plant Genes Involved in Harbouring Symbiotic Rhizobia or Pathogenic Nematodes. *New Phytol.* **2012,** *194,* 511–522.

Dastager, S. G.; Deepa, C. K.; Pandey, A. Isolation and Characterization of Novel Plant Growth Promoting Micrococcus sp NII-**0909** and Its Interaction with Cowpea. *Plant Physiol. Biochem.* **2010,** *48*(12), 987–992.

Davies, P. J. The Plant Hormones: Their Nature, Occurrence, and Functions. In *Plant Hormones—Biosynthesis, Signal Transduction, Action;* 3rd ed., Davies, P. J., Ed.; Kluwer: Dordrecht, The Netherlands, **2010,** pp 1–15.

Davison, J. Plant Beneficial Bacteria. *Biotechnology.* **1988,** *6,* 282–286.

Day, S. D.; Bassuk, N. L. A Review of the Effects of Soil Compaction and Amelioration Treatments on Landscape. *J. Arboricult.,* **1994,** *20*(1), 9–17.

Dechorgnat, J.; Patrit, O.; Krapp, A.; Fagard, M.; Daniel-Vedele, F. Characterization of the *Nrt2.6* Gene in *Arabidopsis thaliana*: A Link with Plant Response to Biotic and Abiotic Stress. *PLoS One.* **2012,** *7*(8), e42491. doi:10.1371/journal.pone.0042491.

Dennis, P. G.; Miller, A. J.; Hirsch, P. R. Are Root Exudates More Important Than Other Sources of Rhizodeposits in Structuring Rhizosphere Bacterial Communities. *FEMS Microbiol. Ecol.* **2010,** *72,* 313–327.

Dohrmann, A. B.; Kuting, M.; Junemann, S.; Jaenicke, S.; Schluter, A.; Tebbe, C. C. Importance of Rare Taxa for Bacterial Diversity in the Rhizosphere of Bt- and Conventional Maize Varieties. *ISME J.* **2013,** *7,* 37–49.

Dokić, L.; Savić, M.; Narančić, T.; Vasiljević, B. Metagenomic Analysis of Soil Microbial Communities. *Arch. Biol. Sci. Belgrade.* **2010,** *62*(3), 559–564.

Drenovsky, R. E.; Elliot, G. N.; Graham, K. J.; Scow, K. M. Comparison of Phospholipids Fatty Acid (PLFA) and Total Soil Fatty Acid Methyl Esters (TSFAME) for Characterizing Soil Microbial Communities. *Soil Biol. Biochem.* **2004,** *36,* 1793–1800.

Dunbar, J.; Ticknor, L. O.; Kuske, C. R. Assessment of Microbial Diversity in Four Southwestern United States Soils by 16S rRNA Gene Terminal Restriction Fragment Analysis. *Appl. Environ. Microbiol.* **2000,** *66,* 2943–2950.

Eilers, K. G.; Debenport, S.; Anderson, S.; Fierer, N. Digging Deeper to Find Unique Microbial Communities: The Strong Effect of Depth on the Structure of Bacterial and Archaeal Communities in Soil. *Soil Biol. Biochem.* **2012,** *50,* 58–65.

Evans, A. The Feeding of the Nine Billion. *Environment and Development.* [online] 2013. http://www.chathamhouse.org/sites/default/files/public/Research/Energy (accessed May 6, 2013).

Fakruddin, M. D.; Mannan, K. S. B. Methods for Analyzing Diversity of Microbial Communities in Natural Environments. *Ceylon J. Sci. (Bio. Sci.)* **2013,** *42*(1), 19–33.

Fisher, M. M.; Triplett, E. W. Automated Approach for Ribosomal Intergenic Spacer Analysis of Microbial Diversity and Its Application to Freshwater Bacterial Communities. *Appl. Environ. Microbiol.* **1999,** *65,* 4630–4636.

Frostegard, A.; Tunlid, A.; Baath, E. Phospholipid Fatty Acid Composition, Biomass, and Activity of Microbial Communities from Two Soil Types Experimentally Exposed to Different Heavy Metals. *Appl. Environ. Microbiol.* **1993,** *59,* 3605–3617.

Garland, J. L.; Mills, A. L. Classification and characterization of heterotrophic microbial communities on the basis of patterns of community-level-sole-carbon-source utilization. *Appl. Environ. Microbiol.* **1991,** *57,* 2351–2359.

Garrity, G. M.; Bell, J. A.; Lilburn, T. G. Taxonomic Outline of Prokaryotes. In *Bergey's Manual of Systematic Bacteriology;* 2nd Ed., Garrity, G. M., Ed.; Springer: New York, 2003, pp 1–399.

Gasser, I.; Müller, H.; Berg, G. Ecology and Characterization of Polyhydroxyalkanoate-Producing Microorganisms on and in Plants. *FEMS Microbiol. Ecol.* **2009,** *70,* 142–150.

Ghazanfar, S.; Azim, A.; Ghazanfar, M. A.; Anjum, M. I.; Begum, I. Metagenomics and Its Application in Soil Microbial Community Studies: Biotechnological Prospects, *J. Anim. Plant Sci.* **2010,** *6*(2), 611–622.

Giller, K. E.; Beare, M. H.; Lavelle, P.; Izac, A. M. N.; Swift, M. J., Agricultural Intensification, Soil Biodiversity and Agroecosystem Function. *Appl. Soil Ecol.* **1997,** *6,* 3–16.

Glick, B. R. Phytoremediation: Synergistic Use of Plants and Bacteria to Cleanup the Environment. *Biotechnol. Adv.* **2001,** *21*(3), 83–393.

Grace, J. B. The Factors Controlling Species Density in Herbaceous Plant Communities: An Assessment. In *Perspectives in Plant Ecology, Evolution and Systematics;* Edwards, P. J., ed.; Urban and Fischer Verlag: Jena, Germany, 1999; pp 1–28.

Gray, E. J.; Smith, D. L. Intracellular and Extracellular PGPR: Commonalities and Distinctions in the Plant-Bacterium Signaling . *Soil Biol. Biochem.* **2005,** *37,* 395–412.

Greene, E. A.; Voordouw, G. Analysis of Environmental Microbial Communities by Reverse Sample Genome Probing. *J. Microbiol. Meth.* **2003,** *53,* 211–219.

Grime, J. P. *Plant Strategies and Vegetation Processes;* Wiley: London, 1979.

Handelsman, J.; Rondon, M. R.; Brady, S. F.; Clardy, J.; Goodman, R. M. Molecular Biological Access to the Chemistry of Unknown Soil Microbes: A New Frontier for Natural Products. *Chem. Biol.* **1998,** *5*(10), 245–249.

Houlden, A.; Timms-Wilson, T. M.; Day, M. J.; Bailey, M. J. Influence of Plant Developmental Stage on Microbial Community Structure and Activity in the Rhizosphere of Three Field Crops. *FEMS Microbiol. Ecol.* **2008,** *65,* 193–201.

Janusauskaite, D.; Kadziene, G.; Auskalniene, O. The Effect of Tillage System on Soil Microbiota in Relation to Soil Structure. *Pol. J. Environ. Stud.* **2013,** *22*(5), 1387–1391.

Karlen, D. L.; Mausbach, M. J.; Doran, J. W.; Cline, R. G.; Harris, R. F.; Schuman, G. E. Soil Quality: A Concept, Definition, and Framework for Evaluation (A Guest Editorial). *Soil Sci. Soc. Am. J.* **1997,** *61,* 4–10.

413

Kauppinen, J.; Pelkonen, J.; Katila, M. J. RFLP Analysis of *Mycobacterium malnroense* Strains using Ribosomal RNA Gene Probes: An Additional Tool to Examine Intraspecies Variation. *J. Microbiol. Meth.* **1994**, *19*, 261–267.

Khan, A. G. Role of Soil Microbes in the Rhizosphere of Plants Growing on Trace Metal Contaminated Soils in Phytoremediation. *J. Trace Elem. Med. Biol.* **2005**, *18*, 355–364.

Kirk J. L.; Beaudette L. A.; Hart, M.; Moutoglis P.; Klironomos, J. N.; Lee, H.; Trevors, J. T. Methods of Studying Soil Microbial Diversity. *J. Microbiol. Meth.* **2004**, *58*, 169–188.

Lemanceau, P.; Mazurier, S.; Avoscan, L.; Robin, A.; Briat, J. F. Reciprocal Interactions between Plants and Fluorescent *Pseudomonas* in Relation to Iron in the Rhizosphere. In *Molecular Microbial Ecology of the Rhizosphere;* de Bruijn, F. J., Ed.; Wiley Blackwell: Hoboken, NJ, USA, 2013, pp 1181–1189.

Lim, H. K.; Chung, E. J.; Kim, J. C.; Choi, G. J.; Jang, K. S.; Chung, Y. R.; Cho, K. L.; Lee, S. W. Characterization of a Forest Soil Metagenome Clone That Confers Indirubin and Indigo Production on *Escherichia coli*. *Appl. Environ. Microbiol.* **2005**, *71*, 7768–7777.

Liu, W. T.; Marsh, T. L.; Cheng, H.; Forney, L. J. Characterization of Microbial Diversity by Determining Terminal Restriction Fragment Length Polymorphisms of Genes Encoding 16S rRNA. *Appl. Environ. Microbiol.* **1997**, *63*, 4516–4522.

Lukow, T.; Dunfield, P. F.; Liesack, W. Use of the T-RFLP Technique to Assess Spatial and Temporal Changes in the Bacterial Community Structure within an Agricultural Soil Planted with Transgenic and Non-transgenic Potato Plants. *FEMS Microbiol. Ecol.,* **2000**, *32*, 241–247.

Lynch, J. M. *Soil Biotechnology: Microbiological Factors in Crop Productivity;* Blackwell: Oxford, 1983.

MacNeil, I. A.; Tiong, C. L.; Minor, C.; August, R. P.; Grossman, T. H.; Loiacono, K. A.; Lynch, B. A.; Phillips, T.; Narula, S.; Sundaramoorthi, R.; Tyler, A.; Aldredge, T.; Long H.; Gilman, M.; Holt, D.; Osburne M. S. Expression and Isolation of Antimicrobial Small Molecules from Soil DNA Libraries. *J. Mol. Microbiol. Biotechnol.* **2001**, *3*, 301–308.

Malik, S.; Beer, M.; Megharaj, M.; Naidu, R. The Use of Molecular Techniques to Characterize the Microbial Communities in Contaminated Soil and Water. *Environ. Int.* **2008**, *34*, 265–276.

Maxam, A. M.; Gilbert, W. A New Method for Sequencing DNA. Proc. Nat. Acad. Sci., **1977**, *74*, 560–564.

McInnes, A.; Thies, J. E.; Abbott, L. K.; Howieson, J. G. Structure and Diversity among Rhizbial Strains, Populations and Communities – A Review. *Soil Biol. Biochem.* **2004**, *36*, 1295–1308.

McNabb, D. H.; Startsev, A. D. Effects of Compaction on Aeration and Morphology of Boreal Forest Soils in Alberta, Canada. *Can. J. Soil Sci.* **2009**, *89*(1), 45–46.

Mendes, R.; Kruijt, M.; de Bruijn, I.; Dekkers, E.; van der Voort, M.; Schneider, J. H.; Piceno, Y. M.; DeSantis, T. Z.; Andersen, G. L.; Bakker, P. A.; et al. Deciphering the Rhizosphere Microbiome for Disease-Suppressive Bacteria. *Science.* **2011**; *332*, 1097–1100.

Metzker, M. L. Emerging Technologies in DNA Sequencing. *Genome Res.* **2005**, *15*, 1767–1776.

Metzker, M. L. Sequencing Technologies – The Next Generation. *Nat. Rev. Genet.* **2010**, *11*, 31–46.

Muyzer, G. DGGE/TGGE, a Method for Identifying Genes from Natural Ecosystems. *Curr. Opin. Microbiol.* **1999**, *2*, 317–322.

Palleroni, N. J. Structure of the Bacterial Genome. In *Handbook of New Bacterial Systematic;* Goodfellow, M.; O'Donnell, A. G., Eds. Academic Press: London, 1993, 57–151.

Peters, S.; Koschinsky, S.; Schwieger, F.; Tebbe, C. C. Succession of Microbial Communities During Hot Composting as Detected by PCR-Single-Strand-Conformation Polymorphism Based Genetic Profiles of Small Subunit rRNA Genes. *Appl. Environ. Microbiol.* **2000**, *66,* 930–936.

Prosser, J. I. Molecular and Functional Diversity in Soil Micro-organisms. *Plant Soil.* **2002**, *244,* 9–17.

Ranjard, L.; Poly, F.; Lata, J. C.; Mougel, C.; Thioulouse, J.; Nazaret, S. Characterization of Bacterial and Fungal Soil Communities by Automated Ribosomal Intergenic Spacer Analysis Fingerprints: Biological and Methodological Variability. *Appl. Environ. Microbiol.* **2001**, *67,* 4479–4487.

Rao, N. S. *Soil Microorganisms and Plant Growth;* 4th edn. Oxford and IBH Publishing: New Delhi, India, 2005, p 407.

Rastogi, G.; Sani, R. Molecular Techniques to Assess Microbial Community Structure, Function, and Dynamics in the Environment; Springer: New York, 2011, pp 29–57.

Revillas, J. J.; Rodelas, B.; Pozo, C.; Martinez-Toledo, M. V.; Gonzalez, L. J. Production of B-Group Vitamins by Two Azotobacter Strains with Phenolic Compounds as Sole Carbon Source under Diazotrophic and Adiazotrophic Conditions. *J. Appl. Microbiol.* **2000**, *89,* 486–493.

Reynolds, H. L.; Packer, A.; Bever, J. D.; Clay, K. Grassroots Ecology: Plant–Microbe–Soil Interactions as Drivers of Plant Community Structure and Dynamics. *Ecology.* **2003**, *84,* 2281–2291.

Rivas, R.; García-Fraile, P.; Velázquez, E. Taxonomy of Bacteria Nodulating Legumes. *Microbiol. Insights.* **2009**, *2,* 51–69.

Sanger, F.; Coulson, A. R. A Rapid Method for Determining Sequences in DNA by Primed Synthesis with DNA Polymerase. *J. Mol. Biol.* **1975**, *94,* 441–448.

Schmidt, S. K.; Gier, M. J. Dynamics of Microbial Populations in Soil: Indigenous Microorganisms Degrading 2,4-Dinitrophenol. *Microb. Eco.* **1989**, *18*(3), 285–296.

Schramm, A.; Larsen, L. H.; Revsbech, N. P.; Ramsing, N. B.; Amann, R.; Schleifer, K. H. Structure and Function of a Nitrifying Biofilm as Determined by *In Situ* Hybridization and the Use of Microelectrodes. *Appl. Environ. Microbiol.,* **1996**, *62,* 4641.

Shoebitz, M.; Ribaudo, C. M.; Pardo, M. A.; Cantore, M. L.; Ciampi, L.; Curá, J. A. Plant Growth Promoting Properties of a Strain of *Enterobacter ludwigii* Isolated from *Lolium perenne* Rhizosphere. *Soil Biol. Biochem.* **2009**, *41*(9), 1768–1774.

Silva, A. P.; Babujia, L. C.; Matsumoto, L. S.; Guimaraes, M. F.; Hungaria, M. Bacterial Diversity under Different Tillage and Crop Rotation Systems in an Oxisol of Southern Brazil. *Open Agric. J.* **2013**, *7* (Supp 11-M6), 40–47.

Smith, C. J.; Osborn, A. M. Advantages and Limitations of Quantitative PCR (Q- PCR)-based Approaches in Microbial Ecology. *FEMS Microbiol. Ecol.* **2009**, *67,* 6–20.

Suleiman, A.; Manoeli, L.; Boldo, J.; Pereira, M.; Roesch, L. Shifts in Soil Bacterial Community after Eight Years of Land-Use Change. *Syst. Appl. Microbiol.* **2013**, *36,* 137–144.

Tiedje, J. M.; Asuming-Brempong, S.; Nusslein, K.; Marsh, T. L.; Flynn, S. J. Opening the Black Box of Soil Microbial Diversity. *Appl. Soil Ecol.* **1999**, *13,* 109–122.

Tonin, C.; Vandenkoornhuyse, P.; Joner, E. J.; Straczek, J.; Leyval, C. Assessment of Arbuscular mycorrhizal Fungi Diversity in the Rhizosphere of *Violoa calaminaria* and Effect of These Fungi on Heavy Metal Uptake by Clover. *Mycorrhiza.* **2001**, *10,* 161– 168.

Torsvik, V.; Daae, F. L.; Sandaa, R. A.; Ovreas, L., Review Article: Novel Techniques for Analysing Microbial Diversity in Natural and Perturbed Environments. *J. Biotechnol.* **1998**, *64*, 53–62.

Torsvik, V.; Goksøyr, J.; Daae, F. L. High Diversity in DNA of Soil Bacteria. *Appl. Environ. Microbiol.* **1990**, *56*, 782–787.

Uroz, S.; Ioannidis, P.; Lengelle, J.; Cébron, A.; Morin, E.; Buée, M.; Martin, F. Functional Assays and Metagenomic Analyses Reveals Differences between the Microbial Communities Inhabiting the Soil Horizons of a Norway Spruce Plantation. *PLoS One.* **2013**, *8*, e55929.

Van Elsas, J. D.; Frois-Duarte, G.; Keijzer-Wolters, A.; Smit, E. Analysis of the Dynamics of Fungal Communities in Soil via Fungal-Specific PCR of Soil DNA Followed by Denaturing Gradient Gel Electrophoresis. *J. Microbiol. Meth.* **2000**, *43*, 133–151.

Wardle, D. A.; Giller, K. E. The Quest for a Contemporary Ecological Dimension to Soil Biology. *Soil Biol. Biochem.* **1996**, *28*, 1549–1554.

Watson, G. W.; Kelsey, P. The Impact of Soil Compaction on Soil Aeration and Fine Root Density of Quercus Palustris. *Urban Forestry Urban Greening.* **2006**, *4*(2), 69–74.

Whittaker, R. H. Evolution and Measurement of Species Diversity. *Taxon.* **1972**, *21*, 213–251.

Yao, H.; He, Z.; Wilson, M. J.; Campbell, C. D. Microbial Biomass and Community Structure in a Sequence of Soils with Increasing Fertility and Changing Land Use. *Microb. Ecol.* **2000**, *40*, 223–237.

Zaidi, S.; Usmani, S.; Singh, B. R.; Musarrat, J. Significance of *Bacillus subtilis* Strains SJ-101 as a Bioinoculant for Concurrent Plant Growth Promotion and Nickel Accumulation in *Brassica juncea*. *Chemosphere.* **2008**, *64*, 991–997.

Zelles, L.; Bai Q. Y. Fractionation of Fatty Acids Derived from Soil Lipids by Solid Phase Extraction and Their Quantitative Analysis by GC-MS. *Soil Biol. Biochem.* **1993**, *25*, 495–507.

Zhuang, X. L.; Chen, J.; Shim, H.; Bai, Z. New Advances in Plant Growth-Promoting Rhizobacteria for Bioremediation. *Environ. Int.* **2007**, *33*, 406–413.

CHAPTER 15

POPULATION DIVERSITY OF *FUSARIUM* SPP. AND ITS INTERACTION WITH PLANT GROWTH PROMOTING RHIZOBACTERIA

S. HARISH[*], R. MANIKANDAN, D. DURGADEVI, and
T. RAGUCHANDER

Department of Plant Pathology, Centre for Plant Protection Studies, Tamil Nadu Agricultural University, Coimbatore 641003, India

[*]*Corresponding author: sankarshari@rediffmail.com*

CONTENTS

ABSTRACT

The ascomycete fungal pathogen *Fusarium* spp. (Teleomorph stage: *Gibberella* spp.), causal agent of vascular wilts in major crops, is a challenging pathogen worldwide. This disease causes economic yield loss either directly or through diverse fungal mycotoxins, which constitute a significant threat to the human health. The diversity and dynamics of *Fusarium* spp. exist in the rhizosphere, which varies with the soil environment, soil climate, and presence of soil microorganisms. During the recent years, plant growth promoting rhizobacteria (PGPR) is being exploited for sustainable agriculture in various parts of the world. Rhizobacteria inhabit plant roots and exert a beneficial effect on plant growth besides reducing the disease incidence. In nature, interactions between the pathogenic and beneficial microorganisms take place, which decide the existence of the pathogen in the rhizosphere region. Interaction of PGPR with *Fusarium* spp. in the rhizosphere may lead to expression of defense related genes in the plants, which can counteract the pathogen infection. On the other hand, interaction with nematodes may aggravate the infection process, which causes more damage to the crop. This review helps in understanding the dynamics and existence of *Fusarium* spp. in the soil and their interaction with the microorganism, which explore the possibility of identifying new proteins or genes in host–pathogen interaction.

15.1 INTRODUCTION

Plants are incessantly challenged with deleterious pathogens which include fungi, bacteria, viruses, insects, and nematodes. While many of these organisms evolved to infect aerial plant parts, such as leaves, stems, flowers, and fruits others target below-ground organs, such as roots and tubers. Specific pathogens target the vascular system composed of xylem vessels, tracheary elements which transports water and minerals that are absorbed by the roots to the photosynthetic organs, and phloem elements, the living tissue that transports organic photosynthetic products. The phloem is rich in sugars, and most vascular pathogens colonize the nutrient-poor xylem vessels. Vascular wilt pathogens are among the most destructive plant pathogens which cause losses to economically important crop plants. Vascular wilt diseases occur worldwide and affect annual crops as well as woody perennials, thus not only affecting food and feed production, but also natural ecosystems. Most of the symptoms caused by vascular wilt pathogens develop in acropetal direction: from bottom to top. Epinasty is the primary disease symptom, followed by flaccidity, chlorosis, vascular browning, and necrosis of the terminal leaflets.

The genus *Fusarium* is considered as one of the most adaptable genera in the fungal kingdom. The most common species, *Fusarium oxysporum* causes vascular wilt disease in a wide variety of economically important crops (Beckman, 1987). It consists of both pathogenic and non-pathogenic strains (Gordon & Martyn, 1997). The strains of *F. oxysporum* are able to stay alive for long periods in soil and enter into the roots inducing either root-rots or wilts. Many other strains can penetrate into the roots, but do not cause disease (Olivain et al., 2004). They are host specific and based on the plant species they infect; they are classified into more than 150 *formae speciales* (Armstrong & Armstrong, 1984; Baayen et al., 2000). Managing the wilt disease of *Fusarium* through chemical means is cost expensive and causes environmental hazards (Mahesh et al., 2010). Hence, the most environmentally safe and effective method of control is the use of biocontrol microorganisms. The population dynamics of *Fusarium* in the soil varies with the soil environment and also the microorganism present in the soil. In this review, a detail understanding of *Fusarium* spp. in the soil with the biotic and abiotic factors will be dealt in detail.

15.2 VASCULAR FUNGAL PATHOGENS—DIVERSITY AND DYNAMICS

Fusarium oxysporum is an omnipresent broad-based soil-borne plant pathogen. The taxonomy of *Fusarium* is based on morphological characteristics, which include the colony color, size, and shape of macroconidia, the presence or absence of microconidia and the formation and character of chlamydospores (Moss & Smith, 1984). In culture, *F. oxysporum* produces colorless to pale yellow mycelium that turns pink or purple with age. With the exception of grasses and most tree crops, few of the widely cultivated crops are not hosts to pathogenic form of *F. oxysporum* (Armstrong & Armstrong, 1984). Virulence has been an extremely useful characteristic for differentiating isolates of *F. oxysporum* into *formae speciales*. The pathogen is distinct in symptomology, epidemiology, and cultivar susceptibility (Vakalounakis, 1996). *F. oxysporum* teleomorph is unknown; hitherto DNA sequence-based phylogenetic analyses place this complex unambiguously in the *Gibberella* clade, close to the *G. fujikuroi* species complex (O'Donnell et al., 2004). Another approach in understanding the diversity in *Fusarium* spp. is the analysis of vegetative compatibility grouping based on the heterokaryotic formation and mycelial interactions (Leslie, 1993; Puhalla, 1985). VCGs are a convenient tool in determining the number of genetically distinct individuals within a population and to understand whether strains with unusual variants have a clonal origin. This has been studied with respect to *Fusarium moniliforme* to know that the mutations at the *pall* and *fuml* loci were of clonal origin (Desjardins et al., 1992). Many pathogens have been characterized based on the internal transcribed spacer (ITS) and intergenic spacer (IGS) region which will be appropriate for studying pathogen identification (Singh et al., 2006). For identification and genetic characterization of *Fusarium* spp., restriction fragment length polymorphism (RFLP) (Gupta et al., 2010), amplified fragment length polymorphism (AFLP) (Stewart et al., 2006), random amplified polymorphic DNA (RAPD) (Manulis et al., 1994; Wright et al., 1996), RFLP-PCR of the intergenic spacer region of ribosomal DNA (IGS rDNA) (Srinivasan et al., 2010) and inter simple sequence repeat (ISSR) were carried out by various research groups (Lin et al., 2012; Tanyolaç & Akkale, 2010). Srinivasan et al. (2010) documented that the phylogenetic analysis of the IGS sequence is very useful for studying the genetic makeup of populations of *F. oxysporum*. Mes et al. (1999) screened two races of *F. oxysporum* f. sp. *lycopersici* for vegetative compatibility and characterized them using RAPD analysis, and found that the RAPD profiles coincided with the vegetative compatibility groups. Do Amaral et al. (2013) reported that three races of *F. oxyspoum* f. sp. *lycopersici* were analyzed by RAPD, ISSR, and restriction fragment length polymorphism of intergenic spacer (RFLP-IGS). Using the marker IGS, restriction with *HinfI*, *RsaI*, *AluI*, and *HaeIII* produced polymorphism information enabling the identification and genetic characterization among isolates of three races of the pathogen. Attitalla et al. (2004) evaluated isozyme analysis, mtDNA-RFLP and high performance liquid chromatography (HPLC) to differentiate two morphologically indistinguishable *formae speciales* of *F. oxysporum lycopersici* and *radicis-lycopersici*. Although HPLC produced distinct profiles for non-pathogenic and pathogenic isolates, the direct mtDNA-RFLP technique proved to be an efficient diagnostic tool for routine differentiation of *lycopersici* and *radicis-lycopersici* isolates. The cultural, morphological diversity and pathogenicity studies along with molecular methods involving the use of polymerase chain reaction will help in better resolution of genetic variation between the strains thus help in understanding the genetic diversity of *Fusarium* spp.

15.3   FACTORS INFLUENCING THE POPULATION DYNAMICS

15.3.1   ABIOTIC AND BIOTIC FACTORS

In the soil, *Fusarium* species are able to persist as mycelium, chlamydospore, and conidia (Booth, 1971). The infection cycle of *F. oxysporum* is initiated by the germination of spores in the soil in response to an undetermined signal within the host root exudates. Upon germination, infective hyphae adhere to the root surface and penetrate the root epidermis directly without the formation of any distinctive structure (Rodriguez-Galvez & Mendgen, 1995). The mycelium then advances inter- or intra-cellularly through the root cortex, until it reaches the xylem vessels and enters them through the pits (Bishop & Cooper, 1983b). At this point, the fungus switches to an endophytic mode of host colonization, during which it remains exclusively within the xylem vessels, using them as avenues to rapidly colonize the host. At this stage, the characteristic wilt symptoms appear as a result of severe water stress, which ultimately lead to complete plant death. Upon plant death, the pathogen grows on the plant surface where it produces chlamydospores that are dispersed into the soil for a second cycle of infection (Bishop & Cooper, 1983b). The low percentage of *F. oxysporum* may be attributed to the soil type as soil type can influence the population structure of *F. oxysporum*, as the species is particularly adapted to a specific abiotic environment or other environmental factor (Edel et al., 2001). Factors such as nematode, soil compaction, crop rotation history, soil pH, herbicide and hail injury, iron chlorosis, and soil type may be important for the development of *Fusarium* root rot (Kremer & Means, 2009; Zhang et al., 2010). Soil type is likely to influence the structure on microbial communities. Generally, sandy texture soil has lower water holding capacity. Highest percentage of *Fusarium* isolates was recovered from sandy loam soil. *F. solani* was found in all the four soil types that is, sandy loam, silty clay loam, silty loam, and silty clay soils in the study conducted by Latiffah et al. (2007). Temperature had a significant influence on the population level of all test *Fusarium* species (Abawi & Barker, 1983). *Fusarium* is a common species in cool to temperate regions of the world (Windels et al., 1988). Saremi and Burgess (2000) reported that the population of *Fusarium sarnbucinum* increased by 25% at cool (13–18°C) temperatures. There was noticeable reduction of population at moderate (19–25°C) and especially at warm (25–30°C) temperatures. Latiffah et al. (2007) isolated four *Fusarium* species *viz.*, *F. solani, F. semitectum, F. equiseti*, and *F. oxysporum* in silty loam soil with pH 3.9. *Fusarium* species were present in acidic condition ranging from pH

3.8 to 6.4 and most of the *F. solani* were isolated in acidic condition of pH 3.8–3.9. Soil moisture induced changes in the soil microbial population, and that in turn affected the *Foc* propagules. Burke et al. (1972) reported that the density of *F. solani* f. sp. *phaseoli*, declined with soil depth. Latiffah et al. (2007) isolated four *Fusarium* species namely, *F. solani*, *F. semitectum*, *F. equiseti*, and *F. oxysporum* at high moisture content. Larkin et al. (1993) reported that conducive soils had high levels of *F. oxysporum* f. sp. *niveum* because populations of general bacteria, actinomycetes, and fluorescent pseudomonads were found to be more in suppressive soils than in conducive soils. *Fusarium* survival on crop residues is affected by soil microbial antagonists (Weller, 2002). The incorporation of green manures increased the density and diversity of microbes in soils which in turn had an effect on pathogen survival (Pérez-Artés et al., 2008). Bailey and Lazarovits (2003) reported that the residue management and organic amendments suppressed the soil-borne pathogens. Kremer and Means (2009) determined the impact of soil moisture content on root colonization of glyphosate-treated soybean by *Fusarium* species. They found that highest levels of *Fusarium* colonization were associated with glyphosate treatment, at the highest soil moisture level. Elevated concentrations of glyphosate delayed decomposition of plant residues, suggesting *Fusarium* preferentially utilized glyphosate as a nutrient source before plant residues (Leslie et al., 1995).

Hoper et al. (1995) found a positive correlation between pH and soil suppressiveness, soils with higher pH being more suppressive toward *Fusarium* wilts. Soil pH influences plant disease infection and development directly by effects on the soil-borne pathogen and populations of microorganisms and indirectly through availability of soil nutrients to the plant host (Ghorbani et al., 2008). Increasing soil pH or calcium levels may be beneficial for disease management in many other crops (Sullivan, 2001). A direct correlation between adequate calcium level, and/or higher pH, and decreasing levels of *Fusarium* occurrence has been established for a number of crops, such as tomato, cotton, and melons (Jones et al., 1989).

15.3.2 ROOT NEMATODE INTERACTION WITH THE POPULATION DYNAMICS OF FUSARIUM

Among various factors involved in the initiation and development of plant diseases, there is a possible association of more than one microorganism with the host plant at a given time. Plant-parasitic nematodes often play a major role in disease interactions. The soil-borne fungi constitute a significant portion of the soil microflora; some of them are recognized plant patho-gens and others are not normally pathogenic under prevailing conditions (Rizvi & Yang, 1996). Atkinson (1892) first reported an interaction between a plant-parasitic nematode and a soil-borne plant pathogenic fungus, which revealed that *Fusarium* wilt of cotton was more severe in soil co-infested with the root-knot nematode (*Meloidogyne* spp.) and *F. oxysporum* than in soil infested with only *F. oxysporum*. The combination of nematode and fungus often results in a synergistic interaction. Back et al. (2000) reported three types of synergistic interactions between fungi and nematodes that can affect the host. Positive or synergistic association between nematode and pathogen occurs when their association results in plant damage exceeding the sum of the individual damage of both pathogens. Conversely, an inter-action between both pathogens results in plant damage less than expected from the sum of both individually, is described as antagonistic. An addi-tive association occurs when nematodes and fungi cause plant damage that equates to the addition of individual damage by both pathogens. Interactions between *Meloidogyne* spp. and *Fusarium* wilt pathogens have been studied and documented in several host crops. Minton and Minton (1963) studied the histopathology of sites infected with *M. incognita* and *F. oxysporum* f. sp. *vasinfectum* on cotton. Sheela and Venkitesan (1990) found that the simulta-neous inoculation of *M. incognita* and *Fusarium* spp. led to the suppression of vine growth in black pepper. Khan and Husain (1989) reported the interac-tion of *M. javanica* and *F. oxysporum* f. sp. *ciceri* on some chickpea cultivars. Lamondia (1992) found that *M. hapla* or *Globodera tabaccum* predispose the tobacco plants to *Fusarium* wilt. In banana, simultaneous or sequential inoculation of root-knot nematode and *Fusarium* sp. caused drastic reduc-tion in the growth parameters than inoculation of either of the pathogens (Jonathan & Rajendran, 1998). Bertrand et al. (2000) reported that corky-root disease has a complex etiology, and emphasized the dominant role of *M. arabicida* as a predisposing agent to subsequent invasion by *F. oxys-porum*. Synergistic relationships between root-knot nematode *Meloidogyne* spp. and *Fusarium* sp. have been reported by several authors (Agbenin & Erinle, 2001; Bertrand et al., 2000; Sharma & Nene, 1990). Jain and Jitendra (2010) observed the difference in wilt appearance of tomato plants inocu-lated either individually, concomitantly, or sequentially with *M. incognita* and *F. oxysporum* f. sp. *lycopersici*. The close proximity of infection court of the nematode and the fungi probably enhanced the possibility of exchange of toxic metabolites from one feeding site to the other thus interfering in the establishment of normal host–parasite relationship. Multiplication rate of nematode was poor in the presence of fungi because of tissue destruction caused by fungi much before the completion of nematode life cycle. Studies were carried out to find the combined effect of root-knot nematode, *M. javanica* and wilt pathogen, *F. udum* in ten wilt resistant/tolerant accessions of pigeonpea. Presence of *M. javanica* with *F. udum* increased wilting from 8 to 33% in KPL-44, 15 to 50% in AWR-74/15, 25 to 50% in ICP 8859 and ICPL 89049, and 15 to 50% in ICP 12745, whereas, in other five accessions wilting did not increase much in presence of nematodes (Bansa et al., 2004; Singh et al., 2006). Ramyabharathi (2014) reported that maximum disease incidence of 95.25% in gerbera was recorded in the treatment with *M. incognita* inoculation 15 d before *F. oxysporum* inoculation. Significant reduction in shoot length, shoot weight, root length, root weight, flower stalk length, flower diameter, number of flowers per plant was noticed in inoculation of *M. incognita* 15 d prior to inoculation of *F. oxysporum*. The same treatment recorded higher gall index of 4.0 with juvenile population of 321.50 and 43.25 females per gram of root.

15.3.3   INTERACTION OF PGPR WITH FUSARIUM SPP.

For plant–microbe interaction, rhizosphere is the major zone which provides a platform for colonization of different microorganisms either beneficial or harmful. Due to secretion of root exudates in the rhizosphere region, higher level of microbial activity exist in the region which may either result in associative, symbiotic, neutralistic or parasitic interactions depending upon nature of the soil and the type of microorganism proliferating in the rhizosphere zone (Hayat et al., 2010). Plant roots offer a niche for the proliferation of soil bacteria that thrive on root exudates and cell lysates. Association of beneficial free-living soil bacteria in the plant root system is usually referred to as plant growth promoting rhizobacteria (PGPR) (Kloepper, 1993). It prevents the deleterious effects from phytopathogenic organisms due to production of antagonistic substances or by inducing resistance against plant pathogens (Glick, 1995). PGPR induce systemic resistance (ISR) through fortifying the physical and mechanical strength of cell wall as well as changing physiological and biochemical reaction of host leading to synthesis of defense chemicals against challenge inoculation of pathogens. Defense reaction occurs due to accumulation of PR proteins (chitinase and β-1,3 glucanases), chalcone synthase, phenylalanine ammonia lyase, peroxidase, phenolics, callose, lignin, and phytoalexins. Burkhead et al. (1994) reported that *P. cepacia* B37W produced pyrrolnitrin antibiotic that was inhibitory to *F. sambucinum*. Kloepper et al. (1992) showed that five of six rhizobacteria that induced systemic resistance in cucumber exhibited both external and internal root colonization. Seed treatment of radish with resistance inducing *P. fluorescens* strain WCS374 reduced *Fusarium* wilt in naturally infested field soil up to 50% (Leeman et al., 1995). Elad and Baker (1985) reported that *Pseudomonas* strains of A1, BK1, TL3B1, and B10 established a siderophore synthesis to inhibit the growth of *F. oxysporum* at *in vitro* conditions. Benhamou et al. (1996) reported that treatment of pea seeds with *P. fluorescens* strain 63-28 resulted in formation of structural barriers, *viz.,* cell wall apposition (papillae) and deposition of newly formed callose and accumulation of phenolic compounds at the site of penetration of invading hyphae of *F. oxysporum* f. sp. *pisi*. Podile and Laxmi (1998) reported that seed bacterized with *B. subtilis* AF1 showed an increase in PO activity from the day six of challenge inoculation with pathogen *F. oxysporum* f. sp. *udum*.

A combination of seed and soil application of fluorescent pseudomonads for soil-borne disease management was reported in chickpea, pigeonpea, and black gram (Jayashree et al., 2000). Fluorescent pseudomonads isolated from suppressive soil were able to induce suppressiveness in conducive soil for crown and root rot caused by *F. oxysporum* f. sp. *radicis lycopersici* (Lemanceau & Alabouvette, 1991). Similarly, application of endophytic bacteria by stem injection reduced the vascular wilt caused by *F. oxysporum* f. sp. *vasinfectum* (Chen et al., 2000). Seed treatment with *P. fluorescens* strain 63-28 restricted the growth of *F. oxysporum* f. sp. *lycopersici* in tomato (M'Piga et al., 1997). *P. fluorescens* isolate Pf1 was found to protect tomato plants from wilt disease caused by *F. oxysporum* f. sp. *lycopersici* (Ramamoorthy et al., 2002). Vivekananthan et al. (2004) reported the foliar application of fluorescent pseudomonad strain FP7 strongly reduced the incidence of anthracnose disease in mango. Application of *Pseudomonas* sp. as seed treatment (10 g/kg seed) followed by soil application (2.5 kg/ha) against root rot effectively enhanced higher plant growth and reduced root rot incidence in black gram (Jayashree et al., 2000). *P. fluorescens* CHA0 strain promoted the germination and plant growth of tomato and effectively controlled *F. oxysporum* f. sp. *lycopersici* (Ardebili et al., 2010). *P. fluorescens* stimulated the activity of peroxidase (PO) in chilli plants inoculated with *F. solani* and gradually increased these activities as compared to untreated control plants (Sundaramoorthy et al., 2012). Higher level expression of defense protein β-1,3-glucanase was noticed in tomato plants treated with Pf1 liquid formulation against *Fusarium* wilt pathogen. Ramyabharathi (2011) reported that production of iturin antibiotic in *Bacillus subtilis* (EPCO 16) is involved in the suppression of *F. oxysporum* f. sp. *lycopersici* in tomato. The liquid formulation of *Bacillus* strains *viz.*, EPCO16 and FZB24 increase the PO activity and induction of more isoforms against *F. oxysporum* f. sp. *lyco-persici* in tomato plants (Elanchezhian, 2012). An investigation into the performance of biocontrol agents and fungicide in the control of grapevine wilt caused by *F. moniliforme* was done by Karunakaran et al. (2003). Shan-mugam et al. (2010) reported that strain mixtures of *Burkholderia atro-phaeus* and *B. cepacia*, when applied through corm dressing and through soil application in the form of talc-based formulations, were inhibitory to the growth of *F. oxysporum* f. sp. *gladioli*. Protective effect in chilli was observed due to the application of rhizobacterial strain *P. fluorescens* (Pf1) against *Fusarium* disease (Sundaramoorthy et al., 2012). An antagonistic effect of six isolates of *Pseudomonas* against *F. oxysporum* f. sp. *ciceris* in chickpea under *in vitro* and *in vivo* conditions was documented (Karimi et al., 2012). Liquid formulation of *P. fluorescens*, Pf1 exhibited higher induc-tion of defense enzymes and reduced the incidence of tomato *Fusarium* wilt disease (Manikandan & Raguchander, 2014). Enhancement of resistance and retardation of *Colletotrichum musae* was observed in banana plants treated with water in oil based PGPR formulation of *P. fluorescens* (FP7) (Mohammed Faisal et al., 2014). Application of *P. fluorescens* liquid formu-lation through drip system significantly reduced the *Fusarium* wilt complex in banana under field conditions (Selvaraj et al., 2014). Thus, interaction of *Fusarium* spp. and PGPR results in the induction of defense enzymes and PR proteins which exhibit antagonistic activity.

15.4 PROTEOMICS APPROACH FOR STUDYING TWO/THREE-WAY INTERACTION

A better understanding of fungal biology, pathogenicity, and plant fungus interactions is needed for efficient and sustainable disease management. It can be obtained by employing various methodological approaches, one of them being proteomics. Proteomics has become a powerful tool for providing important information about pathogenicity and virulence factors, thus opening up new possibilities for crop disease diagnosis and crop protec-tion. Proteomics technology helps in the identification and quantification of all the proteins, study of protein–protein interactions that affect the various complex pathways and networks and structural characterization. Identifying the proteins helps in getting a complete picture of the proteome under study. Proteomic methods, including two-dimensional electrophoresis (2-DE), multi-dimensional liquid chromatography (LC), mass spectrometry (MS), and other derived methodological versions, allow separation, quantification, and identification of thousands of proteins in complex mixtures. Global proteome analysis has become an important approach in gene expression analysis, particularly for organisms with limited genomic resources, since database similarity search allows a high success.

15.4.1 PROTEOMICS IN FUSARIUM SPP.

In particular, application of proteomics in phytopathogenic fungi, helps to understand the differentially expressed proteins involved in altered growth conditions or development stages as well as for the identification of proteins of fungal mycelia (Taylor et al., 2008; Fernandez et al., 2009), secretome (Wang et al., 2005) and subproteomes (Kim et al., 2004; Aisif et al., 2006). Rep et al. (2004) found that tomato *F. oxysporum* pathosystem required a 12 kDa cysteine-rich effector protein for root invasion by the fungus which was identified through proteomic approach. Secreted proteins are of particular interest in phytopathogenic fungi due to their importance in plant–pathogen interactions. Proteomics has also been used successfully in the analysis of metabolic pathways such as mycotoxin synthesis by *Fusarium graminearum*. When it was cultured with a medium containing glucose expressed the cell wall degradation proteins and identified using proteomic study (Phalip et al., 2005; Taylor et al., 2008). Using LC-MS/MS, 229 fungal proteins, mostly glycoside hydrolases and proteases were identified in the secretome of *F. graminearum* during growth on synthetic media. Analyses of proteins secreted by *F. graminearum* during growth in different media and during infection of wheat crop revealed that the differentially expressed proteins could be important in the interaction in plants (Paper et al., 2007). Comparative proteomics of fungal species and strains has contributed to the understanding of the fungal infection process host range, pathogenicity, metabolisms, stress response, signal transduction, and identification of candidate virulence proteins (Xu et al., 2007).

15.4.2 TWO WAY INTERACTION—FUSARIUM SPP. AND HOST

The major contribution of proteomics in crop protection is the identification of both fungal effectors possibly facilitating infection and triggering plant defense and host proteins or biomarkers possibly conferring enhanced resistance, which require subsequent functional analysis of corresponding genes to establish new strategies for disease control. *F. graminearum* synthesizes trichothecene mycotoxins during plant host attack to facilitate spread of the pathogen in wheat and maize (Harris et al., 1999). The analysis of proteomes of mature grains of susceptible barley infected by *F. graminearum* showed that the fungus-induced degradation of the grain proteome and increased *Fusarium* infection depended on the availability of nitrogen amount (Yang et al., 2010). So far, several *F. graminearum* genes relative to mycotoxin production, signal transduction, metabolism, and growth have been analyzed in detail to examine their roles in the virulence and pathogenicity (reviewed by Kazan et al., 2012), but the targets discovered on the basis of the outcomes of proteomics, which may be essential for fungal infection, have not been well investigated. *Fusarium graminearum*, a devastating pathogen of wheat, maize and other cereals, when inoculated, 84 fungal secreted proteins were identified using 1DE and 2DE, followed by MS analyses (Phalip et al., 2005). With respect to host resistance, proteomics has identified many host proteins in response to *F. graminearum*, the majority of which are often involved in primary metabolism, defense, and stress-related responses. However, the most frequently identified host proteins have not been fully investigated, except the PR proteins (e.g., chitinase, β-1,3-glucanase, and thaumatin-like protein), in terms of downstream characterization of their functional roles in enhanced resistance. Houterman et al. (2007) investigated the molecular details of the interaction between the xylem-colonizing plant-pathogenic fungus *Fusarium oxysporum* and tomato in which 33 different proteins were identified whereas 16 tomato proteins were found in the xylem sap for the first time. Amongst these, proteins were peroxidases, chitinases, polygalacturonase and a subtilisin-like protease might be associated with the full virulence of *F. oxysporum*. Wongpiaa and Lomthaisong (2010) investigated the changes of protein profile in resistant (Mae Ping 80) and susceptible (Long Chilli 455) chilli cultivars upon *F. oxysporum* infection using 2DE and MS techniques. The results revealed that 10 protein spots were differentially expressed in the resistant cultivar with five up-regulated, four down-regulated and one supplementary protein, while 15 up-regulated, 11 down-regulated and 11 supplementary protein spots were found in the susceptible cultivar. Some of the induced proteins such as peroxidase, serine/threonine protein kinase, Cu/Zn superoxide dismutase, NADPH HC toxin reductase, and 1-aminocyclopropane-1-carboxylate synthase 3 are involved in plant defense mechanism. Two dimensional analyses of resistant banana root cells challenged with *Fusarium oxysporum* f. sp. *cubense* expressed beta-1,3-glucanase and chitinase compared with the control (Ying et al., 2013). *Fusarium oxysporum* f. sp. *fragariae* induced 79 differentially expressed proteins in resistant strawberry plants and was mainly involved in primary, secondary and protein metabolism, stress and defense responses, antioxidant and detoxification mechanisms, and hormone biosynthesis (Fang et al., 2013). These defense proteins expressed during the interaction were cloned and analyzed for their resistance against infection by the pathogen. Transgenic wheat expressing a α-1-purothionin, a thaumatin-like protein 1, a β-1,3-glucanase (MacKintosh et al., 2007), a class II chitinase (Shin et al., 2011), an antifungal plant defensin (Li et al., 2011), a pectin methylesterase inhibitor (Volpi et al., 2011), a polygalacturonase-inhibiting protein (Ferrari et al., 2012), a lactoferrin (Han et al., 2012), a *Arabidopsis thaliana* NPR1 (Makandar et al., 2006), or a truncated form of the yeast ribosomal protein L3 (Di et al., 2010) enhanced resistance to FHB under greenhouse conditions.

15.4.3 THREE-WAY INTERACTION—FUSARIUM SPP., HOST & PGPR

Most of the studies in plant–microbe interaction focus in the molecular changes in the plant related to pathogen attack and/or plant response to pathogen (Suzuki et al., 2004). However, the influence of biocontrol agent on the interactions between a plant and a pathogen has not yet been investigated in detail by using proteomics. The tripartite interaction will help to elucidate new genes which are expressed in the plant during the infection process. Several signal molecules and defense factors have been identified in plants upon plant microbe interaction (Canovas et al., 2004). Nevertheless, the molecular bases of multiple-player systems that may produce beneficial effects on plant health are largely unknown. Application of proteomics technique in three-way interaction of *P. fluorescens* treated rice plants challenged with sheath blight pathogen unravels the mechanism of resistance against *R. solani* (Saveetha, 2009; Karthiba, 2008). Senthil (2013) identified the *Chaetomium globosum* mediated defense protein and their mechanism against chilli damping-off disease through proteomic approach. Cotton defense responses were induced by application of PGPR against *Fusarium oxysporum* f. sp. *vasinfectum* (Dowd et al., 2004). This proteomics technique on the application of biocontrol agents and their interaction with *Fusarium* spp. helps to understand the molecular insights involved in biologically mediated resistance in plants. Ramamoorthy et al. (2002) reported the induction of 46 kDa chitinase protein in tomato expressed due to application of *P. fluorescens* when challenge inoculated with *F. oxysporum*. f. sp.

lycopersici in tomato. *P. fluorescens* induced the highest activity of polyphenol oxidase (PPO) in banana roots against *Fusarium* wilt disease (Akila et al., 2011). In tomato plants, higher level expression of chitinase protein was noticed upon treated with *P. fluorescens* (Manikandan & Raguchander, 2015). Manikandan (2015) used the proteomic approach in tripartite interaction of *Pseudomonas fluorescens*–tomato–*Fusarium oxysporum* f. sp. *lycopersici* and concluded that *Pseudomonas fluorescens* significantly expressed defense related proteins against wilt disease. From the tripartite interactions, 17 proteins were found significantly expressed and they were involved in improving the plant growth, replication, plant defense against stress, antioxidants, interacting with calcium ions and manganese, transcription, CO_2 fixation in photosynthesis, cutin deposition in the cell walls, systemic resistance, synthesis of nucleosides, defense signaling, activation of phenylpropanoid pathway, function in abiotic stress tolerance, disease resistance, and protein synthesis. Thus the three-way interaction will help to understand the mechanism of resistance by the biological control agent against the vascular pathogens.

15.5 CONCLUSION

Population dynamics of vascular pathogens varies with respect to soil environment *viz.*, soil moisture, soil temperature, soil pH, and soil microorganisms and interaction between these factors will decide the population density of the pathogen in the soil and its survival. Diversity of *Fusarium* exists in the rhizosphere and various molecular tools have been utilized nowadays to understand the dynamics of the pathogens in the soil. Interaction of PGPR with *Fusarium* has induced several defense genes in the plants which can counteract the pathogen infection. On the other hand, interaction with nematodes may enhance the infection process which causes more damage to the crop. Thus, studying the interaction of *Fusarium* spp. with nematodes or PGPR will pay a way to unravel new genes responsible for pathogenecity/ virulence produced by the pathogen and the defense genes produced by the host to counteract the pathogen. Proteomics has become an indispensable tool for understanding molecular and cellular mechanisms in plant–microbe interactions. Host genome sequencing and bioinformatics tools will help to identify novel elements in the host pathogen interaction especially with reference to vascular wilt pathogens. The new proteins or genes will need to thoroughly characterize for their role during the interaction in the plant or in the soil. The functional analysis of the identified proteins or genes in specific hosts is required to elucidate their roles in pathogenicity, virulence, and host resistance. These genes can be cloned and introgressed into the plant for developing resistance against the pathogen. Thus, the identified proteins or genes can be further exploited for their resistance in crop plants against vascular pathogens.

KEYWORDS

- diversity
- *Fusarium*
- plant growth promoting rhizobacteria

REFERENCES

Abawi, G. S.; Barker, K. R. Effects of Cultivar, Soil Temperature and Population Levels of *Meloidogyne incognita* on Root Necrosis and Fusarium Wilt of Tomatoes. *Phytopathol.* **1984,** *74,* 433–438.

Ardebili, Z. O.; Ardebili, N. O.; Seyed Mohammad, M. H. Physiological Effects of *Pseudomonas fluorescens* CHA0 on Tomato (*Lycopersicon esculentum* Mill.) Plants and its Possible Impact on *Fusarium oxysporum* f. sp. *lycopersici. Aus. J. Crop Sci.* **2010,** *5*(12), 1631–1638.

Armstrong, G. M.; Armstrong, J. K. Formae Specials and Races of *Fusarium oxysporum* Causing Wilt Disease. In *Fusarium: Diseases, Biology and Taxonomy;* Nelson, P. E.; Tousson, T. A.; Cook, R. J., Eds.; Pennsylvania State University Press, 1984, pp 391–399.

Atkinson, G. F. Some Diseases of Cotton. *Alabama Polytechnical Institute of Agriculture Experimental Station Bulletin.* **1892,** *41,* 61–65.

Baayen, R. P.; O'Donnell, K.; Bonants, P. J. M.; Cigelnik, E.; Kroon, L. P. N. M.; Roebroeck, E. J. A.; Waalwijk, C. Gene Genealogies and AFLP Analyses in the *Fusarium oxysporum* Complex Identify Monophyletic and Nonmonophyletic Formae Specials Causing Wilt and Rot Disease. *Phytopathology.* **2000,** *90*(8), 891–900.

Back, M. A.; Jenkinson, P.; Haydock, P. P. J. The Interaction between Potato Cyst Nematodes and *Rhizoctonia solani* Diseases in Potatoes. In: *Proceedings of the Brighton Crop Protection Conference, Pests And Diseases;* British Crop Protection Council: Farnham, UK, 2000, 503–506.

Bailey, K. L.; Lazarovits, G. Suppressing Soil-Borne Diseases with Residue Management and Organic Amendments. *Soil Till. Res.* **2003,** *72,* 169–180.

Bansa, S.; Ali, S. S.; Naimuddin; Askary, T. H. Combined Effect of *Fusarium udum* and *Meloidogyne javanica* on Wilt Resistant Accessions of Pigeon Pea. *Ann. Plant Prot. Sci.* **2004,** *12*(1), 130–133.

Beckman, C. H. *The Nature of Wilt Diseases of Plants;* The American Phytopatological Society: St. Paul, MN, 1987.

Benhamou, N.; Kloepper, J. W.; Quadt-Hallman, A.; Tuzun, S. Induction of Defense Related Ultrastructural Modifications in Pea Root Tissues Inoculated with Endophytic Bacteria. *Plant Physiol.* **1996,** *112,* 919–929.

Bertrand, B.; Nunez, C.; Sarah, J. L. Disease Complex in Coffee Involving *Meloidogyne arabicida* and *Fusarium oxysporum. Plant Pathol.* **2000,** *49,* 383–388.

Bishop, C. D.; Cooper, R. M. An Ultrastructural Study of Root Invasion in Three Vascular Wilt Diseases. *Physiol. Mol. Plant Pathol.* **1983a,** *22,* 15–27.

Bishop, C. D.; Cooper, R. M. An Ultrastructural Study of Vascular Colonization in 3 Vascular Wilt Diseases. 1. Colonization of Susceptible Cultivars. *Physiol. Plant Pathol.* **1983b,** *23,* 323–343.

Booth, C. The Present Status of *Fusarium* Taxonomy. *Ann. Rev. of Phytopathol.* **1975,** *13,* 83–93.

Burke, D. W.; Holmes, L. D.; Barker, A. W. Distribution of *Fusarium solani* f. sp. *phaseoli* and Bean Roots in Relation to Tillage and Soil Compaction. *Phytopathology,* **1972,** *62,* 550–554.

Chen, C.; Belanger, R. R.; Benhamou, N.; Paullitz, T. C. Defense Enzymes Induced in Cucumber Roots by Treatment with Plant-Growth Promoting Rhizobacteria (PGPR). *Physiol. Mol. Plant Pathol.* **2000,** *56,* 13–23.

Di, R.; Blechl, A.; Dill-Macky, R.; Tortora, A.; Tumer, N. E. Expression of a Truncated Form of Yeast Ribosomal Protein L3 in Transgenic Wheat Improves Resistance to *Fusarium* Head Blight. *Plant Sci.* **2010,** *178,* 374–380.

Do Amaral, D.; De Almeida, C. M. A.; Malafaia, C. B.; Da Silva, M. L. R. B.; De Menezes Lima, M.; Da Silva, M. V. Identification of Races 1, 2 and 3 of *Fusarium oxysporum* f. sp. *lycopersici* by Molecular Markers. *Afr. J. Microbiol. Res.* **2013,** *7*(20), 2324–2331.

Dowd, C.; Wilson, I. W.; Mcfadden, H. Gene Expression Profile Changes in Cotton Root and Hypocotyl Tissues in Response to Infection with *Fusarium oxysporum* f. sp. *vasinfectum. Mol. Plant. Microbe Interact.* **2004,** *17,* 654–667.

Ferrari, S.; Sella, L.; Janni, M.; De Lorenzo, G.; Favaron, F.; D'Ovidio, R. Transgenic Expression of Polygalacturonase-Inhibiting Proteins in *Arabidopsis* and Wheat Increases Resistance to the Flower Pathogen *Fusarium graminearum. Plant Biol.* **2012,** *14,* S31–S38.

Glick, B. R. The Enhancement of Plant Growth by Free-Living Bacteria. *Can. J. Microbiol.* **1995,** *41,* 109–117.

Gordon, T. R.; Martyn, R. D. The Evolutionary Biology of *Fusarium oxysporum. Ann. Rev. Phytopathol.* **1997,** *35,* 11–128.

Gupta, V. K.; Misra, A. K.; Gaur, R. K. Growth Characteristics of *Fusarium* spp. Causing Wilt Disease in *Psidium guajava* L. in India. *J. Plant Prot. Res.* **2010,** *50*(4), 452–462.

Han, J.; Lakshman, D. K.; Galvez, L. C.; Mitra, S.; Baenziger, P. S.; Mitra, A. Transgenic Expression of Lactoferrin Imparts Enhanced Resistance to Head Blight of Wheat Caused by *Fusarium graminearum. BMC Plant Biol.* **2012,** *12,* 33. doi:10.1186/1471-2229-12-33.

Harris, L. J.; Desjardins, A. E.; Plattner, R. D.; Nicholson, P.; Butler, G.; Young, J. C.; Weston, G.; Proctor, R. H.; Hohn, T. M. Possible Role of Trichothecene Mycotoxins in Virulence of *Fusarium graminearum* on Maize. *Plant Dis.* **1999,** *83,* 954–960.

Hayat, R.; Safdar Ali, S.; Amara, U.; Khalid, R.; Ahmed, I. Soil Beneficial Bacteria and their Rolein Plant Growth Promotion: A Review. *Ann. Microbiol.* **2010,** *60,* 579–598.

Jain, A.; Jitendra, M. **2010**. Pathogenicity of *Fusarium oxysporum* and *Meloidogyne incognita* and Cumulative Effects on Tomato. *Ann. Plant Prot. Sci.* **2010,** *18*(1), 34–42.

Jayashree, K.; Shanmugam, V.; Raghuchander, T.; Ramanathan, A.; Samiyappan, R. Evaluation of *Pseudomonas fluorescens* (Pf-1) against Blackgram and Sesame Root Rot Disease. *J. Biol. Control.* **2000,** *14,* 55–61.

Jonathan, E. I.; Rajendran, G. Interaction of *Meloidogyne incognita* and *Fusarium oxysporum* f. sp. *cubense* on Banana. *Nematol. Mediter.* **1998,** *26,* 9–12.

Karimi, K.; Amini, J.; Harighi, B.; Bahramneja, B. Evaluation of Biocontrol Potential of *Pseudomonas* and *Bacillus* Spp. against *Fusarium* Wilt of Chickpea. *Aus. J. Crop Sci.* **2012,** *6*(4), 695–703.

Kazan, K.; Gardiner, D. M.; and Manners, J. M. On the Trail of a Cereal Killer: Recent Advances in *Fusarium graminearum* Pathogenomics and Host Resistance. *Mol. Plant Pathol.* **2012,** *13,* 399–413.

Khan, T. A.; Husain, S. I. Relative Resistance of Six Cowpea Cultivars as Affected by the Concomitance of Two Nematodes and a Fungus. *Nematol. Mediterr.* **1989,** *17,* 39–41.

Kloepper, J. W. Plant-Growth-Promoting Rhizobacteria as Biological Control Agents. In *Soil Microbial Ecology;* Metting, F.B., Jr. Ed.; Marcel Dekker: New York, **1993;** pp 255–273.

Kremer, R. J.; Means, N. E. Glyphosate and Glyphosate-Resistant Crop Interactions with Rhizosphere Microorganisms. *Euro. J. Agron.* **2009,** *31,* 153–161.

Lamondia, J. A. Predisposition of Broadleaf Tobacco to Fusarium Wilt by Early Season Infection with Tobacco Cyst Nematodes. *J. Nematol.* **1992,** *24*(3), 425–431.

Latiffah, Z.; Mohd Zariman, M.; Baharuddin, S. Diversity of *Fusarium* Species Ii Cultivated Soils in Penang. *Malays. J. Microbiol.* **2007,** *3*(1), 27–30.

Lemanceau, P.; Alabouvette, C. Biological Control of *Fusarium* Diseases by Fluorescent Pseudomonads and Non-Pathogenic *Fusarium*. *Crop Protect.* **1991,** *10,* 279–286.

Leslie, S. B.; Israeli, E.; Lighthart, B.; Crowe, J. H.; Crowe, L.M. Trehalose and Sucrose Protect Both Membranes and Proteins in Intact Bacteria during Drying. *Appl. Environ. Microbiol.* **1995,** *61,* 3592–3597.

Li, Z.; Zhou, M.; Zhang, Z.; Ren, L.; Du, L.; Zhang, B. Expression of a Radish Defensin in Transgenic Wheat Confers Increased Resistance to *Fusarium graminearum* and *Rhizoctonia cerealis*. *Funct. Integr. Genomics.* **2011,** *11,* 63–70.

Lin, Y.; Fuping, L.; Shanshan, L.; Xiao, W.; Ruixuan, Z.; Ziqin, L.; Hui, Z. ISSR Analysis of *Fusarium oxysporum* Schl. In Hebei Province. *Procedia. Environ. Sci.* **2012,** *12,* 1237–1242.

M'Piga, P.; Bélanger R. R.; Paulitz, T. C.; Benhamou, N. Increased Resistance to *Fusarium oxysporum* f. sp. *radices lycopersici* In Tomato Plants Treated with the Endophytic Bacterium *Pseudomonas fluorescens* Strain 63–28. *Physiol. Mol. Plant Pathol.* **1997, 50,** 301–320.

Mackintosh, C. A.; Lewis, J.; Radmer, L. E.; Shin, S.; Heinen, S. J.; Smith, L. A. et al. Overexpression of Defense Response Genes in Transgenic Wheat Enhances Resistance to *Fusarium* Head Blight. *Plant Cell Rep.* **2007,** *26,* 479–488.

Mahesh, M.; Saifulla, M.; Sreenivasa, S.; Shashidhar, K. R. Integrated Management of Pigeonpea Wilt Caused by *Fusarium udum* Butler. *Eur. J. Biol. Sci.* **2010,** *2*(1), 1–7.

Makandar, R.; Essig, J. S.; Schapaugh, M. A.; Trick, H. N.; Shah, J. Genetically Engineered Resistance to *Fusarium* Head Blight in Wheat by Expression of *Arabidopsis* NPR1. *Mol. Plant Microbe Interact.* **2006,** *19,* 123–129.

Manikandan, R.; Raguchander, T. *Fusarium oxysporum* f. sp. *lycopersici* Retardation through Induction of Defensive Response in Tomato Plants Using a Liquid Formulation of *Pseudomonas fluorescens* (Pf1). *Eur. J. Plant. Pathol.* **2014,** *140,* 469–480.

Minton, N. A.; Minton, E. B. Infection Relationship between Meloidogyne Incognita Acrita and *Fusarium oxysporum* f. sp. *vasinfectum* in Cotton. *Phytopathology.* **1963,** *53,* 624 (Abstr.)

Mohammed Faisal, P.; Nagendran, K.; Prema Ranjitham, T.; Karthikeyan, G.; Raguchander, T. ;Prabakar, K. Development and Evaluation of Water-In-Oil Formulation of *Pseudomonas fluorescens* (FP7) against *Colletotrichum musae incitant* of Anthracnose Disease in Banana. *Eur. J. Plant. Pathol.* **2014,** *138,* 167–180.

Moss, M. O.; Smith, J. E. *The Applied Mycology of Fusarium;* Cambridge University Press: New York, 1984.

O'Donnell, K.; Ward, T. J.; Geiser, D. M.; Kistler, C.; Aoki, T. Genealogical Concordance between the Mating Type Locus and Seven other Genes Supports Formal Recognition of Nine Phylogenetically Distinct Species within *Fusarium graminearum* Clade. *Fungal Genet. Biol.* **2004,** *41,* 600–623.

Olivain, C.; Alabouvette, C.; Steinberg, C. Production of a Mixed Inoculum of *Fusarium oxysporum* Fo47 and *Pseudomonas fluorescens* C7 to Control *Fusarium* Diseases. *Biocontrol Sci. Technol.* **2004,** *14,* 227–238.

Pérez-Artés, E.; Roncero, M. I. G.; Jiménez-Díaz, R. M. Restriction Fragment Length Polymorphism Analysis of the Mitochondrial DNA of *Fusarium oxysporum* f. sp. *ciceris. J. Phytopathol.* **2008,** *143*(2), 105–109.

Ramamoorthy, V.; Raguchander, T.; Samiyappan, R. Induction of Defense-Related Proteins in Tomato Roots Treated with *Pseudomonas fluorescens* Pf1 and *Fusarium oxysporum* f. sp. *lycopersici. Plant Soil.* **2002,** *239,* 55–68.

Rodriguez-Galvez, E.; Mendgen, K. The Infection Process of *Fusarium oxysporum* in Cotton Root Tips. *Protoplasma.* **1995,** *189,* 61–72.

Saveetha, K. Interactive Genomics and Proteomics of Plant Growth Promoting Rhizobacteria (PGPR) For the Management of Major Pests and Diseases in Rice. Ph.D. (Ag.) Thesis, TNAU, Coimbatore, India, 2009, p 126.

Selvaraj, S.; Ganeshamoorthi, P.; Raguchander, T.; Seenivasan, N.; Anand, T.; Samiyappan, R. Evaluation of a Liquid Formulation of *Pseudomonas fluorescens* Against *Fusarium oxysporum* f. sp. *cubense* and *Helicotylenchus multicinctus* in Banana Plantation. *Biocontrol.* **2014,** *59,* 345–355.

Shanmugam, V.; Kanoujia, N.; Singh, M.; Singh, S.; Prasad, R. Biocontrol of Vascular Wilt and Corm Rot of Gladiolus Caused by *Fusarium oxysporum* f. sp. *gladioli* Using Plant Growth Promoting Rhizobacterial Mixture. *Crop Protect.* **2011,** *30*(7), 807–813.

Shin, K. H.; Kamal, A. H. M.; Cho, K.; Choi, J. S.; Jin, Y.; Paek, N. C. Defense Proteins are Induced in Wheat Spikes Exposed to *Fusarium graminearum. Plant Omics.* **2011,** *4,* 270–277.

Singh, B. P.; Saikia. R.; Yadav, M.; Singh, R.; Chauhan, V. S.; Arora, D. K. Molecular Characterization of *Fusarium oxysporum* f. sp. *ciceri* Causing Wilt of Chickpea. *Afr. J. Biotechnol.* **2006,** *5,* 497–502.

Srinivasan, K.; Gilardi, G.; Spadaro, D.; Gullino, M. L.; Garibaldi, A. Molecular Characterization through IGS Sequencing of Formae Speciales of *Fusarium oxysporum* Pathogenic on Lamb's Lettuce. *Phytopathol. Mediterr.* **2010,** *49,* 309–320.

Sundaramoorthy, S.; Raguchander, T.; Ragupathi, N.; Samiyappan, R. Combinatorial Effect of Endophytic and Plant-Growth Promoting Rhizobacteria against Wilt Disease of *Capsicum annum* L. Caused by *Fusarium solani. Biol. Control.* **2012,** *60,* 59–67.

Tanyolaç, B.; Akkale, C. Screening of Resistance Genes to Fusarium Root Rot and Fusarium Wilt Diseases in F3 Family Lines of Tomato (*Lycopersicon esculentum*) Using RAPD and Caps Markers. *Afr. J. Biotechnol.* **2010,** *9,* 2727–2730.

Vakalounakis, D. J. Root and Stem Rot of Cucumber Caused by *Fusarium oxysporum* f. sp. *radicis-cucumerinum* f. sp. Nov. *Plant Dis.* **1996,** *80,* 313–316.

Vivekananthan, R.; Ravi, M.; Ramanathan, A.; Samiyappan, R. Lytic Enzymes Induced by *Pseudomonas fluorescens* and other Biocontrol Organisms Mediate Defence against the Anthracnose Pathogen in Mango. *World J. Microbiol. Biotechnol.* **2004,** *20,* 235–244.

Volpi, C.; Janni, M.; Lionetti, V.; Bellincampi, D.; Favaron, F.; D'Ovidio, R. The Ectopic Expression of a Pectin Methyl Esterase Inhibitor Increases Pectin Methyl Esterification and Limits Fungal Diseases in Wheat. *Mol. Plant Microbe Interact.* **2011,** *24,* 1012–1019.

Wang, Y.; Ohara, Y.; Nakayashiki, H.; Tosa, Y.; Mayama, S. Microarray Analysis of the Gene Expression Profile Induced by the Endophytic Plant Growth-Promoting Rhizobacteria, *Pseudo- monas fluorescens* FPT9601-T5 in *Arabidopsis. Mol. Plant Microbe Interact.* **2005,** *18,* 385–396.

Weller, D. M.; Raaijmakers, J. M.; Gardener, B. B. M.; Thomashow, L. S. Microbial Populations Responsible for Specific Soil Suppressiveness to Plant Pathogens. *Ann. Rev. Phytopathol.* **2002,** *40,* 309–348.

Zhang, H.; Kim, M. S.; Krishnamachari, V.; Payton, P.; Sun, Y.; Grimson, M.; Farag, M. A.; Ryu.; C. M., Allen, R.; Melo, I. S.; Pare, P. W. Rhizobacterial Volatile Emissions Regulate Auxin Homeostasis and Cell Expansion in *Arabidopsis. Planta.* **2007,** *226,* 839–851.

Zhang, Z.; Zhang, J.; Wang, Y.; Zheng, X. Molecular Detection Of *Fusarium oxysporum* f. sp. *niveum* and *Mycosphaerella melonis* In Infected Plant Tissues And Soils. *FEMS Microbiol. Lett.* **2005,** *249,* 39–46.

CHAPTER 16

DIVERSITY AND VARIATION IN MAJOR INSECT TRANSMITTED VIRUSES INFECTING VARIOUS CROPS

MOHAMMAD ANSAR

Department of Plant Pathology, Bihar Agricultural University, Sabour 813210, India

Corresponding author: ansar.pantversity@gmail.com

CONTENTS

ABSTRACT

Increasing numbers of review published over the last 15 years in plant virus evolution. In modern era, research in virus evolution has been view from a molecular, to a certain extent than population point of view, There is a requirement for work focusing at process associated in evolution. In evolutionary biology, recognizing the processes that establish the evolution of pathogens is an important part with outcomes viral disease for the management. The genetic exchange *viz.*, recombination, and re-assortment of genomic segments and mutation are the ways that incite evolution. However, the genetic construct of the virus population will depend upon evolutionary stress as selection and unintended genetic drift. The largest section of viruses infecting plants are RNA-viruses, they have greater mutation rates, in fact high rates of recombination as well. The recombination events have been shown to act substantial role in the speciation of plant viruses, predominantly in few taxa. Information provided by the genetic structure analysis of viral populations, figure out the genetic stability of most RNA viruses of plant in spite of their high potential to variation. Assessment of nucleotide sequence data from isolates of various origins has permitted the analysis of the genetic structure for several plant virus species over most of their geographic series. In series of evolution in viral pathogen and adoptability to a geographic area interaction between virus and vector always played a significant role. By virtue of Coexistence, virus and vector population achieves greater survival and dispersal ability. This depends upon the condition in which their genetic and physiological construct from a well-suited vector–virus association. As a result, coevolution between insect-vector and viruses must significantly affect the variability, epidemics, and even emergence of new viruses.

16.1   INTRODUCTION

Studies of plant virus comprehension may not be adequate for explaining evolution in nature. Anthropogenic actions such as crop domestication, long distance movement, and natural habitats exploitation, appear to influence plant virus spread and evolution in cultivated conditions. Small information is known about plant virus diversity, host specificity, and evolution under natural environment where human influence is limited. Our main goal under this study is to enhance the knowledge of diversity and evolution of plant viruses in natural conditions. We address the goal with specific themes that include the evolution in viruses of plants, with reference to infecting some native plants and the phylogenetic relationship. The plant viral populations are genetically diverse in nature. In all living entity, reproduction may result in the creation of individuals that differ genetically from their parents; that known as mutants or variants. In the population of an organism, frequency distribution of genetic variation may change with time, and this progression is called evolution. The study of the genetic structure and evolution in populations is a vital area of biology. In fact the plant viral pathogens, acquaintance of their evolution are essential to the expansion of efficient and sustainable control strategies. However, these commonly not succeed due to evolution of the viral population; defeating of resistance genes by resistance-breaking viruses is the most evident example. The introduction of diagnostic techniques that allows rapid determination of nucleotide sequences, and required to explore the possibilities and risks of new control strategies, like virus-resistant transgenic crops. The matter has lead to a recent increase in the number of publications focusing on variability and evolution of plant viruses, and reviews on related subjects (García-Arenal et al., 2001; Gibbs et al., 1999; Roossinck, 1997).

16.2 GENETIC VARIATION IN VIRAL PATHOGEN

Viral genomes are the highest evolving entities in biological system, mainly because of their small replication time and the large extent of progeny released per infected cell. Evolution occurs by several means: Random mutation, recombination, re-assortment, gene amplification/reduction; and results in quasispecies and altered interfering genomes. Virus genomes depicted a higher mutation rate than cellular organisms. Certainly, viral polymerases have a higher mutation rate than cellular polymerases, mostly are RNA dependent RNA polymerases. Furthermore, virus replication process often does not support cellular compensation mechanisms based on double strand repair. This high mutation rate makes often more deficient than infectious genomes, but it allows to evolve rapidly through natural selection.

Generation of genetic variation by mistake happened during the genome replication of viruses. Basically two prime errors illustrated so far (a) mutation and (b) recombination.

Mutation is the progression that creates differences between nucleotides sequence incorporated into the progeny strand during nucleic acid replication. Initial source of variation in populations through mutation, therefore, the interest in approximated rate taking place (Drake et al., 1998). The differentiation between mutation rate and observed mutation frequency in the analyzed population, both variables may differ greatly, as an unfamiliar fraction of the produced mutants is deleterious and will be abolished from the population by assortment. According to the replication strategy and previous background of the virus, relationship between mutation frequency and mutation rate may also be varied (Drake et al., 1998). An estimated rate of mutation for lytic RNA viruses infecting mammals are in the range of 10^{-4}–10^{-5} miss merge per nucleotide per replication round, resulting about one error (0.76, 96% confidence interval 0.18–1.07) per genome per replication cycle (Drake & Holland, 1999). In a recent estimation based on the detection of mutants lethal for cell-to-cell movement of RNA virus *Tobacco mosaic virus* (TMV), provides an estimate of the mutation rate per genome of 0.10–0.13, which is the same order of degree as estimates for the lytic RNA viruses (Malpica et al., 2002). A large portion (69%) of the mutations was additions and deletions, and half of them involved from three to several bases. A ratio of base replacement to insertions and deletions had only been explained earlier for a *Retrovirus* (Pathak & Temin, 1990). An additional outstanding characteristic is a large fraction of the mutants (35%) multiple mutants, a characteristic not common homology with any other reported mutational range. The data explain that majority of mutations in TMV, and probably in other plant RNA viruses, are not of an adaptive character, and supported the view that high mutation rates of RNA viruses are owing to the necessitate for rapid replication of their chemically unstable RNA-genome to a certain extent than an evolutionary approach (Drake & Holland, 1999). The rate of mutation for ds-RNA phage ø 6 similar to that of TMV (Chao et al., 2002), and this might also with genomes of ds-RNA plant viruses. Mutation rates an array of magnitude lesser for retroviruses than for RNA viruses. The viruses having large ds-DNA genomes, mutation rates per genome are around 0.003 per replication round (Drake et al., 1998). If these values applied to the small ss-DNA plant viruses, the consequences are not known, for which no approximation of mutation rate is present.

Recombination is the process by which segment of genetic information switched between the nucleotide strands of different genetic variants during the process of replication. Therefore, recombination results exchange of genetic information. Populations of various RNA and DNA plant virus sequence analysis provide evidence that recombination may be a great source of evolutionary variations, and it might be predominantly essential for certain virus groups (García-Arenal et al., 2001). Recombination may result in vivid changes in the biological property of the virus at the population level, with major epidemiological events as well as emergence of resistance-breaking strains or the attaining of broader host range (Legg & Thresh,

2000; Monci et al., 2002). The major concerns of gene flow from transgenic plants with pathogen-derived resistances to viruses and on to virus populations have resulted in new approach to analyze the system of recombination and their response in evolution (Tepfer, 2002). In RNA viruses, recombination frequencies depend on the level of sequence homology between the sequences, the distance among the markers used to detect the recombinants, and the presence of hot spots recombination can be as high as mutation rates (Legg & Thresh, 2000). Re-assortment of genomic segments in viruses with genome segmentation may by the genetic exchange, a process also called pseudo-recombination. Re-assortment takes place in natural plant virus populations, and it may take a part as major function in virus evolution (White et al., 1995). The impact of re-assortment on virus biology may be staged, although its function in the genetic structure of virus populations has not been well investigated. Mainly indication is for selection against re-assortants, and for co-adaptation of genomic segments (Garcia-Arenal et al., 2001). In a different source of genetic variation, involvement with new nucleic acid molecules may change the viral pathogenicity and be another source of variation. These events may happen from associations between viruses, and among the viruses and satellites, which are frequent in phytoviruses.

Examples: The presence of satellite RNAs (sat-RNAs) of *Cucumber Mosaic Virus* (*CMV*) strains that causes systemic necrosis in tomato (Garcia-Arenal & Palukaitis, 1999), or the role of *Groundnut Rosette Virus* sat-RNA on the infection process and epidemiology of groundnut rosette disease (Naidu et al., 1999).

Diverse grouping of tomato infecting Brazilian begomoviruses having greater occurrence of coinfection, there is a possibility of recombination between diverse tomato infecting begomovirus species. Resulting recombinant strains might presently be causative to the novel genotypes. This report describes the genetic diversity and recombination between tomato begomoviruses. It is likely ($p = 4.7 \times 10^{-30}$) the isolate DF-BR3 a recombinant (Fig. 16.1a). Although it is apparent that nucleotides −137 to 364 (position 0 at the origin of virion strand replication) originated from a sequence resembling isolate Pa-05 (black region and right tree in Figure 16.1 (b), the remaining of sequence originate from a virus closely similar to isolate DFM (white region and left tree). In parental and recombinant viruses the genetic distances are very small (Fig. 16.1b), representing that the recombination result has occur quite recently. The recombination events obviously (p values $< 10^{-4}$) also detected in four other sequences: TMoLCV, ToCMoV, ToRMV, and TGMV. As expected from the distance analyses of DFM and

TMoLCV, it is clear that part of the replication-associated protein (Rep) and intergenic region (IR) of TMoLCV has a recombinant origin ($p = 1.6 \times 10^{-10}$) with DFM similar to a parental sequence (Alice Kazuko et al., 2006).

Isolate	Rep	IR	CP	Region	Minor parent (Black)	Major parent (White)	Methods	p-value
DF3–BR3				-137/+364	Pa5	DFM	RGBMCS	4.7×10^{-30}
TMoLCV				-528/-28	Unknown	DFM	RGBMCS	1.6×10^{-10}
ToCMoV				-589/-261	DFM	Unknown	GBMCS	3.2×10^{-5}
				-140/+39	DFM	Unknown	RGBMCS	3.0×10^{-9}
ToRMV				+89/+435	ToCMoV	ToSRV	RGBMCS	7.2×10^{-26}
TGMV				-794/-469	Unknown	ToRMV	MC	7.1×10^{-5}

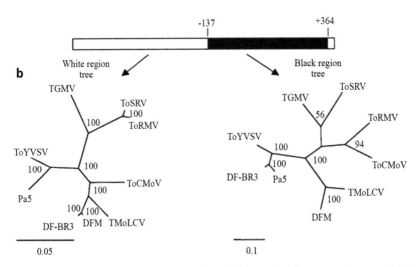

FIGURE 16.1 (a) Schematic representation of the analyzed genome fragment is shown above the recombinant region maps with the positions of partial Rep, partial CP and complete IR sequences indicated. (b) Phylogenetic evidence that isolate DF-BR3 is a recombinant virus. While nucleotides between –137 and +364 of the DF-BR3 genome are derived from a virus closely related to PA-05, the remainder of the analyzed DF-BR3 sequence most closely resembles that of DFM. Numbers associated with branches indicate the percentage of 1,000 bootstrap replicates supporting the existence of that branch in the phylogeny (Source: Pesq. agropec. bras. Brasília, v.41, n.8 (2006)).

16.3 GENETIC VARIATION AMONG PLANT VIRUSES: ANALYSIS AND FACTS

To analyze the genetic variation of phytoviruses, different approach may be used. The variants characterized primarily by differences in biological properties such as the expression of symptoms they caused in different plants, their host series (Fig. 16.2), or possessions of vector transmission as presented in Figure 16.3. The techniques developed to allow the characterization of properties of the virus other than those associated to its relationship with the host and vectors resulted in a dramatic change of outlook. These systems are often more perceptive and reproducible, and permit the typification of more isolates than bioassays. Moreover, these techniques allow the typification of characters that could be neutral; therefore, they would be suitable to analyze the genetic construction of viral populations.

FIGURE 16.2 Symptomatic diversity in plants incited by whitefly transmitted begomoviruses.

Techniques that allowed characterization of viral structural protein(s), like electrophoretic separation of virions and coat protein(s) (CP) subunits, peptide mapping, amino acid composition and sequence analysis of the

CP, and its immunology, as well as monoclonal antibodies and epitope mapping. At present most likely, molecular skills allowing the analysis of the virus genome became accessible. The preference of a given analytical procedure should depend on the object of the analysis, sensitivity and cost of the techniques as well. On the other hand, differences between techniques that provide only qualitative data that is, to facilitate which can be used to recognize variants, and individuals that give information that can be used to quantify how much different the identified variants, that is, to genetic distance estimate. The estimation of genetic distances can originate from data on the amino acid composition of the viral proteins, or by using polyclonal and monoclonal antibodies from serological comparisons (Van Regenmortel, 1982). Genetic distances can also be estimated from restriction fragment length polymorphism (RFLP) analyses (Nei & Tajima, 1981) and ribonuclease T1 finger print (Moya, 1993). The nucleotide sequences analyses of viral genes provide the complete data both to identify genetic variants and to estimate the genetic distance among them. The evidence for genetic variation of viruses, isolates causing different symptoms possibly will get from the similar virus source. It was earlier observed, change in virus traits by serial passages in different host plant species that called host adaptation (Yarwood, 1979). The adaptation, construe as the selection of present variants in the original population of virus or newly generate by it. Biological cloning by single-lesion passage did not abolish heterogeneity in TMV showed by molecular analysis, the same as newly developed variants could take place by mutation (Garcia-Arenal et al., 1984). The analysis of populations takes by the multiplication of inocula resulting from biologically active cDNA clones, RNA virus population heterogeneity showed further (Aldahoud et al., 1989; Ambros et al., 1998; Kurath &Palukaitis, 1990). Therefore, mutants evolve during virus multiplication from biologically or molecularly cloned inocula shown in early and recent works, and laboratory virus isolate always a heterogeneous population included different variants. Such population structure first illustrated in *Tobacco mild green mosaic virus* (*TMGMV*) (Rodrguez-Cerezo & Garcia-Arenal, 1989), typically consists of a main genotype in addition a set of small variants newly generated by mutation. It had been reported earlier for RNA viruses of bacteria and animal in respect of genetic structure, which documented a quasi species and had been linked with the RNA-dependent RNA polymerases error rates.

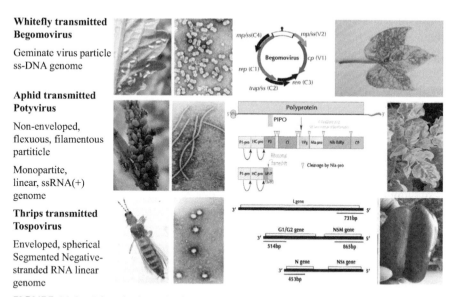

Whitefly transmitted Begomovirus

Geminate virus particle
ss-DNA genome

Aphid transmitted Potyvirus

Non-enveloped, flexuous, filamentous partiticle

Monopartite, linear, ssRNA(+) genome

Thrips transmitted Tospovirus

Enveloped, spherical
Segmented Negative-stranded RNA linear genome

FIGURE 16.3   Diversity in major insect-vector transmitted virus: Genomic organization, associated vector and symptomatic diversity.

16.4   DIVERSITY IN MAJOR VECTOR BORNE VIRUS DISEASES

16.4.1   *WHITEFLY TRANSMITTED BEGOMOVIRUS*

Bean golden mosaic virus, begomoviruses under genus *Begomovirus*, family Geminiviridae, represents a cluster of plant viruses that make use of gene flow provide by recombination (Chatchawankanphanich & Maxwell, 2002; Monci et al., 2002; Padidam et al., 1999; Preiss & Jeske, 2003; Zhou et al., 1997). The virus transmitted in nature by the whitefly *Bemisia tabaci* (Hemiptera: Aleyrodidae) and have geminate icosahedral virions that encapsidated circular single-stranded (ss) DNA genomes (Stanley et al., 2005). Begomoviruses have generally bipartite genomes consisting two DNA components, DNA-A and DNA-B. The DNA-A encodes a Rep, the coat protein (CP), replication enhancer protein (REn), and transcription activator protein (TrAP) that contributes to the regulation and replication of gene expression. The DNA-B encodes proteins that are essential for movement of virus in plants. Open reading frames (ORFs) are structured bi-directionally in both genome components, divided by an IR that having key elements for the replication and transcription of the viral genome, as well as origin of replication (Hanley-Bowdoin et al., 2000). Single genomic component that resembles DNA-A, have been reported in several begomoviruses, among them most of the viruses associated with yellow leaf curl disease in tomato (Moriones & Navas-Castillo, 2000).

Repeated mixed infections in begomoviruses from various ancestry make weed plant species central point for the emergence of new viruses by recombination, which is observable fact for evolution and speciation in Geminiviridae family (Seal et al., 2006). Begomoviruses have emerged worldwide during the last two decades, as a consequence of the insect vector *B. tabaci* spread (Rybicki & Pietersen, 1999), causing infection to a wide-ranging plant species, among some of the agriculturally important crops (Chatchawankanphanich & Maxwell, 2002; Czosnek & Laterrot, 1997; Ribeiro et al., 2003; Stanley et al., 2005; Stonor et al., 2003). Infection reported in early 1990s, associated with the incidence of *Tomato yellow leaf curl Sardinia virus* (TYLCSV). It was later beginning of *Tomato yellow leaf curl virus* (TYLCV) strains, which provided the substrate for contacts, and extend to new host species (Morilla and co-workers, 2003, 2005; Navas-Castillo et al., 1999). Reservoir plants can play a vital role in the emergence of plant virus epidemics by acting of native species (Hull, 2002). Investigation to know the genetic structure and dynamics of *Begomovirus* populations in wild species and potential effects on epidemics of cultivated species are limited and have less information (Frischmuth et al., 1997; Jovel et al., 2004; Ooi et al., 1997; Roye et al., 1997; Sanz et al., 2000).

In an earlier report of Phaneendra et al. (2012), *Tomato leaf curl New Delhi virus* (ToLCNDV) associated with the leaf curl of pumpkin (*Cucurbita moschata*) in Northern India. The disease characterized by coat protein specific primers, indicated the association of a begomovirus with the disease. The sequence comparison and phylogenetic analysis of the entire DNA genome revealed the uniqueness of the virus as ToLCNDV. Sequence similarity analyses were performed by comparing the sequence of other begomoviruses in the database (http://www.ncbi.nlm.nih.gov/nucleotide). Total of 33 isolates of tomato and cucurbits infecting begomoviruses used, the complete DNA-A sequence shared maximum nucleotide sequence identity of 98.1% with ToLCNDV-Tai; (DQ 169056). The highest sequence identity of IR had 99.6% with ToLNDV-Svr (U15015). In comparison between individual ORFs encoding proteins, they found 94.8–99.2% sequence identity with the isolate of ToLCNDV-ND. Based on the complete nucleotide sequences of DNA-A, the phylogenetic tree of virus isolate from pumpkin clustered with 12 ToLCNDV isolates as presented in Figure 16.4.

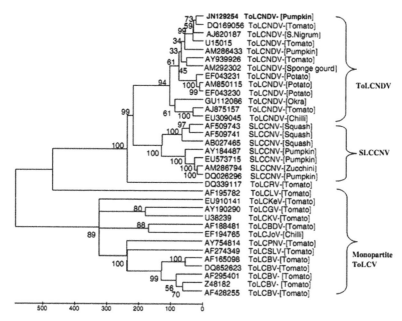

FIGURE 16.4 Parsimonious tree of ToLCNDV-[Pum: IARI: 06] and other related begomovirus isolates based on complete DNA-A sequence. The begomovirus acronyms used are:

ToLCNDV: *Tomato leaf curl New Delhi virus,* ToLCRV: *Tomato leaf curl Rajasthan virus,* ToLCGV: *Tomato leaf curl Gujarat virus,* ToLCBV: *Tomato leaf curl Bangalore virus,* ToLCKV: *Tomato leaf curl Karnataka virus,* ToLCBDV: *Tomato leaf curl Bangladesh virus,* ToLCLV: *Tomato leaf curl Laos virus,* ToLCSLV: *Tomato leaf curl Sri Lanka virus,* ToLCPNV: *Tomato leaf curl Pune virus,* ToLCKeV: *Tomato leaf curl Kerala virus,* ToLCJoV: *Tomato leaf curl Joydebpur virus,* SLCCNV: *Squash leaf curl China virus* (Source: Chigurupati Phaneendra, "Tomato leaf curl New Dehli virus is associated with pumpkin leaf curl: A new disease in Northern India," in Indian J. Virol. 23(1):42–45, 2016. With permission from Springer.)

The geminivirus biodiversity, inciting cassava mosaic in India explores using PCR to amplify the specific gene of *Indian cassava mosaic virus* (ICMV) and *Sri Lankan cassava mosaic virus* (SLCMV). The results depict both ICMV and SLCMV found in mosaic affected cassava; ICMV geographically confined to certain regions, while SLCMV present extensively. In PCR-RFLP analysis, a high 40% share of the samples exhibit novel patterns other than ICMV-type and SLCMV-type patterns. Another diverse geminivirus infecting cassava documented from Sri Lanka, refers to SLCMV, which had greatly lower sequence similarity to ICMV. SLCMV having monopartite genome, which apparently incarcerate the DNA-B of ICMV followed the occurrence of recombination (Saunders et al., 2002). Sequence analysis of cassava-infecting geminivirus DNAs from India showed some of them to be SLCMV and a few to be ICMV. In relative field distribution of ICMV and SLCMV, 38% showed the presence of ICMV alone, 55% showed the presence of SLCMV alone, and 7% showed the presence of both the viruses. In southern districts of Kerala like, Kollam, Pathanamthitta, and Thiruvananthapuram, only SLCMV was detected, but in the northern districts (Kozhikode and Malappuram) ICMV was predominant, however, with slight occurrence of SLCMV. In central districts of the state like Ernakulam, Kottayam, and Idukki, both the viruses were reported, SLCMV being widespread (Patil et al., 2005). Both the viruses were present to same extent in Tamil Nadu state. Detection of mixed infections with both the viruses was also recorded in a few samples from the state. Observation of mosaic disease confined in few plots in Andhra Pradesh and analysis of samples pointed out the occurrence of SLCMV solely (Fig. 16.5).

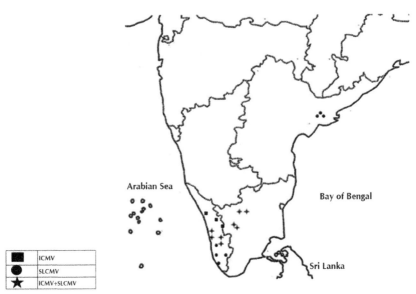

FIGURE 16.5 Diversity of cassava mosaic virus in Southern part of India indicating the presence of ICMV, SLCMV, and dual infection at various regions.

Among most devastating viral diseases of tomato, leaf curl is important that affects crop globally in tropical and temperate regions (Pend et al., 2010; Hanssen et al., 2010; Moriones & Navas-Castillo, 2000). The disease is caused by a complex of viral species under *Begomovirus* genus, and majorities are in monopartite. The first report of disease in the Jordan Valley (1939) and large populations of whiteflies which influence widely

(Avidov, 1944). The big hurdle in tomato production in the Middle East was the spread of TYLCV since the 1970s (Hanssen et al., 2010). The virus was reported in new areas including, Mediterranean Basin, the Far East, the Caribbean, North America, and Australia since mid 1990s (Navas-Castillo et al., 2011). In Italy during late 1980s, another *Begomovirus*; *Tomato yellow leaf curl Sardinia virus* (TYLCSV), causing TYLCD (Credi et al., 1989) was detected, and after shortest period of time, was found in Spain (Moriones et al., 1993). Since then, TYLCSV caused important epidemics in tomato crop in both countries. Mixed infection of TYLCSV and TYLCV led to the emergence of two recombinant viruses, *Tomato yellow leaf curl Malaga virus* (TYLCMaV) (Monci et al., 2002) and *Tomato yellow leaf curl Axarquia virus* (TYLCAxV) (Garcia-Andres et al., 2006). These recombinant viruses possess broader host range than either parent virus. The parental TYLCV quickly displaced by the TYLCMaV and caused epidemics in Southern Spain (Garcia-Andres et al., 2007; Monci et al., 2002; Navas-Castillo et al., 1999). The TYLCSV restricted to Mediterranean countries, however, TYLCV had wide reach (Fig. 16.6). With the conformity of historical records, the phylogeographic analysis advocated that TYLCV most likely came to passes in the Middle East in 1930s–1950s, and spread of evolved Israel and mild strains of TYLCV at continental level began in the 1980s.

Tomato leaf curl disease (ToLCD) is a common disease found in tomato all over India. ToLCD was first reported in Northern India by Vasudeva and Sam Raj (1948), and subsequently from Central India (Varma, 1959) and Southern India (Govindu, 1964; Sastry & Singh, 1973). Several closely related begomoviruses associated with this disease have been cloned and sequenced from India. Two isolates from New Delhi, *Tomato leaf curl New Delhi virus*-Severe (ToLCNDV-Severe) and *Tomato leaf curl New Delhi virus*-Mild (TOLCNDV-Mild), with bipartite genomes, sharing 94% identity in the DNA-A component (with identical DNA-B components), were reported by Padidam et al. (1999), both of which contained an extra ORF (AV3). Four additional isolates, three from Bangalore (Chatchawankan-phanich et al., 1993) and one from Lucknow (Srivastava et al., 1995) were also reported, which indicated that those from North India had a bipartite genome whereas those from South had monopartite genome (like the ones reported from Australia and Taiwan; Muniyappa et al., 2000). DNA sequences of two more isolates, ToLCV-Ban-2 and ToLCV-Ban-4 were reported from Bangalore, sharing 91% identity with each other. In addition, genomic sequences of two additional isolates, which were named *Tomato leaf curl Bangalore virus*, ToLCBV reported subsequently (Kirthi et al., 2002). Later, a new *Begomovirus Tomato leaf curl Gujarat virus* (ToLCGV) was reported from Varanasi, North India and shown to be infectious on the natural host (Chakraborty et al., 2003).

FIGURE 16.6 Geographical distribution of whitefly-transmitted viruses causing diseases in tomato. Circle a, c, e, and f represent the presence of *Tomato yellow leaf curl virus*. Circle b stand for *Tomato yellow leaf curl Sardinia virus. Tomato leaf curl virus* shown in circle d.

16.4.2 *APHID TRANSMITTED POTYVIRUS*

Potyvirus is a genus of viruses in the family *Potyviridae*. There are currently 158 species in this genus including the type species *Potato virus Y* (ICTV, 2015). Potyviruses reported in ~30% of the currently known plant viruses. Similar to begomoviruses, associated virus may cause major losses in agricultural and horticultural crops. The aphids species (> 200) transmit the potyviruses and mainly from the subfamily *Aphidinae* (genera *Macrosiphum* and *Myzus*). The viruses are non-enveloped, flexuous, filamentous 720–850 nm long and 12–15 nm in diameter. Monopartite with linear, ss-RNA(+) genome of 10 kb in size. The 3' terminus has a poly (A) tract and 5' terminus has a genome-linked protein (VPg). The virion RNA is infectious and provides as both the genome and viral messenger RNA. The genomic RNA is encoded into polyproteins which are subsequently processed by the action of three viral-encoded proteinases into functional products. The expression of P3N-PIPO by a -1 ribosomal frameshifting from the P3 ORF and most likely acts as a movement protein. (Vijayapalani et al., 2012) Approximately, more than 4000 potyvirus sequences are available in database of Genbank. Most of them are from the 3-terminal region of the genome that encodes the CP gene. The 59-terminal region of the CP gene is often repetitive, however, that encoding the core and C-terminus of the coat protein gene is not, and appears to have evolved in a consistent hierarchical approach by point mutations and by homologous recombination intermittently (Ward et al., 1995). The trees having closely related topologies were found from every regions of the *Potyvirus* genome, excluding variable 59-terminal region and the 59-terminal region of the coat protein gene with maximum likelihood, parsimony, or distance methods. When the genome was alienated by windows of fixed length rather than gene by gene, topology did not alter (Adams et al., 2005). Phylogenies calculated from coherently evolving coat protein (c-CP) region of 227 sequences, corresponding to all genera of the potyvirids, set all potyviruses in a dense cluster having amazingly uniform radial branch length. The cluster of potyviruses consistently linked in all taxonomies by a small branch to the *Rymovirus* cluster. The rymoviruses were associated by much longer branches to the other genera of *Potyvirid* as presented in Figure 16.7. A phylogenetic study found that

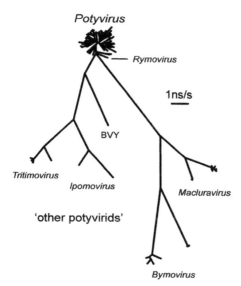

FIGURE 16.7 Unrooted phylogenetic tree showing the relationships of the cCP gene sequences of 227 potyvirids including 183 potyviruses. The 'other potyvirids' include species of *Macluravirus* (maclura mosaic, *Narcissus* latent, and *Cardamom mosaic viruses*), *Bymovirus* (barley mild mosaic, barley yellow mosaic, oat mosaic, wheat spindle streak mosaic and wheat yellow mosaic viruses), *Tritimovirus* (brome streak mosaic, oat necrotic mottle and wheat streak mosaic viruses), *Ipomovirus* (cucumber vein yellowing and sweet potato mild mottle viruses) and *Rymovirus* (ryegrass mosaic, *Agropyron* mosaic and *Hordeum* mosaic viruses). Blackberry virus Y (BVY) an unassigned potyvirid species. The tree intended from the cCP regions using the PhyML program with the HKY+I+G model. (Source: Gibbs et al, "The Prehistory of Potyviruses: Their Initial Radiation Was during the Dawn of Agriculture," 2008, http://journals.plos.org/plosone/article?id=10.1371/journal.pone.0002523.)

radiative deviation of *Potyvirus* occurred around 6,600 years ago that kept in touch with the early development of cultivation (Gibbs et al., 2008). Studies on potyviruses from Australia, which has a of history around two and half centuries in agricultural development, demonstrate that 18 crop infecting potyvirus species less diverse in each other and similar to potyviruses phylogenetically from other part of the world. Nevertheless, *Potyvirus* members found only in Australia have been isolated from either native plants or weeds (Gibbs et al., 2008).

Example: *Hardenbergia mosaic virus,* infecting the native legume (*Hardenbergia comptoniana*), causes no damage but expresses severe symptoms to many cultivated legumes. In the same way, infection of *Passion fruit woodiness virus* in native *Passiflora aurantia* is not damaging. However, cultivated passion fruits (*Passiflora edulis, P. foetida, P. caerulea*) show virulence (Webster et al., 2007).

16.4.3 DIVERSITY AND VARIATION IN LUTEOVIRUSES

Luteovirus is a genus, falls under family Luteoviridae. At present eight species in this genus including the type species viz, Barley yellow dwarf virus-PAV. Several plants serve as natural hosts. The geographical distribution of Luteoviruses is extensive, virus mostly infecting plants via transmission by aphid vectors. The virus replicates within the plant host cell only, and there is no evidence within the vector. The 'luteovirus' come up from the Latin luteus, which means 'yellow'. As a result of infection by the virus, symptomatic yellowing of the plant showed.

The members of Luteoviridae family limited to Brassicaceae, Fabaceae, Poaceae, Solanaceae, and causes severe loss to cultivated crops globally (Hull, 2001). In this group, recombination in the polymerase region of the genome seems to be the main event leading to emergence of new species and virulence evolution in different host taxa (Gibbs & Cooper, 1995). Phylogeny of the group-based analysis of homologous regions, *Pea enation mosaic virus* (Family Luteoviridae, Genus *Enamovirus*) and two additional unassigned members under Luteoviridae family (*Soybean dwarf virus* and *Sugarcane yellow leaf virus*) prove recombination between progenitors associated to *Luteovirus* and *Polerovirus* genera (Moonan et al., 2000). Further, phylogeny of Luteoviridae confirmed that intrafamily and extrafamily recombinations in RNA are most important events causing evolved infectivity in cultivated group members. *Barley yellow dwarf virus* (BYDV) demonstrated a complex trend of virulence in strongly related taxa which reflects diverse adaptive condition in various host plants (Remold, 2002).

16.4.4 THRIPS TRANSMITTED TOSPOVIRUS

Tospoviruses under family *Bunyaviridae*, genus *Tospovirus* are enveloped isometric RNA viruses having a tripartite genome small (S), medium (M), and large (L) segments of ss-RNA. The viruses under the genus are solely transmitted by thrips (Thysanoptera) in a propagative manner. Moreover, it is one of the important plant virus groups infecting important crops extensively across the world (Pappu et al., 2009). Over the last decades, damaging species of thrips have become increasingly shifted in fruit, vegetable, and ornamental crops globally. Incursion of new species into countries or major growing areas, crop production intensification, and the capacity of thrips to develop resistance against systemic insecticides may be the reasons for the same. The transmission ability of several thrips species to the tospoviruses or the tomato spotted wilt group of plant viruses has showed an obvious increase in the world agriculture scenario (Table 16.1). Due to their economic impact on a diverse range of important crops in the Indian subcontinent, a number of the research works have been taken up. However, the diseases caused by tospoviruses are major problems in several other countries of the subcontinent like Bangladesh, Nepal, Pakistan, and Sri Lanka (Raja & Jain, 2006; Pappu et al., 2009; Persley et al., 2006). There is significant limiting factor for the sustainable production due to emergence of diseases caused by tospoviruses. The viruses, *Peanut bud necrosis virus* (PBNV) (Ghanekar et al., 1979; Satyanarayana and co-workers, 1996), *Peanut yellow spot virus* (PYSV) (Satyanarayana and co-workers, 1998), *Watermelon bud necrosis virus* (WBNV) (Jain et al., 1998), *Iris yellow spot virus* (IYSV) (Ravi et al., 2006), and *Capsicum chlorosis virus* (CaCV) (Kunkalikar and co-workers, 2007, 2010) have been recognized. The PBNV, CaCV, and IYSV correspond to top member and PYSV and WBNV are undertaken as tentative species of the genus *Tospovirus* (Fauquet et al., 2005). The viruses like PBNV, WBNV, CaCV, and IYSV were detected in various grown vegetable crops. However, the natural host range of PYSV is not clearly known. The geographical increment of these viruses has been documented in recent years.

TABLE 16.1 Geographical Distribution of Tospovirus Species Infecting Different Host Plants.

Tospovirus species	Geographical distribution	Host plants
Groundnut (peanut) bud necrosis virus (GBNV)	India and South-east Asia	Peanut, other grain legumes, and weed species
Groundnut ringspot virus (GRSV)	South America, South Africa	Peanut , tomato
Impatiens necrotic spot virus (INSV)	USA, West and South Europe, New Zealand, Japan	Ornamentals, peanut, potato, capsicum, weed species
Groundnut Yellow spot virus (GYSV)	India, Thailand	Peanut
Tomato chlorotic spot virus (TCSV)	South America	Tomato, sweet pepper
Tomato spotted wilt virus (TSWV)*	Worldwide	Many hosts among crop, weed and ornamental species
Watermelon silver mottle, virus (WSMoV)	Japan, Taiwan	Watermelon, other cucurbits, tomato
Zucchini lethal chlorosis virus (ZLCV)	Brazil	Zucchini
Capsicum chlorosis virus (CaCV)*	Australia, Thailand, Taiwan, China	Capsicum, tomato, peanut, Hoya (wax flower), gloxinia
Chrysanthemum stem necrosis virus (CNSV)	Brazil	Chrysanthemum
Iris yellow spot virus (IYSV)*	Australia, Brazil, Israel, Japan, The Netherlands, USA	Iris, leek, onion
Melon yellow spot virus (MYSV)	Taiwan, Japan, Thailand	Melon, groundnut
Groundnut chlorotic fan-spot virus	Taiwan	Peanut
Watermelon bud necrosis virus (WBNV)	India	Watermelon
Tomato yellow fruit ring Virus	Iran	Tomato
Calla lily chlorotic spot virus (CCSV)	Taiwan	Calla lilies (*Zantedeschia* spp.)

*Occurrence in Australia

Source: Thrips and Tospovirus: A Management Guide (The State of Queensland, Department of Primary Industries and Fisheries, 2007)

16.4.4.1   CO-EVOLUTION BETWEEN THRIPS AND TOSPOVIRUS

Although, tospoviruses mechanically transmitted under experimental conditions, dispersal, and survivability in nature depend on channel to plants by thrips vectors. The coexistence of virus and vector population derives the survival of TSWV and dispersal ability that is dependent upon a condition in which their genetic and physiological framework forms a well-suited vector–virus interaction. In every phase of the infection cycle, environment and host relations with the virus and the insect are also greatly influenced. Consequently, coevolution among thrips and tospoviruses must significantly sway the observed variability among isolates, epidemics, and even appearance of new tospovirus. Higher error rate inherent in RNA replication by the viral RNA-dependent RNA polymerase (RdRp) (Crotty et al., 2001) influences tremendous genetic variability in *Tospovirus* population and the genomic segment re-assorted between diverse virus isolates in plant kingdom (Qiu et al., 1998). Slight information is available about the genome of thrips, but majority of vector species are characterized by salient morpohological diversity, which implies genetically variable population (Mound, 1996). Undoubtedly tospovirus and thrips population convene the basic necessities for rapid coevolution.

Example: Coevolution between tospovirus and their vector, the altered status of *Thrips tabaci* as a vector of TSWV. This species transmitted all known isolates of TSWV worldwide 40 years back (Sakimura, 1962). However, this vector species appears to be incapable of transmitting new TSWV isolates (Mau et al., 1990; Wijkamp et al., 1995). A novel virus–vector relationship also arose, perhaps as result of coevolutionary actions. For example, *T. palmi* can now transmit newly emergent tospoviruses in potato (Ansar et al., 2015) and *Frankliniella. bispinosa* emerged as vector of TSWV (Webb et al., 1997) while vector of *Irish yellow spot virus*, *T. tabaci* emerged (Gera et al., 1998).

In the formation of these new vector–virus relationship, defined mechanism involved, and reproductive strategies in thrips play significant roles as well (Chatzivassiliou, 2002). Mixed viral infection in many host plants and multiple thrips species provides platform for genetic exchange. The re-assortment of RNA segment form thrips transmissible or resistance-breaking isolates could provide non-transmissible isolates with characteristics of resistance breaking and new transmission behavior. Moreover, indications supported the impression that re-assortment in mixed infection may possibly contribute to the emergence of isolate with a selection of new characteristics as well as emergence of new species of a vector like *T. tabaci*.

KEYWORDS

- adaptation
- diversity
- evolution
- insect-vector
- transmission

REFERENCES

Adams, M. J.; Antoniw, J. F.; Fauquet, C. M. Molecular Criteria for Genus and Species Discrimination Within the Family Potyviridae. *Arch. Virol.* **2005,** *150,* 459–479.

Aldahoud, R.; Dawson, W. O.; Jones, G. E. Rapid, Random Evolution of the Genetic Structure of Replicating Tobacco Mosaic Virus Populations. *Intervirol.* **1989,** *30,* 227–233.

Alice Kazuko Inoue-Nagata, Martin, D. P.; Leonardo, S.; Leonardo De Britto Giordano, Isabel Cristina Bezerra; Antonio Carlos De Ávila. New Species Emergence via Recombination among Isolates of the Brazilian Tomato Infecting Begomovirus Complex. *Pesq. Agropec. Bras. Brasília.* **2006,** *41*(8), 1329–1332.

Ambros, S.; Hernandez, C.; Desvignes, J. C.; Flores, R. Genomic Structure of Three Phenotypically Different Isolates of Peach Latent Mosaic Viroid: Implications of the Existence of Constraints Limiting the Heterogeneity of Viroid Quasi species. *J. Virol.* **1998,** *72,* 7397–7406.

Ansar, M.; Akram, M.; Singh, R. B.; Pundhir, V. S. Epidemiological Studies of Stem Necrosis Disease in Potato Caused by Groundnut Bud Necrosis Virus. *Indian Phytopath.* **2015,** *68*(3), 321–325.

Avidov, H. Z. *Tobacco Whitefly in Israel;* Hassadeh (in Hebrew): Tel Aviv, 1944; pp 1–33.

Chao, L., Rang, C. U.; Wong, L. E. Distribution of Spontaneous Mutants and Inferences about the Replication Mode of the RNA Bacteriophage Ø6. *J. Virol.* **2002,** *76,* 3276–3281.

Chakraborty, S.; Pandey, P. K.; Banerjee, M. K.; Kalloo, G.; Fauquet, C. M. A New Begomovirus Species Causing Tomato Leaf Curl Disease in Varanasi, India. *Plant Dis.* **2003,** *87,* 313.

Chatchawankanphanich, O.; Maxwell, D. P. Tomato Leaf Curl Karnataka Virus from Bangalore, India, Appears to be a Recombinant Begomovirus. *Phytopathol.* **2002,** *92,* 637–645.

Chatzivassiliou, E. K.; Peters, D.; Katis, N. I. The Efficiency by Which Thrips Tabaci Populations Transmit Tomato Spotted Wilt Virus Depends on Their Host Preference and Reproductive Strategy. *Phytopathol.* **2002,** *92,* 603–609.

Credi, R.; Betti, L.; Canova, A. Association of a Geminivirus with a Severe Disease of Tomato in Sicily. *Phytopath. Medit.* **1989,** *28,* 223–226.

Crotty, S.; Cameron, C. E.; Andino, R. RNA Virus Error Catastrophe: Direct Molecular Test by Using Ribavirin. *Proc. Natl. Acad. Sci U.S.A.* **2001,** *98,* 6895–6900.

Czosnek, H.; Laterrot, H. A Worldwide Survey of Tomato Yellow Leaf Curl Viruses. *Arch. Virol.* **1997,** *142,* 1391–1406.

Drake, J. W.; Charlesworth, B.; Charlesworth, D.; Crow, J. F. Rates of Spontaneous Mutation. *Genetics.* **1998,** *148,*1667–1686.

Drake, J. W.; Holland, J. J. Mutation Rates among RNA Viruses. *Proc. Natl. Acad. Sci. U.S.A.* **1999,** *96,* 13910–13913.

Fauquet, C. M.; Mayo, M. A.; Maniloff, J.; Desselberger, U.; Ball, L. A. *Virus Taxonomy, VIIIth Report of ICTV*; Academic Press: New York, 2005; pp 1259.

Frischmuth, T.; Engel, M.; Lauster, S.; Jeske, H. Nucleotide Sequenc Evidence for the Occurrence of Three Distinct Whitefly-Transmitted Sida Infecting Bipartite Geminiviruses in Central America. *J. Gen. Virol.* **1997,** *78,* 2675–2682.

Garcia-Andres, S.; Accotto, G. P.; Navas-Castillo, J.; Moriones, E. Founder Effect, Plant Host, and Recombination Shape the Emergent Population of Begomoviruses that Cause the Tomato Yellow Leaf Curl Disease in the Mediterranean Basin. *Virol.* **2007,** *359,* 302–312.

Garcia-Andres, S.; Monci, F.; Navas-Castillo, J.; Moriones, E. Begomovirus Genetic Diversity in the Native Plant Reservoir *Solanum nigrum*: Evidence for the Presence of a New Virus Species of Recombinant Nature. *Virol.* **2006,** *350,* 433–442.

Garcia-Arenal, F.; Fraile, A.; Malpica, J. M. Variability and Genetic Structure of Plant Virus Populations. *Annu. Rev. Phytopathol.* **2001,** *39,* 157–186.

Garcia-Arenal, F.; Palukaitis, P. Structure and Functional Relationships of Satellite RNAs of Cucumber Mosaic Virus. *Curr. Top. Microbiol. Immunol.* **1999,** *239,* 37–63.

Garcia-Arenal, F.; Palukaitis, P.; Zaitlin, M. Strains and Mutants of Tobacco Mosaic Virus are Both Found in Virus Derived from Single-Lesion-Passaged Inoculum. *Virol.* **1984,** *132,* 131–137.

Gera, A.; Kritzman, A.; Cohen, J.; Raccah, B. Tospovirus Infecting Bulb Crops in Israel. In *Recent Progress in Tospovirus and Thrips Research;* Peters, D., Goldbach, R., Eds.; Wageningen: The Netherlands, 1998; pp 86–87.

Ghanekar, A. M.; Reddy, D. V. R.; Lizuka, N.; Amin, P. W.; Gibbons, R. W. Bud Necrosis of Groundnut (*Arachis hypogaea*) in India Caused by Tomato Spotted Wilt Virus. *Ann. Appl. Biol.* **1979,** *93,* 173–179.

Gibbs, A. J.; Ohshima, K.; Phillips, M. J.; Gibbs, M. J. The Prehistory of Potyviruses: Their Initial Radiation was During the Dawn of Agriculture. *PLoS One.* **2008,** *3,* 1–11.

Gibbs, A. J.; Keese, P. L.; Gibbs, M. J.; Garcia-Arenal, F. Plant Virus Evolution: Past, Present and Future. In *Origin and Evolution of Viruses;* Domingo, E., Webster, R., Holland, J. J., Eds. Academic Press: San Diego, CA, 1999; pp 263–285.

Gibbs, A. J.; Cooper, J. I. A Recombinational Event in the History of Luteoviruses Probably Induced by Base-Paring between Genomes of Two Distinct Viruses. *Virol.* **1995,** *206,* 1129–1132.

Hanley-Bowdoin, L.; Settlage, S. B.; Orozco, B. M.; Nagar, S.; Robertson, D. Geminiviruses: Models for Plant DNA Replication, Transcription, and Cell Cycle Regulation. *Crit. Rev. Biochem. Mol. Biol.* **2000,** *35,* 105–140.

Hanssen, I. M.; Lapidot, M.; Thomma, B. P. Emerging Viral Diseases of Tomato Crops. *Mol. Plant Microbe Interact.* **2010,** *23,* 539–548.

Hull, R. *Matthews Plant Virology*; Academic Press: San Diego, 2001.

Hull, R. *Matthew's Plant Virology*, 4th ed.; Academic Press: San Diego, 2002.

ICTV . Virus Taxonomy: 2014 Release (accessed June 15, 2015).

Jain, R. K.; Pappu, H. R.; Pappu, S. S.; Krishanareddy, M.; Vani, A. Watermelon Bud Necrosis Tospovirus is a Distinct Virus Species Belonging to Serogroup IV. *Arch. Virol.* **1998**, *143*, 1637–1644.

Jovel, J.; Reski, G.; Rothenstein, D.; Ringel, M.; Frischmuth, T.; Jeske, H. Sida Micrantha Mosaic is Associated with a Complex Infection of Begomoviruses Different from Abutilon Mosaic Virus. *Arch. Virol.* **2004**, *149*, 829–841.

Kunkalikar, S. R.; Sudarsana, P.; Rajagopalan, P.; Zehr, U. B.; Ravi, K. S. Biological and Molecular Characterization of Capsicum Chlorosis Virus Infecting Chili and Tomato in India. *Arch. Virol.* **2010**, *155*, 1047–1057.

Kunkalikar, S. R.; Sudarsana, P.; Rajagopalan, P.; Zehr, U. B.; Naidu, R. A.; Ravi, K. S. First Report of Capsicum Chlorosis Virus in Tomato in India. *Plant Health Prog.* doi: 10.1094/PHP-2007-1204-01-BR.

Kurath, G.; Palukaitis, P. Serial Passage of Infectious Transcripts of a Cucumber Mosaic Virus Satellite RNA Clone Results in Sequence Heterogeneity. *Virol.* **1990**, *173*, 8–15.

Legg, J.; Thresh, J. M. Cassava Mosaic Virus Disease in East Africa: A Dynamic Disease in a Changing Environment. *Virus Res.* **2000**, *71*, 135–149.

Malpica, J. M.; Fraile, A.; Moreno, I.; Obies, C. I.; Drake, J. W.; García Arenal, F. The Rate and Character of Spontaneous Mutation in an RNA Virus. *Genetics.* **2002**, *162*, 1505–1511.

Mau, R. F. L.; Bauista, R.; Cho, M. M.; Ullman, D. E.; Gusukuma, L. Factors Affecting the Epidemiology of TSWV in Field Crops: Comparative Virus Acquisition Efficiency of Vectors and Suitability of Alternate Host to *Frankliniella occidentalis* (Pergande). In *Virus-Thrips-Plant Interaction of Tomato Spotted Wilt Virus,* Proceedings of United States Department of Agriculture Workshop 87, April 18–19, 1990; ARS: Beltsville, MD, 1990; pp 21–27.

Monci, F.; Sanchez-Campos, S.; Navas-Castillo, J.; Moriones, E. A Natural Recombinant Between the Geminiviruses Tomato Yellow Leaf Curl Sardinia Virus and Tomato Yellow Leaf Curl Virus Exhibits a Novel Pathogenic Phenotype and is Becoming Prevalent in Spanish Populations. *Virol.* **2002**, *303*, 317–326.

Moonan, F.; Molina, J.; Mirkov, T. E. Sugarcane Yellow Leaf Virus: An Emerging Virus that has Evolved by Recombination Between Luteoviral and Poleroviral Ancestors. *Virol.* **2000**, *269*, 156–171.

Morilla, G.; Antúnez, C.; Bejarano, E. R.; Janssen, D.; Cuadrado, I. M. A New Tomato Yellow Leaf Curl Virus Strain in Southern Spain. *Plant Dis.* **2003**, *87*, 1004.

Morilla, G.; Janssen, D.; García-Andrés, S.; Moriones, E.; Cuadrado, I. M.; Bejarano, E. R. Pepper (Capsicum Annuum), is a Dead-End Host for Tomato Yellow Leaf Curl Virus (TYLCV). *Phytopathol.* **2005**, *95*, 1089–1097.

Moriones, E.; Navas-Castillo, J. Tomato Yellow Leaf Curl Virus, an Emerging Virus Complex Causing Epidemics Worldwide. *Virus Res.* **2000**, *71*, 123–134.

Moriones, E.; Arno, J.; Accotto, G. P.; Noris, E.; Cavallarin, L. First Report of Tomato Yellow Leaf Curl Virus in Spain. *Plant Dis.* **1993**, *77*, 953.

Mound, L. A. The Thysanoptera Vector Species of Tospoviruses. *Acta. Hortic.* **1996**, *431*, 298–309.

Moya, A.; Rodríguez-Cerezo, E.; García-Arenal, F. Genetic Structure of Natural Populations of the Plant RNA Virus Tobacco Mild Green Mosaic Virus. *Mol. Biol. Evol.* **1993**, *10*, 449–456.

Muniyappa, V.; Venkatesh, H. M.; Ramappa, H. K.; Kulkarni, R. S.; Zeidan, M.; Tarba, C. Y.; Ghanim, M.; Czosnek H. Tomato Leaf Curl Virus from Bangalore (Tolcv-Ban4): Sequence

Comparison with Indian to LCV Isolates, Detection in Plants and Insects, and Vector Relationships. *Arch. Virol.* **2000**, *145*(8), 1583–1598.

Naidu, R. A.; Kimmins, F. M.; Deom, C. M.; Subrahmanyam, P.; Chiyembekeza, A. J.; Van Der Merwe. J. A. Groundnut Rosette: A Virus Disease Affecting Groundnut Production in Sub-Saharan Africa. *Plant Dis.* **1999**, *83*, 700–709.

Navas-Castillo, J.; Fiallo-Olive, E.; Campos, S. S. Emerging Virus Diseases Transmitted by Whiteflies. *Annu. Rev. Phytopathol.* **2011**, *49*, 15.1–15.30.

Navas-Castillo, J.; Sánchez-Campos, S.; Díaz, J. A.; Sáez, E.; Moriones, E. Tomato Yellow Leaf Curl Virus-Is Causes a Novel Disease of Common Bean and Severe Epidemics in Tomato In Spain. *Plant Dis.* **1999**, *81*, 19–32.

Nei, M.; Tajima, F. DNA Polymorphism Detectable by Restriction Endonucleases. *Genetics.* **1981**, *97*, 145–163.

Ooi, K.; Ohshita, S.; Ishii, I.; Yahara, T. Molecular Phylogeny of Geminivirus Infecting Wild Plants in Japan. *J. Plant. Res.* **1997**, *110*, 247–257.

Padidam, M.; Sawyer, S.; Fauquet, C. M. Possible Emergence of New Geminiviruses by Frequent Recombination. *Virol.* **1999**, *265*, 218–225.

Pappu, H. R.; Jones, R. A. C.; Jain, R. K. Global Status of Tospovirus Epidemics in Diverse Cropping Systems: Successes Achieved and Challenges Ahead. *Virus Res.* **2009**, *141*, 219–236.

Pathak, V. K.; Temin, H. M. Broad Spectrum of *In Vivo* Forward Mutations. Hypermutations. and Mutational Hotspots in a Retroviral Shuttle Vector after a Single Replication Cycle: Substitutions, Frameshifts, and Hypermutations. *Proc. Natl. Acad. Sci. U.S.A.* **1990**, *8*, 6019–6023.

Patil, B. L.; Rajasubramaniam, S.; Bagchi, C.; Dasgupta, I. Both Indian Cassava Mosaic Virus and Sri Lankan Cassava Mosaic Virus are Found in India and Exhibit High Variability as Assessed by PCR-RFLP, *Arch. Virol.* **2005**, *150*, 389–397.

Persley, D. M.; Thomas, J. E.; Sharman, M. Tospoviruses–An Australian Perspective. *Aust. Plant Pathol.* **2006**, *35*, 161–180.

Phaneendra, C.; Rao, K. R.; Jain, R. K.; Mandal, B. Tomato Leaf Curl New Delhi Virus is Associated with Pumpkin Leaf Curl: A New Disease in Northern India. *Indian J. Virol.* **2012**, *23*(1), 42–45.

Preiss, W.; Jeske, H. Multitasking in Replication is Common Among Geminiviruses. *J. Virol.* **2003**, *77*, 2972–2980.

Qiu, W. P.; Geske, S. M.; Hickey, C. M.; Moyer, J. W. Tomato Spotted Wilt Tospovirus Genome Reassortment and Genome Segment-Specific Adaptation. *Virol.* **1998**, *244*, 186–94.

Raja, P.; Jain, R. K. Molecular Diagnosis of Groundnut Bud Necrosis Virus Causing Bud Blight of Tomato. *Indian Phytopathol.* **2006**, *59*, 359–362.

Ravi, K. S.; Kitkaru, A. S.; Winter, S. Iris Yellow Spot Virus in Onions: A New Tospovirus Record Form India. *Plant Pathol.* **2006**, *55*, 288.

Remold, S. K. Unapparent Virus Infection and Host Fitness in Three Weedy Grass Species. *J. Ecol.* **2002**, *90*, 967–977.

Ribeiro, S. G.; Ambrozevicius, L. P.; Avila, A. C.; Bezerra, I. C.; Calegario, R. F.; Fernandes, J. J.; Lima, M. F.; De Mello, R. N.; Rocha, H.; Zerbini, F. M. Distribution and Genetic Diversity of Tomato-Infecting Begomoviruses in Brazil. *Arch. Virol.* **2003**, *148*, 281–295.

Rodrguez-Cerezo, E.; Garcıa-Arenal, F. Genetic Heterogeneity of the RNA Genome Population of the Plant Virus U5-TMV. Virol. **1989**, *170*, 418–423.

Roossinck, M. J. Mechanisms of Plant Virus Evolution. *Annu. Rev. Phytopathol.* **1997**, *35,* 191–209.

Roye, M. E.; Mclaughlin, W. A.; Nakhla, M. K.; Maxwell, D. P. Genetic Diversity Among Geminiviruses Associated with the Weed Species *Sida* spp., *Macroptilium lathyroides*, and *Wissadula amplissima* from Jamaic. *Plant Dis.* **1997**, *81,* 1251–1258.

Rybicki, E. P.; Pietersen, G. Plant Virus Problems in the Developing World. *Adv. Virus Res.* **1999**, *53,* 127–175.

Sakimura, K. The Present Status of Thrips-Borne Viruses. In *Biological Transmission of Disease Agents;* Maramorosch, K., Ed.; Academic Press: New York, 1962; pp 33–40.

Sanz, A. I.; Fraile, A.; Garcia-Arenal, F.; Zhou, X. P.; Robinson, D. J.; Khalid, S.; Butt, T.; Harrison, B. D. Multiple Infection, Recombination and Genome Relationships Among Begomovirus Isolates Found in Cotton and Other Plants in Pakistan. *J. Gen. Virol.* **2000**, *81,* 1839–1849.

Sastry, K. S. M.; Singh, S. J. Assessment of Loss in Tomato by Tomato Leaf Curl Virus. *Ind. J. Mycol. Plant Pathol.* **1973**, *27,* 274–297.

Satyanarayana, T.; Gowda, S.; Lakshminaryan Reddy, K.; Metals, E.; Dawson, W. O. ; Reddy, D. V. R. Peanut Yellow Spot Virus is a Member of a New Serogroup of Tospovirus Genus Based on Small (S) RNA Sequence and Organization. *Arch. Virol.* **1998**, *143,* 353–364.

Satyanarayana, T.; Mitchell, S. E.; Reddy, D. V. R.; Brown, S.; Kresovich, S.; Jarret, R.; Naidu, R. A.; Damski, J. W. The Complete Nucleotide Sequence and Genome Organization of the M RNA Segment of Peanut Bud Necrosis Tospovirus and Comparison with other Tospoviruses. *J. Gen. Virol.* **1996**, *77,* 2347–2352.

Saunders, K.; Nazeera, S.; Mali, V. R.; Malathi, V. G.; Briddon, R.; Markham, P. G.; Stanley, J. Characterisation of Sri Lankan Cassava Mosaic Virus and Indian Cassava Mosaic Virus: Evidence for Acquisition of a DNA B Component by Amonopartite Begomovirus. *Virol.* **2002**, *293,* 63–74.

Seal, S. E.; Van Den Bosch, F.; Jeger, M. J. Factors Influencing Begomovirus Evolution and their Increasing Global Significance: Implications for Sustainable Control. *Crit. Rev. Plant Sci.* **2006**, *25,* 23–46.

Srivastava, K. M.; Hallan, V.; Raizada, R. K.; Chandra, G.; Singh, B. P.; Sane, P. V. Molecular Cloning of Indian Tomato Leaf Curl Virus Genome Following a Simple Method of Concentrating the Supercoiled Replicative Form of Viral DNA. *J. Virol. Methods.* **1995**, *51*(2–3), 297–304.

Stanley, J.; Bisaro, D. M.; Briddon, R. W.; Brown, J. K.; Fauquet, C. M.; Harrison, B. D.; Rybicki, E. P.; Stenger, D. C. Geminiviridae. In *Virus Taxonomy, VIIIth Report of the ICTV;* Fauquet, C. M., Mayo, M. A., Maniloff, J., Desselberger, U., Ball, L. A. Eds.; Elsevier/ Academic Press: London, 2005; pp 301–326.

Stonor, J.; Hart, P.; Gunther, M.; De Barro, P.; Rezaian, M. A. Tomato Leaf Curl Geminivirus in Australia: Occurrence, Detection, Sequence Diversity and Host Range. *Plant Pathol.* **2003**, *52,* 379–388.

Tepfer, M. Risk Assessment of Virus-Resistant Transgenic Plants. *Annu. Rev. Phytopathol.* **2002**, *40,* 467–491.

Van Regenmortel, M. H. V. *Serology and Immunochemistry of Plant Viruses.* Academic Press: New York, 1982.

Vijayapalani, P.; Maeshima, M.; Nagasaki-Takekuchi, N.; Miller, W. A. Interaction of the Trans-Frame Potyvirus Protein P3N-PIPO with Host Protein Pcap1 Facilitates Potyvirus Movement. *PLoS Pathog.* **2012**, *8*(4). doi: 10.1371/Journal.ppat.1002639.

Varma, J. P. Tomato Leaf Curl: ICAR Proceedings, Seminar on diseases of horticultural plants (Simla), 1959, pp 182–200.

Vasudeva, R. S.; Sam Raj, J. A Leaf Curl Disease of Tomato. *Phytopathology.* **1948,** *38,* 364–369.

Ward, C. W.; Weiller, G.; Shukla, D. D.; Gibbs, A. J. Molecular Systematics of the Potyviridae, the Largest Plant Virus Family. In *Molecular Basis of Virus Evolution;* Gibbs, A. J., Calisher, C. H., Garcia-Arenal, H., Eds.; Cambridge University Press: Cambridge. 1995; pp 477–500.

Webb, S. E.; Kok-Yokomi, M. L.; Tsai, J. H. Evaluation of *Frankliniella bispinosa* as a Potential Vector to Tomato Spotted Wilt Virus. *Phytopathol.* **1997,** *87,* 102.

Webster, C. G.; Coutts, B. A.; Jones, R. A. C.; Jones, M. G. K.; Wylie, S. J. Virus Impact at the Interface of an Ancient Ecosystem and a Recent Agro ecosystem: Studies on Three Legume-Infecting Potyviruses in the Southwest Australian Floristic Region. *Plant Pathol.* **2007,** *56,* 729–742.

White, P. S.; Morales, F. J.; Roossinck, M. J. Interspecific Reassortment in the Evolution of a Cucumovirus. *Virol.* **1995,** *207,* 334–337.

Wijkamp, I.; Almarza, R.; Goldbach, R.; Peters, D. Distinct Levels of Specificity in Thrips Transmission of Tospoviruses. *Phytopathol.* **1995,** *85,* 1069–74.

Yarwood, C. E. Host Passage Effects with Plant Viruses. *Adv. Virus Res.* **1979,** *25,* 169–190.

Zhou, X. P.; Liu, Y. L.; Calvert, L.; Munoz, C.; Otim-Nape, G. W.; Robinson, D. J.; Harrison, B. D. Evidence that DNA-A of a Geminivirus Associated with Severe Cassava Mosaic Disease in Uganda has Arisen by Interspecific Recombination. *J. Gen. Virol.* **1997,** *78,* 2101–2111.

INDEX

Printed and bound by CPI Group (UK) Ltd, Croydon, CR0 4YY

23/10/2024

01777701-0017